POROUS MEDIA: THEORY AND EXPERIMENTS

# POROUS MEDIA: THEORY AND EXPERIMENTS

*Edited by*

R. DE BOER

*University of Essen, Germany*

Reprinted from *Transport in Porous Media*
Volume 34, Nos. 1–3 (1999)

SPRINGER-SCIENCE+BUSINESS MEDIA, B.V.

**A C.I.P. catalogue record for this book is available from the Library of Congress.**

ISBN 978-94-010-5939-8 ISBN 978-94-011-4579-4 (eBook)
DOI 10.1007/978-94-011-4579-4

---

*Printed on acid-free paper.*

# Table of Contents

# Preface

The EUROMECH Colloquium 366, 'Porous Media – Theory and Experiments' was held at the Bildungszentrum für die Entsorgungs- und Wasserwirtschaft GmbH B·E·W, Essen, Germany, from 23 to 27 June 1997.

The goal of EUROMECH 366 was the presentation of recent findings in the macroscopic porous media theory (mixture theory restricted by the volume fraction concept) concerning general concepts and special investigations in the theoretical as well as the experimental field. Herein, numerical results requiring new solution strategies were also to be included. Moreover, foundations of fundamental statements in the macroscopic porous media theory (e.g. the effective stress principle for incompressible and compressible constituents by micromechanic investigations) were welcome.

Emphasis was placed upon the need to bring together scientists from various branches where porous media theories play a dominant role, namely from theoretical mechanics, agriculture, biomechanics, chemical engineering, geophysics and soil mechanics as well as from petroleum energy and environmental engineering.

More than 80 people from 12 different countries expressed their interest in the Colloquium, and finally, 58 took part in the meeting presenting 42 papers. Among the talks were seven principal lectures given by leading scientists in the a.m. fields invited by the organizers.

As Chairman of EUROMECH 366, I would like to thank the co-chairmen and all of my co-workers from the Institute of Mechanics, FB 10, University of Essen, for their help in organizing the Colloquium, in particular, Dr.-Ing. W. Walther, Priv.-Doz. Dr.-Ing. J. Bluhm and Dipl.-Ing. S. Breuer. Moreover, I would like to express my gratitude to the sponsors who financially supported the meeting – Fördervereinigung für die Stadt Essen e.V., A. Sutter GmbH, Essen, Robert-Bosch-Stiftung, Stuttgart, Bildungszentrum für die Entsorgungs- und Wasserwirtschaft GmbH B·E·W, Essen, Universität Essen.

The Editor of *Transport in Porous Media* kindly offered to publish a special issue of the Journal with selected new papers reflecting the contents of the talks at the conference. Originally, more than 30 authors intended to contribute to the special issue. After the peer-reviewing process, 20 articles were approved for publication. It is my wish to thank the reviewers for their careful examination of the manuscripts. The large number of accepted papers is a testament to the high quality of all those submitted.

This special issue of *Transport in Porous Media* with its 20 articles will provide, without any doubt, a valuable overview of recent developments in the field of porous media, on the theoretical as well as the experimental level, and it will be helpful in the realization of a consistent macroscopic porous media theory.

R. DE BOER
*Chairman*

# On the Role of the Interface Mechanical Interaction in a Gravity-Driven Shear Flow of an Ice-Till Mixture

T. WU and K. HUTTER
*Department of Mechanics, Technische Universität Darmstadt, Hochschulstraße 1, D-64289 Darmstadt, Germany*

(Received: 18 December 1996; in final form: 8 October 1997)

**Abstract.** The ice-till mixtures at the base of glaciers and ice sheets play a very important role in the movement of the glaciers and ice sheets. This mixture is modelled as an isothermal flow which is overlain by a layer of pure ice. In this model, ice is treated as usual as a very viscous fluid with a constant true density, while till, which is assumed to consist of sediment and bound (that is, moving with the sediment) interstitial water and/or ice, is also assumed in a first approximation to behave such as a fluid. For an isothermal flow below the melting point the water component can be neglected. Therefore, only the mass and momentum balances for till and ice are needed. To complete the model, no-slip and stress-free boundary conditions are assumed at the base and free-surface, respectively. The transition from the till-ice mixture layer to the overlying pure ice layer is idealized in the model as a moving interface representing in the simplest case the till material boundary, at which jump balance relations for till and ice apply. The mechanical interactions are considered in the mixture basel layer, as well as at the interface via the surface production. The interface mechanical interaction is supposed to be only a function of the volume fraction jump across the interface. In the context of the thin-layer approximation, numerical solutions of the lowest-order form of the model show a till distribution which is reminiscent to the ice-till layer in geophysical environment.

**Key words:** ice-till mixture, thermodynamic consistent formulation, gravity shear flow, interface interaction, numerical solutions.

## 1. Introduction

Observations in bore-holes drilled into glaciers till the base (Engelhardt *et al.*, 1978) have shown that, close to the rock bed, the glacier ice is increasingly contaminated by sediment. In the extreme case, a basal layer consisting predominantly of sediment and bound (that is, moving with the sediment) water forms a so-called till, with up to approximately 85% till possible (Wu *et al.*, 1996), the remainder consisting of ice, free water and/or possibly cavities. The sediment part of this till is likely eroded from the basal rock surface by the moving ice and incorporated into the near-bottom ice as the glacier or ice sheet moves. Assuming, on the scale of the entire base of a glacier or ice sheet, that the free water constituent is negligible, a glacier or ice sheet can be idealized as consisting of a relatively thick pure ice layer riding on top of a relatively thin ice-till mixture layer at the base.

In general, the sediment part of the till consists of various-sized particles (for example clay, silt, sand, gravel, boulders, and so on) which together constitute a *granular material*. In a first approximation treated in this work, we ignore the (higher-order) effects due to the volume-fraction-gradient dependence of the till stress tensor, and treat the till constituent as viscous. In other words, the effects of volume fraction gradients and friction between the sediment grains on the stress are assumed negligible, something that is perhaps in fact not very realistic for high sediment concentrations; on the other hand, interstitial water could have a lubricating effect, such that the effect of intergranular friction is reduced. Note in addition that the study of creeping flow of soil containing various amounts of water has shown that the assumption of viscous behaviour can account for observed deformations both in the laboratory and in the field (Hutter and Vulliet, 1985; Vulliet and Hutter, 1988a, b). On this basis, we follow Hutter *et al.* (1994), and Svendsen *et al.* (1996) in treating the ice-till system as a mixture of two constant true density viscous fluids. In addition, we assume for simplicity that this mixture is saturated, that is, the mixture volume is always equal to that of till plus ice in the mixture, that is, no cavities or voids arise during the flow. Analogous to the constraint of constant density in a fluid, this constraint is maintained by a pressure, the so-called saturation pressure.

Since the motion of the layer is extremely slow, Stokes flow is assumed. The till and ice momentum balances reduce then to force balances between the constituent stresses, gravitation and the momentum exchange forces. The boundary conditions at the base are assumed to be no-slip for both constituents, while the free surface is assumed to be stress free. As such, we neglect the process of entrainment of sediment at the base into the flow, and concentrate on the mechanical aspects of the basal layer. The interface between the till and overlying pure ice moves with the till material velocity, and is nonmaterial with respect to the ice. As such, the jump balance relations for both till and ice apply at this interface. In addition, as recognized by Hutter *et al.* (1994), ice and till can interact mechanically with each other at this interface, just as they do via, for example, drag forces in the bulk. The interface mechanical interactions were introduced into this model. Using the constant mechanical interaction Wu *et al.* (1996) show that it is predominantly the thickness of the basal (mixture) layer that is influenced by the ice-till momentum interaction. In the current work the interface mechanical interaction is assumed to be only dependent on the volume fraction jump at the interface. Using a relatively simple form of this interaction, we investigate qualitatively the effect of this interface interaction between ice and till on the mechanical behaviour and thickness of the lower layer, modelled as a till-ice mixture.

To begin with, the governing field equations, along with boundary and transition conditions, are formulated in Section 2. Next, the mechanical interactions in the bulk as well as at the interface are introduced into the model (Section 3). After that, a scaling analysis for gravity-driven shear flow down an inclined curved surface is carried out and the governing equations are then simplified (Section 4). Lastly, numerical solutions of the lowest-order form of the resulting model are obtained and their implications for the mechanical behaviour and thickness of the basal layer are discussed in Section 5.

## 2. Governing Equations and Boundary Conditions

Consider the isothermal flow of an ice-till mixture overlain by a pure ice layer down an inclined plane (Figure 1). As discussed above, the lower layer is modelled as a mixture of two very viscous, constant true density fluids, so that the constituents true density $\hat{\rho}_\alpha$ is constant and the partial density $\rho_\alpha = \nu_\alpha \hat{\rho}_\alpha$ varies only with the till volume fraction $\nu_\alpha$, where the index $\alpha$ is T for till and I for ice. In this case the volume fraction $\nu_\alpha$ is the independent variable, instead of the partial density. Assuming no mass exchange between ice and till, the mass and momentum balances for till and ice in the mixture layer then take the forms

$$\begin{aligned} &\frac{\partial \nu_\alpha}{\partial t} + \operatorname{div}(\nu_\alpha \mathbf{v}_\alpha) = 0, \\ &\operatorname{div} \mathbf{T}_\alpha + \hat{\rho}_\alpha \nu_\alpha \mathbf{g} + \mathbf{m}_\alpha = \mathbf{0}. \end{aligned} \tag{2.1}$$

Here, $\mathbf{v}_\alpha$ and $\mathbf{T}_\alpha$ represent the partial velocity and constituent partial stress for till and ice with the index T and I, respectively; $\mathbf{g}$ is the vector of gravity acceleration. The mechanical interaction force between till and ice can be simply written as $\mathbf{m}_\mathrm{T} = -\mathbf{m}_\mathrm{I} = \mathbf{m}$. Since Stokes flow was assumed, the accelaration term in the momentum balance was neglected. In addition, the lower mixture layer is assumed to be saturated, i.e.,

$$\nu_\mathrm{T} + \nu_\mathrm{I} = 1. \tag{2.2}$$

In other words, we assume that no 'cavities' or 'voids' can arise in this layer during its flow. Defining $\nu = \nu_\mathrm{T}$, we then have $\nu_\mathrm{I} = 1 - \nu$. The relation (2.2) is a saturation constraint which leads to the loss of an independent variable, one of the volume fractions. Such a constraint is analogous to the classical constant density

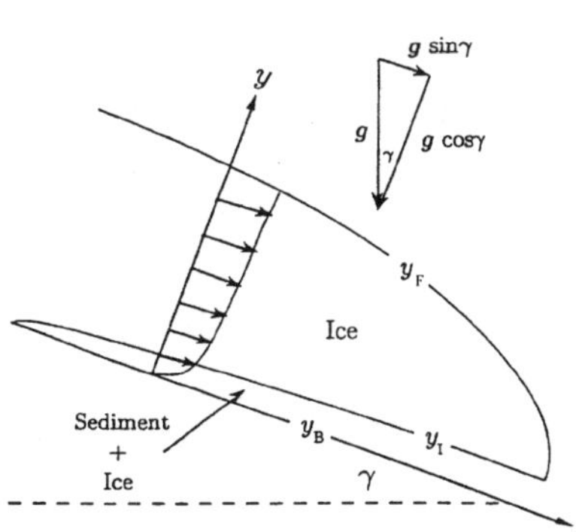


*Figure 1.* Two-dimensional model geometry. $y_\mathrm{B}$, $y_\mathrm{I}$, and $y_\mathrm{F}$ represent the positions of the base, interface, and free surface of the system as a function of $x$ and $t$.

('incompressibility') constraint for a fluid, in which case the mass density is lost as such a variable, to be 'replaced' by the pressure maintaining the constraint as a new unknown. By analogy, in the current mixture context, this new unknown pressure $p$ replaces the volume fraction lost via the saturation constraint (2.2) as an unknown in the model, and maintains this constraint. Hence, we refer to $p$ as the saturation pressure. As in the classical incompressible fluid case, $p$ enters the balance relations via the stress constitutive relation; in addition, in the current mixture context, the till-ice momentum interaction also depends on $p$ (see, e.g., Svendsen and Hutter, 1995, and below). In the upper pure ice layer, the governing equations take the standard forms

$$\begin{aligned} &\operatorname{div} \mathbf{v}_{\mathrm{I}} = \mathbf{0}, \\ &\operatorname{div} \mathbf{T}_{\mathrm{I}} + \hat{\rho}_{\mathrm{I}}\, \mathbf{g} = \mathbf{0}. \end{aligned} \tag{2.3}$$

Since ice and till are modeled as viscous fluids, the constitutive relation of the partial stress is

$$\mathbf{T}_\alpha = -p_\alpha \mathbf{I} + \mu_\alpha\, \mathbf{D}'_\alpha, \tag{2.4}$$

where $p_\alpha$ represents the constituent (equilibrium) pressure, $\mathbf{D}'_\alpha$ its deviatoric part of the symmetric part of the constituent velocity gradient $\nabla \mathbf{v}_\alpha$, and $\mu_\alpha$ the shear viscosity. The second part on the right hand side is the viscous (non-equilibrium) part of $\mathbf{T}_\alpha$, in which the bulk viscosities of the till and ice constituents are neglected. The constituent pressures are related to the mixture saturation pressure $p$ by the constitutive assumption

$$p_\alpha = \nu_\alpha\, p \tag{2.5}$$

in the context of (2.2). Clearly, we have $p_{\mathrm{I}} \neq p_{\mathrm{T}}$ in general. The mechanical interaction will be discussed in Section 3.

To complete the basic formulation the boundary conditions are specified. The free surface $y = y_{\mathrm{F}}(x, t)$ is assumed stress free, i.e.,

$$\mathbf{T}_{\mathrm{I}}\mathbf{n} = \mathbf{0}, \tag{2.6}$$

where $\mathbf{n}$ is the unit normal to the surface. At the base $y = y_{\mathrm{B}}(x, t)$ , no-slip conditions for ice and till are assumed, i.e.,

$$\mathbf{v}_\alpha = \mathbf{0}. \tag{2.7}$$

Here we neglected the process of entrainment of sediment into the basal layer. Since the upper layer contains no till, the interface $y = y_{\mathrm{I}}(x, t)$ is material with respect to till. That means the surface velocity is the same as the till velocity. The mass and momentum jump conditions are

$$\begin{aligned} &[\![\rho_\alpha(\mathbf{v}_\alpha - \mathbf{v}_{\mathrm{T}})]\!] \cdot \mathbf{n} = 0, \\ &\rho_\alpha^{\pm}\{(\mathbf{v}_\alpha^{\pm} - \mathbf{v}_{\mathrm{T}}) \cdot \mathbf{n}\}\, [\![\mathbf{v}_\alpha]\!] - [\![\mathbf{T}_\alpha]\!]\mathbf{n} = \boldsymbol{\varpi}_\alpha, \end{aligned} \tag{2.8}$$

where $(2.8)_1$ was used to write $(2.8)_2$, $[\![\psi]\!] = \psi^+ - \psi^-$ (+ for the upper, pure-ice layer, and − for the till-ice layer), and $\boldsymbol{\varpi}_\alpha$ represents a constituent interaction force conceptually analogous to mechanical interaction as given in (2.1), but one on the interface between the lower till-ice layer and the upper pure ice layer, rather than in the bulk (see Hutter *et al.*, 1994). In the till case ($\alpha$ = T), the first term in $(2.8)_2$ vanishes; as for ice, dimensional considerations show it to be negligible. Consequently, we have

$$[\![\mathbf{T}_\alpha]\!]\mathbf{n} = -\boldsymbol{\varpi}_\alpha. \tag{2.9}$$

In general, the sum of the constituent interface mechanical interaction terms is zero, i.e., $\boldsymbol{\varpi}_\mathrm{T} + \boldsymbol{\varpi}_\mathrm{I} = \mathbf{0}$, insuring momentum balance in the mixture as a whole at the interface. For a justification of this in terms of physics, see Wu *et al.*, (1996).

If the velocity profile is determined up to the free surface, a further kinematic statement on the velocity jump at the interface is needed. For an interface between two single viscous fluids, continuity of the tangential component of the velocity field would be required, i.e., $[\![\mathbf{v} \times \mathbf{n}]\!] = \mathbf{0}$, representing a no-slip condition. Since we have two components on one side, and only one on the other side, it seems inappropriate to demand $[\![\mathbf{v}_\alpha \times \mathbf{n}]\!] = \mathbf{0}$, because for till this would imply that its tangential velocity component at the interface vanishes, while that for ice does not. Consequently, a proper generalization of the no-slip condition for the situation is to assume the continuity of the tangential *barycentric velocity* across the interface, i.e.,

$$[\![(\nu\mathbf{v}_\mathrm{T} + (1-\nu)\mathbf{v}_\mathrm{I}) \times \mathbf{n}]\!] = \mathbf{0}, \tag{2.10}$$

where the barycentric velocity (mixture velocity) is defined via $\mathbf{v} = \nu\mathbf{v}_\mathrm{T} + (1-\nu)\mathbf{v}_\mathrm{I}$. Because of no till on the positive side of the interface, i.e., $\nu^+ = 0$, this reduces to the simple form

$$(\mathbf{v}_\mathrm{T} - \mathbf{v}_\mathrm{I})^- \times \mathbf{n} = \mathbf{0}, \tag{2.11}$$

which expresses the no-slip condition when $\nu^- > 0$. Physically, the no-slip condition requires also the continuity of the tangential ice velocity across the interface, i.e.,

$$[\![\mathbf{v}_\mathrm{I}]\!] \times \mathbf{n} = \mathbf{0}. \tag{2.12}$$

## 3. Mechanical Interactions

The constitutive relations for $\mathbf{T}_\alpha$ have been formulated in the last section. In order to solve the equations, we need the constitutive relations for the mechanical interaction $\mathbf{m}$ in the bulk, as well as $\boldsymbol{\varpi}_\alpha$ at the interface. In general, the constituent mechanical interaction rate density $\mathbf{m}$ in the lower layer will depend on the corresponding independent fields and their gradients, i.e.,

$$\mathbf{m} = \hat{\mathbf{m}}(\nu, p, \mathbf{v}_\mathrm{T}, \mathbf{v}_\mathrm{I}, \nabla\nu, \nabla p, \nabla\mathbf{v}_\mathrm{T}, \nabla\mathbf{v}_\mathrm{I}). \tag{3.1}$$

Subjecting (3.1) to the frame-indifference reduces this to the form

$$\mathbf{m} = \hat{\mathbf{m}}(\nu, p, \mathbf{v}_d, \nabla\nu, \nabla p, \mathbf{D}_{\mathrm{T}}, \mathbf{D}_{\mathrm{I}}, \mathbf{W}_d), \tag{3.2}$$

where $\mathbf{D}_\alpha$ is the symmetric part of $\nabla\mathbf{v}_\alpha$, $\mathbf{W}_\alpha$ its skew-symmetric part, $\mathbf{v}_d = \mathbf{v}_{\mathrm{I}} - \mathbf{v}_{\mathrm{T}}$ the relative velocity, and $\mathbf{W}_d = \mathbf{W}_{\mathrm{I}} - \mathbf{W}_{\mathrm{T}}$ the relative spin. In this work, the influence of $\mathbf{D}_{\mathrm{T}}$, $\mathbf{D}_{\mathrm{I}}$ and $\mathbf{W}_d$ on the interaction is neglected. Lastly, the constituent mechanical interaction rate density $\mathbf{m}$ is assumed to take the simplest possible form, i.e., the linear form

$$\mathbf{m} = m_{\mathbf{v}}\,\nu(1-\nu)\,\mathbf{v}_d + m_\nu\,(\nabla\nu) + m_p\,(\nabla p)\,, \tag{3.3}$$

where in this case the coefficients $m_{\mathbf{v}}$, $m_\nu$ and $m_p$ are at most functions of $\nu$ and $p$; in particular, the dependence of $m_p$ on $\nu$ must be such that $\mathbf{m}$ vanishes as $\nu$ goes to 0 or 1. Note that the constituent mechanical interaction coefficient $m_{\mathbf{v}}$ associated with drag interactions is proportional to the inverse of the so-called Darcy permeability parameter. The assumption $m_p = 0$ was used here as done by Drew and Segel (1971) and MacKenzie (1984). As for $m_\nu$, it has the unit of pressure, and the simplest form $m_\nu = \delta p$ is taken here. As shown in the general thermodynamic formulation of Svendsen and Hutter (1995) for a mixture of isotropic viscoelastic materials, this last coefficient takes the usual form $m_\nu = p$ (i.e., $\delta = 1$; e.g., MacKenzie, 1984) only when one assumes that the mixture inner free energy* depends negligibly on (in the current context) $\nu$, something that will not in general be the case. Beyond this, Hutter *et al.* (1994) show that, for such a choice (i.e., $\delta = 1$), there exist no non-trivial solutions for the steady shear flow of a mixture of two viscous fluids overlain by a single viscous fluid layer in the context of a simple parallel-slab geometry. The influence of the parameter $\delta \in (0, 1)$ was studied by Svendsen *et al.* (1996). Finally Wu *et al.* (1996a) interpret $\boldsymbol{\varpi}_\alpha$ as the limit of $\int m_\nu \nabla\nu\,\mathrm{d}v$ over a pillbox like volume embracing the singular surface where $\nu^+ = 0,\ \nu^- \neq 0$.

The presence of a non-zero constituent interface interaction force $\boldsymbol{\varpi}_\alpha$ in (2.9) can be motivated as follows: Consider a pillbox with its upper surface in the pure ice and its lower surface in till-ice mixture, with both of these surfaces parallel to the interface. Assume for simplicity a state at rest with no shear stresses. Surface forces from the pillbox sides can be ignored (or vanish in this simple case when the force balance in the direction of the pillbox thickness is considered, which we do here for the sake of simplicity). With the top as the positive side, and the bottom as the negative side, (2.4) and (2.9) then imply

$$\begin{aligned}
&(1-\nu^-)p^- - p^+ = \varpi_{\mathrm{I}} && \text{ice,}\\
&\nu^- p^- = \varpi_{\mathrm{T}} && \text{till,}\\
&p^- - p^+ = 0 && \text{mixture,}
\end{aligned} \tag{3.4}$$

with $\varpi_\alpha = \boldsymbol{\varpi}_\alpha \cdot \mathbf{n}$. Now, since $p^- = p^+$ follows from $(3.4)_3$, and $p^+ \neq 0$, setting $\boldsymbol{\varpi}_\alpha = \mathbf{0}$ in $(3.4)_2$ implies $\nu^- = 0$, that is, no jump of till volume fraction across

* This is the part of the mixture free energy not depending on diffusion.

the interface in this case. As such, for a given value of $p^-$, $(3.4)_2$ couples the 'magnitude' of the jump in the till volume fraction (i.e., $[\![\nu]\!] = \nu^-$) to that of the till-ice mechanical interaction $\varpi_\alpha$ at the interface; in particular, the larger the till-ice mechanical interaction at the interface, the larger the jump in till volume fraction. Consequently, $\boldsymbol{\varpi}_\alpha$ clearly defines the sharpness of the interface between the till below and pure ice above, with $\boldsymbol{\varpi}_\alpha = \mathbf{0}$ corresponding to no interface (in the sense of a jump in the till volume fraction) at all. Observations (e.g., Engelhardt *et al.*, 1978) indicate that indeed the interface between the pure ice above and the till below is rather abrupt, or sharp; in other words, till is concentrated in the basal layer. In any case, we would also expect $\boldsymbol{\varpi}_\alpha \neq \mathbf{0}$ on physical grounds; indeed, the steep gradient in the till volume fraction from the bottom to the top of the pillbox induces mechanical interaction (forces) between the till and ice which in the limit as the pillbox thickness goes to zero become corresponding interface interactions which are non-zero in general.

Constitutive relations for the interface mechanical interaction $\boldsymbol{\varpi}_\alpha$ are needed. Since the interface between the pure ice layer and the till-ice mixture layer is a singular surface, some of the independent fields may not be continuous and have a jump, and their gradients may not exist near the interface. So we assume that $\boldsymbol{\varpi}_\alpha$ depend on the jumps of the corresponding independent fields, i.e.,

$$\boldsymbol{\varpi}_\alpha = \hat{\boldsymbol{\varpi}}_\alpha \left([\![\nu]\!], [\![p]\!], [\![\mathbf{v}_\mathrm{T}]\!], [\![\mathbf{v}_\mathrm{I}]\!]\right). \tag{3.5}$$

## 4. Dimensional Analysis and Simplification

In this section, we adapt the above balance and constitutive relations to the two-dimensional, parallel-sided slab idealization of a glacier or ice sheet dealt with in this work (see Figure (1)). To this end, we work with the following scalings:

$$\begin{gathered}[x] = L, \quad [y] = H, \quad [t] = T, \quad [u_\alpha] = U, \quad [v_\alpha] = V, \\ [\mathbf{S}_\mathrm{T}] = \hat{\rho}_\mathrm{T} g H \,\sin\gamma, \; [\mathbf{S}_\mathrm{I}] = \hat{\rho}_\mathrm{I} g H \,\sin\gamma, \; [p] = \hat{\rho}_\mathrm{I} g H \,\cos\gamma, \; [\boldsymbol{\varpi}_\alpha] = \iota_\alpha.\end{gathered} \tag{4.1}$$

Here, $L$, $H$, $U$, $V$ and $T$ represent a typical length, thickness of the two-layer system, $x$-velocity, $y$-velocity, and time, respectively, chosen such that $T = L/U = H/V$.

Substituting (4.1) into (2.1)–(2.5), and introducing the thin-layer approximation $\varepsilon = H/L \ll 1$, we obtain the following non-dimensionalized component forms of the model relations for the lower mixture layer:

$$\begin{aligned}&-\nu p_{,y} + (1-\delta) p\,\nu_{,y} = \mathcal{R}\nu, \\ &(1-\nu) p_{,y} + (1-\delta) p\,\nu_{,y} = -(1-\nu), \\ &\tau_{\mathrm{T},y} = -[\nu + \mathcal{M}\nu(1-\nu)u], \\ &\tau_{\mathrm{I},y} = -(1-\nu) + \mathcal{R}\mathcal{M}\nu(1-\nu)u, \\ &u_{\mathrm{T},y} = 2\mathcal{G}_\mathrm{T}\tau_\mathrm{T}, \qquad u_{\mathrm{I},y} = 2\mathcal{G}_\mathrm{I}\tau_\mathrm{I},\end{aligned} \tag{4.2}$$

to O(1) in $\varepsilon$, where now, and in the rest of this work, *all variables are non-dimensional.* In these last relations appear the non-dimensional quantites $\mathcal{R} = \hat{\rho}_{\rm T}/\hat{\rho}_{\rm I}$, $\mathcal{M} = m_{\rm v}U/\hat{\rho}_{\rm T}g\sin\gamma$, $\mathcal{G}_{\rm T} = H^2\hat{\rho}_{\rm T}g\sin\gamma/\mu_{\rm T}U$ and $\mathcal{G}_{\rm I} = H^2\hat{\rho}_{\rm I}g\sin\gamma/\mu_{\rm I}U$; in addition, we have introduced $\tau_\alpha = T_{\alpha xy}$ (for more details, see Hutter *et al.*, 1994). Similarly, the boundary conditions from (2.6)–(2.7), (2.9) become

$$p = 0 \quad \text{and} \quad \tau_{\rm I} = 0 \quad \text{at} \quad y = y_{\rm F}(x,t), \tag{4.3}$$

$$u_\alpha = 0 \quad \text{and} \quad v_\alpha = 0 \quad \text{at} \quad y = y_{\rm B}(x,t), \tag{4.4}$$

$$\begin{aligned} &p^- = p^+, \qquad \tau_{\rm T}^- = \mathcal{Q}_{\rm T}\,\varpi_{{\rm T}x} \quad \text{at} \quad y = y_{\rm I}(x,t),\\ &\nu^- p^- = \mathcal{P}_{\rm T}\,\varpi_{{\rm T}y}, \qquad \tau_{\rm I}^+ - \tau_{\rm I}^- = \mathcal{Q}_{\rm I}\,\varpi_{{\rm I}x}, \end{aligned} \tag{4.5}$$

with $\mathcal{Q}_{\rm T} = \iota_{\rm T}/\hat{\rho}_{\rm T}gH\sin\gamma$, $\mathcal{Q}_{\rm I} = \iota_{\rm I}/\hat{\rho}_{\rm I}gH\sin\gamma$, and $\mathcal{P}_{\rm T} = \iota_{\rm T}/\hat{\rho}_{\rm I}\,gH\cos\gamma$. The dimensionless equations in the upper layer are

$$p_{,y} = -1, \quad \tau_{{\rm I},y} = -1, \quad u_{{\rm I},y} = 2\mathcal{G}_{\rm I}\tau_{\rm I}, \tag{4.6}$$

from which we obtain

$$p(y) = -(y-1), \quad \tau_{\rm I}(y) = -(y-1), \quad u_{\rm I}(y) = u_{\rm I}(1) - \mathcal{G}_{\rm I}(1-y)^2, \tag{4.7}$$

where $y_{\rm F} = 1$ and the boundary conditions (4.3) were used. From $(4.5)_2$, we infer that if the till volume fraction $\nu^-$ does not vanish, the interface till-ice interaction force $\varpi_{{\rm T}y}$ in the $y$-direction must differ from zero.

Using the thin layer approximation $\varepsilon = H/L \ll 1$, the velocity component $v_\alpha = {\rm O}(\varepsilon)$ is negligible compared with the tangential velocity $u_\alpha = {\rm O}(1)$. The kinematical conditions at the interface (2.11) and (2.12) yield the dimensionless relation for the velocity

$$u_{\rm I}^+ = u_{\rm I}^- = u_{\rm T}^-. \tag{4.8}$$

With (4.4) we can integrate the velocities from the base to the interface, and then use (4.8) and $(4.7)_3$ to obtain the ice velocity at the free surface.

In this work, we consider the interface mechanical interaction in the tangential direction of the interface, i.e.,

$$\varpi_{{\rm T}x} = 0 \quad \text{and} \quad \varpi_{{\rm I}x} = 0, \tag{4.9}$$

which imply the ice shear stress is continuous across the interface and the till shear stress vanishes, i.e.,

$$\tau_{\rm T}^- = 0 \quad \text{and} \quad \tau_{\rm I}^+ = \tau_{\rm I}^-, \tag{4.10}$$

from $(4.5)_{2,4}$. With this assumption only the constitutive relation for $\varpi_{{\rm T}y}$ is needed. Because of $(4.5)_1$ and (4.8), (3.5) is reduced to

$$\varpi_{{\rm T}y} = \hat{\hat{\varpi}}_{{\rm T}y}(\nu^-, u_{\rm T}), \tag{4.11}$$

where the velocity components $v_\alpha$ have been omitted for the thin layer approximation. Further simplification in (4.11) is to neglect the dependence on the velocity and to take the simple power form

$$\varpi_{Ty} = B\,(\nu^-)^\beta, \tag{4.12}$$

where $\beta$ and $B$ are material parameters. Therefore, $(4.5)_2$ can be written as

$$\nu^- p^- = B\,\mathcal{P}_T(\nu^-)^\beta, \tag{4.13}$$

Since both of $\nu^-$ and $p^-$ are nonnegative, we have $B \geqslant 0$. For $B = 0$, we have $\nu^- p^- = 0$ and all the mechanical interactions vanish. This case was discussed in Svendsen *et al.* (1996), so that we assume here $B > 0$. Physically one can imagine that the mechanical interaction at the interface increases if the volume fraction jump increases. So $\beta \geqslant 0$ can be assumed in (4.13).

In the next section, these quantities are prescribed, and their effect on the thickness and mechanical behaviour of the basal layer is quantitatively investigated in some simple cases.

## 5. Numerical Solutions and Discussion

The analytical solution in the upper ice layer have been obtained in (4.7) and can be used as boundary conditions for the boundary value problem in the lower mixture layer if we know the interface position $y_I$. In this section we will concentrate on the numerical solutions in the ice-till layer.

Relations $(4.2)_{1,2}$ can be rewritten in the forms

$$(1-\delta)p\nu_{,y} = (1-\mathcal{R})\nu(1-\nu), \qquad p_{,y} = -1 + (1-\mathcal{R})\nu. \tag{5.1}$$

Since $\nu = 0.85$ is the largest physically reasonable value (see Wu *et al.*, 1996) for the till volume fraction, we shall impose the boundary condition $\nu = 0.85$ at $y = y_B$. At the interface we have the boundary conditions from $(4.5)_2$, $(4.7)_1$ and (4.13),

$$p^- = 1 - y_I \quad \text{and} \quad (\nu^-)^{1-\beta} = \frac{B}{1-y_I}. \tag{5.2}$$

In general, we have three boundary conditions $\nu = 0.85$ at $y_B = y$ and (5.2) with unknown $y_I$ for the equation system (5.1), which turns out to be a boundary value problem and can be solved numerically.

In addition, restricting attention for simplicity to the case of constant viscosities, i.e., $\mathcal{G}_\alpha =$ constant, $(4.2)_{3\text{–}6}$ can be combined to obtain the single relation

$$u_{d,yy} = 2\,[\mathcal{G}_I(1-\nu) - \mathcal{G}_T\nu] - 2\,\mathcal{M}\nu(1-\nu)(1+\mathcal{R})\,u_d \tag{5.3}$$

for $u_d = u_I - u_T$. From $(4.4)_1$ and (4.6) the boundary conditions for $u_d$ at the base and at the interface are simply

$$u_d(y = y_B) = 0 \quad \text{and} \quad u_d(y = y_I) = 0. \tag{5.4}$$

Thus, for $u_d$ we have a two-point-boundary value problem, which can be solved numerically using a shooting technique. The shear stress $\tau_\alpha$ is then obtained via numerical integration of $(4.2)_{3,4}$ with the boundary conditions (4.10) and the solutions of $u$ and $\nu$. Finally, the velocity $u_\alpha$ is obtained by integrating $(4.2)_{5,6}$ with $(4.4)_1$.

For the numerical solutions, we chose $H = 1000\,\mathrm{m}$, $\mathcal{R} = 2.7$, $\mathcal{G}_\mathrm{I} = 1$, $\mathcal{G}_\mathrm{T} = 1$, $\mathcal{M} = 1$ and $\mathcal{P}_\mathrm{T} = 1$; the first of these is an appropriate order of magnitude value for the thickness of an ice sheet, and the other values are chosen for simplicity. In particular, note that $\mathcal{G}_\mathrm{T}/\mathcal{G}_\mathrm{I} = \mathcal{R}\,\mu_\mathrm{I}/\mu_\mathrm{T}$, so that these assumptions correspond to assuming that the viscosity of the 'granular' till is about 3 times that of ice. As shown by Hutter *et al.* (1994), one obtains only trivial solutions of (5.1) when $\delta = 1$, that is, a single-layer, single-constituent 'mixture'. Svendsen *et al.* (1996) investigated the effects of varying $\delta$ for $\nu^- = 0$ (at the interface) on the till volume fraction, pressure, shear stress and velocity profiles in a single layer mixture model for a glacier or ice sheet. In particular, they showed that the most realistic till volume fraction profiles arise for the case $\delta \sim 0.95$. Wu *et al.*, (1996) use this value of $\delta$, and vary $\nu^-$, which is tantamount to vary $\varpi_{Sy}$ in the context of $(4.5)_2$, and is easier to implement numerically. It can be interpreted as the special case with $\beta = 0$ in this paper (see below).

Here the value of $\delta = 0.95$ is also used and the influence of the interface mechanical interaction force with parameters $\beta$ and $B$ is investigated. In the last section we motivated the choices $B > 0$ and $\beta \geqslant 0$. For $\beta = 0$, the mechanical interaction force at the interface is the predetermined parameter $B$. The numerical solution of the boundary value problem (5.1) gives us $\nu^-$ and $y_\mathrm{I}$. This is the same as if the volume fraction at the interface would be known (see Wu *et al.*, 1996). In that work $\nu^-$ varies, and then the mechanical interaction (here $B$) and $y_\mathrm{I}$ are obtained through the numerical solution of Equations (5.1). When $\beta = 1$, $(5.2)_2$ yields $y_I = 1 - B$, that is, the interface position is known. With the boundary conditions $\nu = 0.85$ at $y_\mathrm{B} = y$ and $(5.2)_1$ Equations (5.1) can be then solved numerically. If one knows $\nu^-$, the interface mechanical interaction can be determined through (4.13).

Now we will show the numerical results in the two cases $\beta \in (0, 1)$ and $\beta > 1$:

- For $\beta \in (0, 1)$, using (5.2), we have

$$y_\mathrm{I} = 1 - B\,(\nu^-)^{\beta-1} \quad \text{for} \quad 1 \geqslant y_I \geqslant 0, \tag{5.5}$$

which implies $B < 1$, since $(\nu^-)^{\beta-1} > 1$. If $\nu^-$ decreases, $B$ should decrease significantly. Figure 2 shows the till volume fraction for $B = 0.05$, $\beta = 0.5, 0.9$, $B = 0.1$ and $\beta = 0.9$. In addition for reasons of comparison, the profiles of the volume fraction for $B = 0$ are given in Figure 2(a). For the same value of $B$, if $\beta$ increases, the jump of volume fraction $\nu^-$ decreases, and the lower till layer becomes thinner. A sharp interface can be seen for $\beta = 0.5$ and $B = 0.05$. With the same value of $\beta$ and an increasing value of $B$, $\nu^-$ increases and the lower till layer becomes thicker. The volume fraction profiles show that if $\beta \lesssim 1$ with

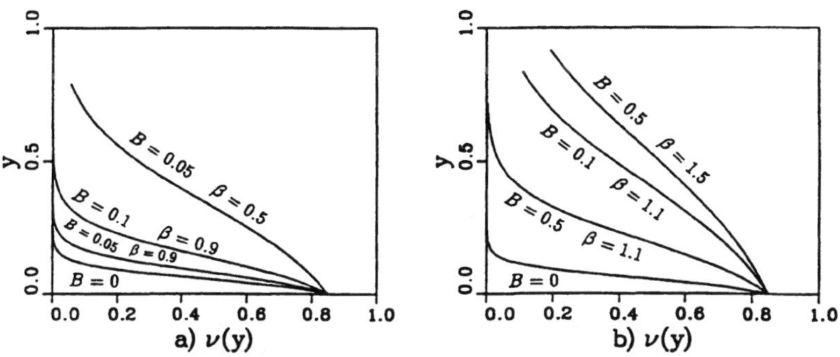


*Figure 2.* Till volume fraction for $\beta \in (0, 1)$ (a) and $\beta > 1$ (b); profiles for different values of $B$. For these calculations, we chose $H = 1000\,\mathrm{m}$, $\mathcal{R} = 2.7$, $\mathcal{G}_\mathrm{I} = 1$, $\mathcal{G}_\mathrm{T} = 1$, $\mathcal{M} = 1$ and $\mathcal{P}_\mathrm{T} = 1$ (see text).

a relatively small $B$, e.g., $B = 0.05$, we obtain a two layer solution similar to the solution with $B = 0$. Mathematically, from (5.2) we have in this case for $\beta \lesssim 1$,

$$\lim_{\nu \to 0} (\nu^-)^{1-\beta}(1 - y_I) = B \to 0, \tag{5.6}$$

a result consistent with a one layer solution with $B = 0$, $\nu^- = 0$ and $y_\mathrm{I} = 1$ (see Svendsen *et al.*, 1996). Then, the dimensionless interface mechanical interaction force $\varpi_{\mathrm{T}y} = B\,(\nu^-)^\beta \to 0$ vanishes.

- For $\beta > 1$, Figure 2(b) shows the till volume fraction profiles for $\beta = 1.1$, $B = 0.5, 0.1$, $\beta = 1.5$, $B = 0.5$, and $B = 0$. For a fixed value of $\beta$, if $B$ increases, the jump of the till volume fraction $\nu^-$ decreases, and the lower till layer becomes thinner. With the same $B$ and an increasing value of $\beta$, $\nu^-$ increases and the till layer (mixture) becomes thicker. The results are just opposite to those of the last case. For $\beta \gtrsim 1$, a two layer solution similar to the last case can also be obtained, but the parameter $B$ (e.g. $B = 0.5$) is much larger than in the last case. With a relatively small $B$, e.g., $B = 0.1$, a sharp interface appears. Mathematically, it is totally different from the last case. Since $\beta \gtrsim 1$, we have to rewrite $(5.2)_2$, i.e.,

$$\frac{(\nu^-)^{\beta-1}}{(1 - y_\mathrm{I})} = \frac{1}{B}. \tag{5.7}$$

With $B \to \infty$, we have $(\nu^-)^{\beta-1} \to 0$ and $y_\mathrm{I} \neq 1$, The first means $\nu^- \to 0$ and the second leads to a real two layer solution which is qualitatively similar to the observations (i.e., Engelhardt *et al.*, 1987). The interface mechanical interaction force $\varpi_{\mathrm{T}y} = B\,(\nu^-)^\beta$ may not vanish for very large $B$. This is consistent with the result that the nonvanishing interface interaction leads mathematically to a real two layer distribution.

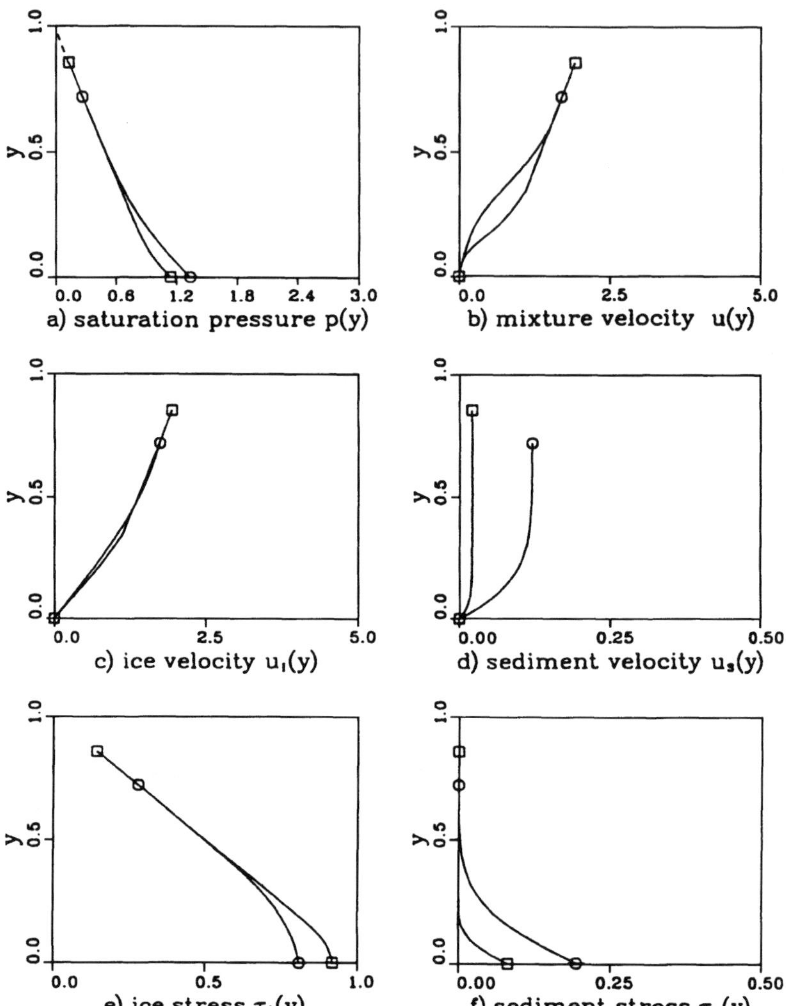


*Figure 3.* Saturation pressure (a) mixture velocity (b) ice velocity (c) till velocity (d) ice shear stress (e) and till shear stress (f) profiles with the parameters $\beta = 0.9$, $B = 0.05$ (curves with squares) and $\beta = 1.1$, $B = 0.5$ (curves with circles). For these calculations, we chose $H = 1000\,\text{m}$, $\mathcal{R} = 2.7$, $\mathcal{G}_\text{I} = 1$, $\mathcal{G}_\text{T} = 1$, $\mathcal{M} = 1$ and $\mathcal{P}_\text{T} = 1$ (see text).

The saturation pressure, mixture velocity, ice velocity, till velocity, ice shear stress and till shear stress in the lower mixture layer ($0 < y < y_\text{I}$) of the two-layer system for the two cases $\beta = 0.9$, $B = 0.05$ and $\beta = 1.1$, $B = 0.5$ are shown in Figures 3(a)–(f). The pressure profiles vary in the till layer in correspondence with the varying till volume fraction. In the upper part of the mixture layer the pressure profiles are almost linear, just like the linear pressure profile in the upper ice layer (dashed line in Figure 3(a)). The form of the velocity and stress profiles, that is, the qualitative behaviour of the solutions, do not differ significantly from the related results for $B = 0$ (see Svendsen *et al.*, 1996).

The main purpose of the current work has been to investigate the effect of till-ice interaction processes, both in the bulk and on the till-ice interface. As shown in this and previous works (Svendsen *et al.*, 1996; Wu *et al.*, 1996a, b) the bulk and interface momentum interactions between till and ice have a dramatic influence on this distribution, as well as on the thickness of the till layer. In this work we have used the interface interaction in the form (4.12) and found that, if the parameter $\beta$ is near unity, one can obtain the till distribution qualitatively similar to that of the observations. Especially, with $\beta \gtrsim 1$, the interface production may not vanish, even though the jump $\nu^-$ is very small. This effect creates a promising two layer solution, which is qualitatively relevant to the ice-till layer distribution observed in the real situations.

## References

1. Drew, D. A. and Segel, L. A.: 1971, Averaged equations for two-phase flows, *Stud. Appl. Math. L*, 205–231.
2. Engelhardt, H. F., Harrison, W. D. and Kamb, B.: 1978, Basal sliding conditions at the glacier bed as revealed by borehole photography, *J. Glaciol.* **20**, 469–508.
3. Hutter, K. and Vulliet, L.: 1985, Gravity-driven slow creeping flow of a thermoviscous body at elevated temperatures, *J. Thermal Stresses* **8**, 99–138.
4. Hutter, K., Jöhnk, K. and Svendsen, B.: 1994, On interfacial transition conditions in two-phase gravity flow, *Z. Angew. Math. Phys.* **45**, 746–762.
5. MacKenzie, D.: 1984, The generation and compaction of partially molten rock, *J. Petrol.* **25**, 713–765.
6. Svendsen, B. and Hutter, K.: 1995, On the thermodynamics of a mixture of viscous fluids with constraints, *Int. J. Engng. Sci.* **33**, 2021–2054.
7. Svendsen, B., Wu, T., Jöhnk, K. and Hutter, K.: 1996, On the role of mechanical interactions in the steady-state gravity flow of a two-constituent mixture down an incline plane, *Proc. R. Soc. London A* **452**, 1189–1205.
8. Vulliet, L. and Hutter, K.: 1988a, A continuum model for natural slops in slow movement, *Geotechnique* **38**, 199–217.
9. Vulliet, L. and Hutter, K.: 1988b, Set of constitutive models for soil under slow movement, *J. Geotech. Engng. ASCE* **114**, 1022–1041.
10. Wu, T., Jöhnk, K., Svendsen, B. and Hutter, K.: 1996, On the gravity-driven shear flow of an ice-till mixture, *Annal. Glaciol.* **23**, 359–363.
11. Wu, T., Hutter, K. and Svendsen, B.: 1998, On shear flow of a saturated ice-sediment mixture with thermodynamic equilibrium pressure and momentum exchange, *Proc. Royal Soc. London A* **454**, 71–88.

# Finite Deformation Models and Field Performance *

S. ARNOD[1], M. BATTAGLIO[2], N. BELLOMO[3], D. COSTANZO[2],
R. LANCELLOTTA[2] and L. PREZIOSI[3]
[1]*PhD student, Politecnico di Torino, Corso Duca degli Abruzzi 24, Torino, 10129, Italy*
[2]*Dpt. Structural Eng., Soil Mechanics Group, Politecnico di Torino, Corso Duca degli Abruzzi 24, Torino, 10129, Italy*
[3]*Dpt. Mathematics, Politecnico di Torino, Corso Duca degli Abruzzi 24, Torino, 10129, Italy*

(Received: 22 June 1997; in final form: 12 December 1997)

**Abstract.** This paper reports about the derivation of a fully nonlinear model characterized by finite deformations without smallness assumptions. The soil is assumed to be saturated, and no restrictions are introduced on the constitutive laws. Initial boundary value problems are formulated with reference to geotechnical problems, such as consolidation under own weight or sedimentation of solid particles in a quiescent fluid, and back-analyses of field performance of an embankment resting on a soft clay deposit.

**Key words:** consolidation, deformable porous media, mixture theory, test embankment, back analysis, soft clay deposit.

## 1. Introduction

This paper deals with nonlinear consolidation problems characterized by finite strain and material nonlinearity. The reference framework is a recent paper by Lancellotta and Preziosi [1], in which a new mathematical model was proposed within the framework of mixture theory. Due to its derivation without smallness assumptions, this model substantially differs from the nonlinear models available in the literature [2–5], still developed in the framework of small deformation theory. Indeed, nonlinearity of model [1] is not simply induced by constitutive laws on material behaviour. In particular, the soil is assumed to be saturated and fracture-free. No restriction applies to the stress–strain and permeability–void ratio relations, but there is a one-to-one mapping in compression.

The paper consists of three parts. Section 2 is a report on the modelling procedure. A three-dimensional model is derived, by considering balance of mass and momentum for the constituents. Then, in Section 3, both Eulerian and Lagrangian formulations are derived in the one-dimensional case, in view of the application. Finally, Section 4 deals with the back-analysis of field performance of an embankment resting on a soft clay deposit.

---

* Partially supported by MURST, Minister for University and Technological Research and by C.N.R. Contract n.96.03858.ct01.

The pertinent literature and general background can be found in the recent review paper by de Boer [6], which provides an updated state-of-the-art, from pioneer works to recent developments. Engineering motivations and objectives can be found in specialized books on the above topic [7, 8].

## 2. A Three-Dimensional Mathematical Model

As already mentioned, the starting point for the derivation of the model is the mixture theory [9]. In the isothermal case and in absence of chemical reactions or phase changes, the theory of mixtures gives rise to the following system of equations:

$$\begin{aligned} &\frac{\partial \phi_{\mathrm{s}}}{\partial t} + \nabla \cdot (\phi_{\mathrm{s}} \mathbf{v}_{\mathrm{s}}) = 0, \\ &\frac{\partial \phi_{\mathrm{l}}}{\partial t} + \nabla \cdot (\phi_{\mathrm{l}} \mathbf{v}_{\mathrm{l}}) = 0, \\ &\rho_{\mathrm{s}}\phi_{\mathrm{s}} \left( \frac{\partial \mathbf{v}_{\mathrm{s}}}{\partial t} + \mathbf{v}_{\mathrm{s}} \cdot \nabla \mathbf{v}_{\mathrm{s}} \right) = \nabla \cdot \tilde{\mathbb{T}}_{\mathrm{s}} + \rho_{\mathrm{s}}\phi_{\mathrm{s}} \mathbf{g} + \mathbf{m}_{\sigma}, \\ &\rho_{\mathrm{l}}\phi_{\mathrm{l}} \left( \frac{\partial \mathbf{v}_{\mathrm{l}}}{\partial t} + \mathbf{v}_{\mathrm{l}} \cdot \nabla \mathbf{v}_{\mathrm{l}} \right) = \nabla \cdot \tilde{\mathbb{T}}_{\mathrm{l}} + \rho_{\mathrm{l}}\phi_{\mathrm{l}} \mathbf{g} - \mathbf{m}_{\sigma}, \end{aligned} \tag{2.1}$$

where s and l stand for solid and liquid phase, respectively. Moreover, referring to the $p$-constituent:

- $\rho_p$ is the 'true' density, that is the density of the material which is used as $p$-constituent of the mixture, e.g., $\rho_{\mathrm{water}} = 1000\,\mathrm{Kg/m^3}$;
- $\phi_p$ is the volume fraction, i.e., the volume occupied by the $p$-constituent over the total volume. Therefore, assuming saturation, $\phi_{\mathrm{s}} + \phi_{\mathrm{l}} = 1$;
- $\underset{\sim}{\mathbf{v}}_p$ is the velocity;
- $\tilde{\mathbb{T}}_p$ is the partial stress tensor, which describes the behaviour of the $p$-constituent when the other constituent is co-present. The difference between the partial stress $\tilde{\mathbb{T}}_p$ and the stress $\mathbb{T}_p$ of the $p$-constituent taken separately, usually named Reynolds stress, is well emphasized by a comparison with the model obtained using the ensemble average approach [10];
- $\mathbf{g}$ is the gravitational acceleration;
- $\mathbf{m}_\sigma$ is the momentum supply (also named internal body force or interaction force) and is related to the local interactions between the constituents across the interface separating them. It appears in both momentum equations, the third and fourth equations in (2.1), with a different sign, because of the type of interaction (action–reaction) force between constituents.

Adding the first two (continuity) equations yields

$$\nabla \cdot \mathbf{v}_{\mathrm{c}} = \nabla \cdot (\phi_{\mathrm{s}} \mathbf{v}_{\mathrm{s}} + \phi_{\mathrm{l}} \mathbf{v}_{\mathrm{l}}) = 0, \tag{2.2}$$

where $\mathbf{v}_c$ is the so-called composite velocity. In this way, a divergence-free velocity field is identified. In handling the model it is very useful to exploit this property using (2.2) in place of one of the two continuity equations stated in (2.1). Similar calculations can be developed for the momentum equations for the mixture, which can replace either the third or fourth equation in (2.1).

However, in order to obtain a model suitable for engineering applications, the application of Darcy's law and some further assumptions are necessary. In detail, the following ones are proposed:

(i) Inertia of the liquid constituent is neglected;

(ii) The viscous forces for the liquid phase are neglected, so that the stress tensor reduces to a scalar;

(iii) The interaction force depends linearly on $\mathbf{v}_s - \mathbf{v}_l$ through an invertible matrix, which depends on the deformation of the soil skeleton.

$$\mathbf{v}_l - \mathbf{v}_s = -\frac{\mathbb{K}(\mathbb{E}_s)}{\mu\phi_l}(\nabla P_l - \rho_l \mathbf{g}), \tag{2.3}$$

where $\mu$ is the liquid viscosity, $\mathbb{K}$ is the 'effective permeability' tensor, $P_l$ is the pore liquid pressure,

$$\mathbb{E}_s = \tfrac{1}{2}(\mathbb{F}_s\mathbb{F}_s^T - \mathbb{I}) \tag{2.4}$$

is the Lagrangian (finite deformation) strain tensor for the soil skeleton, and $\mathbb{F}_s$ is its deformation gradient;

(iv) Both inertia of the liquid and solid constituent are neglected.

Thus, technical calculations yield the following expression of the three-dimensional model:

$$\begin{aligned}
&\frac{\partial \phi_s}{\partial t} + \nabla \cdot (\phi_s \mathbf{v}_s) = 0,\\
&\nabla \cdot \mathbf{v}_c = 0,\\
&\mathbf{v}_l - \mathbf{v}_s = -\frac{\mathbb{K}(\mathbb{E}_s)}{\mu\phi_l}(\nabla P_l - \rho_l \mathbf{g}),\\
&\nabla \cdot \mathbb{T}_m + \rho_m \mathbf{g} = 0,
\end{aligned} \tag{2.5}$$

where

$$\mathbb{T}_m = \widetilde{\mathbb{T}}_s + \widetilde{\mathbb{T}}_l - \rho_s\phi_s(\mathbf{v}_s - \mathbf{v}_m) \otimes (\mathbf{v}_s - \mathbf{v}_m) - \rho_l\phi_l(\mathbf{v}_l - \mathbf{v}_m) \otimes (\mathbf{v}_l - \mathbf{v}_m), \tag{2.6}$$

is the stress tensor of the mixture considered as a whole (see [9, 10]) and the quantities $\mathbf{v}_s - \mathbf{v}_m$ and $\mathbf{v}_l - \mathbf{v}_m$ are the so-called diffusive velocities.

A basic equation introduced in all consolidation models is Darcy's law, which can be deduced from the momentum equation for the liquid constituent (2.4).

## 3. Eulerian and Lagrangian Formulation in One Space Dimension

Calculations related to several interesting applications, such as the ones dealt with in the sections which follow, can be studied in one space dimension. Therefore, it is now assumed that both flow and strain take place along the vertical direction, and that the medium is isotropic in horizontal planes, so that consolidation occurs along a principal direction of the permeability tensor. In particular, the above model can be rewritten, in one space dimension, as follows:

$$\begin{aligned} &\frac{\partial \phi_s}{\partial t} + \frac{\partial}{\partial z}(\phi_s v_s) = 0, && \frac{\partial v_c}{\partial z} = 0, \\ &v_l - v_s = -\frac{K(\phi_s)}{\mu(1-\phi_s)}\left(\frac{\partial P_l}{\partial z} + \rho_l g\right), && \frac{\partial P_l}{\partial z} + \frac{\partial \sigma'}{\partial z} + \rho_m g = 0, \end{aligned} \tag{3.1}$$

where $\sigma' = \sigma_m - P_l$ is the 'effective stress' and $\sigma_m$ is the total stress in the mixture: both of them are taken as positive in compression.

The above equation has been obtained by also recalling that the continuity equation of the solid constituent can be written in Lagrangian form as

$$\det \mathbb{F}_s = \frac{\phi_s^*}{\phi_s}, \tag{3.2}$$

where $\phi_s^*$ is the volume ratio of the undeformed reference configuration, and that in one-dimension the only nontrivial component of the Lagrangian strain tensor is

$$E_{zz} = \frac{1}{2}\left(\frac{\phi_s^{*2}}{\phi_s^2} - 1\right). \tag{3.3}$$

It follows that the dependence of the permeability tensor $\mathbb{K}$ on $\mathbb{E}_s$ is equivalent in one-dimensional problems to that on $\phi_s$. The same is not true for the stress tensor $\mathbb{T}_m$, which refers to the whole mixture. In this case the possible dependence on the Lagrangian strain tensor involves not only the volume fraction, but also the fluid properties.

The final result is obtained by technical calculations reported in [1] and is stated by the following equation:

$$\frac{\partial e}{\partial t} + C(t)\frac{\partial e}{\partial z} + (1+e)^2\frac{\partial}{\partial z}\left[\frac{1}{1+e}\frac{K(e)}{\mu}\left(\frac{\partial \sigma'}{\partial z} + \frac{\rho_s - \rho_l}{1+e}g\right)\right] = 0\,, \tag{3.4}$$

where $e$ is the void ratio

$$e = \frac{\phi_l}{\phi_s} = \frac{1}{\phi_s} - 1\,, \tag{3.5}$$

and where the constitutive relation for the effective stress $\sigma'$ is still to be specified.

In order to obtain the Lagrangian formulation, recall that the material form of the continuity conditions is given by

$$\frac{\mathrm{d}}{\mathrm{d}t}(\rho_s \phi_s \det \mathbb{F}_s) = 0, \tag{3.6}$$

which in one dimension reduces to

$$\frac{\partial z}{\partial Z} = \frac{\phi_s^*}{\phi_s} = \frac{1+e}{1+e^*}, \tag{3.7}$$

where $e^*$ is the void ratio in the undeformed reference configuration and where the material coordinate $z$ indicates the current position of the particle that in the undeformed reference configuration was identified by the reference coordinate $Z$.

Then, technical calculations and the assumption that the excess pore pressure is related to the flow through Darcy's law, which can be rewritten in a Lagrangian framework as

$$\frac{e}{1+e}(v_l - v_s) = -\frac{k(e)}{\rho_l g}\frac{\partial u}{\partial Z}\frac{\partial Z}{\partial z}, \tag{3.8}$$

yield

$$\frac{\partial e}{\partial t} + (1+e^*)\frac{\partial}{\partial Z}\frac{k(e)}{\rho_l g}\frac{1+e^*}{1+e}\left(\frac{\partial \sigma'}{\partial Z} + \frac{\rho_s - \rho_l}{1+e^*}g\right) = 0. \tag{3.9}$$

*Remark 3.1.* The Lagrangian formulation is often preferred to the Eulerian formulation since it does not involve the $C(t)$ term which appears in Equations (3.4) and (3.6). Furthermore, Equation (3.9) has the advantage that it is defined on a fixed domain, while (3.4) usually involves integration over a time-varying domain.

*Remark 3.2.* Both models have been derived without any smallness assumption regarding the void ratio-effective stress, or the void ratio-permeability relationship, and for this reason represent a general formulation. Anyway, in order to solve them, the above relationships must be specified as 'single-valued' functions. In particular, the fact that the relation is single-valued means that one is focusing on compression problems obtained, for instance, imposing a load on the soil layer. The release of the load may, in fact, in principle, lead to an expansion of the layer, which certainly will not follow the same rule due to anelastic soil behaviour.

The evolution of the system can be obtained solving, by suitable computational methods [12], the initial boundary value problem implemented with the above models. The structure of the equation is such that, besides the initial condition, two boundary conditions, one on each boundary, are needed in order to have a properly formulated problem. In addition, in the Eulerian formulation the quantity $C(t)$ and the evolution equations for the moving boundaries still need to be determined.

The statement of the boundary conditions must be related to the analysis of the physics of the system. The general procedure and the analysis of two specific cases: *consolidation of a draining layer* and *sedimentation and consolidation under own weight* are dealt with in [1].

The quantitative analysis of the above problems, which is developed by generalized collocation methods, can be addressed to the back analysis described in the next section.

## 4. Back Analysis of Field Performance

As an example of application of the above presented mathematical model for soil consolidation problems, the case history of Porto Tolle will be briefly discussed here (see Figure 1).

The area being considered covers a very flat delta plain (see Figure 1(b) and 1(c)), with original elevation ranging between −1 and −2 m.s.l. This grade was raised to elevation 0.00–0.50 m by placing a sand fill during the years 1974–1975. The soil profile (see Figure 1(d)) consists of a stratum of very recent, loose to medium silty sand with embedded thin layers and lenses of soft cohesive and organic sediments. Below this stratum a 20–22 m thick deposit of light brown soft silty-clay is encountered. It contains frequent seams and lenses of fine silty-sand, which never exceed a few centimeters. In its lower part, the presence of organic matter in the silty-clay matrix is quite frequent, and during the borings a number of small pockets of organic gas were encountered. The soft silty-clay deposit terminates abruptly at a depth of 28–30 m in a medium dense sand.

The ground water level in the upper sand stratum ranges between −1 and −1.5 m.s.l.; in the lower sand stratum it exceeds the level found in the upper one by 1.5–2 m. As a consequence, the clay is subjected to a slight upward seepage. The

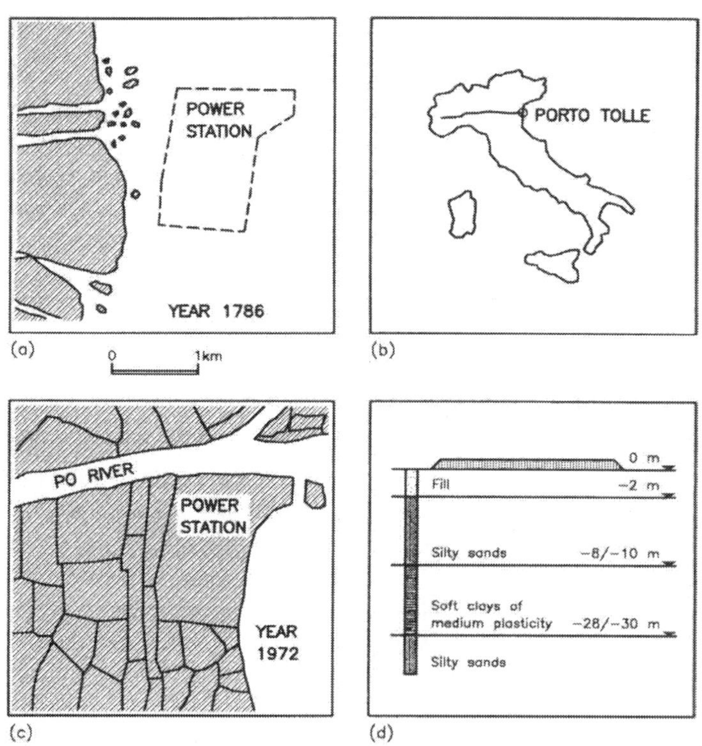


*Figure 1.* Location of Porto Tolle site and soil profile.

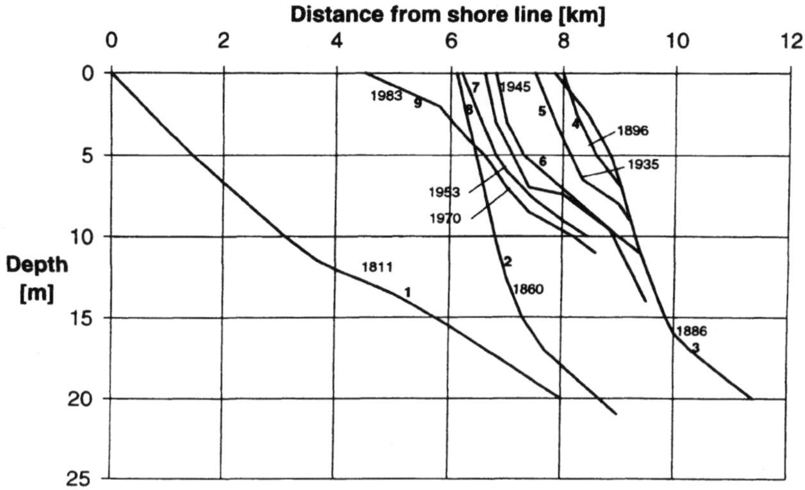


*Figure 2.* Shore history.

local geology indicates the presence of holocene deposits up to 150–200 m thick. Since the soft clay deposits do not exceed 6 000–10 000 years in age, this thickness corresponds to a very high sedimentation rate. In particular, the upper silty sand strata have been deposited within the past 200 years (see Figures 1(a) and 2). This geological information confirms that the soft cohesive deposit is only slightly overconsolidated, due to ground water oscillation.

Nine floating-roof storage tanks were built on this area during the construction of the 2.4 *GW* thermal power plant of ENEL, and, because of the expected large settlements, a full scale investigation was carried out by means of a large test embankment [13], whose behaviour over time was extensively monitored.

The main aim of this study was to investigate the consolidation behaviour of the thick layer of soft clay, under the imposed load of a test embankment. To this purpose, the surface settlements, the vertical displacements of the soil strata up to a depth of 40 m and the excess pore water pressures have been monitored for a period of more than three years, i.e., during and after the construction of the embankment.

The experimental embankment was 200 m × 250 m wide, with a height of 7 m, resulting in a uniform pressure on the soil of 114 kPa. It was built in less than 3 months.

Before the construction of the embankment, several instruments of various types were installed: surface reference plates, to measure the settlement of the ground surface; eight instrumented boreholes (with inductive measuring devices), to measure subsurface soil displacements, and 110 piezometers (62 electrical and 48 closed hydraulic type piezometers), to monitor at various depths in the ground the pore water pressures.

The instrumentation was installed up to a depth of 40 m under the embankment, in order to identify the vertical movements and the pore water pressure increase caused

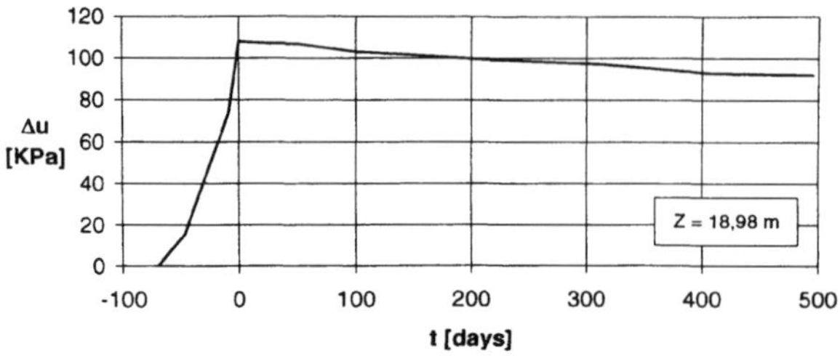


*Figure 3.* An example of excess pore pressure $\Delta u$ dissipation with time $t$.

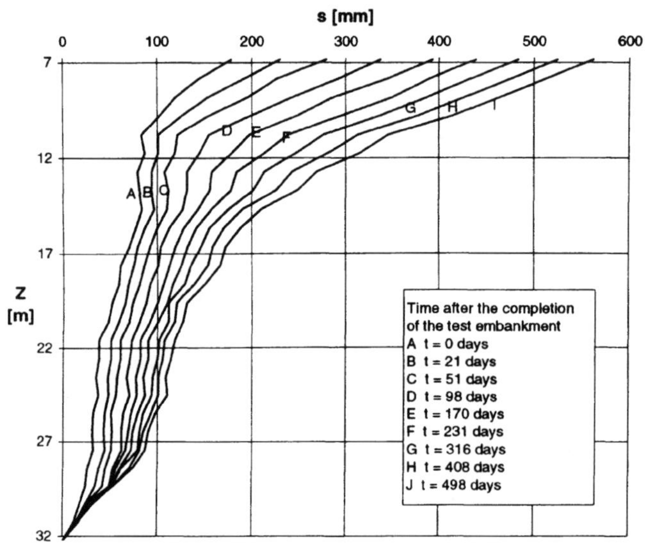


*Figure 4.* Settlement variation with depth at various times (settlement isochrones).

by the consolidation of the clay layer, and, on the other hand, the rapid compression of the sand strata (the upper and the deep layers).

Figure 3 shows an example of the variation over time of the excess pore pressure $\Delta u$, measured at a depth of 19 m; the zero time assumed as a reference is the completion of the embankment. At this time, the maximum value of the pore pressure increment in the clay layer was reached, and thereafter the soil consolidation occurred under the whole weight of the embankment. The progressive increase of the consolidation settlement was simultaneously monitored.

Figure 4 shows settlement values, measured at various depths in the clay layer, during the consolidation process: each line represents an isochrone in terms of settlement.

Figure 5 reports the excess pore water pressure values versus depth, as measured in the clay, after 160 days from the completion of the embankment (isochrone in terms of excess pore pressure). These values, measured at the time $t$ within the thickness $2H$ of the consolidating layer, are interpolated by the shown continuous line.

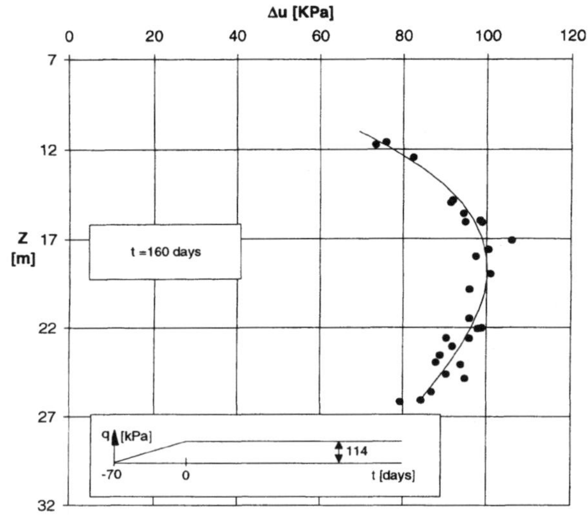


*Figure 5.* An example of measured excess pore pressure $\Delta u$ distribution with depth.

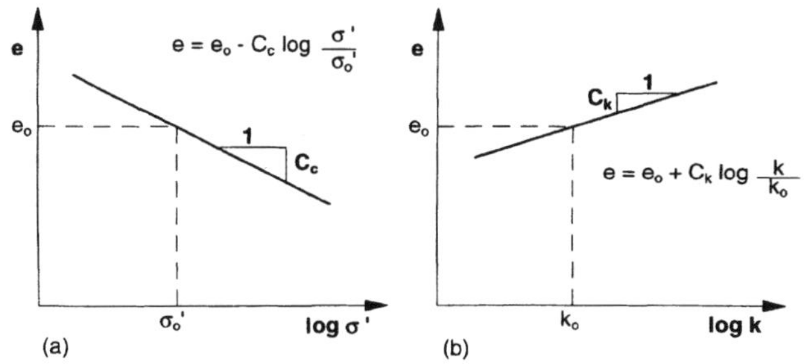


*Figure 6.* Constitutive laws for soft clays: (a) Void ratio $e$, effective stress $\sigma'$; (b) Void ratio $e$, hydraulic conductivity $k$.

The use of Equation (3.4) or (3.9) requires specification of constitutive laws, i.e., changes of void ratio with effective stress, as well as changes of hydraulic conductivity with void ratio, as schematically shown in Figure 6.

In the present analysis the compressibility parameters have been introduced according to the values deduced from laboratory tests (see [4, 15] for more insight into this aspect), whereas the hydraulic conductivity has been back-figured from field performance.

An example of this kind of back analysis is presented in Figure 7, in which a comparison of the calculated excess pore pressure isochrone in the clay layer (continuous line) and the measured values at the same time are shown. The values of $k_0$ (initial value of hydraulic conductivity, corresponding to a value $e_0$ of the void ratio) and $C_k$ (Figure 6) used for the analysis are also shown.

A remarkable increase of meaningfulness can be achieved if the analysis is done only with reference to the field performance of the consolidating clay layer: in this

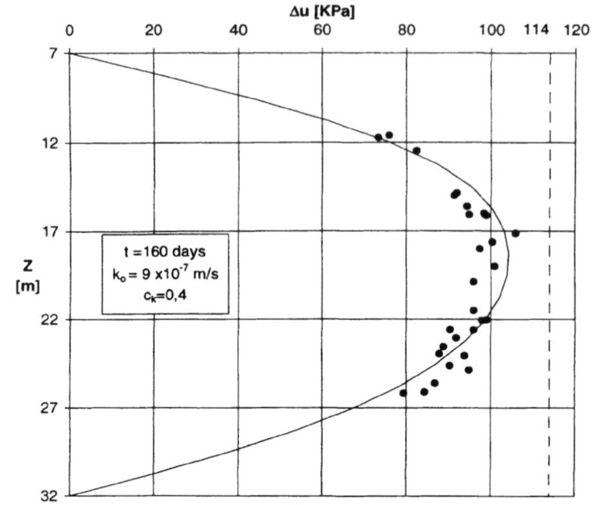


*Figure 7.* Computed versus measured excess pore pressure distribution with depth.

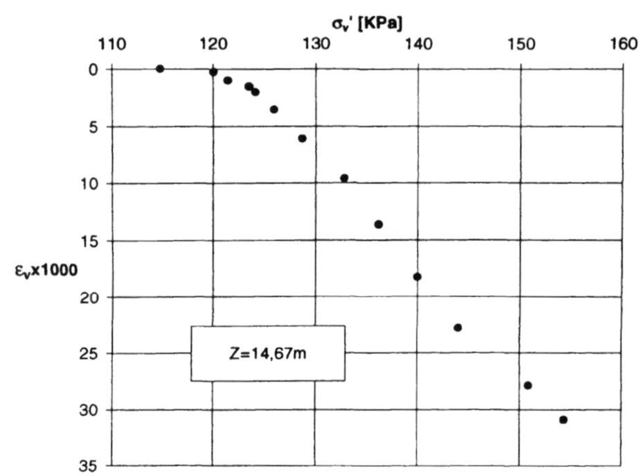


*Figure 8.* Stress–strain curve as obtained from in situ measurement of settlement and excess pore pressure.

case the compressibility parameters are deduced from the knowledge of the excess pore pressure and the associated settlements value.

In this view, experimental and full-scale stress-strain curves of the soil at various depths have been derived, on the basis of the vertical strain values and the corresponding vertical effective stress values.

Figure 8 shows an example of one of these curves, deduced from the pore pressure and vertical strain $\varepsilon_v$ measured at a depth of 14.67 m (upper part of the consolidating layer): the form of this curve points out a stress level corresponding to an abrupt change in the compressibility of the material. This stress value is the preconsolidation vertical stress, i.e., the maximum vertical effective stress ever supported by the soil in the past.

Comparing the values of the preconsolidation stress, evaluated in this way, with those estimated in the laboratory from the compression curves of incremental loading oedometer tests, it can be seen that a good agreement exists, irrespective of the depth within the clay layer.

The results presented in Figure 7 are the first examples of application of mixture theory based models to full scale engineering problems. The model presents a large capability in capturing the most relevant aspects of the problem as presented by the influence of self-weight and full nonlinearity (both geometric and material nonlinearities are taken into account).

## References


1. Lancellotta, R. and Preziosi, L.: *Int. J. Engng. Sci.* **35**(10–11) (1997), 1045–1063.
2. Davis, E. H. and Raymond, G. P.: A non-linear theory of consolidation, *Geotechnique* **15** (1965), 161–173.
3. Janbu, N.: Consolidation of clay layers based on non-linear stress strain, *Proc. VI ICSMFE*, Montreal, 1965, pp. 83–87.
4. Mikasa, M.: *The Consolidation of Soft Clay*, Civil Engineering in Japan, Japanese Society of Civil Engineering, 1965, pp. 21–26.
5. Cornetti, P. and Battaglio, M.: Nonlinear consolidation of soil: modelling and solution techniques, *Math. Comput. Modelling* **20** (1994), 1–12.
6. de Boer, R.: Highlights in the historical development of the porous media theory: toward a consistent macroscopic theory, *Appl. Mech. Rev.* **49** (1996), 201–262.
7. Lancellotta, R.: *Geotechnical Engineering*, Balkema, 1995.
8. Lee, I. K., White, W. and Ingles, O. G.: *Geotechnical Engineering*, Pitman, 1983.
9. Bowen, R. M.: Incompressible porous media models by the use of the theory of mixtures, *Int. J. Engng. Sci.* **18** (1980), 1129–1148.
10. Preziosi, L.: The theory of deformable porous media and its application to composite material manufacturing, *Surveys Math. Indust.* **6** (1996), 167–214.
11. Munaf, D., Wineman, A. S., Rajagopal, K. R. and Lee, D.W.: A boundary value problem in groundwater motion analysis – Comparison of predictions based on Darcy's law and the continuum theory of mixtures, *Math. Models Methods Appl. Sci.* **3** (1993), 231–248.
12. Bellomo, N. and Preziosi, L.: *Mathematical Modelling, Analysis and Scientific Computation*, CRC Press, Boca Raton, 1995.
13. Billotta, E. and Viggiani, C.: Una indagine sperimentale in vera grandezza sul comportamento di un banco di argilla normalmente consolidata, *XII Conv. Naz. Geotecnica*, Cosenza, 1975, pp. 232–240.
14. Skempton, A. W.: The consolidation of clays by gravitational compaction, *Q. J. Geol. Soc. London* **125** (1970), 373–412.
15. Burland, J. B.: On the compressibility and shear strength of natural clays, *Geotechnique* **40** (1990), 329–378.
16. Hegg, U., Jamiolkowski, M. B., Lancellotta, R. and Parvis, E.: Performance of large oil tanks on soft ground, *Proc. Piling and Ground Treatment for Foundations*, T. Telford, London, 1983, pp. 87–94.

# The Peculiarities of Linear Wave Propagation in Double Porous Media

A. A. GUBAIDULLIN and O. YU. KUCHUGURINA
*Tyumen Institute of Mechanics of Multiphase Systems, Siberian Branch of the Russian Academy of Sciences, Taimyrskaya St. 74, Tyumen, 625000 Russia*

(Received: 22 June 1997)

**Abstract.** The features of propagation of longitudinal and transverse waves (LW and TW) in fractured porous medium (FPM) saturated with liquid are investigated by methods of multiphase mechanics. The mathematical model of FPM accounting for inequality of velocities and pressures of liquid in pores and fractures, liquid mass exchange and nonstationary interaction forces is developed. Processes of monochromatic wave propagation are studied. The dispersion relation is obtained and the effect of model parameters on wave propagation is analysed. It is established that one transverse and three longitudinal waves propagate in FPM saturated with liquid. The fastest LW is a deformational wave and the two others are filtrational. Filtrational waves attenuate much stronger than deformational and transverse waves. Distinction of velocities and pressures in liquid in various pore systems provides an explanation for the existence of the two filtrational waves in porous medium with two different characteristic sizes of pores.

**Key words:** fractured porous medium, skeleton, longitudinal and transverse waves.

## 1. Introduction

Consider porous medium with pores of two characteristic sizes which may differ by an order of magnitude or more: primary (intergranular) – rather small pores, and secondary – the large-size pores. The secondary pores represented by fractures and caves are the result of mechanical or chemical action. Large pores which are usually fractures divide medium into blocks while small pores divide it into grains. Volume of liquid in fractures may be much smaller than that in the blocks or the volumes may be of the same order. One of the most typical sample of medium with double porosity, that is, fractured porous medium (FPM) is schematically shown in Figure 1.

Problems regarding motion of fluids in fractured and FPM have previously been under consideration by many researchers. A set of equations governing processes of liquid filtration in media with double porosity was suggested by Barenblatt *et al.* (1960). The processes of nonstationary filtration of liquid and gas were then studied by Nikolayevskii *et al.* (1970, 1984) and Barenblatt *et al.* (1984). In the papers by Barenblatt *et al.* (1960), Nikolayevskii *et al.* (1970, 1984) and Barenblatt *et al.* (1984) the skeleton of the medium was assumed to be immovable. In the paper by

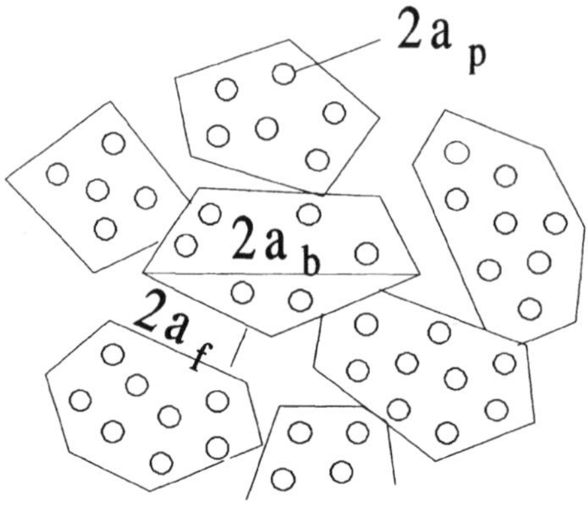


*Figure 1.* Scheme of FPM.

Nikolayevskii (1984) it was suggested to account for dependence of porosity on pressure of liquid but in the further analysis the dependence was neglected. Buyevich (1984) considered FPM as two coexisting homogeneous porous continua. The influence of stressed state of material and liquid pressure on effective parameters of continua was considered and a system of equations other than equations of filtration by Barenblatt *et al.* (1960) was derived.

Processes of filtration of liquid and gas through porous materials were considered by Dinariyev (1990, 1992, 1994) with the use of a fractal model of FPM. In the paper by Saucier (1992), the analytical method is presented which qualitatively describes full permeability of heterogeneous porous medium with different scales of heterogeneity and their different distribution in space. Auriault and Boutin (1992) and Auriault and Royer (1993, 1994) investigated filtration of compressible fluid through FPM using method of asymptotic series. In the paper by Bai *et al.* (1994) the analytical solution of the problem on filtration of homogeneous liquid in fractured medium, which is permeable via blocks and fractures, was considered. In the work by Chen (1994) the equations of flow in double porous medium are derived by means of the spatial averaging method. Unlike traditional models, in these equations intensity of liquid transfer between pores and fractures depends not only on difference of pressures in blocks and fractures, but on their gradients as well.

Wilson and Aifantis developed the theory of consolidation with double porosity (1982) and investigated acoustic wave propagation in fractured porous rock (1984). In their paper (1984) porous medium is assumed to consist of two degrees of porosity and the resulting three-velocity model thus yields three types of longitudinal waves (LWs); one associated with the elastic properties of the matrix material and one each for the fluids in the pore space and the fracture space. Some special limiting cases are studied analytically. Two approaches are considered: one based on Biot's approach utilizing the 'viscodynamic' operator and one based on Aifantis' 'multiporosity theory'. The research is carried out in terms of pressures of liquid in pores and fractures and total stress of the medium, that is, without using the concept of effective stress.

In the work by Egorov *et al.* (1990) mathematical models of acoustic wave propagation in saturated bounded and unbounded porous and fractured porous media were considered. It is assumed in FPM that porous blocks and liquid within the block have the same velocity, but mass exchange between blocks and fractures is accounted for. In the scheme of the two-velocity model used by the authors there are two LWs propagating in the medium. Some estimations of velocity and attenuation of the fast wave are made.

In the papers by Tuncay and Corapcioglu (1996a, b) a theory of wave propagation in FPM is based on the double porosity concept. The macroscopic constitutive relations of elastic porous media saturated by one or two fluids and saturated FPM can be obtained from the constitutive relations developed in the paper. In the simplest case, the final set of governing equations reduce to Biot's equations (1956a, b, 1962). Body waves in FPM saturated by two fluids are studied. The final equations of the four-velocity motion are written in terms of displacements only. The authors show the existence of four compressional and one rotational wave. In the authors' opinion, the second compressional wave arises because of fractures, whereas the fourth compressional wave is associated with the pressure difference between the fluid phases in the porous blocks. The effects of fractures on the phase velocity and attenuation coefficient of body waves are numerically investigated for a fractured sandstone saturated by air and water phases. All compressional waves are diffusive-type waves, that is, highly attenuated, and do not exist at low frequencies.

It should be noted that most of the works considering media with double porosity are devoted to filtration processes, whereas features of wave propagation in such media have not been adequately studied. In the present paper we develop a three-velocity model for FPM. The constitutive equation for effective stress of skeleton is derived; the viscous-inertial effects are accounted in interaction force between liquid and skeleton and in liquid exchange between pores and fractures.

In order to study processes of linear wave propagation we suggest a model of the medium with double porosity which accounts for inequality of velocities and pressures in liquid in systems of pores and fractures and mass exchange between pore systems. We distinguish solid phase (parameters of solid phase have subscript s), liquid in pores (subscript p) and liquid in fractures (subscript f). The subscript l refers to parameters of liquid averaged over the volume of all liquid, or to constant physical parameters of liquid. Characteristic linear sizes of the medium are mean radius of primary pores $a_p$, mean half-width of fractures $a_f$, and half of mean diameter of porous block $a_b$ (see Figure 1).

## 2. Effective Stress in the Medium with Double Porosity. Mass and Momentum Balance Equations

For adequate description of mechanical processes in porous media it is necessary to know effective stress $\sigma_{s*}$. In usual porous medium it is proportional to the difference of true stresses in phases, and total stress in medium is a sum of effective stress $\sigma_{s*}$ and stress in liquid phase $\sigma_l$:

$$\sigma_{s*} = \alpha_s(\sigma_s - \sigma_l), \qquad \sigma = \sigma_{s*} + \sigma_l. \tag{1}$$

Here, $\alpha_s$ is the volume fraction of the solid phase, $\sigma_s$ and $\sigma_l$ are averaged microscopic stresses of solid and liquid phases correspondingly. However, in a double porous medium, pressures in liquid in primary and secondary pores may be different. So averaged stress in liquid needs to be expressed in terms of stress in pores $\sigma_p$ and fractures $\sigma_f$. It results in the following expression:

$$\sigma_{s*} = \alpha_s(\sigma_s - \sigma_l), \qquad \sigma_l = \frac{\alpha_p\sigma_p + \alpha_f\sigma_f}{\alpha_p + \alpha_f}. \tag{2}$$

Here, $\alpha_k$, $\sigma_k$ are volume fraction and averaged microscopic stress in $k$-phase correspondingly ($k$ = s, p, f), $\sigma_l$ is averaged stress in liquid over the volume of all pores; at that total stress in the medium, $\sigma$ is a sum of effective stress of skeleton $\sigma_{s*}$ and averaged stress of liquid phase $\sigma_l$ as well as in usual porous medium, that is

$$\sigma = \sigma_{s*} + \sigma_l. \tag{3}$$

Equations of mass and momentum balance (Nigmatulin, 1978, 1990) with account of (2) for solid skeleton, liquid in pores and fractures are written as

$$\begin{gathered}
\frac{\partial \rho_s}{\partial t} + \nabla^k \rho_s v_s^k = 0, \\
\frac{\partial \rho_p}{\partial t} + \nabla^k \rho_p v_p^k = -q, \qquad \frac{\partial \rho_f}{\partial t} + \nabla^k \rho_f v_f^k = q, \\
\rho_s \frac{\mathrm{d}v_s^k}{\mathrm{d}t} = -\alpha_s \nabla^k p_l + \nabla^j \sigma_{s*}^{jk} + F_p^k + F_f^k, \qquad p_l = \frac{\alpha_p p_p + \alpha_f p_f}{\alpha_p + \alpha_f},
\end{gathered} \tag{4}$$

$$\rho_p \frac{\mathrm{d}v_p^k}{\mathrm{d}t} = -\alpha_p \nabla^k p_p - F_p^k - q v_p^k, \qquad \rho_f \frac{\mathrm{d}v_f^k}{\mathrm{d}t} = -\alpha_f \nabla^k p_f - F_f^k + q v_f^k. \tag{5}$$

Here, $\alpha_k$, $\rho_k$, $v_k$ are volume fraction, reduced density and velocity of $k$-phase correspondingly ($k$ = s, p, f); $p_p$, $p_f$ are pressures in pores and fractures, $\sigma_{s*}$ is the effective stress of the skeleton; $q$ the intensity of mass exchange between pores and fractures, $F_p$, $F_f$ are forces of interaction between the solid phase and liquid in pores and fractures correspondingly. Superscripts are coordinate indices, summation is implied over the repetitive superscripts. It is assumed that averaged viscous stress tensor in liquid is negligible and the viscous effects are considered in the processes of interaction of liquid and skeleton and liquid exchange between pore systems only.

The following equations for true densities of phases $\rho_k^\circ$ are valid:

$$\rho_k = \alpha_k \rho_k^\circ \quad (k = \text{s, p, f}), \qquad \alpha_s + \alpha_p + \alpha_f = 1. \tag{6}$$

Expressions for interaction forces $F_p$, $F_f$ in (5) for harmonic oscillations of frequency $\omega$ are written as

$$\begin{aligned} &F_p = F_{mp} + F_{\mu p} + F_{Bp}, \qquad F_f = F_{mf} + F_{\mu f} + F_{Bf}, \\ &F_{mj} = 0.5\eta_{mj}\alpha_s\alpha_j\rho_l^\circ i\omega(\nu_j - \nu_s), \quad (j = p, f), \\ &F_{\mu j} = \eta_{\mu j}\alpha_s\alpha_j a_j^{-2}\mu_l(\nu_j - \nu_s), \\ &F_{Bj} - \eta_{Bj}\alpha_s\alpha_j a_j^{-1}\sqrt{2\rho_l^\circ\omega\mu_l}(1+i)(\nu_j - \nu_s), \end{aligned} \tag{7}$$

where $F_{mk}$ is the associated mass force, $F_{\mu k}$ is analogous to viscous force of Stokes, $F_{Bk}$ is Basse force due to the nonstationary viscous layer near the interphase boundary; i is an imaginary unit; $\mu_l$ the liquid viscosity, $\eta_{mk}$, $\eta_{\mu k}$, $\eta_{Bk}$ are factors taking into account the structure of the medium.

When pressures $p_p$, $p_f$ are different, the exchange of liquid between systems of pores may take place. When studying processes of filtration of liquid in double porous media, the following expression for $q$ is usually used (Barenblatt *et al.*, 1984):

$$q = \eta_q \frac{\rho_l^\circ k_p}{\mu_l}\frac{p_p - p_f}{a_b^2}, \tag{8}$$

where $k_p$ is the permeability of the primary pore system, $\eta_q$ a factor characterising the geometry of the medium. Formula (8) was obtained for inertialess flow. It should be noted that permeability $k_p$ may be expressed in terms of a factor in stationary viscous force $F_{\mu p}$, namely: $k_p^{-1} = (\alpha_s/\alpha_p)(\eta_{\mu p}a_p^{-2})$. In order to account for nonstationary effects of liquid motion, $\eta_{\mu p}a_p^{-2}$ needs to be replaced by the factor in the full viscous force $(F_{\mu p} + F_{Bp})$

$$\eta_{\mu p}a_p^{-2} + \eta_{Bp}a_p^{-1}\sqrt{\frac{2\rho_l^\circ\omega}{\mu_l}}(1+i),$$

which depends on angular frequency $\omega$. So the final expression for intensity of liquid exchange $q$ is written in the form:

$$q = \eta_q \frac{\alpha_p}{\alpha_s} f(\omega)\frac{\rho_l^\circ}{\mu_l}\frac{p_p - p_f}{a_b^2}, \tag{9}$$

$$f(\omega) = \left(\eta_{\mu p}a_p^{-2} + \eta_{Bp}a_p^{-1}\sqrt{\frac{2\rho_l^\circ\omega}{\mu_l}}(1+i)\right)^{-1}.$$

## 3. Constitutive Equations for Elastic Skeleton and for Phases

In order to obtain constitutive equations for skeleton of double porous medium, we follow the method which was used before for usual porous medium (Nikolayevskii *et al.*, 1970; Nigmatulin, 1978). As was shown in these papers, macroscopic strain tensor of solid phase $\varepsilon_s$ is determined not only by microscopic strain tensor of the skeleton material (write $\mathring{\varepsilon}_s$ for averaged microscopic strain tensor), but by deformation due to displacement of grains relative to each other as well. The latter part of the macroscopic strain $\varepsilon_s$ is named by effective strain and is written as $\varepsilon_{s*}$

$$\varepsilon_s = \mathring{\varepsilon}_s + \varepsilon_{s*}, \tag{10}$$

$$\frac{d}{dt}\varepsilon_s^{kl} = \frac{1}{2}(\nabla^k v_s^l + \nabla^l v_s^k). \tag{11}$$

It would appear reasonable if in double porous medium effective deformation, $\varepsilon_{s*}$ results from displacements of grains relative to each other within the porous blocks as well as from displacement of blocks relative to each other. So, we will write it as follows: $\varepsilon_{s*p}$ and $\varepsilon_{s*f}$. First, assume that the effective strain $\varepsilon_{s*}$ is a sum of $\varepsilon_{s*p}$ and $\varepsilon_{s*f}$:

$$\varepsilon_{s*} = \varepsilon_{s*p} + \varepsilon_{s*f} \tag{12}$$

and

$$\varepsilon_{s*p} = (1-\eta)\varepsilon_{s*}, \qquad \varepsilon_{s*f} = \eta\varepsilon_{s*}, \tag{13}$$

where $\eta$ is some scalar value. In general, the linear relation between $\varepsilon_{s*p}$, $\varepsilon_{s*f}$ and $\varepsilon_{s*}$ is of tensor character, that is,

$$\varepsilon_{s*f}^{kl} = \gamma^{klmn}\varepsilon_{s*}^{mn}, \qquad \varepsilon_{s*p}^{kl} = \varepsilon_{s*}^{kl} - \gamma^{klmn}\varepsilon_{s*}^{mn},$$

but for simplicity we assume that this relation is determined by the scalar value $\eta$.

The following assumption is that effective stress $\sigma_{s*}$ depends linearly on the effective strain $\varepsilon_{s*}$. Then, taking (12) into consideration, $\sigma_{s*}$ can be written as a sum

$$\sigma_{s*} = \sigma_{s*p} + \sigma_{s*f}, \tag{14}$$

where $\sigma_{s*p}$ is the part of $\sigma_{s*}$ which depends linearly on $\varepsilon_{s*p}$ and $\varepsilon_{s*f}$ depends on $\varepsilon_{s*f}$. Analogous to effective stress in usual porous medium (see (1)), we believe each of the stresses $\sigma_{s*p}$, $\sigma_{s*f}$ to be proportional to the difference of averaged stresses in phases with some factor of proportionality:

$$\sigma_{s*p} = \kappa_p(\sigma_s - \sigma_p), \qquad \sigma_{s*f} = \kappa_f(\sigma_s - \sigma_f).$$

So, $\sigma_{s*p}$ is the effective stress between the grains within the block, and $\sigma_{s*f}$ is the effective stress between the porous blocks. Inserting the latter equalities into (14) and taking (2) into account, one can define the factors $\kappa_p$, $\kappa_f$ as

$$\kappa_p = \frac{\alpha_s\alpha_p}{\alpha_p + \alpha_f}, \qquad \kappa_f = \frac{\alpha_s\alpha_f}{\alpha_p + \alpha_f}$$

at that $\kappa_p + \kappa_f = \alpha_s$. Thus,

$$\sigma_{s*p} = \frac{\alpha_s\alpha_p}{\alpha_p + \alpha_f}(\sigma_s - \sigma_p), \qquad \sigma_{s*f} = \frac{\alpha_s\alpha_f}{\alpha_p + \alpha_f}(\sigma_s - \sigma_f). \tag{15}$$

Effective stresses $\sigma_{s*p}$ and $\sigma_{s*f}$ may be interpreted as follows: $\sigma_{s*p}$ is a part of $\sigma_{s*}$ which governs momentum transfer in the solid phase through the contacts between the grains within the porous block while $\sigma_{s*f}$ is a part of $\sigma_{s*}$ which governs momentum transfer through contacts between the blocks. In the limiting cases, when $\alpha_p = 0$ (fractured medium) or $\alpha_f = 0$ (usual porous medium), there $\sigma_{s*p} = 0$, $\sigma_{s*} = \sigma_{s*f}$ or $\sigma_{s*f} = 0$, $\sigma_{s*} = \sigma_{s*p}$, correspondingly.

For slight deformation of medium it is then assumed that each of the relations $\sigma^{kl}_{s*p}(\varepsilon^{mn}_{s*p})$, $\sigma^{kl}_{s*f}(\varepsilon^{mn}_{s*f})$ satisfies the Hook law with some elastic moduli which characterizes the skeleton of the medium with double porosity:

$$\begin{aligned}\sigma^{kl}_{s*p} &= \alpha_s(\tilde{\lambda}_{*p}\varepsilon^{mm}_{s*p}\delta^{kl} + 2\mu_{*p}\varepsilon^{kl}_{s*p}),\\ \sigma^{kl}_{s*f} &= \alpha_s(\tilde{\lambda}_{*f}\varepsilon^{mm}_{s*f}\delta^{kl} + 2\mu_{*f}\varepsilon^{kl}_{s*f}),\end{aligned} \tag{16}$$

here, $\delta^{kl}$ is the Kronecker delta symbol.

Strain tensor $\mathring{\varepsilon}_s$ satisfies the Hook law:

$$\mathring{\varepsilon}^{kl}_s = \frac{1}{2\mu_s}\left(\sigma^{kl}_s - \frac{\lambda_s}{3\lambda_s + 2\mu_s}\sigma^{mm}_s\delta^{kl}\right), \tag{17}$$

where $\lambda_s$, $\mu_s$ are elastic moduli for solid phase material.

Taking into consideration (10), (11), (15), (17) and $\sigma^{kl}_p = -\delta^{kl}p_p$, $\sigma^{kl}_f = -\delta^{kl}p_f$, the Equations in (16) are written as

$$\begin{aligned}\varepsilon^{kl}_{s*p} &= (1-\eta)\left[\varepsilon^{kl}_s - \frac{1}{2\mu_s}\frac{\alpha_p+\alpha_f}{\alpha_s\alpha_p}\sigma^{kl}_{s*p} + \right.\\ &\quad \left. + \frac{\lambda_s}{2\mu_s(3\lambda_s+2\mu_s)}\frac{\alpha_p+\alpha_f}{\alpha_s\alpha_p}\sigma^{mm}_{s*p}\delta^{kl} + \frac{p_p}{3\lambda_s+2\mu_s}\delta^{kl}\right],\\ \varepsilon^{kl}_{s*f} &= \eta\left[\varepsilon^{kl}_s - \frac{1}{2\mu_s}\frac{\alpha_p+\alpha_f}{\alpha_s\alpha_f}\sigma^{kl}_{s*f} + \right.\\ &\quad \left. + \frac{\lambda_s}{2\mu_s(3\lambda_s+2\mu_s)}\frac{\alpha_p+\alpha_f}{\alpha_s\alpha_f}\sigma^{mm}_{s*f}\delta^{kl} + \frac{p_f}{3\lambda_s+2\mu_s}\delta^{kl}\right].\end{aligned} \tag{18}$$

Having inserted (18) into (16) and having solved it for $\sigma_{s*p}$ and $\sigma_{s*f}$, we obtain

$$\begin{aligned}\sigma^{kl}_{s*p} &= (1-\eta)\alpha_s[\lambda_{*p}\varepsilon^{mm}_s\delta^{kl} + 2\mu_{*p}\varepsilon^{kl}_s + \nu_{*p}p_p\delta^{kl}],\\ \sigma^{kl}_{s*f} &= \eta\alpha_s[\lambda_{*f}\varepsilon^{mm}_s\delta^{kl} + 2\mu_{*f}\varepsilon^{kl}_s + \nu_{*f}p_f\delta^{kl}],\end{aligned} \tag{19}$$

where

$$\begin{aligned}\lambda_{*p} &= \tilde{\lambda}_{*p}\left[1 + \mathrm{O}\left(\frac{\tilde{\lambda}_{*p}}{\lambda_s}, \frac{\tilde{\mu}_{*p}}{\mu_s}\right)\right], \qquad \mu_{*p} = \tilde{\mu}_{*p}\left[1 + \frac{\tilde{\mu}_{*p}}{\mu_s}\right]^{-1},\\ \lambda_{*f} &= \tilde{\lambda}_{*f}\left[1 + \mathrm{O}\left(\frac{\tilde{\lambda}_{*f}}{\lambda_s}, \frac{\tilde{\mu}_{*f}}{\mu_s}\right)\right], \qquad \mu_{*f} = \tilde{\mu}_{*f}\left[1 + \frac{\tilde{\mu}_{*f}}{\mu_s}\right]^{-1},\\ \nu_{*p} &= \frac{\lambda_{*p} + 2\mu_{*p}/3}{\lambda_s + 2\mu_s/3}, \qquad \nu_{*f} = \frac{\lambda_{*f} + 2\mu_{*f}/3}{\lambda_s + 2\mu_s/3}.\end{aligned} \tag{20}$$

Effective stress $\sigma_{s*}$ may be written as

$$\sigma_{s*}^{kl} = \alpha_s[\lambda_* \varepsilon_s^{mm} \delta^{kl} + 2\mu_* \varepsilon_s^{kl} + (1-\eta)\nu_{*p} p_p \delta^{kl} + \eta \nu_{*f} p_f \delta^{kl}], \tag{21}$$

where

$$\lambda_* = (1-\eta)\lambda_{*p} + \eta\lambda_{*f}, \qquad \mu_* = (1-\eta)\mu_{*p} + \eta\mu_{*f}.$$

It should be noted that if one derives constitutive equations for the skeleton without taking into account double porosity and inequality of pressures $p_p$, $p_f$ (as it was done in the papers by Nikolayevskii (1970) and Nigmatulin (1978)), one would obtain the formula which contains some averaged elastic moduli of skeleton $\lambda_*$, $\mu_*$ and averaged liquid pressure

$$p_l = \frac{\alpha_p p_p + \alpha_f p_f}{\alpha_p + \alpha_f},$$

$$\sigma_{s*}^{kl} = \alpha_s[\lambda_* \varepsilon_s^{mm} \delta^{kl} + 2\mu_* \varepsilon_s^{kl} + \nu_* p_l \delta^{kl}], \qquad \nu_* = \frac{\lambda_* + 2\mu_*/3}{\lambda_s + 2\mu_s/3}. \tag{22}$$

This formula can also be taken as a constitutive equation for the skeleton but it is somewhat less detailed than (21). If one takes $\lambda_*$, $\mu_*$ as it is written above and puts coefficients of $p_p$, $p_f$ equal in (21) and (22), that is,

$$\nu_* \frac{\alpha_p}{\alpha_p + \alpha_f} = (1-\eta)\nu_{*p}, \qquad \nu_* \frac{\alpha_f}{\alpha_p + \alpha_f} = \eta \nu_{*f},$$

after eliminating $\nu_*$ the scalar value $\eta$ is determined:

$$\eta = \left(1 + \frac{\alpha_p}{\alpha_f} \frac{\nu_{*f}}{\nu_{*p}}\right)^{-1}. \tag{23}$$

Thus, it is necessary to define four elastic constants $\lambda_{*p}$, $\mu_{*p}$, $\lambda_{*f}$, $\mu_{*f}$ in order to study double porous medium with an elastic skeleton. Effective stress $\sigma_{s*}$, strain $\varepsilon_s$ and pressures $p_p$, $p_f$ are related by constitutive equation (21); values $\nu_{*p}$, $\nu_{*f}$, $\eta$ are determined by coefficients $\lambda_{*p}$, $\mu_{*p}$, $\lambda_{*f}$, $\mu_{*f}$ and volume fractions $\alpha_p$, $\alpha_f$. One may note that at $\alpha_p = 0$ or $\alpha_f = 0$ it is $\eta = 1$ or $\eta = 0$, correspondingly, and (21) coincides with the constitutive equation for usual porous medium with elastic moduli $\lambda_{*f}$, $\mu_{*f}$ or $\lambda_{*p}$, $\mu_{*p}$.

Constitutive equations for phases are used in the following form:

$$p'_s = \frac{K_s \rho_s^{\circ\prime}}{\rho_{s0}^{\circ}}, \qquad \left(K_s = \lambda_s + \frac{2\mu_s}{3}\right),$$
$$p'_p = \frac{K_l \rho_p^{\circ\prime}}{\rho_{l0}^{\circ}}, \qquad p'_f = \frac{K_l \rho_f^{\circ\prime}}{\rho_{l0}^{\circ}}, \tag{24}$$

where $K_s$, $K_l$ are the elastic bulk moduli for the solid phase material and liquid, correspondingly (subscript 0 corresponds to the equilibrium value of quantity, primed symbol means the deviation of quantity from its equilibrium value).

## 4. The Equations of Simultaneous Deformations of Phases

To close the set of equations for usual porous medium, either of the two following relations is commonly used:

$$p_{s*} = \alpha_s(p_s - p_l), \qquad p_{s*} = -\tfrac{1}{3}\sigma_{s*}^{mm}, \tag{25}$$

or the equation of variation of porosity $\alpha_s$ which is obtained from (25) and is written as (Nikolayevskii, 1970; Nigmatulin, 1978)

$$\frac{\alpha_s'}{\alpha_s} = (1 - \nu_*)\left(\frac{\rho_s'}{\rho_s} - \frac{p_l'}{K_s}\right). \tag{26}$$

For double porous medium the two closing equations need to be written because there are two pressures in the liquid. The first one may be taken as a relation between pressures resulting from (2):

$$p_{s*} = \alpha_s(p_s - p_l), \tag{27}$$

where

$$p_{s*} = -\frac{\sigma_{s*}^{mm}}{3}, \qquad p_l = \frac{\alpha_p p_p + \alpha_f p_f}{\alpha_p + \alpha_f}.$$

For the second equation the similar relation between pressures $p_s$, $p_p$ within the porous blocks (that is a part of the mixture occupied by solid and liquid in primary pores) may be used. However, it requires the definition of effective pressure within the porous block. It is more convenient to make use of Equation (26). In order to write an equation for the variation of $\alpha_s$ within the porous block, one should replace $p_l'$ by $p_p'$, $\nu_*$ by $\nu_{*p}$, $\alpha_s$ by $\alpha_s/(\alpha_s + \alpha_p)$ and $\rho_s = \alpha_s\rho_s^\circ$ by

$$\frac{\rho_s}{\alpha_s + \alpha_p} = \frac{\alpha_s\rho_s^\circ}{\alpha_s + \alpha_p}$$

because these quantities are referred to the volume of porous block per unit volume of the medium. It results in

$$\frac{\alpha_s'}{\alpha_s} + \frac{\alpha_f'}{1 - \alpha_f}\nu_{*p} = (1 - \nu_{*p})\left(\frac{\rho_s'}{\rho_s} - \frac{p_p'}{K_s}\right). \tag{28}$$

## 5. Propagation of Linear Waves in Fractured Porous Medium

The set of Equations (4), (5), (11), (13), (19), (24), (27) and (28) at given expressions for mass and momentum exchange (7), (9) and elastic constants of skeleton is closed. After linearization it becomes a linear uniform system of equations and it may be used in studying linear wave propagation in FPM.

The solutions of the system that were considered, present one-dimensional monochromatic waves, that is $v_j^x = v_j = V_j\,\exp(i\omega t - ikx)$, $v_j^y = v_j^z = 0$ for LW or

$v_j^z = v_j = V_j \exp(i\omega t - ikx)$, $v_j^x = v_j^y = 0$ for transverse wave (TW) ($j = \mathrm{s, p, f}$; $\omega$ is frequency, $k$ is a complex wave number). Dispersion relation is obtained as a result of substitution of the solution into the set of equations and equating its determinant to zero.

It was found that dispersion relation for LWs is a cubic polynomial in $k^2$ while that for TWs is a linear equation in $k^2$. It means that in the medium three modes of LWs and one TW may propagate. Solving the dispersion dependence $k^{(j)} = k^{(j)}(\omega)$, the velocity $C^{(j)}$ and attenuation decrement $\delta^{(j)}$ of each mode is determined as follows:

$$C^{(j)} = \frac{\omega}{\mathrm{Re}\, k^{(j)}}, \qquad \delta^{(j)} = -\mathrm{Im}\, k^{(j)}, \tag{29}$$

where $j = 1, 2, 3$ refer to LWs and $j = 4$ to TW.

It should be remembered that there are two modes of LW in usual porous medium that are explained for two ways of momentum transfer: by liquid and by solid phase at once. That is why the reasonable question arises: what is responsible for the third wave in double porous medium?

To answer this question we consider wave propagation in FPM, saturated with liquid with an incompressible immovable skeleton. A linearized set of equations governing motion of liquid is written as (at that $v_\mathrm{s} = 0$ and $\alpha_j = \mathrm{const}$)

$$\begin{aligned}
&\frac{\partial \rho_\mathrm{p}}{\partial t} + \rho_{\mathrm{p}0}\nabla^k v_\mathrm{p}^k = -q, \qquad \frac{\partial \rho_\mathrm{f}}{\partial t} + \rho_{\mathrm{f}0}\nabla^k v_\mathrm{f}^k = q,\\
&\rho_{\mathrm{p}0}\frac{\partial v_\mathrm{p}^k}{\partial t} = -\alpha_{\mathrm{p}0}\nabla^k p_\mathrm{p} - F_\mathrm{p}^k, \qquad \rho_{\mathrm{f}0}\frac{\partial v_\mathrm{f}^k}{\partial t} = -\alpha_{\mathrm{f}0}\nabla^k p_\mathrm{f} - F_\mathrm{f}^k,\\
&\rho_j' = \alpha_{j0}\rho_j^{\circ\prime} \quad (j = \mathrm{p, f}), \qquad p_\mathrm{p}' = \frac{K_\mathrm{l}\rho_\mathrm{p}^{\circ\prime}}{\rho_{\mathrm{l}0}^{\circ}}, \qquad p_\mathrm{f}' = \frac{K_\mathrm{l}\rho_\mathrm{f}^{\circ\prime}}{\rho_{\mathrm{l}0}^{\circ}}.
\end{aligned} \tag{30}$$

The set of Equations (30) with the inertial terms in momentum equations being neglected, is equivalent to the equations of Barenblatt and Zheltov, which are commonly used to study processes of filtration in FPM (Barenblatt *et al.*, 1960). The dispersion relation following from (30) is a quadratic equation in $k^2$:

$$\begin{aligned}
&\bar{\xi}^4 - \bar{\xi}^2\left[\left(1 + \frac{\chi}{\alpha_{\mathrm{f}0}}\right)(1 + \alpha_{\mathrm{s}0}\Phi_\mathrm{f}) + \left(1 + \frac{\chi}{\alpha_{\mathrm{p}0}}\right)(1 + \alpha_{\mathrm{s}0}\Phi_\mathrm{p})\right] +\\
&+ \left(1 + \frac{\chi}{\alpha_{\mathrm{f}0}} + \frac{\chi}{\alpha_{\mathrm{p}0}}\right)(1 + \alpha_{\mathrm{s}0}\Phi_\mathrm{f})(1 + \alpha_{\mathrm{s}0}\Phi_\mathrm{p}) = 0,
\end{aligned} \tag{31}$$

where $\bar{\xi} = C_\mathrm{l}k/\omega$, $C_\mathrm{l} = \sqrt{K_\mathrm{l}/\rho_{\mathrm{l}0}^{\circ}}$ is the speed of sound in liquid,

$$\chi = \eta_q \frac{\alpha_{\mathrm{p}0}}{\alpha_{\mathrm{s}0}}\frac{f(\omega)}{a_\mathrm{b}^2}\frac{K_\mathrm{l}}{i\omega\mu_\mathrm{l}},$$

$$\Phi_\mathrm{p} = 0.5\eta_{m\mathrm{p}} + (i\omega t_{\mu\mathrm{p}})^{-1} + \eta_{\mathrm{Bp}}\left(\frac{2}{\eta_{\mu\mathrm{p}}}\right)^{1/2}(i\omega t_{\mu\mathrm{p}})^{-1/2}(1 - i),$$

$$\Phi_{\mathrm{f}} = 0.5\eta_{m\mathrm{f}} + \left(\frac{\eta_{m\mathrm{f}}}{\eta_{m\mathrm{p}}}\right)\left(\frac{a_{\mathrm{p}}}{a_{\mathrm{f}}}\right)^2 (\mathrm{i}\omega t_{\mu\mathrm{p}})^{-1} + \\ + \eta_{\mathrm{Bf}}\left(\frac{2}{\eta_{\mu\mathrm{p}}}\right)^{1/2}\left(\frac{a_{\mathrm{p}}}{a_{\mathrm{f}}}\right)(\mathrm{i}\omega t_{\mu\mathrm{p}})^{-1/2}(1-\mathrm{i}),$$

$$t_{\mu\mathrm{p}} = \frac{\rho_{10}^{\circ} a_{\mathrm{p}}^2}{\eta_{\mu\mathrm{p}}\mu_{\mathrm{l}}}.$$

It shows that there are two LWs propagating in liquid in FPM; their speed and attenuation are determined by mass exchange (it is accounted by term $\chi$) and interaction forces (terms $\Phi_{\mathrm{p}}$, $\Phi_{\mathrm{f}}$).

With interaction forces being neglected (at $\Phi_{\mathrm{p}}, \Phi_{\mathrm{f}} = 0$), but taking into account mass exchange ($\chi \neq 0$), the solutions of the dispersion relation (31) are

$$(\bar{\xi})^2 = \left(\frac{k}{\omega}C_{\mathrm{l}}\right)^2 = 1, \qquad 1 + \frac{\chi}{\alpha_{\mathrm{f}0}} + \frac{\chi}{\alpha_{\mathrm{p}0}}. \tag{32}$$

In this case speed of one mode is equal to speed of sound in liquid $C_{\mathrm{l}}$ and it does not attenuate, but the speed of the other mode is less than $C_{\mathrm{l}}$ and it attenuates when propagating (as $\chi$ is a complex value).

When mass exchange is neglected ($\chi = 0$) but interaction forces are accounted for ($\Phi_{\mathrm{p}}, \Phi_{\mathrm{f}} \neq 0$), solutions of (31) are

$$(\bar{\xi})^2 = \left(\frac{k}{\omega}C_{\mathrm{l}}\right)^2 = 1 + \alpha_{\mathrm{s}0}\Phi_{\mathrm{f}}, \qquad 1 + \alpha_{\mathrm{s}0}\Phi_{\mathrm{p}}, \tag{33}$$

both waves propagate with the velocity less than $C_{\mathrm{l}}$ and attenuate.

When both mass exchange and interaction forces are neglected ($\chi = 0, \Phi_{\mathrm{p}}, \Phi_{\mathrm{f}} = 0$), (31) has the following solutions

$$(\bar{\xi})^2 = \left(\frac{k}{\omega}C_{\mathrm{l}}\right)^2 = 1, 1, \tag{34}$$

and each of the two waves propagates with the speed equal to the speed of sound in liquid $C_{\mathrm{l}}$ and does not attenuate. It is evident that in the latter case the equivalent character of propagation of these two waves is explained by the fact that liquid viscosity is not at all accounted for, though its effect depends highly on the size of heterogeneity in the medium. This is why one should take into consideration the difference between the velocities and pressures of liquid in various pore systems due to different effects of liquid viscosity.

## 6. Calculation Data of Speed and Attenuation of Linear Waves

Using the dispersion relations for longitudinal and transverse waves, velocity and attenuation of each mode were calculated by (29). The material of the solid phase

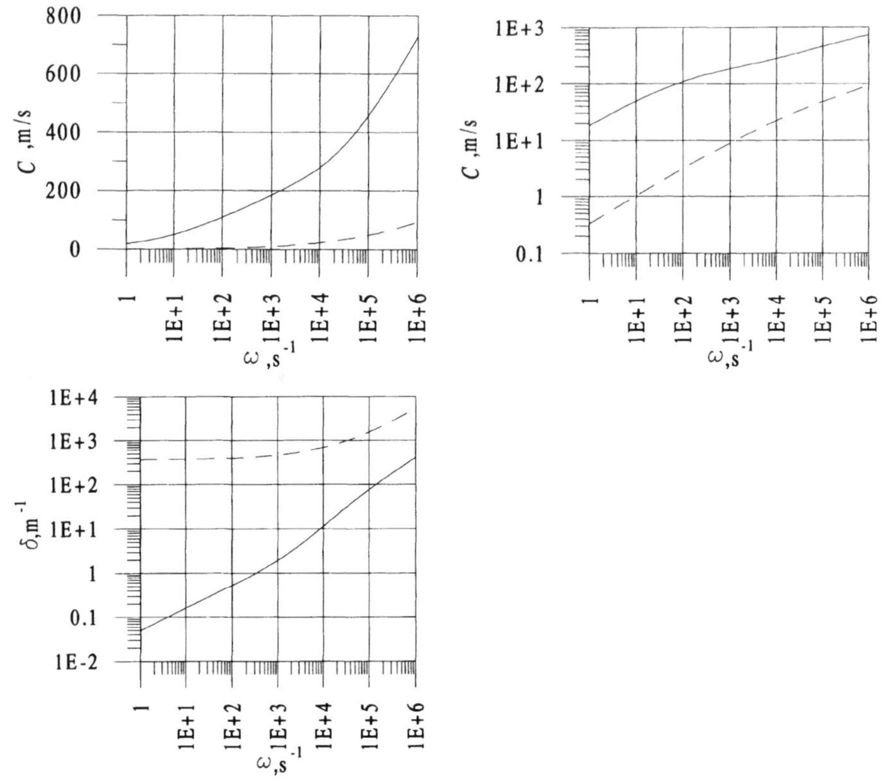


*Figure 2.* Phase velocity (in linear and logarithmic scale) and linear damping decrement of LWs in FPM with incompressible and immovable skeleton.

is quartz and that of liquid is water; initial pressure in the medium $p_0 = 0.1$ MPa, phase parameters are: $\rho^\circ_{s0} = 2500\,\mathrm{kg/m^3}$, $K_s = 5 \times 10^{10}$ Pa, $\rho^\circ_{l0} = 1000\,\mathrm{kg/m^3}$, $K_l = 2.25 \times 10^9$ Pa, $\mu_l = 10^{-3}$ Pa s. The main parameters of FPM are: $\alpha_p = 0.25$, $\alpha_f = 0.01$, $a_p = 0.01$ mm, $a_f = 0.1$ mm, $a_b = 2$ mm.

In Figure 2 phase velocity and damping decrement of linear waves in FPM with incompressible and immovable skeleton are presented. As was pointed out above, two longitudinal waves may propagate in the liquid in such a medium. Speeds of both waves are less at low frequencies; the faster wave (which is shown by the solid line) damps less than the other one (dashed line). In Figure 3 speed and damping decrement of waves in FPM with an account of compressibility and motion of skeleton are shown. Speeds of the faster LW ‘1’ which is a deformational wave and the transverse one ‘4’ depend weakly on frequency; they attenuate much less than waves ‘2’ and ‘3’ which are filtrational ones. Comparing Figures 2 and 3, one can see that waves ‘2’ and ‘3’ in Figure 3 correspond to waves shown by solid and dashed lines in Figure 2. It is evident that deformational and TWs do not exist in the medium with an immovable skeleton. The lower speed of wave ‘2’ in Figure 3, when comparing

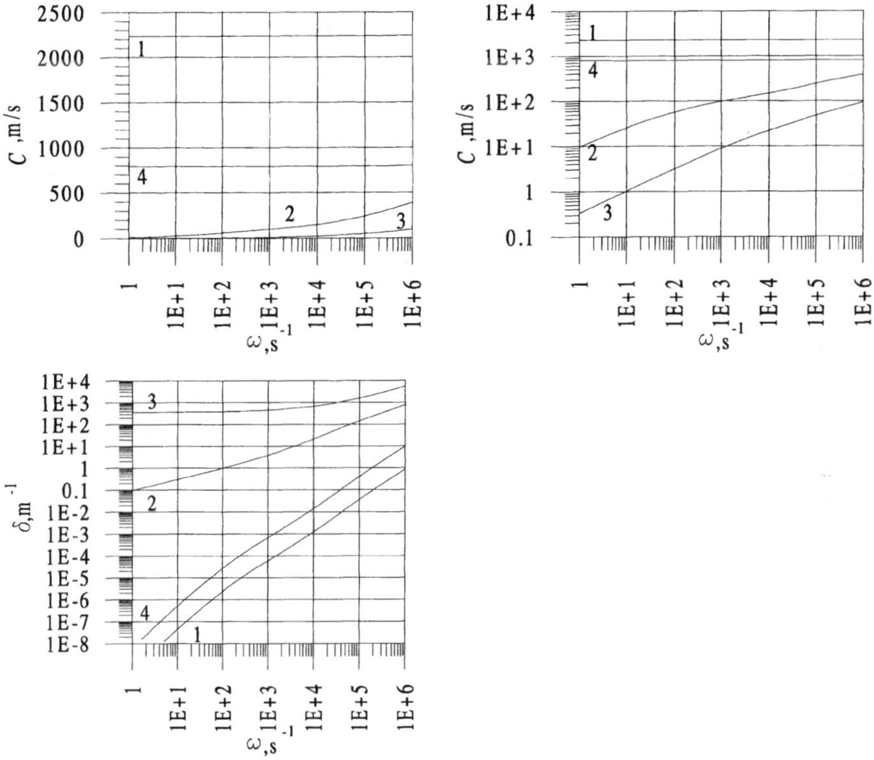


*Figure 3.* Phase velocity (in linear and logarithmic scale) and linear damping decrement of longitudinal (1–3) and transverse (4) waves in FPM.

it with the corresponding one in Figure 2, is explained by effect of compressibility and motion of the skeleton of FPM (this effect is equivalent to the effect of compressibility of walls on sound propagation in narrow tubes, see the monograph by Isakovich, 1973).

Figure 4 illustrates the influence of intensity of mass exchange between pore systems on wave propagation in FPM. One may see that the intensity of mass exchange has a visible effect on filtrational wave '3' only. With the factor of liquid exchange $\eta_q$ growing, the speed of wave '3' decreases and its attenuation increases. Thus, the filtrational wave '3' is more noticeable at small intensity of mass exchange.

The influence of the other parameters of medium and interaction forces on characteristics of wave propagation in the FPM was likewise numerically investigated.

As follows from the analysis of the linearized system of equations, during TW propagation, pressures in liquid in pores as well as in fractures are constant, that is,

$$p_{\mathrm{p}}^{(4)} = p_{\mathrm{f}}^{(4)} = p_0,$$

and the shear components of effective stress tensor $\sigma_{s*}$ vary only.

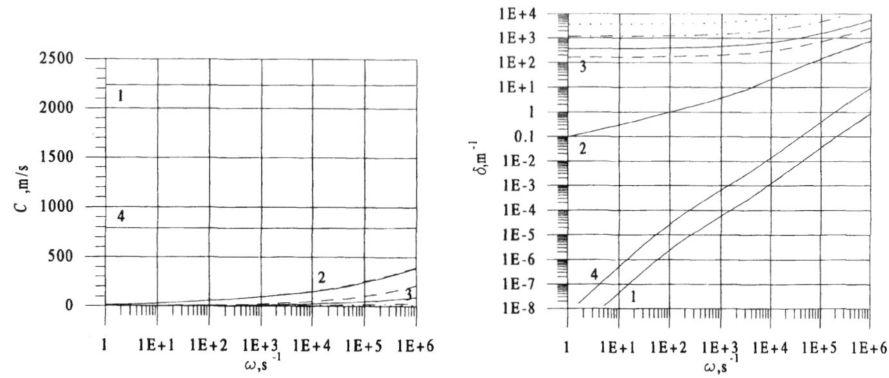


*Figure 4.* Phase velocity and linear damping decrement of longitudinal (1–3) and transverse (4) waves in FPM at various intensity of mass exchange. Dashed, solid, dash-and-dotted and dotted lines correspond to $\eta_q = 0.1, 0.5, 5, 50$.

The character of pressure variation

$$p_{\mathrm{p}}^{(j)} = A_{\mathrm{p}}^{(j)} \exp(\mathrm{i}\omega t - \mathrm{i}k^{(j)}x), \qquad p_{\mathrm{f}}^{(j)} = A_{\mathrm{f}}^{(j)} \exp(\mathrm{i}\omega t - \mathrm{i}k^{(j)}x),$$

averaged pressure in liquid

$$p_{\mathrm{l}}^{(j)} = \frac{\alpha_{\mathrm{p}0} p_{\mathrm{p}}^{(j)} + \alpha_{\mathrm{f}0} p_{\mathrm{f}}^{(j)}}{\alpha_{\mathrm{p}0} + \alpha_{\mathrm{f}0}}$$

and effective longitudinal stress

$$\sigma_{\mathrm{s}*}^{(j)} = A_{\mathrm{s}}^{(j)} \exp(\mathrm{i}\omega t - \mathrm{i}k^{(j)}x)$$

in each mode of the LWs was studied (here $A_{\mathrm{p}}^{(j)}$, $A_{\mathrm{f}}^{(j)}$, $A_{\mathrm{s}}^{(j)}$ are complex amplitudes, $j = 1, 2, 3$, primes are dropped). It is found that in the frequency range considered, namely, at $1\mathrm{s}^{-1} \leqslant \omega \leqslant 10^6\,\mathrm{s}^{-1}$, the following ratios of amplitudes are near to the real constant values:

$$\frac{p_{\mathrm{l}}^{(1)}}{-\sigma_{\mathrm{s}*}^{(1)}} \approx \mathrm{const} > 0, \qquad \frac{p_{\mathrm{p}}^{(1)}}{p_{\mathrm{f}}^{(1)}} \approx 1,$$

$$\frac{p_{\mathrm{l}}^{(2)}}{-\sigma_{\mathrm{s}*}^{(2)}} \approx -1, \qquad \frac{p_{\mathrm{p}}^{(2)}}{p_{\mathrm{f}}^{(2)}} \approx 1, \qquad \frac{p_{\mathrm{l}}^{(3)}}{-\sigma_{\mathrm{s}*}^{(3)}} \approx -1,$$

besides,

$$\mathrm{Re}\left(\frac{p_{\mathrm{p}}^{(3)}}{p_{\mathrm{f}}^{(3)}}\right) < 0.$$

The constant is determined by parameters of undisturbed medium: initial volume fractions, densities, and so on. In particular, for the FPM with the parameters mentioned above the constant is about 2.4. For the third wave, the ratio $p_\mathrm{p}^{(3)}/p_\mathrm{f}^{(3)}$ is not constant at $1\,\mathrm{s}^{-1} \leqslant \omega \leqslant 10^6\,\mathrm{s}^{-1}$, but it is allowable to conclude that pressures $p_\mathrm{p}$, $p_\mathrm{f}$ are not equal. From this it follows that in the first wave both liquid and skeleton are simultaneously compressed or rarefied, with equal pressures in pores and fractures. As wave ‘2’ or ‘3’ propagates, variation of total pressure in the medium is near zero:

$$\sigma^{(2)} = \sigma_{\mathrm{s}*}^{(2)} - p_\mathrm{l}^{(2)} \approx 0, \qquad \sigma^{(3)} = \sigma_{\mathrm{s}*}^{(3)} - p_\mathrm{l}^{(3)} \approx 0.$$

In the second wave as well as in the first one, pressures in liquid in various pore systems are equal. In contrast, in the third wave pressures, $p_\mathrm{p}$, $p_\mathrm{f}$ are different, so that propagation of the wave ‘3’ is accompanied by liquid exchange between pores and fractures. The very strong damping of the third wave means that pressures $p_\mathrm{p}$, $p_\mathrm{f}$ become equal to their equilibrium value in a very short time and, consequently, a process of mass transfer ends. Thus, the difference in velocities and liquid pressures in various pore systems provides an explanation for the third wave in porous medium with two different characteristic sizes of pores.

## 7. Conclusions

The main results and conclusions are as follows:

1. A mathematical model with three velocities and three pressures has been developed for FPM with elastic skeleton;
2. It has been shown that there are three modes of LWs and one TW in the medium; at low frequencies two LWs propagate with slow speed, that is, these two waves are filtrational. Filtrational waves damp much stronger than fast longitudinal (deformational) and transverse waves. Speeds of fast longitudinal ‘1’ and transverse ‘4’ waves are determined mostly by elastic moduli characterizing the skeleton of the medium. Speed and attenuation of faster filtrational wave ‘2’ depends mainly on the interphase viscous force, whereas the speed of slower filtrational wave ‘3’ depends mainly on the intensity of mass exchange between pore systems and its attenuation is determined by both intensity of mass exchange and viscous force between liquid and skeleton.
3. It has been found how pressures and effective stress vary in each mode, namely:

    - in the fast LW ‘1’ both skeleton and liquid are simultaneously compressed or rarefied, with equal pressures in pores and fractures;
    - in LW ‘2’ the skeleton is compressed and the liquid is rarefied (or the skeleton is rarefied and the liquid is compressed), pressures in pores and fractures are equal;
    - in LW ‘3’ the skeleton is compressed and the liquid as a whole is rarefied (or the skeleton is rarefied and the liquid is compressed), with different pressures in pores and fractures;
    - in TW ‘4’ the pressures in liquid are constant and equal to their equilibrium value, whereas the shear components of effective stress of skeleton vary.

4. It has been shown that the third LW in the medium with two characteristic sizes of pores is caused by the difference of velocities and pressures in pores and fractures.

## Acknowledgement

The research described in this publication was made possible in part by Grant No. 98-01-00831 from the Russian Foundation for Basic Research.

## References

Auriault, J.-L. and Boutin, C.: 1992, Deformable porous media with double porosity. Quasi-static. 1: Coupling effects, *Transport in Porous Media* **7**(1), 63–82.

Auriault, J.-L. and Royer, P.: 1993, Ecoulement d'un gaz dans un milieu poreux a double porosite, *C. R. Acad. Sci., Ser. 2*, **317**(4), 431–436.

Bai, M., Ma, Q. and Roegiers, J.-C.: 1994, Dual-porosity behaviour of naturally fractured reservoirs, *Int. J. Numer. Anal. Meth. Geomech.* **18**(6), 359–376.

Barenblatt, G. I., Zheltov, Yu. P. and Kochina, I. N.: 1960, Basic concepts in the theory of seepage of homogeneous liquids in fissured rocks, *PMM* **24**(5), 1286–1303.

Barenblatt, G. I., Entov, V. M. and Ryzhik, V. M.: 1984, *Flow of Liquids and Gases in Natural Seams*, Nedra, Moscow (in Russian).

Biot, M. A.: 1956a, Theory of propagation of elastic waves in a fluid-saturated porous solid: 1. Low-frequency range, *J. Acoust. Soc. Am.* **28**(2), 169–178.

Biot, M. A.: 1956b, Theory of propagation of elastic waves in a fluid-saturated porous solid: 2. Higher frequency range, *J. Acoust. Soc. Am.* **28**(2), 179–191.

Biot, M. A.:1962, Mechanics of deformation and acoustic propagation in porous media, *J. Appl. Soc. Phys.* **33**(4), 1482–1498.

Buyevich, Yu. A.: 1984, Structural and mechanical properties and filtration in elastic fractured porous material, *IFZh* **46**(4), 593–600.

Chen Zhangxin: 1994, Large-scale averaging analysis of single phase flow in fractured reservoirs, *SIAM J. Appl. Math.* **54**(3), 541–659.

Dinariyev, O. Yu.: 1990, Filtration in fractured medium with fractal geometry of fractures, *Doklady AN, MZhG* (5), 66–70.

Dinariyev, O. Yu.: 1992, Flow of liquid and gases in porous media with fractal geometry, *Doklady AN, MZhG* (5), 101–109.

Dinariyev, O. Yu.: 1994, Curve of pressure restore for fractal fractured porous medium, *PMM* **58**(4), 172–175.

Egorov, A. G., Kosterin, A. V. and Skvortsov, E. V.: 1990, *Consolidation and Acoustical Waves in Saturated Porous Media*, Izdatel'stvo Kazanskogo universiteta, Kazan (in Russian).

Isakovich, M. A.: 1973, *General Acoustics*, Nauka, Moscow (in Russian).

Nigmatulin, R. I.: 1978, *Fundamentals of Mechanics of Heterogeneous Media*, Nauka, Moscow (in Russian).

Nigmatulin, R. I.: 1990, *Dynamics of Multiphase Media*, Hemisphere, New York.

Nikolayevskii, V. N., Basniyev, K. S., Gorbunov, A. T. and Zotov, G. A.: 1970, *Mechanics of Saturated Porous Media*, Nedra, Moscow (in Russian).

Nikolayevskii, V. N.: 1984, *Mechanics of Porous and Fractured Media*, Nedra, Moscow (in Russian).

Royer, P. and Auriault J.-L.: 1994, Transient quasi-static gas flow through a rigid porous medium with double porosity, *Transport in Porous Media* **17**(1), 33–57.

Saucier Antoine: 1992, Effective permeability of multifractal porous media, *Physica. A* **183**(4), 381–397.

Tuncay, K. and Corapcioglu, M. Y.: 1996a, Wave propagation in fractured porous media, *Transport in Porous Media* **23**, 237–258.

Tuncay, K. and Corapcioglu, M. Y.: 1996b, Body waves in fractured porous media saturated by two immiscible Newtonian fluids, *Transport in Porous Media* **23**, 259–273.

Wilson, R. K., and Aifantis, E. C.: 1982, On the theory of consolidation with double porosity, *Int. J. Engng. Sci.* **20**(9), 1009–1035.

Wilson, R. K. and Aifantis, E. C.: 1984, A double porosity model for acoustic wave propagation in fractured-porous rock, *Int. J. Engng. Sci.* **22**(8), 1209–1217.

# A Micromechanics-Based Approach to the Failure of Saturated Porous Media

P. DE BUHAN and L. DORMIEUX
*Laboratoire de Mécanique des Solides (URA 317), ENPC-CERCSO, 6 et 8 av. Blaise Pascal, Cité Descartes Champs-sur-Marne, 77455-MARNE-LA-VALLEE, France*

(Received: 22 June 1997; in final form: 13 October 1997)

**Abstract.** This contribution is devoted to the implementation of a homogenization method for deriving the strength or failure properties of a fluid-saturated porous medium, from those exhibited by its individual constituents at the microscopic level. Within this context, a specific attention is paid to the possibility of adopting an effective stress formulation. While the case of a purely cohesive solid matrix provides the first illustrative example where the 'effective stress principle' as originally stated by Terzaghi is fully applicable, the analysis is then particularly focused on porous sandstones, modelled as periodic packings of cemented rigid grains. A closed-form analytical expression is thus obtained for the strength criterion of such rock materials, which proves to be a function of a generalized effective stress formed as a linear combination of the total stress and the pore pressure, as in the case of poroelasticity. It is shown in particular that the key microstructural parameter involved in this formulation is the ratio between the intergranular contact area and the grain cross-section area. A possible extension of the homogenization procedure in order to account for a still more realistic description of the sandstone microstructure is finally outlined.

**Key words:** saturated porous medium, effective stress, failure criterion, homogenization method, sandstone.

## 1. Introduction

Whether the deformability and strength properties of a fluid-saturated porous material can be assessed by means of the so-called 'effective stress' concept, remains a much debated issue both from the fundamental and practical view points. In a recent paper, Lade and de Boer (1997) give a comprehensive historical account of this question, from the original formulation proposed by Terzaghi (1943) for dealing with the consolidation problem of water-saturated soil layers, to the most recent theoretical and experimental developments stimulated by the increasing demand for reliable and predictive design methods in such fields as geotechnical or petroleum engineering.

Expressed in general terms, the effective stress principle refers implicitly to a macroscopic description of the saturated porous medium, modelled as a continuum, at any point of which the skeleton and fluid particles are geometrically coincident (Figure 1(a)). The couple of stress variables associated with this description are the

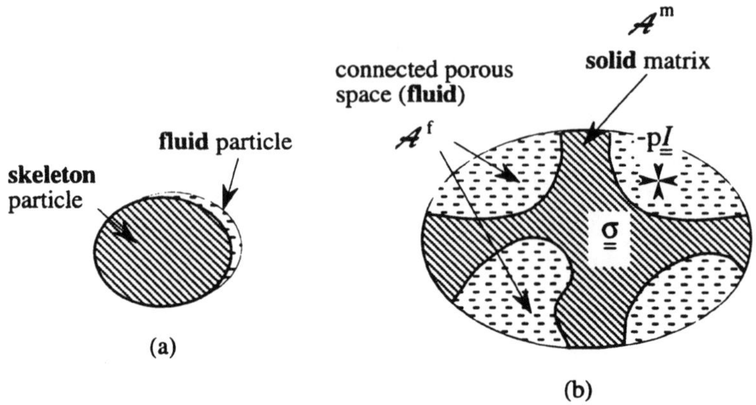


*Figure 1.* Perceptions of a saturated porous medium at the (a) macroscopic and (b) microscopic scales.

total stress $\underline{\underline{\Sigma}}$ and the pore pressure $P^{\star}$, which control the skeleton's behaviour, so that, according to the Terzaghi's postulate, such a behaviour is governed by the sole effective stress defined as $\underline{\underline{\Sigma}} + P\underline{\underline{I}}$.

Within the context of linear poroelasticity, it is classically shown (Biot, 1955; Rice and Cleary, 1976; Thompson and Willis, 1991; Coussy, 1995; Cheng, 1997, among many other contributors), that the skeleton deformations are directly proportional to a generalized effective stress $\underline{\underline{\Sigma}} + P\underline{\underline{B}}$, where $\underline{\underline{B}}$ reduces to $\underline{\underline{I}}$, thus recovering Terzaghi's assumption, in the case of an incompressible solid matrix making up the skeleton at the microscopic scale. The latter result suggests that, resorting to a description of the porous material at the microscopic scale, as shown in Figure 1(b), where the solid and fluid constituents are now geometrically distinct, is liable to produce relevant information as regards to its macroscopic behaviour, and thus provide a rational framework for investigating the question of effective stress formulation.

Such a micromechanical approach will be adopted in this paper in order to examine the range of validity of an effective stress formulation for describing the strength or failure properties of fluid-saturated porous materials, with a special emphasis on sandstones in the second part of the paper. Indeed, contrary to what is usually undisputably accepted even for rock materials (see for instance Jaeger and Cook, 1976, or Goodman, 1980), some recent experiments concerning saturated cemented materials (Pietruszczak *et al.*, 1995; Pietruszczak and Pande, 1995) have revealed significant deviations from the Terzaghi's effective stress concept.

The strength or failure criterion of a fluid-saturated porous medium can always be expressed by means of a condition involving both $\underline{\underline{\Sigma}}$ and $P$ as independent variables, in the form:

$$F(\underline{\underline{\Sigma}}, P) \leqslant 0 \iff \underline{\underline{\Sigma}} \in G(P), \tag{1}$$

$^{\star}$ Unless specified otherwise, macroscopic quantities are denoted by capital letters, while those pertaining to the microscopic description will be represented by small letters.

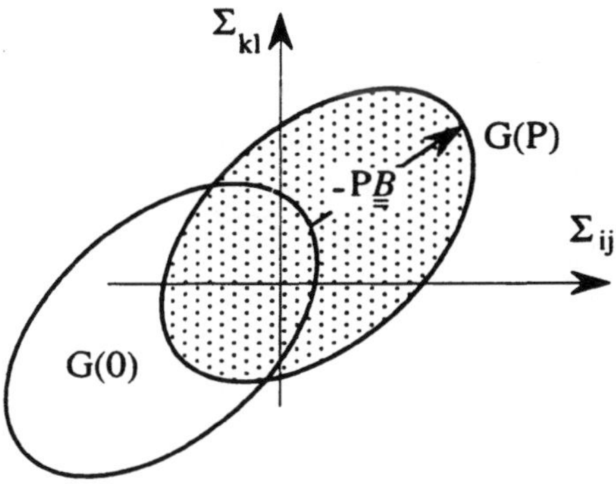


*Figure 2.* Geometrical interpretation of the generalized effective stress principle in the space of total stresses.

where $G(P)$ denotes, for a prescribed value of the pore pressure (drained experiment), the domain of admissible stresses, which can be geometrically represented in the six-dimensional stress space (Figure 2). It follows that a generalized effective stress formulation of the strength criterion would mean that $F(.)$ depends on $\underline{\underline{\Sigma}} + P\underline{\underline{B}}$ only, implying that the domain $G(P) + P\underline{\underline{B}} = \{\underline{\underline{\Sigma}} + P\underline{\underline{B}}, \underline{\underline{\Sigma}} \in G(P)\}$ is independent of the pore pressure $P$, and is therefore equal to $G(0)$, which is the domain relating to the strength properties of the porous material in which the pore pressure is kept equal to zero. $G(P)$ may be deduced from $G(0)$ through a translation equal to $-P\underline{\underline{B}}$. The validity of such an assumption will now be carefully examined and discussed in the light of a homogenization procedure allowing to derive the strength properties of the porous material as a whole, from those exhibited by its individual components (fluid and solid) at the microscopic scale.

## 2. Homogenization Theory Applied to Saturated Porous Media

The purpose of this section is to lay the fundamentals which are necessary for applying the yield design (or limit analysis*) homogenization method (Suquet, 1985; de Buhan, 1986) to saturated porous materials. Such a method relies upon the heuristic assumption that the porous medium may be modelled as a two-phase periodic material when regarded at the microscopic scale. According to this method, it turns out that the determination of the macroscopic strength condition amounts to solving a yield design boundary value problem relating to a unit representative cell of the porous material, denoted by $\mathcal{A}$ in the sequel. For illustrative purpose, the latter will be taken as a cube of unit side, referred to an orthonormal frame $(O\xi_1\xi_2\xi_3)$, with its origin placed at the cube center as shown in Figure 3. The solid matrix which is the microscopic counterpart of the macroscopic concept of skeleton occupies a subdomain $\mathcal{A}^m$ of

* The term 'yield design' will be preferred to 'limit analysis', since the latter refers to elastic perfectly plastic materials, with associated flow rule, while the former seems much more appropriate when dealing with geomaterials such as soils, rocks or concretes, for which no other information than their strength properties are required in the analysis (see Salençon, 1990, for further details).

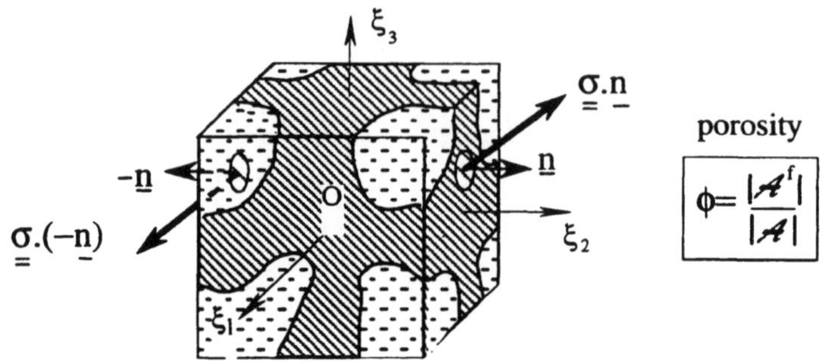


*Figure 3.* A periodic microscopic model for saturated porous materials.

$\mathcal{A}$, whereas the fluid entirely occupies the complementary porous space denoted by $\mathcal{A}^{\mathrm{f}}$, so that its volume fraction defined as $\phi = |\mathcal{A}^{\mathrm{f}}|/|\mathcal{A}|$ is the porosity. It should be emphasized that $\mathcal{A}^{\mathrm{f}}$ represents the connected porous space in which the fluid is allowed to circulate, while $\mathcal{A}^{\mathrm{m}}$ represents the matrix which may possibly contain microvoids, forming an occluded porosity which does not contribute to the fluid mass transport in the porous network.

Under the assumption of quasi-static conditions adopted throughout the paper, inertial effects are neglected. Therefore, a microscopic stress field $\underline{\underline{\sigma}}$ defined in each point $\underline{\xi}$ of $\mathcal{A}$ is said to be statically admissible for the problem, if and only if it satisfies the equilibrium equation with no body forces:

$$\operatorname{div} \underline{\underline{\sigma}}(\underline{\xi}) = 0 \quad \forall\, \underline{\xi} \in \mathcal{A} \tag{2}$$

along with $[\underline{\underline{\sigma}}] \cdot \underline{n} = 0$ across possible stress discontinuity surfaces ($\underline{n}$ is the local unit normal to such a surface), while $\underline{\underline{\sigma}} \cdot \underline{n}$ complies with the antiperiodicity condition, that is takes opposite values on opposite sides of the unit cell, as sketched in Figure 3. The set of such statically admissible stress fields will be denoted by $\mathcal{S}$.

According to the static definition of the macroscopic strength criterion, a state of stress $\underline{\underline{\Sigma}}$ is allowable, if one can exhibit $\underline{\underline{\sigma}} \in \mathcal{S}$ such that

$$\begin{aligned} &\underline{\underline{\Sigma}} = \langle \underline{\underline{\sigma}} \rangle = \phi \langle \underline{\underline{\sigma}} \rangle^{\mathrm{f}} + (1-\phi)\langle \underline{\underline{\sigma}} \rangle^{\mathrm{m}}, && \text{(a)} \\ &\underline{\underline{\sigma}}(\underline{\xi}) \begin{cases} \in g^{\mathrm{m}}(\underline{\xi}) & \forall \underline{\xi} \in \mathcal{A}^{\mathrm{m}}, \\ = -p(\underline{\xi})\underline{\underline{I}} & \forall \underline{\xi} \in \mathcal{A}^{\mathrm{f}}, \end{cases} && \text{(b)} \end{aligned} \tag{3}$$

where $\langle\cdot\rangle$, $\langle\cdot\rangle^{\mathrm{m}}$ and $\langle\cdot\rangle^{\mathrm{f}}$ denote volume averaging over $\mathcal{A}$, $\mathcal{A}^{\mathrm{m}}$ and $\mathcal{A}^{\mathrm{f}}$ respectively, $g^{\mathrm{m}}(\underline{\xi})$ represents the solid matrix strength domain at point $\underline{\xi}$, and $p(\underline{\xi})$ is the local pressure in the fluid (thanks to the quasi-static conditions prevailing, shear stresses due to viscosity are neglected in the present analysis). Combining Equations (2) and (3) yields in particular:

$$\underline{\underline{\Sigma}} = -\phi P \underline{\underline{I}} + (1-\phi)\langle \underline{\underline{\sigma}} \rangle^{\mathrm{m}} \tag{4}$$

since, owing to the equilibrium requirement (2), the fluid pressure is uniform throughout the connected porous space, the corresponding value being identified with the

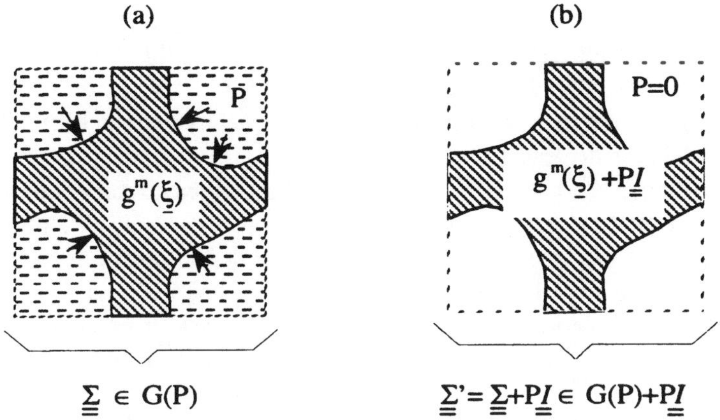


*Figure 4.* Yield design problems attached to the (a) fluid-saturated and (b) 'dry' representative cell of porous material.

macroscopic pore pressure $P$. The definition of $G(P)$ given by (3), where $\underline{\underline{\sigma}} \in \mathcal{S}$, may be more conveniently rewritten as

$$\begin{aligned}
&\underline{\underline{\Sigma}} \in G(P) \quad \text{or} \quad \underline{\underline{\Sigma}}' = \underline{\underline{\Sigma}} + P\underline{\underline{I}} \in G(P) + P\underline{\underline{I}} \\
&\qquad\qquad \Updownarrow \\
&\underline{\underline{\Sigma}}' = (1-\phi)\langle \underline{\underline{\sigma}}' \rangle^{\mathrm{m}}, \quad \underline{\underline{\sigma}}' = \underline{\underline{\sigma}} + P\underline{\underline{I}} \in \mathcal{S} \\
&\underline{\underline{\sigma}}'(\underline{\xi}) \begin{cases} = 0 & \forall \underline{\xi} \in \mathcal{A}^{\mathrm{f}}, \\ \in g^{\mathrm{m}}(\underline{\xi}) + P\underline{\underline{I}} & \forall \underline{\xi} \in \mathcal{A}^{\mathrm{m}}. \end{cases}
\end{aligned} \tag{5}$$

As it is clearly apparent from the above formulation, the determination of $G(P) + P\underline{\underline{I}}$ may be obtained from solving a yield design boundary value problem defined over the representative cell in which the fluid has been completely removed ('dry' porous material), provided the matrix strength domain be altered correspondingly through a translation equal to $P\underline{\underline{I}}$ in the stress space (Figure 4).

## 3. Case of a Purely Cohesive Matrix Material (de Buhan and Dormieux, 1996)

A first important situation is considered in which the constituent solid matrix obeys a 'pressure-insensitive' strength condition, that is such that the matrix failure is exclusively governed by the value of the deviatoric stress. This condition may be expressed in the following form:

$$g^{\mathrm{m}}(\underline{\xi}) = g^{\mathrm{m}}(\underline{\xi}) + \lambda\underline{\underline{I}} \quad \forall \underline{\xi} \in \mathcal{A}^{\mathrm{m}}, \quad \forall \lambda. \tag{6}$$

The well-known Tresca or von Mises criteria provide immediate examples of such a property, which may apply for instance to quartz or silica grains. It could be given a simple geometrical interpretation, by observing that the domain $g^{\mathrm{m}}$ drawn in the space of $\underline{\underline{\sigma}}$ is a cylinder, the generatrices of which are parallel to the axis of isotropic states of stress ('hydrostatic' axis).

It follows immediately from the introduction of (6) into (5), that the domain $G(P) + P\underline{\underline{I}}$ is identical with $G(0)$, so that the determination of the criterion for any value of the pore pressure is readily obtained from that of the 'dry' porous material, thereby providing a full justification of the Terzaghi's effective stress formulation. Two additional comments deserve to be made:

(a) Such a result should be closely connected to the previously mentioned similar property, established in the context of poroelasticity, that is assuming notably that the solid matrix behaves elastically. Indeed, it can be noticed that the solid matrix incompressibility from which the Terzaghi's formulation has been established, is equivalent to pressure-independence, since, in the expression of its constitutive law, the no-volume change kinematic constraint implies that the stress is defined up to an arbitrary isotropic component.

(b) It is important to observe that, unlike the constituent solid matrix, $G(0)$ is not of the purely deviatoric kind, that it exhibits a pressure-dependence. This must obviously be attributed to the presence of empty cavities which constitutes the porous network usually filled by the circulating fluid. Furthermore, making use of the definition of $G(0)$ given by (3) and (4), where $P$ is taken equal to zero, it can be easily proved that $G(0)$ is enclosed in $(1 - \phi)g^{\mathrm{m}}$:

$$G(0) \subset (1 - \phi)g^{\mathrm{m}}, \tag{7}$$

which means that, referring for instance to the case when the solid matrix obeys a Tresca criterion with a cohesion or shear strength equal to $c^{\mathrm{m}}$, the shear strength properties of the dry porous skeleton are at least reduced by a factor equal to the solid volume fraction.

(c) The condition of periodicity, which may appear somewhat artificial, is not essential for deriving the above results concerning the general definition of $G(P)$ on the one hand, the proof that the Terzaghi's effective stress formulation applies in the case of a purely cohesive matrix, on the other hand. Indeed, considering any loading mode of a representative volume element of saturated porous material, in which the loading parameters would represent the components of the total macroscopic stress tensor, the same kind of reasoning as that performed in Sections 2 and 3 would have been possible.

## 4. A Simple Model for Saturated Porous Sandstones

From the microstructural point of view, sandstones may be roughly described as packing of grains in mutual interaction through joints formed during the compaction and cementation process of porous sedimentary rocks, as a result of an intergranular pressure solution mechanism (see, for instance, Guéguen and Palciauskas (1994)). Figure 5 gives the example of such a periodic assemblage, the unit representative cell being a cubic volume containing one single grain in contact with six adjacent others through identical joint interfaces located on the different sides of the cube.

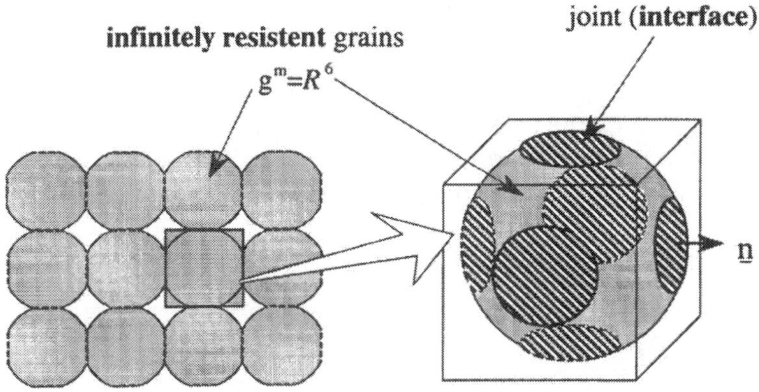


*Figure 5.* Periodic model for porous sandstones and associated representative cubic cell.

Referring to the generic notations introduced in Section 2, the solid domain $\mathcal{A}^m$ now includes the volume of the grain itself, denoted by $\mathcal{A}^g$, along with three joint interfaces, denoted by $\mathcal{J}^\alpha$, $\alpha = 1$ to 3 contained in three orthogonal sides $\mathcal{S}^\alpha$ of the cube of outer unit normals $\underline{e}_\alpha$. $\eta = |\mathcal{J}^\alpha|/|\mathcal{S}^\alpha|$ thus represents the ratio of the contact area between adjacent grains, to the total area of the cube side. It will be assumed in the present analysis that the grain is infinitely resistent, that is, $g^m = \mathcal{R}^6$, whereas the strength properties of the joints will be characterized by a Mohr–Coulomb failure criterion expressed in the form:

$$\underline{T}(\underline{n} = \underline{e}_\alpha) \in g^j(c, \varphi) \Longleftrightarrow |\underline{\tau}| \leqslant c - \sigma \tan \varphi, \tag{8}$$

where $\underline{T}(\underline{n} = \underline{e}_\alpha) = \sigma \underline{e}_\alpha + \underline{\tau}$ is the stress-vector acting upon joint $\mathcal{J}^\alpha$ of normal $\underline{n} = \underline{e}_\alpha$, while $c$ and $\varphi$ denote the cohesion and friction angle respectively. Those parameters will be supposed to remain unaffected by the fluid presence in the remaining porous space $\mathcal{A}^f$. The following property can easily be established concerning the strength domain $g^j(c, \varphi)$ which has been drawn in the $(\sigma, \underline{\tau})$ three-dimensional space (Figure 6):

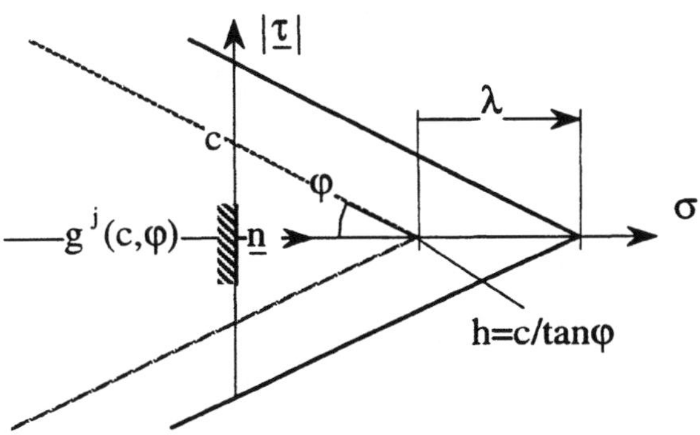


*Figure 6.* Representation of joint failure condition in the $(\sigma, \underline{\tau})$-space.

$$\forall\lambda > -h, \quad g^j(c,\varphi) + \lambda \underline{n} = g^j((1+\lambda/h)c,\varphi) = (1+\lambda/h)g^j(c,\varphi), \tag{9}$$

where $h = c/\tan\varphi$ is the maximum admissible normal stress in tension. This relationship, which will be used later on, means that translating the interface strength domain by an arbitrary algebraic quantity along the $\sigma$-axis, is strictly equivalent to multiplying its cohesion by a factor equal to $(1 + \lambda\tan\varphi/c)$, without altering the friction angle.

The general definition of domain $G(P) + P\underline{\underline{I}}$ given by (5), is therefore particularized as follows:

$$\begin{gathered}
\underline{\underline{\Sigma}}' \in G(P) + P\underline{\underline{I}} \\
\Updownarrow \\
\underline{\underline{\Sigma}}' = (1-\phi)\langle\underline{\underline{\sigma}}'\rangle^{\mathrm{m}}, \quad \underline{\underline{\sigma}}' \in \mathcal{S} \\
\underline{\underline{\sigma}}'(\underline{\xi}) \begin{cases} = 0 & \forall\underline{\xi} \in \mathcal{A}^{\mathrm{f}}, \\ \in g^{\mathrm{m}}(\underline{\xi})(= \mathcal{R}^6) + P\underline{\underline{I}} & \forall\underline{\xi} \in \mathcal{A}^{\mathrm{g}}, \end{cases}
\end{gathered} \tag{10}$$

and

$$\underline{\underline{\sigma}}'(\underline{\xi}).\underline{e}_\alpha \in g^j(c,\varphi) + P\underline{e}_\alpha \qquad \forall\underline{\xi} \in \mathcal{J}^\alpha, \quad \alpha = 1,2,3$$

or using Equation (9) with $\lambda = P > -h$, together with the obvious relation $\mathcal{R}^6 + P\underline{\underline{I}} = \mathcal{R}^6$:

$$\underline{\underline{\Sigma}}' \in G(P) + P\underline{\underline{I}} \Leftrightarrow \left(1 + \frac{P}{h}\right)\underline{\underline{\Sigma}}' \in G(0) \tag{11}$$

which leads to the following key result:

$$G(P) + P\underline{\underline{I}} = \left(1 + \frac{P}{h}\right) G(0). \tag{12}$$

As it clearly appears from the above closed-form expression, the formulation of the strength criterion for a saturated porous sandstone, is not amenable to the Terzaghi's postulate except in the very specific situation where $G(0)$ would remain invariant by any positive homothetic expansion, signifying that it is a cone with its apex at the origin of the stress-space. It will now be proved that the simplified microstructural model adopted for sandstone is not compatible with such a situation. This will be made possible by explicitly determining $G(0)$.

## 5. Determination of $G(0)$

Although it does not reduce to the Terzaghi's formulation, thus providing a palpable example where the effective stress principle in its original form should be dismissed, Equation (12), however, demonstrates that the determination of the failure properties for any value of the pore pressure, only requires being able to specify those exhibited

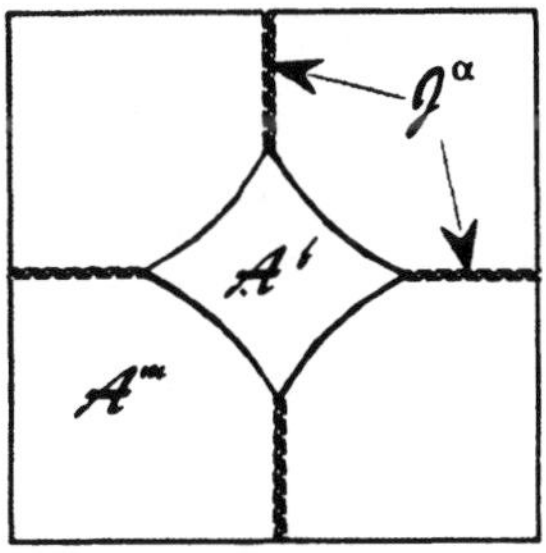


*Figure 7.* Side view of a unit cubic cell of dry porous sandstone.

by the porous material under 'dry' conditions. The static definition of $G(0)$ is given by (10), where $P = 0$, that is expressed in a more compact form:

$$G(0) = \{\underline{\underline{\Sigma}} = (1-\phi)\langle\underline{\underline{\sigma}}\rangle^{\mathrm{m}},\ \underline{\underline{\sigma}} \in \mathcal{S}_0,\ \underline{\underline{\sigma}} \cdot \underline{\mathrm{e}}_\alpha \in g^j(c,\varphi) \text{ on } \mathcal{J}^\alpha\}, \tag{13}$$

where $\mathcal{S}_0$ is the subset of $\mathcal{S}$ formed by the microscopic stress fields which satisfy the stress-free boundary condition on $\partial\mathcal{A}^{\mathrm{m}} \cap \partial\mathcal{A}^{\mathrm{f}}$ (fluid/grain boundary, as is shown in Figure 7, where for the sake of clarity, the unit representative cubic volume has been chosen with its apice coinciding with the grain centers).

## 5.1. KINEMATIC APPROACH TO $G(0)$

Following the same kind of reasoning as that employed for deriving the macroscopic failure properties of jointed blocky rock masses (Bekaert and Maghous, 1996), or masonry-like materials (de Buhan and de Felice, 1997), the kinematic definition of $G(0)$ resulting from the dualism of the static one given by (13) through the virtual work principle, is now introduced. It turns out in particular that, due to the unlimited strength assumption made for the grain solid matrix, the only velocity fields that are worth being considered in the kinematic approach, are such that their restriction to each grain is a rigid body translation, thus inducing velocity jumps at the joint interfaces.

Denoting by $\underline{\mathrm{U}}^\alpha$ the velocity discontinuity calculated when crossing joint $\mathcal{J}^\alpha$ along its normal $\underline{\mathrm{e}}_\alpha$, it can be shown that the corresponding macroscopic strain rate, which appears to be the kinematic variable associated with $\underline{\underline{\Sigma}}$ by duality, simply writes

$$\underline{\underline{\mathrm{D}}} = \sum_{\alpha=1}^{3} \underline{\mathrm{U}}^\alpha \overset{s}{\otimes} \underline{\mathrm{e}}_\alpha, \tag{14}$$

($\underline{\mathrm{a}} \overset{s}{\otimes} \underline{\mathrm{b}}$ denotes the symmetric part of the tensorial product $\underline{\mathrm{a}} \overset{s}{\otimes} \underline{\mathrm{b}}$). Furthermore, it is proved that the support function of $G(0)$, defined for any value of $\underline{\underline{\mathrm{D}}}$ by

$$\Pi(\underline{\underline{\mathrm{D}}}) = \sup_{\underline{\underline{\Sigma}}}\{\underline{\underline{\Sigma}} : \underline{\underline{\mathrm{D}}};\ \underline{\underline{\Sigma}} \in G(0)\} \tag{15}$$

may also be calculated as

$$\Pi(\underline{\underline{D}}) = \operatorname*{Min}_{\underline{U}^\alpha} \left\{ \sum_{\alpha=1}^{3} \int_{\mathcal{J}^\alpha} \pi(\underline{e}_\alpha; \underline{U}^\alpha)\, \mathrm{d}S; \ \text{ with } (\underline{\underline{D}} \text{ of the form (14)} \right\}, \tag{16}$$

where $\pi(\underline{e}_\alpha; \underline{U}^\alpha)$ is the support function of $g^j(c, \varphi)$, defined in the same way as $\Pi(\underline{\underline{D}})$ for $G(0)$ by

$$\pi(\underline{U}^\alpha; \underline{e}_\alpha) = \operatorname*{Sup}_{\underline{T}(\underline{e}_\alpha)} \{\underline{T}(\underline{e}_\alpha).\underline{U}^\alpha;\ \underline{T}(\underline{e}_\alpha) \in g^j(c, \varphi)\}, \tag{17}$$

which on account of (8) yields

$$\pi(\underline{U}^\alpha; \underline{e}_\alpha) = \begin{cases} \dfrac{c}{\tan\varphi}\,(\underline{U}^\alpha.\underline{e}_\alpha) & \text{if } \ \underline{U}^\alpha.\underline{e}_\alpha \geqslant |\underline{U}^\alpha| \sin\varphi, \\ +\infty & \text{otherwise.} \end{cases} \tag{18}$$

The above kinematic constraint imposed on $\underline{U}^\alpha$ corresponds to the fact that the velocity jumps across the joints must be inclined by an angle at least equal to $\varphi$ to the joint interface. Introducing the latter expression into Equation (16), one thus obtains

$$\Pi(\underline{\underline{D}}) = \eta \frac{c}{\tan\varphi} \operatorname*{Min}_{\underline{U}^\alpha} \left\{ \sum_{\alpha=1}^{3} \underline{U}^\alpha.\underline{e}_\alpha; \underline{\underline{D}} = \sum_{\alpha=1}^{3} \underline{U}^\alpha \overset{s}{\otimes} \underline{e}_\alpha, \underline{U}^\alpha.\underline{e}_\alpha \geqslant |\underline{U}^\alpha| \sin\varphi \right\}. \tag{19}$$

## 5.2. CHARACTERIZATION OF $G(0)$ AS A CONE-SHAPED CRITERION

Let us introduce $\mathcal{C}$ as the set of $\underline{\underline{D}}$ of the form (14) satisfying conditions $\underline{U}^\alpha.\underline{e}_\alpha \geqslant |\underline{U}^\alpha| \sin\varphi$ . Observing that $\operatorname{tr}\underline{\underline{D}} = \sum_\alpha \underline{U}^\alpha.\underline{e}_\alpha$ for $\underline{\underline{D}} \in \mathcal{C}$, it follows from (18) that

$$\Pi(\underline{\underline{D}}) = \begin{cases} \eta \dfrac{c}{\tan\varphi} \operatorname{tr}\underline{\underline{D}} & \text{if } \ \underline{\underline{D}} \in \mathcal{C}, \\ +\infty & \text{otherwise,} \end{cases} \tag{20}$$

so that the comparison with (15) leads to the following kinematic definition of $G(0)$:

$$\begin{aligned} G(0) &= \bigcap_{\underline{\underline{D}} \in \mathcal{C}} \{\underline{\underline{\Sigma}}; \underline{\underline{\Sigma}} : \underline{\underline{D}} \leqslant \Pi(\underline{\underline{D}})\} \\ &= \bigcap_{\underline{\underline{D}} \in \mathcal{C}} \left\{ \underline{\underline{\Sigma}}; \left( \underline{\underline{\Sigma}} - \frac{\eta c}{\tan\varphi \underline{\underline{I}}} \right) : \underline{\underline{D}} \leqslant 0 \right\}, \end{aligned} \tag{21}$$

since the set of representative points $\underline{\underline{\Sigma}}$ in the stress space, defined by the inequality $(\underline{\underline{\Sigma}} - \eta c/\tan\varphi \underline{\underline{I}}) : \underline{\underline{D}} \leqslant 0$, is a half space bounded by the hyperplane passing through point $(\eta c/\tan\varphi)\underline{\underline{I}}$ located on the hydrostatic axis, and whose outer normal

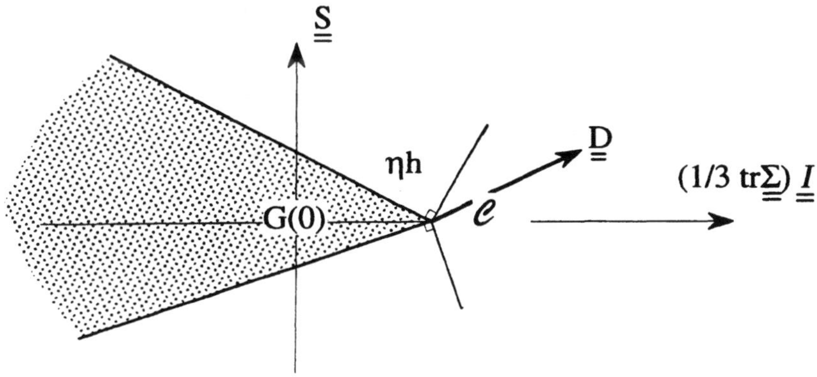


*Figure 8.* Representation of $G(0)$ with its cone of outer normals in the stress-space ($\underline{\underline{S}}$ is the deviatoric component of $\underline{\underline{\Sigma}}$).

is $\underline{\underline{D}}$*. As a result, it then becomes possible to construct $G(0)$ as the convex envelope of such planes, proving that it is a cone of apex $(\eta c/\tan\varphi)\underline{\underline{I}}$, whereas $\mathcal{C}$ is simply the cone of its outer normals (Figure 8).

Besides, it can be easily established that $G(0)$ contains the set of isotropic stresses of the form $\underline{\underline{\Sigma}} = \Sigma\underline{\underline{I}}$, $\Sigma \leqslant \eta c/\tan\varphi$, which constitutes a semi-line located on the hydrostatic axis. This simply results from the fact that all such states of stress verify

$$(\Sigma - \eta c/\tan\varphi)\,\mathrm{tr}\,\underline{\underline{D}} \leqslant 0 \quad \forall \underline{\underline{D}} \in \mathcal{C}, \tag{22}$$

since)

$$\mathrm{tr}\,\underline{\underline{D}} = \sum_{\alpha} \underline{U}^{\alpha}.\underline{e}_{\alpha} \geqslant \sum_{\alpha} |\underline{U}^{\alpha}| \sin\varphi \geqslant 0 \quad \forall \underline{\underline{D}} \in \mathcal{C},$$

which implies from (21) that they do belong to $G(0)$. It should be noted in particular that the homogenized dry porous material exhibits infinite strength properties under isotropic compressive loading. This is obviously due to the initial assumption made concerning the unlimited strength properties of the grains, which might be confirmed by referring to the lower-bound static approach.

Such an approach is outlined in Figure 9(a), which represents a stress distribution obtained by superposing three piecewise constant uniaxial stress fields, that is of the form

$$\underline{\underline{\sigma}}(\underline{\xi}) = \mu \sum_{\alpha=1}^{3} I^{\alpha}(\underline{\xi})\underline{e}_{\alpha} \otimes \underline{e}_{\alpha}, \tag{23}$$

where $\mu$ is a constant, and $I^{\alpha}(\underline{\xi})$ is an indicator function equal to 1 or 0, depending on whether point $\underline{\xi}$ is located or not in the cylindrical volume having joint $\mathcal{J}^{\alpha}$ as a

* Indeed, tensors $\underline{\underline{\Sigma}}$ and $\underline{\underline{D}}$ may be regarded as vectors of $\mathcal{R}^6$, and their doubly contracted product identified with the scalar product of these vectors.

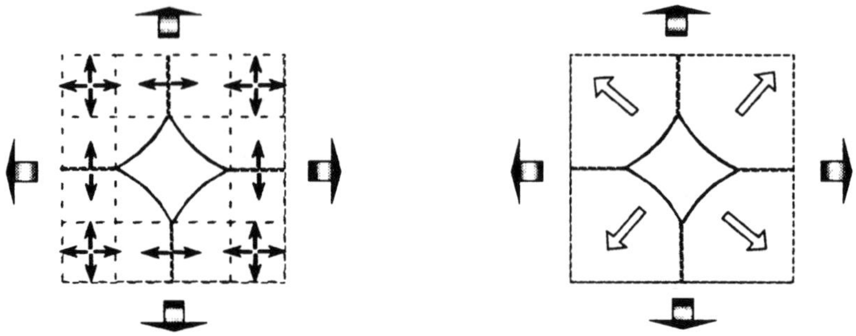

(a) (b)

*Figure 9.* (a) Statics (lower-bound approach) and (b) kinematics (upper-bound approach) of the unit representative cell under isotropic tensile loading.

cross-section. This stress field is obviously statically admissible in the sense that it belongs to $\mathcal{S}_0$, in such a manner that it equilibrates a macroscopic stress equal to

$$\underline{\underline{\Sigma}} = (1-\phi)\,\langle\underline{\underline{\sigma}}\rangle^{\mathrm{m}} = \eta\mu\underline{\underline{I}}. \tag{24}$$

On the other hand, it satisfies the joints failure condition if and only if

$$\underline{\underline{\sigma}}(\underline{\xi}).\underline{e}_\alpha \in g^j(c,\varphi) \quad \forall\underline{\xi} \in \mathcal{J}^\alpha \quad \forall\alpha, \tag{25}$$

that is,

$$\mu \leqslant \frac{c}{\tan\varphi}. \tag{26}$$

It thus confirms that the half line made up of the hydrostatic states of stress of the form $\underline{\underline{\Sigma}} = \Sigma\,\underline{\underline{I}}$, with $\Sigma \leqslant \eta c/\tan\varphi$, is enclosed in $G(0)$. $\eta c/\tan\varphi$ does represent the exact value of the tensile resistance of the dry porous material subjected to isotropic loading conditions. Figure 9(b) pictures the associated failure mechanism corresponding to an isotropic extension strain rate

$$\underline{U}^\alpha = U\underline{e}_\alpha, \quad U > 0, \text{ hence, from (14) } \underline{\underline{D}} = U\underline{\underline{I}}. \tag{27}$$

It finally comes out that the static definition of $G(0)$, which proves to be equivalent to the kinematic one (21), reads as follows:

$$G(0) = \{\underline{\underline{\Sigma}};\ \underline{\underline{\Sigma}}.\underline{e}_\alpha \in \eta g^j(c,\varphi) \quad \forall\alpha\}, \tag{28}$$

thereby clearly demonstrating that the homogenized dry porous material exhibits anisotropic (orthotropic) strength properties arising, as could be expected, from the preferential orientations of joints.

### 5.3. GENERALIZED EFFECTIVE STRESS FORMULATION

Let us come back to the situation where the porous network between the grains is filled with a fluid. Using Equation (12), and taking advantage of the fact that $G(0)$ is a cone-shaped criterion, make it possible to derive a generalized effective stress formulation for the strength properties of the saturated material, similar to that encountered in the context of poroelastic behaviour.

Indeed, $G(0)$ may be additively decomposed as

$$G(0) = \eta\, h\, \underline{\underline{I}} + C(0), \tag{29}$$

where $C(0)$ is a cone with its apex at the origin. Consequently, Equation (12) can be transformed as follows:

$$G(P) + P\underline{\underline{I}} = \left(1 + \frac{P}{h}\right) G(0) = \eta(h + P)\underline{\underline{I}} + C(0) \tag{30}$$

using the fact that $C(0)$ remains invariant by any positive similarity (i.e., $\lambda C(0) = C(0)\ \forall\lambda > 0$), so that finally

$$G(P) + P\underline{\underline{I}} = G(0) + \eta P\underline{\underline{I}}, \quad \text{or} \quad G(P) + (1-\eta)P\underline{\underline{I}} = G(0). \tag{31}$$

One immediately recognizes a generalized effective stress formulation, since the failure condition of the saturated porous sandstone only depends on an effective stress of the form $\underline{\underline{\Sigma}} + BP\underline{\underline{I}}$, with

$$B = 1 - \eta. \tag{32}$$

This important result is worth being commented on.

(a) First, it gives clear evidence that the deviation from the Terzaghi's formulation is all the more significant that the contact areas betwen grains are important, that is the cementation process is more advanced. Conversely, the Terzaghi's postulate is again recovered in the case of negligible contact areas (with respect to the grain size), which has already been demonstrated for any failure condition involving intergranular pointwise contact forces (de Buhan and Dormieux, 1996). This provides a theoretical justification to the widely accepted Terzaghi's principle concerning granular (sands) or particulate (clays) soils.

(b) There is a striking, but not so surprising, similarity between Equation (32) and its poroelastic analogue

$$B = 1 - \frac{K_0}{K_s}, \tag{33}$$

where $K_s$ is the solid matrix elastic bulk modulus (microscopic scale), and $K_0$ the bulk modulus of the drained porous material as a whole (macroscopic scale). Indeed, apart from being defined as the proportion of the joint contact area within the total grain cross-section, $\eta$ may also be interpreted as the ratio between the isotropic

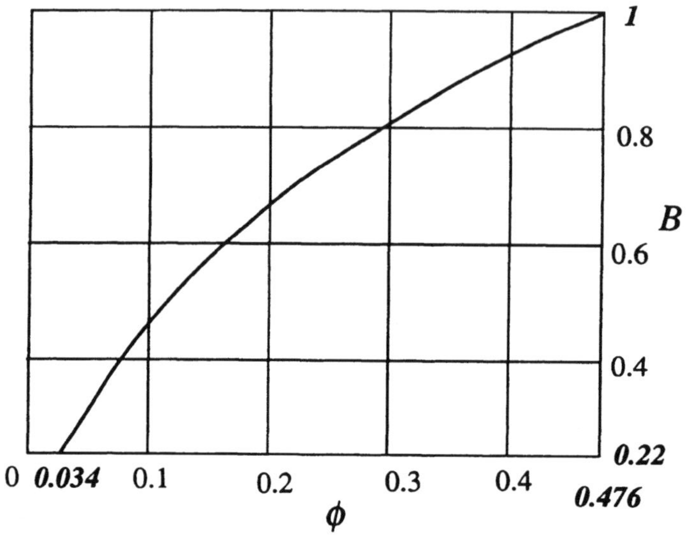


*Figure 10.* Coefficient $B$ vs. porosity $\phi$ for a periodic cubic array of cemented spherical grains.

tensile resistance of the homogenized porous sandstone from which the fluid has been removed, and the tensile strength of the joint itself.

(c) Deviation from the effective stress principle may be indirectly related to porosity. Considering for instance the particular case of cubic arrangement adopted here for describing the sandstone fabric, where the grains are assumed to be spherical, it can be easily established from simple geometrical considerations that

$$\phi = 1 - \frac{\pi}{2} - 3\eta + \frac{8\pi}{3}\left(\frac{\eta}{\pi} + \frac{1}{4}\right)^{3/2}. \tag{34}$$

This relationship holds true for $\eta$ ranging between 0 (pointwise intergranular contacts) and $\pi/4$ (tangent joint interfaces), which corresponds to a porosity varying from $\phi = 3.4\%$ ($\eta = \pi/4$) to $\phi = 47.6\%$ ($\eta = 0$). Figure 10 displays the corresponding variations of $B = 1 - \eta$ as an increasing function of $\phi$.

## 6. Concluding Remarks

Making use of the yield design homogenization method applied to inhomogeneous media exhibiting a periodic microstructure, the strength properties of a fluid-saturated porous material have been specified in two situations: the case of a purely cohesive solid matrix making up the skeleton, where it has been proved that the Terzaghi's formulation is perfectly appropriate, and more significantly as regards the potential applications, the case of sandstones, for which a simplified but representative periodic model has been adopted. A closed-form expression for the strength criterion has been derived in the latter case, which turns out to be of the generalized effective stress type, with the analogue of a Biot coefficient depending on the porosity only.

However, it should be kept in mind that the validity of such a result is subjected to several constitutive assumptions, notably the fact that the failure of the skeleton as a whole is exclusively controlled by that of the joint interfaces between adjoining grains, discarding any particle crushing phenomenon which might involve failure of the grains themselves. More particularly, the choice of a Mohr–Coulomb condition proposed in this paper for describing the strength of these joint interfaces, should be discussed further in the light of the actual physical mechanisms taking place in their formation (Rutter, 1976; Tada *et al.*, 1987).

Moreover, the introduction of a periodic model, that is of a perfectly ordered description, while allowing analytical calculations to be carried through, may appear somewhat unrealistic as regards the actual microstructure of sandstones, which is more answerable to a statistical description. A possible way to overcome such a difficulty would be to resort to homogenization techniques applied to randomly inhomogeneous media. Referring for example more particularly to models developed in the field of polycrystal plasticity (see for instance Zaoui, 1985), the microstructure of a sandstone could be regarded as a 'polycrystal', that is as an assemblage of randomly oriented and distributed 'single crystals', any such individual crystal consisting of a periodic arrangement of cemented grains as described and homogenized in Section 5. Applying such a two-step homogenization procedure would undoubtedly result in smoothing out the artificial anisotropy evidenced at the level of a single crystal, and thus provide a more realistic assessment of the overall strength properties of sandstones. Yet, it should not significantly alter the generalized effective stress formulation deduced from the periodic model.

## References


Bekaert, A. and Maghous, S.: 1996, Three-dimensional yield strength properties of jointed rock mass as a homogenized medium, *Mech. Cohesive-Frictional Mater.* **1**, 1–24.

Biot, M. A.: 1955, Theory of elasticity and consolidation of a porous anisotropic solid, *J. Appl. Phys.* **26** 182–185.

de Buhan, P.: 1986, A fundamental approach to the yield design of reinforced soil structures – Chap. II: Yield design homogenization theory for periodic media (in French), Doctorate Thesis, Paris.

de Buhan, P. and Dormieux, L.: 1996, On the validity of the effective stress concept for assessing the strength of saturated porous materials: a homogenization approach, *J. Mech. Phys. Solids* **44**(10), 1649–1667.

de Buhan, P. and de Felice, G.: 1997, A homogenization approach to the ultimate strength of brick masonry, *J. Mech. Phys. Solids* **45**(7), 1085–1104.

Cheng, A. H.-D.: 1997, Material coefficients of anisotropic poroelasticity, *Int. J. Rock Mech. Min. Sci.* **34**(2), 199–205.

Coussy, O.: 1995, *Mechanics of Porous Continua*, Wiley.

Goodman, R. E.: 1980, *Introduction to Rock Mechanics*, Wiley.

Guéguen, Y. and Palciauskas, V.: 1994, *Introduction to the Physics of Rocks*, Princeton University Press, Princeton, N.J.

Jaeger, J. C. and Cook, N. W. G.: 1976, *Fundamentals of Rock Mechanics*, Chapman & Hall, London.

Lade, P. V. and de Boer, R.: 1997, The concept of effective stress for soil, concrete and rock, *Géotechnique* **47**(1), 61–78.

Pietruszcak, S., Turu, G. and Pande, G. N.: 1995, On the mechanical response of saturated cemented materials – Part I: theoretical considerations, *Int. J. Numer. Anal. Meth. Geomech.* **19**, 555–562.

Pietruszcak, S. and Pande, G. N.: 1995, On the mechanical response of saturated cemented materials – Part II: experimental investigation and numerical simulations, *Int. J. Numer. Anal. Meth. Geomech.* **19**, 563–571.

Rice, J. R. and Cleary, M. P.: 1976, Some basic stress diffusion solutions of fluid saturated elastic porous media with compressible constituents, *Rev. Geophys. Space Phys.* **14**, 227–291.

Rutter, E. H.: 1976, The kinetics of rock deformation by pressure solution, *Phil. Trans. R. Soc. London A* **283**, 203–219.

Salençon, J.: 1990, An introduction to the yield design theory and its application to soil mechanics, *Eur. J. Mech. A/Solids* **9**, 477–500.

Suquet, P.: 1985, Elements of homogenization for inelastic solid mechanics, *Homogenization Techniques for Composite Media, C.I.S.M.*, Springer-Verlag, Udine, Italy, pp. 193–278.

Tada, R., Maliva, R. and Siever, R.: 1987, A new mechanism for pressure solution in porous quartzose sandstone, *Geochemica et Cosmochimica Acta* **51**, 2295–2301.

Terzaghi, K.: 1943, *Theoretical Soil Mechanics*, Wiley.

Thompson, M. and Willis, J. R.: 1991, A reformulation of the equations of anisotropic poroelasticity, *J. Appl. Mech. ASME* **58**, 612–616.

Zaoui, A.: 1985, Approximate statistical modelling and applications, *Homogenization Techniques for Composite Media, C.I.S.M.*, Springer-Verlag, Udine, Italy, pp. 338–397.

# Contributions to Theoretical/Experimental Developments in Shock Waves Propagation in Porous Media

S. SOREK[1,2], A. LEVY[2], G. BEN-DOR[2] and D. SMEULDERS[3]
[1]*J. Blaustein Institute for Desert Research, Water Resources Research Center, Sde Boker Campus, Israel*
[2]*Department of Mechanical Engineering, Pearlstone Center for Aeronautical Studies, Ben-Gurion University of the Negev, Beer Sheva, Israel*
[3]*Faculty of Applied Earth Sciences, Delft University of Technology, Delft, The Netherlands*

(Received: 22 June 1997; in final form: 30 October 1997)

**Abstract.** Macroscopic balance equations of mass, momentum and energy for compressible Newtonian fluids within a thermoelastic solid matrix are developed as the theoretical basis for wave motion in multiphase deformable porous media. This leads to the rigorous development of the extended Forchheimer terms accounting for the momentum exchange between the phases through the solid-fluid interfaces. An additional relation presenting the deviation (assumed of a lower order of magnitude) from the macroscopic momentum balance equation, is also presented. Nondimensional investigation of the phases' macroscopic balance equations, yield four evolution periods associated with different dominant balance equations which are obtained following an abrupt change in fluid's pressure and temperature. During the second evolution period, the inertial terms are dominant. As a result the momentum balance equations reduce to nonlinear wave equations. Various analytical solutions of these equations are described for the 1-D case. Comparison with literature and verification with shock tube experiments, serve as validation of the developed theory and the computer code.

A 1-D TVD-based numerical study of shock wave propagation in saturated porous media, is presented. A parametric investigation using the developed computer code is also given.

**Key words:** saturated and multiphase porous media, mass, momentum and energy balance equations, Forchheimer terms, Compaction waves, 1-D numerical simulation, shock tube experiments.

## 1. Introduction

The expected significance of research concerning the propagation of nonlinear waves in porous media is that it can serve as a tool in designing, for example porous plates covering, say, shelters or aircraft wings enabling the attenuation of the blast intensity. The analysis of changes produced in waves velocities may lead to the identification of different underground fluid pockets. Such modeling can also develop a methodology for cleaning groundwater from trapped contaminants which are localized at specific sites. It can explore the effect of shock waves sweeping over dust layers or sand dunes. Such modeling can also develop a methodology for pressure amplification when shock waves are discharged towards a collision at a focal point.

A porous material as defined by Gibson and Ashby [18] is a material 'made up of an interconnected network of solid strata or plates which forms the edges and faces of cells'. In general, porous materials can be divided into flexible or rigid. We use the term *rigid* in the loose relative sense which is common when applied to porous materials in general, and foams, in particular, that is the strain does not exceed a few percent. Unlike flexible porous materials which can undergo very large deformations (the strain can reach values larger than 0.9) even when moderate forces act on them, rigid porous materials do not exhibit usually large deformations. Many researchers have investigated experimentally the interaction of planar shock waves with flexible porous materials, e.g., Gelfand *et al.* [17], Gvozdeva *et al.* [21], Henderson *et al.* [24], Skews [46], Skews *et al.* [47] and Ben-Dor *et al.* [10]. Experimental investigation of the head-on collision of planar shock waves with rigid porous materials were reported by Sandusky and Liddiard [45] and Levy *et al.* [30]. In all the above-mentioned studies the pores of the porous material were air saturated. The interaction of planar shock waves with water saturated or unsaturated porous materials was reported by Zaretsky and Igra [53] for flexible polyurethane foams and van der Grinten *et al.* [20] for solid porous materials which were made by gluing together sand particles.

In recent years, the authors were involved in developing the theoretical basis and experimental verifications for wave motion in multiphase deformable porous media ([7–11, 26, 27, 29–35, 38, 48–52]). These were formulated in the form of macroscopic balance equations of mass, momentum and energy for compressible Newtonian fluids within a thermoelastic solid matrix.

In general, two type of approaches have been employed by researches for treating the above-mentioned interactions.

### 1.1. THE SINGLE PHASE APPROACH

The basic idea in the single phase approach is to replace the solid matrix $(\,)_s$, and the fluid $(\,)_f$, occupying the pores, by a single fictitious phase. This approach has been applied in two different ways. The first, known as the *bulk approach* is appropriate for both flexible and rigid porous materials. The second, known as the *pseudo-gas approach*, is appropriate to gas saturated weak flexible porous materials.

#### 1.1.1. *The Bulk Approach*

In this approach, the porous material is assumed to be a single phase whose properties are derived from the properties of the individual phases comprising it. The Lagrangian governing equations describing the head-on collision of planar shock waves with flexible foams were developed by Mazor *et al.* [38]. Numerical simulations based on this model were conducted by Ben-Dor *et al.* [10]. Relatively good agreement was evident when the numerical predictions were compared to experimental results.

### 1.1.2. *The Pseudo-Gas Approach*

Gelfand *et al.* [17] proposed a pseudo-gas approach, in which the gaseous phase and the foam matrix are treated as a homogeneous pseudo-gas with new properties. As a result of this approach, the interaction problem is reduced to the refraction of a planar shock wave at a gaseous interface (that is the interface separating the pure gas and the pseudo gas). Consequently, the flow field can be calculated using gas dynamic relations with the boundary condition that the pressures and the velocities in the flow states in either side of the gas/foam interface are equal.

Gvozdeva and Faresov [22] adopted Gelfand *et al.*'s [17] pseudo-gas model and presented an analytical model for the calculation of various parameters in the flow field (for example, the compaction wave velocity, the compaction induced flow velocity, etc.).

A similar approach was adopted by Li and Ben-Dor [35]. However, their model was limited to analytical predictions of the flow field in the vicinity of the gas/foam interface immediately after the front edge of the foam was struck head-on by the planar shock wave.

## 1.2. THE MULTIPHASE APPROACH

In this approach, the porous medium is considered as a multiphase in which the various phases interact with each other. A detailed description of this approach was given by Baer and Nunziato [5]. A one-dimensional two-phase analysis of air as the fluid phase was presented by Baer [3] and Powers *et al.* [44]. In addition to numerical solutions, Baer [3], Powers *et al.* [44] and Levy *et al.* [31] also presented simplified analytical models for calculating the jump conditions across compaction waves in rigid porous materials.

Biot's [12] work on compression waves propagation was probably the first one to employ the fundamentals of porous media mechanics. A large number of papers have appeared in the literature following Biot's pioneering work. Among the recent ones are those by Smeulders *et al.* [48], Degrande and de Roeck [14] and Nigmatulin and Gubaidullin [41]. Basically these refer to microscopic representations of the phases' balance equations within the framework of the theory of mixtures. Most refer to linear waves which take place when momentum dissipation terms dominate. As an example, Attenborough [1] presents a theory dealing with the motion of sound waves through an ideal saturated porous matrix with parallel cylindrical pores. Using microscopic physical parameters, he extended this to account for randomly distributed pores. An extensive literature survey of similar approaches is given by Corapcioglu [13].

A series of shock tube experiments in which the interaction of weak planar shock waves with low density flexible foams was conducted by Skews [46]. Baer [4] presented an analytical model, based on the continuum mixture theory for describing the shock-foam (porous material) interaction phenomenon. His numerical simulations were found to agree quite well with Skews [46] experiments. For these experiments, better resolutions were obtained by Olim *et al.* [43] who considered the matrix as dust particles. As a result, the problem was replaced by the well known problem of calculating the relaxation zone in dusty shock waves (for details see Igra and Ben-Dor [26]). In the momentum interaction they replaced the classical expressions for the drag force by Darcy–Forchheimer forces. Olim *et al.* [43] noted correctly that this approach can be justified only for infinitely weak foams in which the internal stresses are negligibly small.

Macroscopic mass, momentum and energy balance equations for the fluid phase and the solid matrix were formulated on the basis of *representative elementary volume* (REV) concepts by Bear and Bachmat [6]. Based on these, Bear and Sorek [7] developed the dominant macroscopic forms of the mass and momentum balance equations following an abrupt pressure impact in saturated porous materials under isothermal conditions. It was shown that during a certain time period, due to the domination of momentum inertial terms, the fluid momentum balance equation conforms to nonlinear wave forms.

This initiated the establishment of the macroscopic theoretical basis for wave motion in multiphase deformable porous media.

Bear *et al.* [8] and Sorek *et al.* [49] elaborated on these for the case of thermoelastic porous media, describing the theoretical basis for obtaining displacement and shock waves, respectively. Following Sorek *et al.* [49], Levy *et al.* [32] introduced additional Forchheimer terms and obtained a variety of nonlinear wave equation forms. These together with the development of the evolving balance equations following an abrupt simultaneous change of the fluid temperature and pressure, are the major novel theoretical aspects when compared with Nikolaevskij [42].

Based on the findings of Powers *et al.* [44], these macroscopic equations were simplified by Levy *et al.* [31] who proposed analytical expressions for calculating the jump conditions across compaction waves in rigid porous materials. The predictions of the model developed by Levy *et al.* [31] were found to be better than those of Powers *et al.* [44] when they were compared to the experimental results of Sandusky and Liddiard [45].

Krylov *et al.* [27] presented a 1-D simple analytical solution of the macroscopic isothermal case. The wave equation was transformed to Euler's equation describing the motion of a 'new' fluid with properties related to the fluid which actually occupied the pores of the porous material. A similar analytical solution was presented by Sorek *et al.* [51] for the nonisothermal case. In that study they presented a method leading to generalized fluid density, pressure and temperature. Using their generalized properties Euler's equation was rewritten as the 1-D expression for the analytical solution of the fluid's equation of motion. To obtain this, the porosity, the matrix strain and temperature of the solid were developed as explicit functions of pressure.

The coupled changes between solute's mass balance equation and the 'new' fluid wave equation, were developed theoretically and solved analytically for a 1-D case by Sorek and Levy-Hevroni [50] and Sorek [52] for nondeformable and deformable, respectively, saturated porous media.

In the following we will present the development leading to the macroscopic nonlinear wave equations for a fluid in saturated deformable and thermoelastic porous media.

## 2. Phase Balance Equations

As a starting point for analyzing the motion of a phase, that is the spatial and temporal changes that it may undergo as a result of various excitations, Bear and Bachmat [6] describe the general form for a microscopic balance equation for an extensive quantity in a single phase continuum. The microscopic balance equations for each phase are then transformed by *averaging* rules, Bachmat and Bear [2] and Bear and Bachmat [6], into macroscopic equivalents using REV concepts.

### 2.1. MACROSCOPIC BALANCE EQUATIONS

In a saturated porous medium, a compressible Newtonian fluid and a thermoelastic solid matrix, the phases' macroscopic balance equations, neglecting their dispersive flux terms, read:

*The fluid mass balance equation:*

$$\frac{\partial}{\partial t}(\phi\bar{\rho}_{\mathrm{f}}^{\mathrm{f}}) = -\nabla\cdot(\phi\bar{\rho}_{\mathrm{f}}^{\mathrm{f}}\overline{\mathbf{V}_{\mathrm{f}}}^{\mathrm{f}}). \tag{1}$$

*The solid mass balance equation:*

$$\frac{\partial}{\partial t}[(1-\phi)\bar{\rho}_{\mathrm{s}}^{\mathrm{s}}] = -\nabla\cdot[(1-\phi)\bar{\rho}_{\mathrm{s}}^{\mathrm{s}}\overline{\mathbf{V}_{\mathrm{s}}}^{\mathrm{s}}], \tag{2}$$

where $\bar{e}_{\alpha}^{\alpha}$ denotes the average of an intensive variable $e_\alpha$, over the $\alpha$-phase ($\alpha \equiv \mathrm{f}, \mathrm{s}$) within the REV, $\rho$ denotes the mass density of the phase, $\phi$ denotes the porosity and $\mathbf{V}$ denotes the mass weighted velocity of the phase.

The general form of the momentum balance equation for the $\alpha$ phase, assuming that the micoroscopic solid–fluid interface $\mathcal{S}_{\mathrm{fs}}$ is material to mass flux, reads

$$\begin{aligned}
&\frac{\partial}{\partial t}[\theta_\alpha(\bar{\rho}_{\alpha}^{\alpha}\overline{\mathbf{V}_\alpha}^{\alpha} + \overline{\mathbf{d}_{\alpha}^{M}}^{\alpha})] \\
&\quad = -\nabla\cdot[\theta_\alpha(\bar{\rho}_{\alpha}^{\alpha}\overline{\mathbf{V}_\alpha}^{\alpha}\overline{\mathbf{V}_\alpha}^{\alpha} - \overline{\boldsymbol{\sigma}_\alpha}^{\alpha} + \overline{\mathbf{d}_{\alpha}^{M}}^{\alpha}\overline{\mathbf{V}_\alpha}^{\alpha} + \overline{(\rho_\alpha\overset{\circ}{\mathbf{V}}_\alpha)\overset{\circ}{\mathbf{V}}_\alpha}^{\alpha})] + \\
&\quad\quad + \frac{1}{\mathcal{U}_0}\int_{\mathcal{S}_{\mathrm{fs}}}\boldsymbol{\sigma}_\alpha\cdot\boldsymbol{\xi}\,\mathrm{d}s + \theta_\alpha\overline{\rho_\alpha\mathbf{F}_\alpha}^{\alpha},
\end{aligned} \tag{3}$$

in which $\theta_\alpha$ denotes the volume fraction of the $\alpha$ phase, $\boldsymbol{\sigma}_\alpha$ denotes its stress tensor, $\mathbf{F}_\alpha$ denotes its body force vector, $\mathcal{U}_0$ denotes the REV volume, $\boldsymbol{\xi}$ denotes a unit vector perpendicular to the REV interface and $\overline{\mathbf{d}_{\alpha}^{M}}^{\alpha}$ ($\equiv \overline{\overset{\circ}{\rho}_\alpha\overset{\circ}{\mathbf{V}}_\alpha}^{\alpha}$) denotes the variations vector (within the volume of the $\alpha$ phase in REV) from the specific momentum of the $\alpha$

phase. In obtaining the macroscopic momentum balance equations, we neglect $\overline{\mathbf{d}_{\alpha}^{M}}^{\alpha}$ (i.e., $\left|\bar{\rho}_{\alpha}^{\alpha}\overline{\mathbf{V}_{\alpha}}^{\alpha}\right| \gg \left|\overline{\mathbf{d}_{\alpha}^{M}}^{\alpha}\right|$), as well as the dispersive flux of momentum $(\overline{(\rho_{\alpha}\mathring{\mathbf{V}}_{\alpha})\mathring{\mathbf{V}}_{\alpha}}^{\alpha})$ compared with the advective one $(\bar{\rho}_{\alpha}^{\alpha}\overline{\mathbf{V}_{\alpha}}^{\alpha}\overline{\mathbf{V}_{\alpha}}^{\alpha})$.

To evaluate the value of $\overline{\mathbf{d}_{\alpha}^{M}}^{\alpha}$, we need to solve separately the momentum balance equation of the *deviation terms*. Concerning the fluid, this reads

$$\frac{\partial}{\partial t}(\phi\overline{\mathbf{d}_{\mathrm{f}}^{M}}^{\mathrm{f}}) = -\nabla\cdot\left[\phi\left(\overline{\mathbf{d}_{\mathrm{f}}^{M}}^{\mathrm{f}}\bar{\mathbf{V}}_{\mathrm{f}}^{\mathrm{f}} + \overline{(\rho_{\mathrm{f}}\mathring{\mathbf{V}}_{\mathrm{f}})\mathring{\mathbf{V}}_{\mathrm{f}}}^{\mathrm{f}}\right)\right]. \tag{4}$$

We note that the first RHS term in (4), represents an advective flux of this momentum balance equation. To solve for $\overline{\mathbf{d}_{\alpha}^{M}}^{\alpha}$ in (4), we need first to solve the macroscopic mass and momentum balance equations from which we will estimate $\overline{\mathbf{V}_{\mathrm{f}}}^{\mathrm{f}}$, the phase macroscopic velocity. Moreover, as an approximation of (4), we can assume that the second RHS term (the dispersive flux of the momentum) is negligible in comparison to the advective one.

Unlike Bear and Bachmat [6], for high fluid velocities and based on the microscopic momentum balance equation of the fluid at $\mathcal{S}_{\mathrm{fs}}$, the solid–fluid interface, we assume

$$\left|\left(\rho_{\mathrm{f}_j}\frac{\partial V_{\mathrm{f}_j}}{\partial t} - \frac{\partial \tau_{\mathrm{f}_{ij}}}{\partial x_i}\right)\xi_j\right| \ll \left|\left(\frac{\partial P}{\partial x_j} + \rho_{\mathrm{f}} g\frac{\partial Z}{\partial x_j} + \rho_{\mathrm{f}} V_{\mathrm{f}_j}\frac{\partial V_{\mathrm{f}_j}}{\partial x_i}\right)\xi_j\right| \tag{5}$$

in which $\tau_{f_{ij}}$ denotes the shear stress tensor of the Newtonian fluid, $P$ denotes its pressure, $\mathbf{g}$ denotes the gravity acceleration vector, $Z$ denotes the altitude parallel to $\mathbf{g}$. We then obtain,

*The fluid momentum balance equation:*

$$\begin{aligned}
\phi\bar{\rho}_{\mathrm{f}}^{\mathrm{f}}\frac{D_{\mathrm{f}}}{Dt}(\overline{\mathbf{V}_{\mathrm{f}}}^{\mathrm{f}}) &= -\phi\mathbf{T}_{\mathrm{f}}^{*}(\nabla\bar{P}^{\mathrm{f}} + \bar{\rho}_{\mathrm{f}}^{\mathrm{f}} g\nabla Z) - \frac{c_{\mathrm{f}}}{2\Delta_{\mathrm{f}}^{2}}\frac{\bar{\rho}_{\mathrm{f}}^{\mathrm{f}}}{\phi}\tilde{\mathbf{F}}\mathbf{q}_{\mathrm{r}}\mathbf{q}_{\mathrm{r}} + \\
&\quad + \bar{\rho}_{\mathrm{f}}^{\mathrm{f}}(\mathbf{I} - \mathbf{T}_{\mathrm{f}}{}^{*})(\nabla\mathbf{q}_{\mathrm{r}} + \phi\nabla\overline{\mathbf{V}_{\mathrm{s}}}^{\mathrm{s}})\cdot\overline{\mathbf{V}_{\mathrm{s}}}^{\mathrm{s}} + \\
&\quad + (\bar{\mu}_{\mathrm{f}}^{\mathrm{f}} + \overline{\lambda''_{\mathrm{f}}}^{\mathrm{f}})[\nabla(\nabla\cdot\mathbf{q}_{\mathrm{r}} + \phi\nabla\cdot\overline{\mathbf{V}_{\mathrm{s}}}^{\mathrm{s}}) - (\nabla\cdot\overline{\mathbf{V}_{\mathrm{f}}}^{\mathrm{f}})\nabla\phi] + \\
&\quad + \bar{\mu}_{\mathrm{f}}^{\mathrm{f}}\left[\nabla\cdot(\nabla\mathbf{q}_{\mathrm{r}} + \phi\nabla\overline{\mathbf{V}_{\mathrm{s}}}^{\mathrm{s}}) - \frac{c_{\mathrm{f}}}{\Delta_{\mathrm{f}}^{2}}\boldsymbol{\alpha}\mathbf{q}_{\mathrm{r}}\right],
\end{aligned} \tag{6}$$

where $\mathrm{D}_{\alpha}/\mathrm{D}t \equiv \partial/\partial t + \mathbf{V}_{\alpha}\cdot\nabla$ denotes the *material derivative*, which expresses the rate of change from the point of view of an observer traveling at the velocity $\mathbf{V}_{\alpha}$, $\bar{\mu}_{\mathrm{f}}^{\mathrm{f}}$ and $\overline{\lambda''_{\mathrm{f}}}^{\mathrm{f}}$ denote, respectively, the first and second fluid viscosities, $c_{\mathrm{f}}$ denotes a shape factor, $\boldsymbol{\alpha}$ denotes a tensor associated with the matrix directional cosines, $\Delta_{\mathrm{f}}$ denotes the hydraulic radius of the pore spaces, $\mathbf{I}$ denotes a unit tensor, $\mathbf{q}_{\mathrm{r}}(\equiv \phi\bar{\mathbf{V}}_{\mathrm{r}})$ denotes the specific flux related to the relative velocity $\bar{\mathbf{V}}_{\mathrm{r}}(\equiv \overline{\mathbf{V}_{\mathrm{f}}}^{\mathrm{f}} - \overline{\mathbf{V}_{\mathrm{s}}}^{\mathrm{s}})$ and $\tilde{\mathbf{F}}$ denotes the third rank Forchheimer tensor.

Note that the Forchheimer term originated from a microscopic inertial term at the solid–fluid interface (see also Du Plessis and Masliyan [15] and Du Plessis *et al.* [16]). This is in contradiction to the assumption made by Hsu and Cheng [25], Nield [39, 40], and Olim *et al.* [43] who refer to it as a drag force.

*The porous medium momentum balance equation*

By summing the macroscopic momentum balance equations for the fluid and for the thermoelastic solid phases, the rate of momentum exchange across the solid–fluid interface is eliminated and we obtain the momentum balance equation for the porous medium as a whole in the form

$$\begin{aligned}\bar{\rho}_{\mathrm{b}}\frac{\mathrm{D_s}}{\mathrm{D}t}(\overline{\mathbf{V}_{\mathrm{s}}}^{\mathrm{s}}) + \phi\bar{\rho}_{\mathrm{f}}^{\mathrm{f}}\frac{\mathrm{D_r}}{\mathrm{D}t}(\bar{\mathbf{V}}_{\mathrm{r}}) &= -\phi\bar{\rho}_{\mathrm{f}}^{\mathrm{f}}(\bar{\mathbf{V}}_{\mathrm{r}}\nabla\cdot\overline{\mathbf{V}_{\mathrm{s}}}^{\mathrm{s}} + \overline{\mathbf{V}_{\mathrm{s}}}^{\mathrm{s}}\nabla\cdot\mathbf{V}_{\mathrm{r}}) + \nabla\cdot\bar{\boldsymbol{\sigma}}_{\mathrm{t}} + \\ &\quad + \bar{\rho}_{\mathrm{b}}g\boldsymbol{\nabla}Z + \frac{1}{\mathcal{U}_0}\int_{\mathcal{S}_{\mathrm{fs}}}(\boldsymbol{\sigma}_{\mathrm{f}} - \boldsymbol{\sigma}_{\mathrm{s}})\cdot\boldsymbol{\xi}_{\mathrm{f}}\,\mathrm{d}S, \qquad (7)\end{aligned}$$

where $\bar{\rho}_{\mathrm{b}}$ denotes the *bulk density*, given by

$$\bar{\rho}_{\mathrm{b}} \equiv \phi\bar{\rho}_{\mathrm{f}}^{\mathrm{f}} + (1-\phi)\bar{\rho}_{\mathrm{s}}^{\mathrm{s}}, \qquad (8)$$

the *total bulk stress* $\bar{\boldsymbol{\sigma}}_{\mathrm{t}}$ is given by

$$\bar{\boldsymbol{\sigma}}_{\mathrm{t}} \equiv \phi\bar{\boldsymbol{\sigma}}_{\mathrm{f}}^{\mathrm{f}} + (1-\phi)\bar{\boldsymbol{\sigma}}_{\mathrm{s}}^{\mathrm{s}} = \bar{\boldsymbol{\sigma}}_{\mathrm{s}}' + \bar{\boldsymbol{\sigma}}_{\mathrm{f}}^{\mathrm{f}}, \qquad (9)$$

in which $\bar{\boldsymbol{\sigma}}_{\mathrm{s}}'$ denotes the effective stress tensor for the thermoelastic solid matrix in which its constitutive relation is assumed to be similar to its microscopic one,

$$\bar{\boldsymbol{\sigma}}_{\mathrm{s}}' = 2\overline{\mu_{\mathrm{s}}'}^{\mathrm{s}}\boldsymbol{\varepsilon}_{sk} + [\overline{\lambda''_{\mathrm{s}}}^{\mathrm{s}}\varepsilon_{sk} - \bar{\eta}(\overline{T_{\mathrm{s}}}^{\mathrm{s}} - \overline{T_{\mathrm{s0}}}^{\mathrm{s}})]\mathbf{I} \qquad (10)$$

$\overline{T_\alpha}^{\alpha}$ denotes the $\alpha$-phase temperature, $\overline{\mu_{\mathrm{s}}'}^{\mathrm{s}}$, $\overline{\lambda_{\mathrm{s}}''}^{\mathrm{s}}$ and $\bar{\eta}$ denote the Lame's constants for a thermoelastic solid matrix, $\boldsymbol{\varepsilon}_{sk}$ and $\varepsilon_{sk}$ denote, respectively, its strain tensor and volumetric strain associated with $\mathbf{w}_{\mathrm{s}}$, its displacement vector, and defined by

$$\boldsymbol{\varepsilon}_{sk} = \tfrac{1}{2}[\boldsymbol{\nabla}\mathbf{w}_{\mathrm{s}} + (\boldsymbol{\nabla}\mathbf{w}_{\mathrm{s}})^{T}]; \quad \varepsilon_{sk} = \nabla\cdot\mathbf{w}_{\mathrm{s}}. \qquad (11)$$

In writing the macroscopic energy balance equations for the fluid and the solid phases, we assume linear thermodynamics, omit external energy sources and the energy associated with the fluid shear tensor $\bar{\boldsymbol{\tau}}_{\mathrm{f}}^{\mathrm{f}}$ (viscous dissipation) is assumed to be negligible. We thus obtain:

*The fluid energy balance equation:*

$$\begin{aligned}&\frac{\partial}{\partial t}\left\{\phi\bar{\rho}_{\mathrm{f}}^{\mathrm{f}}\left[C_{\mathrm{f}}\overline{T_{\mathrm{f}}}^{\mathrm{f}} + \frac{(\overline{V_{\mathrm{f}}}^{\mathrm{f}})^2}{2}\right]\right\} \\ &\quad = \mathbf{T}_{\mathrm{s}}^{*}\bar{P}^{\mathrm{f}}\overline{\mathbf{V}_{\mathrm{s}}}^{\mathrm{s}}\cdot\nabla\phi - \mathbf{T}_{\mathrm{f}}^{*}\nabla\cdot(\phi\bar{P}^{\mathrm{f}}\overline{\mathbf{V}_{\mathrm{f}}}^{\mathrm{f}}) - \alpha^{*\mathrm{H}}(\overline{T_{\mathrm{f}}}^{\mathrm{f}} - \overline{T_{\mathrm{s}}}^{\mathrm{s}}) + \\ &\quad + \overline{\mathbf{V}_{\mathrm{s}}}^{\mathrm{s}}\cdot\left[\frac{c_{\mathrm{f}}}{2\Delta_{\mathrm{f}}^2}\frac{\bar{\rho}_{\mathrm{f}}^{\mathrm{f}}}{\phi}\tilde{\mathbf{F}}\,\mathbf{q}_{\mathrm{r}}\mathbf{q}_{\mathrm{r}} + \bar{\rho}_{\mathrm{f}}^{\mathrm{f}}(\mathbf{I} - \mathbf{T}_{\mathrm{f}}^{*})(\boldsymbol{\nabla}\mathbf{q}_{\mathrm{r}} + \phi\boldsymbol{\nabla}\overline{\mathbf{V}_{\mathrm{s}}}^{\mathrm{s}})\cdot\overline{\mathbf{V}_{\mathrm{s}}}^{\mathrm{s}}\right] -\end{aligned}$$

$$
- \nabla \cdot \left\{ \phi \bar{\rho}_{\mathrm{f}}^{\mathrm{f}} \left[ C_{\mathrm{f}} \overline{T_{\mathrm{f}}}^{\mathrm{f}} + \frac{(\overline{V_{\mathrm{f}}}^{\mathrm{f}})^2}{2} \right] \overline{\mathbf{V}_{\mathrm{f}}}^{\mathrm{f}} \right\} +
$$

$$
+ \nabla \cdot [\phi \mathbf{D}^{*\mathrm{H}} \boldsymbol{\nabla} (C_{\mathrm{f}} \bar{\rho}_{\mathrm{f}}^{\mathrm{f}} \overline{T_{\mathrm{f}}}^{\mathrm{f}}) + \phi \bar{\lambda}_{\mathrm{f}}^{\mathrm{f}} \boldsymbol{\nabla} \overline{T_{\mathrm{f}}}^{\mathrm{f}}], \tag{12}
$$

where $C_\alpha$ denotes the constant specific heat at constant volume for the fluid phase ($\alpha \equiv f$) or at constant strain for the solid matrix ($\alpha \equiv \mathrm{s}$), $\overline{\lambda_\alpha}^\alpha$ denotes the thermal conductivity for the $\alpha$-phase, $\alpha^{*\mathrm{H}}$ denotes the heat transfer coefficient and $\mathbf{D}^{*\mathrm{H}}$ denotes the dispersive heat tensor for the fluid.

*The solid energy balance equation:*

$$
\frac{\partial}{\partial t} \left\{ (1-\phi) \bar{\rho}_{\mathrm{s}}^{\mathrm{s}} \left[ C_{\mathrm{f}} \overline{T_{\mathrm{s}}}^{\mathrm{s}} + \frac{(\overline{V_{\mathrm{s}}}^{\mathrm{s}})^2}{2} \right] \right\}
$$

$$
= -\alpha^{*\mathrm{H}} (\overline{T_{\mathrm{s}}}^{\mathrm{s}} - \overline{T_{\mathrm{f}}}^{\mathrm{f}}) - \mathbf{T}_{\mathrm{s}}^{*} \bar{P}^{\mathrm{f}} \overline{\mathbf{V}_{\mathrm{s}}}^{\mathrm{s}} \cdot \boldsymbol{\nabla} \phi + \mathbf{T}_{\mathrm{s}}^{*} \nabla \cdot [(1-\phi) \bar{P}^{\mathrm{f}} \overline{\mathbf{V}_{\mathrm{s}}}^{\mathrm{s}}] - \nabla \cdot (\boldsymbol{\sigma}_{\mathrm{s}}' \overline{\mathbf{V}_{\mathrm{s}}}^{\mathrm{s}}) -
$$

$$
- \overline{\mathbf{V}_{\mathrm{s}}}^{\mathrm{s}} \cdot \left[ \frac{c_{\mathrm{f}}}{2\Delta_{\mathrm{f}}^2} \frac{\bar{\rho}_{\mathrm{f}}^{\mathrm{f}}}{\phi} \tilde{\mathbf{F}} \mathbf{q}_{\mathrm{r}} \mathbf{q}_{\mathrm{r}} + \bar{\rho}_{\mathrm{f}}^{\mathrm{f}} (\mathbf{I} - \mathbf{T}_{\mathrm{f}}^{*}) (\boldsymbol{\nabla} \mathbf{q}_{\mathrm{r}} + \phi \boldsymbol{\nabla} \overline{\mathbf{V}_{\mathrm{s}}}^{\mathrm{s}}) \cdot \overline{\mathbf{V}_{\mathrm{s}}}^{\mathrm{s}} \right] -
$$

$$
- \boldsymbol{\nabla} \cdot \left\{ (1-\phi) \bar{\rho}_{\mathrm{s}}^{\mathrm{s}} \left[ C_{\mathrm{s}} \overline{T_{\mathrm{s}}}^{\mathrm{s}} + \frac{(\overline{V_{\mathrm{s}}}^{\mathrm{s}})^2}{2} \right] \overline{\mathbf{V}_{\mathrm{s}}}^{\mathrm{s}} - (1-\phi) \overline{\lambda_{\mathrm{s}}}^{\mathrm{s}} \boldsymbol{\nabla} \overline{T_{\mathrm{s}}}^{\mathrm{s}} \right\}. \tag{13}
$$

The set of balance equations (1), (2), (6), (7), (12) and (13) together with the definitions (8)–(11) can be solved for the macroscopic flow properties, given appropriate values for the various expressions representing the effects of small-scale features on the averaged large-scale flow. From hereafter we will refer to the macroscopic equations only and, accordingly, the $\overline{()}^\alpha$ designator will be omitted.

## 3. Dominant Forms of the Balance Equations Evolving After Abrupt Simultaneous Changes in Fluid Pressure and Temperature

In what follows we will describe the form of the dominant balance equations in a saturated porous media, obtained after abrupt simultaneous changes (e.g., in a well) in fluid pressure and temperature. Our objective is to analyze the order of magnitude of the flux terms appearing in the obtained macroscopic balance equations in order to eliminate nondominating ones. To achieve this goal, a dimensional analysis is carried out in which all the terms are compared with those associated with the rate of fluid pressure and temperature changes. In doing so, we follow Bear and Sorek [7], Sorek *et al.* [49] and Levy *et al.* [32]. Similar investigation can also be found in Sorek [52] in the case of abrupt change in pressure together with the problem of solute transport.

### 3.1. PERIOD OF UNIFORM DISTRIBUTIONS OF PRESSURE, STRESS AND TEMPERATURE

At the instant of abrupt change in the fluid's pressure and temperature, the nondimensional factors associated with these rates are of order one while all others are much smaller than one. This yields an approximate fluid mass balance equation in the form,

$$\left(\beta_{\mathrm{P}}\frac{\partial P}{\partial t}+\beta_{\mathrm{T}}\frac{\partial T_{\mathrm{f}}}{\partial t}\right)+V_{\mathrm{f}_j}\left(\beta_{\mathrm{P}}\frac{\partial P}{\partial x_j}+\beta_{\mathrm{T}}\frac{\partial T_{\mathrm{f}}}{\partial x_j}\right)=0 \tag{14}$$

for a compressible fluid ($\rho_{\mathrm{f}}=\rho_{\mathrm{f}}(P,T_{\mathrm{f}})$), denoting $\beta_{\mathrm{P}}$ and $\beta_{\mathrm{T}}$ as its pressure and temperature compressibilities, respectively. The solid matrix is deformable ($\phi=\phi(P)$) and assuming $|\partial\phi/\partial t|\gg|\mathbf{V}_{\mathrm{s}}\cdot\nabla\phi|$:

The fluid's momentum balance equation becomes

$$\phi T^*_{\mathrm{f}_{ij}}\frac{\partial P}{\partial x_j}=0. \tag{15}$$

The porous media momentum balance equation reads

$$\nabla\boldsymbol{\sigma}'_{\mathrm{s}}=(1-\phi)\rho_{\mathrm{s}}\,g\nabla Z. \tag{16}$$

The fluid energy balance equation is given by

$$|\mathbf{V}_{\mathrm{f}}+\mathbf{V}_{\mathrm{s}}|\nabla T_{\mathrm{f}}\cdot\nabla T_{\mathrm{f}}=0. \tag{17}$$

From (15) and (17) we note that P and $T_{\mathrm{f}}$ have constant distributions. Thus, this together with (14) yields

$$\beta_{\mathrm{P}}\frac{\partial P}{\partial t}=-\beta_{\mathrm{T}}\frac{\partial T_{\mathrm{f}}}{\partial t}. \tag{18}$$

Note that in view of (14) and (18), although the fluid is considered compressible, during this first time period it behaves as if it was incompressible, i.e.,

$$\rho_{\mathrm{f}}=\text{const.}\qquad\text{at}\quad t=0+. \tag{19}$$

The solid energy balance equation becomes

$$(1-\phi)\rho_{\mathrm{s}}C_{\mathrm{s}}\frac{\mathrm{D_s}T_{\mathrm{s}}}{\mathrm{D}t}=0, \tag{20}$$

hence, $T_{\mathrm{s}}$ remains constant along characteristic curves defined by

$$\frac{\mathrm{D_s}\mathbf{x}_{\mathrm{s}}}{\mathrm{D}t}=\mathbf{V}_{\mathrm{s}}, \tag{21}$$

where $\mathbf{x}_{\mathrm{s}}$ denotes the position vector along the characteristic path. The pressure and the fluid temperature impulse propagate without spatial attenuation over a characteristic, $()_c$, distance $L^{\mathrm{P}}_c$ defined by

$$L^{\mathrm{P}}_c\cong\mathrm{O}\left(\frac{t^{\mathrm{P}}_c}{\rho_{\mathrm{f}}\,\beta_{\mathrm{P}}\,V^{\mathrm{f}}_c}\right). \tag{22}$$

Hence, with the increase in time ($t = 0+$) at the onset of the fluid's pressure and temperature, throughout the $L_c^{\mathrm{P}}$ distance the fluid density will remain constant. This constitutes a linear ratio between the change of the fluid's pressure and temperature, both of which are uniformly distributed in space. The effective matrix stress is vertically uniformly distributed and constant elsewhere and its temperature and rate of change are related to its velocity.

### 3.2. PERIOD OF NONLINEAR WAVE PROPAGATION

After the first time period, it was concluded (Levy *et al.* [32]) that the fluid and the solid mass balance equations, remain in the form of (1) and (2), respectively.

During this time period, the fluid momentum balance equation conforms to

$$\phi\rho_{\mathrm{f}}\frac{\mathrm{D_f}}{\mathrm{D}t}(V_{\mathrm{f}_i}) + \phi T^*_{\mathrm{f}_{ij}}\left(\frac{\partial P}{\partial x_j} + \rho_{\mathrm{f}}\, g\frac{\partial Z}{\partial x_j}\right) + \frac{c_{\mathrm{f}}}{2\Delta_{\mathrm{f}}^2}\tilde{F}_{ijk}\,\phi\rho_{\mathrm{f}}\, V_{\mathrm{r}_j} V_{\mathrm{r}_k} = 0. \tag{23}$$

We subtract the fluid momentum balance equation from the porous medium momentum balance equation and assume that $|(1-\phi)\,\rho_{\mathrm{s}}| \gg |\phi\rho_{\mathrm{f}}(T_{\mathrm{s}}^* - T_{\mathrm{f}}^*)|$.

The obtained solid momentum balance equation, reads

$$(1-\phi)\rho_{\mathrm{s}}\frac{\mathrm{D_s}}{\mathrm{D}t}(V_{\mathrm{s}_i}) + (1-\phi)\left(T^*_{\mathrm{s}_{ij}}\frac{\partial P}{\partial x_j} + \rho_{\mathrm{s}}\, g\frac{\partial Z}{\partial x_i}\right) - \\ - \frac{\partial\sigma'_{\mathrm{s}_{ij}}}{\partial x_j} - \frac{c_{\mathrm{f}}}{2\Delta_{\mathrm{f}}^2}\tilde{F}_{ijk}\,\phi\rho_{\mathrm{f}}\, V_{\mathrm{r}_j} V_{\mathrm{r}_k} = 0. \tag{24}$$

The approximate energy balance equations conform to:

*The fluid energy balance equation*

$$\phi\rho_{\mathrm{f}}\left[C_{\mathrm{f}}\frac{\mathrm{D_f}}{\mathrm{D}t}(T_{\mathrm{f}}) + V_{\mathrm{f}_i}\frac{\mathrm{D_f}}{\mathrm{D}t}(V_{\mathrm{f}_i})\right] + T^*_{\mathrm{f}_{ij}}\left[\frac{\partial}{\partial x_i}(\phi P V_{\mathrm{f}_j}) - P\, V_{\mathrm{s}_j}\frac{\partial\phi}{\partial x_i}\right] + \\ + \frac{c_{\mathrm{f}}}{2\Delta_{\mathrm{f}}^2}\tilde{F}_{ijk}\,\phi\rho_{\mathrm{f}}\, V_{\mathrm{r}_j} V_{\mathrm{r}_k} V_{\mathrm{s}_i} = 0. \tag{25}$$

*The solid energy balance equation*

$$(1-\phi)\rho_{\mathrm{s}}\left[C_{\mathrm{s}}\frac{\mathrm{D_s}}{\mathrm{D}t}(T_{\mathrm{s}}) + V_{\mathrm{s}_j}\frac{\mathrm{D_s}}{\mathrm{D}t}(V_{\mathrm{s}_j})\right] + \\ + T^*_{\mathrm{s}_{ij}}\left[\frac{\partial}{\partial x_i}(1-\phi)P V_{\mathrm{s}_j} + P\, V_{\mathrm{s}_j}\frac{\partial\phi}{\partial x_i}\right] - \\ - \frac{\partial\sigma'_{\mathrm{s}_{ij}}}{\partial x_i} V_{\mathrm{s}_j} - \frac{c_{\mathrm{f}}}{2\Delta_{\mathrm{f}}^2}\tilde{F}_{ijk}\,\phi\rho_{\mathrm{f}}\, V_{\mathrm{r}_i} V_{\mathrm{r}_k} V_{\mathrm{s}_j} = 0. \tag{26}$$

On the basis of nondimensional scalars of the same order we note that the time span during which (23) to (26) are valid is given by

$$t_c^{\mathrm{P}} \cong \mathrm{O}\left(\frac{\beta_{\mathrm{P}}\, P_c\, V_c^{\mathrm{f}}}{g}\right). \tag{27}$$

During this period of evolution, inertial terms dominate and we note the emergence of nonlinear wave equations.

Actually, two more evolution stages are reported in Bear and Sorek [7] and Levy *et al.* [32]. During the third time period, the dissipative effects in the balance equations arise and become of the same order of magnitude as the inertial terms. During the fourth time period, the dissipative terms are dominating over the inertial ones. The various forms of the dominant balance equations during the third and fourth time periods are beyond the scope of this manuscript.

## 4. Analytical Solutions

### 4.1. ACROSS STRONG COMPACTION WAVES IN GAS SATURATED RIGID POROUS MEDIA

Based on experimental results and some additional simplifying assumptions, we show that the general macroscopic equations governing the flow field which is developed in a gas saturated rigid porous medium domain can be simplified to a form which enabled us to develop two analytical models for calculating the jump conditions across strong compaction waves.

Predictions obtained by these two simplified analytical models were compared to the experimental results of Sandusky and Liddiard [45] and to predictions of the model proposed by Powers *et al.* [44]. Fairly good to excellent agreements are evident.

In view of (1)–(13), we can write the *1-D mass, momentum and energy balance equations of the porous medium as a whole*. This we do by combining the fluid and solid balance equations, for each of the balance equations, noting that for the 1-D case $T_f^* = T_s^* = 1$. Dimensional analysis of these combined balance equations showed that the gaseous-phase terms were negligible in comparison to the solid-phase terms. Thus, the *1-D porous medium balance equations of mass, momentum and energy* could be approximated by those of the solid-phase ones. These, upon dropping the s subscript (e.g., $\theta \equiv \theta_s = 1 - \phi$), read respectively

$$\frac{\partial}{\partial t}(\theta\rho) + \frac{\partial}{\partial x}(\theta\rho V) = 0, \tag{28}$$

$$\frac{\partial}{\partial t}(\theta\rho V) + \frac{\partial}{\partial x}(\theta\rho V^2 + \theta\sigma) = 0, \tag{29}$$

$$\frac{\partial}{\partial t}\left[\theta\rho\left(\mathcal{I} + \frac{V^2}{2}\right)\right] + \frac{\partial}{\partial x}\left[\theta\rho V\left(\mathcal{I} + \frac{V^2}{2} + \frac{\sigma}{\rho}\right)\right] = 0. \tag{30}$$

Based on findings of Baer [3] and Powers *et al.* [44] that the velocity of compaction waves is almost constant, one can further reduce (28)–(30) by applying the well known Galilean transformation (Ben-Dor [9]), i.e.,

$$\frac{\mathrm{d}}{\mathrm{d}\xi}(\theta\rho u) = 0, \tag{31}$$

$$\frac{\mathrm{d}}{\mathrm{d}\xi}(\theta\rho u^2 + \theta\sigma) = 0, \tag{32}$$

$$\frac{\mathrm{d}}{\mathrm{d}\xi}\left(\mathcal{I} + \frac{u^2}{2} + \frac{\sigma}{\rho}\right) = 0, \tag{33}$$

where $u \equiv V - \mathcal{D}$, $\xi \equiv x - \mathcal{D}t$ and $\mathcal{D}$ denotes the constant velocity of the compaction wave.

The above set (31)–(33), consists of three equations and five ($\phi$, $\rho$, $u$, $\sigma$, $\mathcal{I}$) unknowns. Following Powers *et al.* [44], we add the *caloric equation for the solid phase*

$$\mathcal{I} + \frac{\sigma}{\rho} = \frac{\gamma}{(\gamma - 1)}\frac{\sigma + \mathcal{S}}{\rho}, \tag{34}$$

where $\gamma$ and $\mathcal{S}$ are parameters that define Tait's equation of state. The value of $\gamma$ is chosen to match the Hugoniot data. It is analogous to the specific heat ratio for equation of state of an ideal gas, $\mathcal{S}$ is the non-ideal solid parameter. When $\mathcal{S} = 0$, the equation of state is referred to as being ideal. An additional equation can be based on empirical relations (see, for example, the *compaction equation* in Powers *et al.* [44]). However, the parameters of this may need to be correlated with each separate experiment. Instead, we will prescribe one of the unknowns.

In what follows, we describe two simplified models. In the first, we assume that the solid phase density remains constant while in the second, post compaction wave porosity $\phi$, is close to zero. The relations, as obtained from these two models, are presented in terms of $X_{21} \equiv X_2/X_1$ where X denotes a property of the solid phase (e.g., $\theta$, $\sigma$, $u$, etc.), and the indices 1 and 2 denote location ahead and behind the compaction wave, respectively.

#### 4.1.1. *Model I: Post Compaction Wave for Small Porosity Values*

It can be assumed that the solid phase starts to be compressed only when its volume fraction $\theta_2$ approaches unity. Hence, when $\theta_2$ is sufficiently smaller than unity, the solid phase can be assumed to be incompressible ($\rho_2 = \rho_1 \equiv \rho$). Under these assumptions (31)–(33), together with (34), can be integrated to read

$$\theta_1\mathcal{D} = \theta_2(\mathcal{D} - u), \tag{35}$$

$$\theta_1\rho\mathcal{D}^2 + \theta_1\sigma_1 = \theta_2\rho(\mathcal{D} - u)^2 + \theta_2\sigma_2, \tag{36}$$

$$\frac{\gamma}{(\gamma - 1)}\frac{\sigma_1 + \mathcal{S}}{\rho} + \frac{\mathcal{D}^2}{2} = \frac{\gamma}{(\gamma - 1)}\frac{\sigma_2 + \mathcal{S}}{\rho} + \frac{(\mathcal{D} - u)^2}{2}, \tag{37}$$

(35)–(37) can be combined to

$$\sigma_{21} = \frac{\Lambda + 1/\theta_{21}}{\Lambda + \theta_{21}}, \tag{38}$$

where

$$\Lambda \equiv \frac{1+\gamma}{1-\gamma}. \tag{39}$$

Note the similarity between (38) and the well known Rankine–Hugoniot equation which relates the pressure to density ratios across shock waves in gases (e.g., Glass and Hall [19], p. 66). When rewriting (38) as a function of $\theta_{21}$, it becomes evident that $\theta_{21} \to -\Lambda$ when $\sigma_{21} \to \infty$ and that $\theta_{21} = 1$ when $\sigma_{21} = 1$.

In order to relate the normal stress ratio across the compaction wave to the compaction wave, $M_s (\equiv \mathcal{D}/c_1)$, Mach number, we use Powers *et al.* [44] to express $c_1$ the sound speed in the solid ($c = \pm\sqrt{\partial\sigma/\partial\rho}$), ahead of the compaction wave. Combining (35), (36) and (38) yields

$$M_s = \left[\frac{(\Lambda-1)\theta_{21}}{\gamma(1+\bar{\mathcal{S}})(\Lambda+\theta_{21})}\right]^{1/2}, \tag{40}$$

where $\bar{\mathcal{S}} = \mathcal{S}/\sigma_1$.

The nondimensional velocity of the solid phase, $U_{21} \equiv u_2/c_1$, can be obtained by combining (35) and (40),

$$U_{21} = \frac{2\left[(1+\bar{\mathcal{S}})\,M_s^2 - 1\right]}{(1+\gamma)(1+\bar{\mathcal{S}})M_s}. \tag{41}$$

We note from (41) that $U_{21}$ depends on the non-ideal behavior of the material, also $U_{21} = 0$ when $\sigma_{21} = 1$.

### 4.1.2. *Model II: Post Compaction Wave for Porosity Very Close to Zero*

In the case of $\theta \to 1$, the assumption that the solid phase is incompressible (i.e., $\rho_1 \neq \rho_2$) can no longer be valid. We thus substitute $\rho_1$ and $\rho_2$ in (35)–(37), to obtain

$$\rho_{21} = \frac{-\Lambda\,\sigma_{21} + (1-\Lambda)\bar{\mathcal{S}} + \theta_{12}}{-\Lambda + (1-\Lambda)\bar{\mathcal{S}} + \theta_{21}\,\sigma_{21}}. \tag{42}$$

From (42) we note that $\rho_{21} \to -\Lambda/\theta_{21}$ when $\sigma_{21} \to \infty$ as well as that $\rho_{21} = 1$ when $\sigma_{21} = 1$ and $\theta_{21} = 1$. It is also evident that for the incompressible solid (i.e., $\rho_{21} = 1$), (42) conforms to (38).

Following a procedure similar to the one outlined earlier, the compaction wave Mach number can be calculated from

$$M_s^2 = \frac{(\Lambda-1)\left[(1+\bar{\mathcal{S}})\theta_{21}\,\rho_{21} - (1+\theta_{21}\,\bar{\mathcal{S}})\right]}{\gamma(1+\bar{\mathcal{S}})(1-\theta_{12}\,\rho_{12})(\theta_{21}\,\rho_{21}+\Lambda)}. \tag{43}$$

The solid nondimensional velocity $U_{21}$ can be calculated from

$$U_{21} = \sqrt{\frac{(\Lambda-1)(1-\theta_{12}\,\rho_{12})\left[(1+\bar{\mathcal{S}})\theta_{21}\,\rho_{21} - (1+\theta_{21}\,\bar{\mathcal{S}})\right]}{\gamma(1+\bar{\mathcal{S}})(\theta_{21}\,\rho_{21}+\Lambda)}}. \tag{44}$$

By virtue of (44), $U_{21} = 0$ when $\sigma_{21} = 1$ and $\theta_{21} = 1$.

*Table I.* Comparison of analytical models with experiments of head-on collision of planar shock waves with rigid porous materials

| | Sandusky and Liddiard Experimental Data (1985) | Powers *et al.*'s Model (1989) | Present Model I |
|---|---|---|---|
| $\theta_2$ | 0.91 | 0.83 | 0.83 |
| $\mathcal{D}$ | 300 | 340 | 365 |
| $(\theta\sigma)_2$ | $3.4 \times 10^7$ | $4.5 \times 10^7$ | $4.74 \times 10^7$ |
| $\rho_{21}$ | not available | not available | 1* |

*Table II.* Comparison of analytical models with experiments of head-on collision of planar shock waves with rigid porous materials

| | Sandusky and Liddiard Experimental Data (1985) | Powers *et al.*'s Model (1989) | Present Model I | Present Model II |
|---|---|---|---|---|
| $\theta_2$ | 0.95 | 0.97 | 1 | 1** |
| $\mathcal{D}$ | 370 | 410 | 365 | 370 |
| $(\theta\sigma)_2$ | $5.9 \times 10^7$ | $6.2 \times 10^7$ | $5.71 \times 10^7$ | $5.8 \times 10^7$ |
| $\rho_{21}$ | not available | 1901/1900 | 1* | 1901/1900 |

$T_1 = 300\,\mathrm{K}, \quad \rho_1 = 1900\,\mathrm{Kg/m^3}, \quad C_s = 1500\,\mathrm{J/(Kg\,K)}$
$\mathcal{S} = 8.976 \times 10^6\,\mathrm{m^2/s^2}, \quad \gamma = 5, \quad \theta_1 = 0.73, \quad u_1 = 100\,\mathrm{m/s}$
(*) The model assumes an incompressible solid.
(**) This was imposed as an input.

### 4.1.3. *Results and Discussion*

The experimental results of Sandusky and Liddiard [45], the predictions of the empirical model proposed by Powers *et al.* [44] and those of our simplified analytical models are shown in Tables I and II, for two cases of different initial conditions.

In the first experiment, the post compaction wave volume fraction of the solid phase was sufficiently less than unity, i.e., $\theta_2 = 0.91$, and hence, only Model I could be used.

In the second experiment, the post compaction wave volume fraction of the solid phase was 0.95. First we applied Model I from which we obtained $\theta_2 \approx 1$. For this reason we decided to apply Model II.

Note that both Models I and II are not as good as the empirical model of Powers *et al.* [44] in predicting $\theta_2$, yet both are much better in predicting $\mathcal{D}$ and $(\theta\sigma)_2$.

We thus conclude:

1. Model II could be practically applied for $\theta_2 \geqslant 0.95$.
2. Using the compaction equation is much more essential when $\theta_2$ is sufficiently less than unity. Dropping this equation when $\theta_2 \to 1$ seems to be justified.
3. The simplified analytical models are sufficient for estimating the post compaction wave properties behind strong compaction waves propagating in air saturated rigid porous materials.

### 4.2. WEAK WAVES IN GAS SATURATED RIGID POROUS MEDIA

In the following we describe a 1-D, horizontal direction, simple analytical solution for the fluid wave equation. The assumptions used are:

- The solid phase preserves its volume.
- The microscopic solid–fluid interfaces are material with respect to the momentum of both phases.
- There are no external energy sources.
- The conductive and dispersive heat fluxes of the phases is negligible in comparison with their advective heat flux.
- The rate of heat transferred between the fluid and solid phases is negligible.

Based on this conceptual model, we obtain the 1-D balance equations.

We consider the traveling wave propagating, say, in the positive direction of the $x$-axis. In such a wave all the quantities are functions of $(x - ct)$, where $c$ denotes the linear speed of wave propagation. Thus, the porosity may be represented by the real part of a complex expression

$$\phi = \Re e[A\, \mathrm{e}^{-\mathrm{i}(\omega t - kx)}], \tag{45}$$

where $\omega$ denotes the wave frequency and $k$ denotes a displacement factor per unit length. Actually, we can write,

$$c = \left|\frac{\omega}{k}\right| \gg |V_{\mathrm{s}}|\,. \tag{46}$$

By virtue of (45) and (46), we can confirm a previous assumption, namely,

$$\left|\frac{\partial(1-\phi)}{\partial t}\right| \gg \left|V_{\mathrm{s}}\frac{\partial(1-\phi)}{\partial x}\right|. \tag{47}$$

Furthermore, the nondimensional form of (47) can be viewed in terms of Strouhal number $\mathrm{St}^e[\equiv L_c^e/(t_c^e\, V_c^\alpha)]$. Letting $\alpha \equiv \mathrm{s}$ and $e \equiv \phi$, (47) will dictate $\mathrm{St}^\phi \gg 1$. Similarly, letting $e \equiv \mathrm{w_s}$ we may assume

$$\mathrm{St}^{\mathrm{w_s}} \equiv \frac{L_c^{\mathrm{w_s}}}{t_c^{\mathrm{w_s}}\, V_c^{\mathrm{s}}} \gg 1. \tag{48}$$

We can therefore write

$$V_{\mathrm{s}} = \frac{\mathrm{D_s w_s}}{\mathrm{D}t} \cong \frac{\partial \mathrm{w_s}}{\partial t}. \tag{49}$$

In view of (11) the definition for the volumetric strain: (47) and (49), we obtain

*The solid mass balance equation:*

$$\frac{1}{(1-\phi)}\frac{\partial \phi}{\partial t} = \frac{\partial V_{\mathrm{s}}}{\partial x} = \frac{\partial \varepsilon_{\mathrm{s}k}}{\partial t} \tag{50}$$

Thus by virtue of (50), $V_{\mathrm{s}}$ and $\varepsilon_{\mathrm{s}k}$ may be represented in a way similar to (45). Hence, their spatial and temporal changes read

$$\left|\frac{\omega}{k}\right| |\varepsilon_{\mathrm{s}k}| = |V_{\mathrm{s}}|\,. \tag{51}$$

In view of (46) and (51) we conclude that,

$$|\varepsilon_{\mathrm{s}k}| \ll 1. \tag{52}$$

Integrating (50) in view of (52) we obtain

$$\frac{1-\phi}{1-\phi_0} = \exp(-\varepsilon_{\mathrm{s}k}) \cong 1 - \varepsilon_{\mathrm{s}k}, \tag{53}$$

in which we assume that $\varepsilon_{\mathrm{s}k\,(x,0)} = 0$ and where $\phi_0$ $(= \phi(x,0))$ denotes an initial porosity.

Concerning the fluid's mass and momentum balance equations, we may define 'new' fluid quantities $(\,)_{\mathrm{n}}$, by the following,

$$\rho_{\mathrm{n}} = \phi\rho_{\mathrm{f}} \tag{54}$$

and

$$\frac{\partial P_{\mathrm{n}}}{\partial x} = T_{\mathrm{f}}^{*}\phi\frac{\partial P}{\partial x}. \tag{55}$$

In view of (54) and (55) we rewrite, respectively, the fluid's mass and momentum balance equations in a form equivalent to Euler's equation describing the propagation of simple (Riemann) fluid waves. These balance equations become

*The fluid mass balance equation:*

$$\frac{\mathrm{D_f}\rho_{\mathrm{n}}}{\mathrm{D}t} + \rho_{\mathrm{n}}\frac{\partial V_{\mathrm{f}}}{\partial x} = 0. \tag{56}$$

*The fluid momentum balance equation:*

$$\frac{\mathrm{D_f} V_{\mathrm{f}}}{\mathrm{D}t} + \frac{1}{\rho_{\mathrm{n}}}\frac{\partial P_{\mathrm{n}}}{\partial x} = 0. \tag{57}$$

Following Landau and Lifshitz [28], and the requirement that the fluid velocity is a function of its density, we can on the basis of (56) or (57) write the pressure wave equation of the 'new' fluid

$$\frac{\partial P_n}{\partial t} + (V_n \pm c_n)\frac{\partial P_n}{\partial x} = 0, \tag{58}$$

where $V_n$ and $c_n$ denote the 'new' fluid velocity and its sound velocity, respectively, given by

$$V_n = \pm \int_0^{P_n} \frac{dP_n}{\rho_n c_n} \qquad c_n = \pm \sqrt{\frac{dP_n}{d\rho_n}}. \tag{59}$$

Note that with the absence of heat exchange between the phases, we regard the motion as being adiabatic for the solid and the fluid. In such a case the specific entropy $s_f$, of the fluid remains constant (i.e., isentropic). Actually, when we assume that the processes of energy dissipation are negligible, we refer to *ideal* fluids, the motion of which must necessarily be adiabatic. By subtracting from the fluid's energy balance equation its mass balance multiplied by $s_f$, we obtain

*The fluid energy balance equation:*

$$\frac{D_f s_f}{Dt} = 0. \tag{60}$$

*The solid momentum balance equation:*

$$\frac{D_s V_s}{Dt} - \frac{\lambda_s''}{\rho_s(1-\phi)}\frac{\partial^2 w_s}{\partial x^2} + \frac{\eta}{\rho_s(1-\phi)}\frac{\partial T_s}{\partial x} + \frac{T_s^*}{\rho_s}\frac{\partial P}{\partial x} = 0. \tag{61}$$

*The solid energy balance equation:*

$$(1-\phi)\rho_s C_s \frac{D_s T_s}{Dt} + \eta T_s \frac{\partial V_s}{\partial x} = 0. \tag{62}$$

4.2.1. *The Isothermal Case*

By virtue of (48) and (49) for the isothermal case we rewrite (61) as

*The solid momentum balance equation:*

$$\frac{\partial^2 w_s}{\partial t^2} - c_s^2 \frac{\partial^2 w_s}{\partial x^2} + \frac{T_s^*}{\rho_s}\frac{\partial P}{\partial x} = 0, \tag{63}$$

where $c_s$ denotes the speed of propagation of the displacement wave in the solid phase given by

$$c_s^2 \equiv \frac{\lambda_s''}{\rho_s(1-\phi)}. \tag{64}$$

In view of (63) considering $M_s(\equiv V_s/c_s)$ the solid phase Mach number, we assume that

$$\mathrm{St}_s^w M_s^2 \ll 1, \tag{65}$$

which in view of (48) implies the motion of a *weak wave* ($M_s \ll 1$) and that the solid inertia in (63) is negligible so that the solid momentum balance equation can be approximated by

$$-\frac{\lambda_s''}{(1-\phi)}\frac{\partial \varepsilon_{sk}}{\partial x} + T_s^* \frac{\partial P}{\partial x} = 0. \tag{66}$$

Substituting (53) into (66) when considering (52), we integrate (66) for $\varepsilon_{sk}$ and obtain

$$\varepsilon_{sk} = (1-\phi_0)\zeta T_s^* \quad \text{in which} \quad \zeta \equiv \frac{P-P_0}{\lambda_s''} \ll 1, \tag{67}$$

where $\varepsilon_{sk} = 0$ at $P = P_0$. Actually we note that $\zeta$ in (67) is negligible for many practical means for example when considering solid porous materials such as SiC or $Al_2\,O_3$. We relate this to the fact that $\Delta P\ (\equiv P - P_0)$ is of many orders smaller than $\lambda_s''$ as exemplified in the case of a head-on reflection of shock waves (Liepmann and Roshko [37]).

In view of (53), (67) and the compatibility relation in (11), we obtain

$$\phi = \phi_0 + \zeta(1-\phi_0)^2 T_s^* \cong \phi_0, \qquad \mathrm{w}_s = \frac{1-\phi_0}{\lambda_s''} T_s^* \int_{-\infty}^{x} (P - P_0)\,\mathrm{d}x. \tag{68}$$

Based on our findings for the solid phase we can relate the fluid quantities with those of the 'new' fluid. Substituting (68) into (54), by virtue of (67), (55) and assuming that $P_n(x, 0) = 0$ we obtain

$$\rho_n \cong \phi_0\,\rho_f, \qquad P_n \cong \phi_0(P - P_0)T_f^*. \tag{69}$$

In view of (54), (55), (64) and (68), we expand $c_n$ of (59) and the compressibility $\beta_n$ of the 'new' fluid can be approximated. These now become

$$c_n^2 \cong T_f^* c_f^2, \qquad \beta_n \cong \frac{\beta_P}{T_s^*[1 - \lambda_s''/(\rho_s\, c_s^2)]}, \tag{70}$$

in which the fluid compressibility $\beta_P\ (\equiv (1/\rho_f)\,\mathrm{d}\rho_f/\mathrm{d}P = 1/(\rho_f\, c_f^2))$ is related to its sound velocity. For typical solid density ($\rho_s = 2000\,\mathrm{Kg/m^3}$) and solid sound velocity ($c_s = 7000\,\mathrm{m/s}$), we always have $\lambda_s''/(\rho_s\, c_s^2) > 1$, while $T_s^*/(\beta_P\,\lambda_s'') \ll 1$. We also note that for $\beta_P > 0$ we obtain $\beta_n < 0$.

### 4.2.2. *The Nonisothermal Case*

Let us first start with the solid phase and assume that

$$\mathrm{St}^{T_s} \equiv \frac{\mathrm{L}_c^{T_s}}{t_c^{T_s}\, V_c^s} \gg 1. \tag{71}$$

Considering (11), (53) and (71), we rewrite (62) the solid energy balance equation in the form

$$(1-\phi_0)\rho_s\, C_s \frac{\partial T_s/\partial t}{T_s} + \eta \frac{\partial \varepsilon_{sk}/\partial t}{(1-\varepsilon_{sk})} = 0, \tag{72}$$

from which, by virtue of (52), we obtain,

$$T_s = T_{s_0}\left[1 - \frac{\eta\varepsilon_{sk}}{(1-\phi_0)\rho_s\, C_s}\right], \tag{73}$$

where $T_{s_0}$ denotes the solid temperature at $\varepsilon_{sk} = 0$.

In view of (11), (48), (49), (53), (65) and (73), we approximate (61) the solid momentum balance equation, to read

$$\frac{\lambda_{\mathrm{eff}}}{(1-\phi_0)(1-\varepsilon_{sk})}\frac{\partial}{\partial x}(1-\varepsilon_{sk}) = T_s^* \frac{\partial P}{\partial x} \tag{74}$$

in which

$$\lambda_{\mathrm{eff}} \equiv \left[1 + \frac{T_{s_0}\alpha_s^2\lambda_s''}{(1-\phi_0)\rho_s\, C_s}\right]\lambda_s'' \cong \lambda_s'', \tag{75}$$

where $\alpha_s$ denotes the thermal expansion coefficient (related to $\eta$ the thermal stress coefficient, i.e., $\eta = \alpha_s\, \lambda_s''$). The approximation in (75) is possible for thermoelastic rigid materials (e.g., SiC or $Al_2O_3$) and when considering the appropriate order of magnitudes, i.e., $\lambda_s'' \cong 400\,\mathrm{GPa}$, $T_{s_0} = 300\,\mathrm{K}$, $\alpha_s = 5\times 10^{-6}\,\mathrm{K}^{-1}$, $C_s = 1000\,\mathrm{J/Kg/K}$, $\rho_s = 2000\,\mathrm{Kg/m^3}$ and $0 < \phi_0 < 1$.

Solving (74) in view of (52) and assuming that $P = P_0$ at $\varepsilon_{sk} = 0$ yields

$$\varepsilon_{sk} \cong \frac{(1-\phi_0)T_s^*}{\lambda_{\mathrm{eff}}}(P - P_0). \tag{76}$$

In view of (75) and (76) when substituted into (73), for different values of $\rho_s$ it is evident that practically

$$T_s \cong T_{s_0}. \tag{77}$$

When substituting (76) into (53) we obtain

$$\phi = \phi_0 + \frac{(1-\phi_0)^2 T_s^*}{\lambda_{\mathrm{eff}}}(P - P_0). \tag{78}$$

Accordingly, using (76) and referring to (11), the compatibility equation, we write,

$$w_s = \frac{1-\phi_0}{\lambda_{\text{eff}}} T_s^* \int_{-\infty}^{x} (P - P_0)\,\mathrm{d}x. \tag{79}$$

Concerning the fluid phase, we need to address six variables ($\phi$, $\rho_f$, $P$, $T_f$, $V_f$, $s_f$). For these we have three balance equations, (56), (57), (60), and (78) a relation for the porosity. We thus add constitutive laws assuming that these macroscopic forms are similar to their microscopic counterparts while the coefficients (to be determined experimentally) are different in value from those of the microscopic ones. For an *ideal fluid* we can thus add the relation

$$\left(\frac{P}{P_0}\right)\left(\frac{\rho_{f_0}}{\rho_f}\right)^{C_f/C_P} = \exp\left(\frac{s_f - s_{f_0}}{C_f}\right), \tag{80}$$

where $\rho_{f_0}$ and $s_{f_0}$ are reference values obtained when $P = P_0$ and $C_P$ denotes the fluid specific heat at constant pressure. We also write the equations of state for a perfect gas in the form

$$\rho_f = \frac{P}{RT_f}, \tag{81}$$

in which $R$ denotes the specific gas constant.

Substituting (78) into (55), integrating, and in view of (80) we obtain

$$P_n = \left[A + \tfrac{1}{2}B(P - P_0)\right](P - P_0) \cong A(P - P_0), \tag{82}$$

where

$$A \equiv \phi_0 T_f^* \quad \text{and} \quad B \equiv \frac{(P - P_0)^2 T_f^* T_s^*}{\lambda_{\text{eff}}} \ll 1. \tag{83}$$

We note that in view of the approximation expressed in (75), B is practically negligible and $P_n$ is approximated, accordingly. We can rewrite $\phi$ in terms of $P_n$, this in view of (78), (82) and (83) becomes

$$\phi = \frac{\sqrt{A^2 + 2BP_n}}{T_f^*} \cong \phi_0. \tag{84}$$

We note that the approximate forms of (82) and (84) are in line with the isothermal case. By virtue of (54), (82) and (83), we can also rewrite the constitutive relation (80) in the form

$$\exp\left(\frac{s_f - s_{f_0}}{C_f}\right) = \left(\frac{\sqrt{A^2 + 2BP_n} - A}{BP_0}\right)\left(\frac{\sqrt{A^2 + 2BP_n}}{T_f^*}\frac{\rho_{f_0}}{\rho_n}\right)^{C_P/C_f}$$

$$\cong \left(1 + \frac{1}{A}\frac{P_n}{P_0}\right)\left(\phi_0 \frac{\rho_{f_0}}{\rho_n}\right)^{C_P/C_f}. \tag{85}$$

Altogether, we now face four ($\rho_n$, $V_f$, $P_n$, $s_f$) variables. To solve for these variables, we have three balance equations (56), (57), (60) and a constitutive relation (85). It is noteworthy to find an expression for the temperature $T_n$, of the generalized fluid. This will also allow us to write the generalized state function commencing from (81).

We start by introducing the relation for the generalized enthalpy $h_n(P_n, s_f)$. Using the definition for the enthalpy we postulate that

$$dh_n = T_n\, ds_f + \frac{dP_n}{\rho_n}, \tag{86}$$

from which we have

$$\frac{1}{\rho_n} = \frac{\partial h_n}{\partial P_n} \quad \text{and} \quad T_n = \frac{\partial h_n}{\partial s_f}. \tag{87}$$

In view of (54), (55), (86) and (87) we obtain

$$h_n(P_n, s_f) = T_f^* h(P, s_f) \tag{88}$$

and

$$T_n = T_f^* T_f. \tag{89}$$

The equation of state of the generalized fluid may now be obtained from (54), (82)–(84), (89) when substituted into (81)

$$\rho_n = \frac{\sqrt{A^2 + 2BP_n}\left(BP_0 - A + \sqrt{A^2 + 2BP_n}\right)}{BRT_n} \cong \frac{AP_0}{RT_n}. \tag{90}$$

Hence, we note that the density of the generalized fluid becomes dependent only on the temperature of this fluid.

In what follows we will describe the roll of Forchheimer terms by comparing shock tube experiments and numerical simulation of compaction waves traveling through a 1-D saturated thermoelastic porous medium.

## 5. Numerical Solution for Compaction Waves in a 1-D Saturated Thermoelastic Porous Medium

During the second time interval, compaction wave equations are formed. The 1-D balance equations of mass, momentum and energy of the fluid ((1), (23) and (25), respectively) and of the solid ((2), (24) and (26), respectively) when written in a vector form (Levy *et al.* [34]) become

$$\frac{\partial \mathbf{U}}{\partial t} + \frac{\partial \mathbf{F}}{\partial x} = \mathbf{Q}. \tag{91}$$

The variables vector **U**, is defined by

$$\mathbf{U} = [r_f, r_s, m_f, m_s, E_f, E_s]^T, \tag{92}$$

where

$$r_{\mathrm{f}} \equiv \phi\rho_{\mathrm{f}}, \quad m_{\mathrm{f}} \equiv r_{\mathrm{f}} V_{\mathrm{f}}, \quad E_{\mathrm{f}} \equiv r_{\mathrm{f}} e_{\mathrm{f}} = r_{\mathrm{f}}\left(C_{\mathrm{f}} T_{\mathrm{f}} + \frac{V_{\mathrm{f}}^2}{2}\right),$$
$$r_{\mathrm{s}} \equiv (1-\phi)\rho_{\mathrm{s}}, \quad m_{\mathrm{s}} \equiv r_{\mathrm{s}} V_{\mathrm{s}}, \quad E_{\mathrm{s}} \equiv r_{\mathrm{s}} e_{\mathrm{s}} = r_{\mathrm{s}}\left(C_{\mathrm{s}} T_{\mathrm{s}} + \frac{V_{\mathrm{s}}^2}{2}\right), \tag{93}$$

$e_\alpha$ denotes the total energy per mass unit of the $\alpha$ phase. The flux vector **F**, is given by

$$\mathbf{F} = \begin{bmatrix} m_{\mathrm{f}} \\ m_{\mathrm{s}} \\ \dfrac{m_{\mathrm{f}}^2}{r_{\mathrm{f}}} + T_{\mathrm{f}}^* \phi P \\ \dfrac{m_{\mathrm{s}}^2}{r_{\mathrm{s}}} - \sigma_{\mathrm{s}}' + \left(1 - \phi T_{\mathrm{f}}^*\right) P \\ \dfrac{m_{\mathrm{f}}}{r_{\mathrm{f}}}\left(E_{\mathrm{f}} + T_{\mathrm{f}}^* \phi P\right) \\ \dfrac{m_{\mathrm{s}}}{r_{\mathrm{s}}}[E_{\mathrm{s}} - \sigma_{\mathrm{s}}' + (1 - \phi T_{\mathrm{f}}^*) P] \end{bmatrix}. \tag{94}$$

The source vector **Q**, is given by

$$\mathbf{Q} = \begin{bmatrix} 0 \\ 0 \\ T_{\mathrm{f}}^* P \dfrac{\partial \phi}{\partial x} - \tilde{F} r_{\mathrm{f}} \left|\dfrac{m_{\mathrm{f}}}{r_{\mathrm{f}}} - \dfrac{m_{\mathrm{s}}}{r_{\mathrm{s}}}\right| \left(\dfrac{m_{\mathrm{f}}}{r_{\mathrm{f}}} - \dfrac{m_{\mathrm{s}}}{r_{\mathrm{s}}}\right) \\ -T_{\mathrm{f}}^* P \dfrac{\partial \phi}{\partial x} + \tilde{F} r_{\mathrm{f}} \left|\dfrac{m_{\mathrm{f}}}{r_{\mathrm{f}}} - \dfrac{m_{\mathrm{s}}}{r_{\mathrm{s}}}\right| \left(\dfrac{m_{\mathrm{f}}}{r_{\mathrm{f}}} - \dfrac{m_{\mathrm{s}}}{r_{\mathrm{s}}}\right) \\ \dfrac{m_{\mathrm{s}}}{r_{\mathrm{s}}}\left[T_{\mathrm{f}}^* P \dfrac{\partial \phi}{\partial x} - \tilde{F} r_{\mathrm{f}} \left|\dfrac{m_{\mathrm{f}}}{r_{\mathrm{f}}} - \dfrac{m_{\mathrm{s}}}{r_{\mathrm{s}}}\right| \left(\dfrac{m_{\mathrm{f}}}{r_{\mathrm{f}}} - \dfrac{m_{\mathrm{s}}}{r_{\mathrm{s}}}\right)\right] \\ \dfrac{m_{\mathrm{s}}}{r_{\mathrm{s}}}\left[-T_{\mathrm{f}}^* P \dfrac{\partial \phi}{\partial x} + \tilde{F} r_{\mathrm{f}} \left|\dfrac{m_{\mathrm{f}}}{r_{\mathrm{f}}} - \dfrac{m_{\mathrm{s}}}{r_{\mathrm{s}}}\right| \left(\dfrac{m_{\mathrm{f}}}{r_{\mathrm{f}}} - \dfrac{m_{\mathrm{s}}}{r_{\mathrm{s}}}\right)\right] \end{bmatrix}, \tag{95}$$

in which $\tilde{F}$ denotes the Forchheimer coefficient for an isotropic solid matrix. The pressure $P$, is prescribed by the equation of state

$$\phi P = (\gamma - 1)\left(E_{\mathrm{f}} - \frac{m_{\mathrm{f}}^2}{2 r_{\mathrm{f}}}\right), \tag{96}$$

where $\gamma$ denotes the specific heat capacity ratio of the fluid phase.

The porosity $\phi$, and the skeleton strain $\varepsilon_{sk}$, can be expressed as

$$\phi = 1 - \frac{r_{\mathrm{s}}}{\rho_{\mathrm{s}}}, \qquad \varepsilon_{sk} = 1 - \frac{r_{\mathrm{s}}}{r_{\mathrm{s}0}}. \tag{97}$$

In view of (10) we can express the constitutive relation for the effective stress $\sigma_s'$, in the form

$$\sigma_s' = E_\varepsilon \left( \frac{r_{s0} - r_s}{r_{s0}} \right) - E_T \left( \frac{E_s}{r_s} - \frac{E_{s0}}{r_{s0}} - \frac{m_s^2}{2r_s^2} + \frac{m_{s0}^2}{2r_{s0}^2} \right), \tag{98}$$

in which $E_\varepsilon (\equiv \lambda_s'' + \mu_s')$, $E_T (\equiv \eta / C_s)$ are the one-dimensional macroscopic Lame coefficients for a thermoelastic solid and the subscript $(\,)_{s0}$ denotes the solid properties at an initial state.

The set (91) of six partial differential governing equations defined by (92), (94) and (95) are to be solved for six unknowns namely, $r_f, r_s, m_f, m_s, E_f$ and $E_s$. Consequently, in principle given the appropriate conditions, the set is complete and can be solved. However, due to the complexity of the equations, an analytical solution of the set is not feasible. Instead, a numerical solution has been performed.

Note that in writing (91) we assumed that the gradients of the porosity were very small and therefore they might be written as source terms in the source vector **Q**, given by (95). As a result the source vector **Q**, contains derivative terms and thus affects the character of the equations (hyperbolic rather than elliptic-hyperbolic). This is very convenient for numerical purposes and is correct only when the gradients of the porosity are very small as is the case in this study. Applying this procedure removes the ill-posedness associated with the elliptic-hyperbolic character of the equations which is usually manifested in highly oscillatory solutions as the mesh is refined.

The numerical scheme for solving (91) is based on an upwind *TVD* shock-capturing scheme, originally developed by Harten [23], which was extended to solve the problem of compaction waves propagation in saturated thermoelastic porous media.

## 6. Experimental Study

The head-on collision of planar shock waves with rigid porous materials was investigated experimentally in order to validate the predictions of the physical model and the numerical code. The experiments were conducted in a 75 mm × 75 mm shock tube. The incident shock wave Mach number range was $1.2 \leqslant M_i \leqslant 1.7$, the initial pressures and temperatures throughout the experimental study were about 830 mbar and 288 K, respectively.

The rigid porous samples were made of silicon carbide (SiC) and alumina ($Al_2O_3$). The SiC manufactured porous material had either 10 or 20 pores per inch and the $Al_2O_3$ manufactured porous material had either 30 or 40 pores per inch. The porosity of these porous materials as well as the initial length $L_0$, of the samples which were cut from these materials are given in Table III. The porosity $\phi$, was simply calculated from $\phi = 1 - \rho_b / \rho_s$, where $\rho_b$ is the bulk density of the porous material and $\rho_s$ the material density of the skeleton of which the porous material is made. Note that the above expression is limited to gas saturated porous materials as was the case in the present study. The cross-section dimensions of the samples

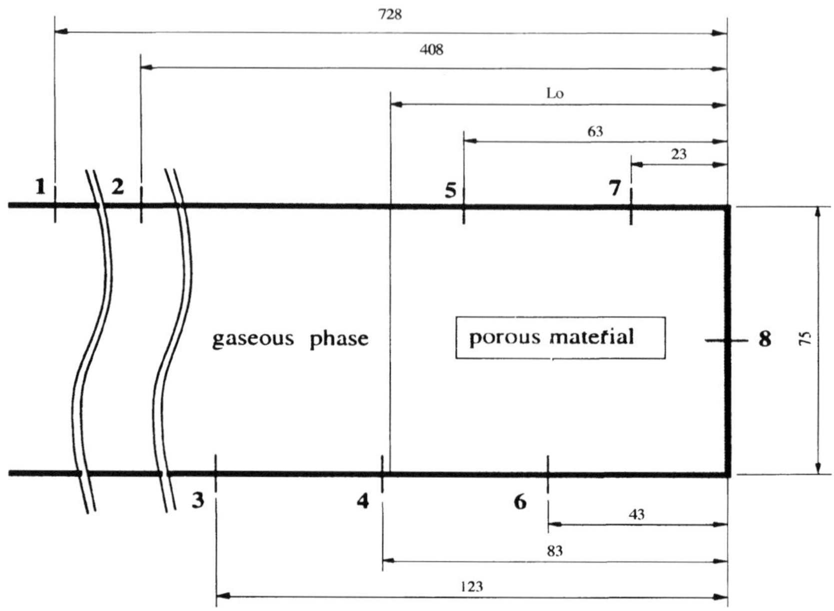


*Figure 1.* Schematic illustration of the experimental set-up and the location of the pressure transducers (numbered 1 to 8).

were 75 mm × 75 mm and hence, as a result of the head-on collision with the shock wave they underwent a uni-axial strain (also known as tri-axial stress) compression. Twelve experiments were conducted with each sample. The samples were mounted at the end of the driven section of the shock tube in such a way that their back edges were supported by the shock-tube end-wall.

A schematic illustration of the experimental set-up used during the course of the experimental study is shown in Figure 1. Seven pressure transducer ports were used in each experiment. The pressure transducers in ports 1 and 2 were used to measure the incident shock wave velocity. Pressure transducers were mounted in ports 4, 5, 6, 7 and 8 for the samples having an initial length up to (inclusive) 83 mm (see Table III). For the alumina sample with an initial length of 99 mm the pressure transducer of port 4 was moved to port 3. This was done in order to have a measurement of the pressure changes caused by the shock wave which reflected backwards from the front edge of the porous material.

## 7. Comparison Between the Numerical Predictions and the Experimental Results

In order to validate the physical model and the numerical code, the governing equations were solved numerically for different samples and a variety of initial conditions and compared to experimental results. In order to solve the governing equations and compare the solution to the experimental results, the various parameters which appear in the physical model namely: the macroscopic Lame coefficients for a thermoelastic solid $E_\varepsilon$ and $E_T$, the Forchheimer factor $\tilde{F}$, the fluid tortuosity $T_f^*$, and the intrinsic density of the solid matrix $\rho_s$, which appear in the physical model had to be estimated for each sample. Based on the properties of the materials of which the samples were made of and the fact that the upper limit of the elastic stress reduces when the porosity increases, in a $(1-\phi)^2$ manner, Gibson and Ashby [19], the magnitudes of the macroscopic Lame coefficients $E_\varepsilon$ and $E_T$, were estimated to be identical for all the samples:

$$E_\varepsilon = 380 \times 10^7 \text{ Pa} \quad \text{and} \quad E_T = 26.207 \text{ Kg/m}^3.$$

The intrinsic density of the solid matrix for all the samples (SiC and $Al_2O_3$), as provided by the manufacturer, was

$$\rho_s = 2000 \pm 60 \text{ Kg/m}^3.$$

The tortuosity factors $T_f^*$, for the various samples were found by estimating the ratio between the speed of sound of the air inside the porous medium $a_f$ and that in pure air $a$ (Li *et al.* [36]),

$$\left(\frac{a_f}{a}\right)^2 = \frac{\left(1 - T_f^* + \gamma T_f^*\right) T_f^*}{\gamma}. \tag{99}$$

The Forchheimer factors $\tilde{F}$, for the various samples where found experimentally. In these experiments, each sample was mounted in a pipe and the pressure drop across it was measured as a function of the air flow rate. The pressure drop was found to be a parabolic function of the air velocity. This was done by assuming that the pressure drop depends linearly on the length of the sample. The values of the tortuosity and the Forchheimer factors, as obtained experimentally for the various samples, are presented in Table III.

In order to validate the physical model, (91) was solved numerically for different samples and initial conditions and compared to experimental results. Table IV represents the initial conditions of 18 different experiments. The comparison between the

*Table III.* The Forchheimer ($\tilde{F}$) and the tortuosity ($T_f^*$) factors, porosity ($\phi$) and sample length ($L_0$) for the different samples

| Sample material | Sample type | Pores per inch (ppi) | $\tilde{F}$ (1/m) | $T_f^*$ | $\phi$ | $L_0$ (mm) |
|---|---|---|---|---|---|---|
| Silicon carbide | I | 10 | 300 | 0.7 | $0.728 \pm 0.016$ | 40, 60, 81 |
| (SiC) | II | 20 | 500 | 0.7 | $0.745 \pm 0.001$ | 41, 62, 83 |
| Alumina | III | 30 | 900 | 0.75 | $0.814 \pm 0.010$ | 48, 93 |
| ($Al_2O_3$) | IV | 40 | 1800 | 0.75 | $0.821 \pm 0.007$ | 50, 99 |

*Table IV.* The initial conditions for the various experiments

| Test number | Sample material | Sample length [mm] | Initial pressure (KPa) | Initial temperature (K) | Shock Mach number-$M_i$ |
|---|---|---|---|---|---|
| 1 | | 40 | 83.04 | 290.0 | 1.378 |
| 2 | SiC | 60 | 83.11 | 290.5 | 1.385 |
| 3 | | 81 | 83.41 | 290.5 | 1.377 |
| 4 | 10 ppi | 81 | 83.71 | 289.5 | 1.543 |
| 5 | | 81 | 83.70 | 290.5 | 1.734 |
| 11 | | 48 | 82.84 | 292.0 | 1.374 |
| 12 | $Al_2O_3$ | 93 | 83.41 | 290.5 | 1.377 |
| 13 | 30 ppi | 93 | 83.36 | 291.0 | 1.539 |
| 14 | | 93 | 83.12 | 291.5 | 1.744 |
| 15 | | 50 | 83.09 | 292.0 | 1.374 |
| 16 | $Al_2O_3$ | 50 | 83.08 | 292.5 | 1.545 |
| 17 | 40 ppi | 50 | 83.10 | 292.5 | 1.741 |
| 18 | | 99 | 83.16 | 290.5 | 1.377 |

predictions of the numerical simulations and the experimental results, for all the cases appearing in Table IV. In what follows, we will present two of these comparisons the experiments of which are listed in Table IV as test numbers 1 and 13.

Figures 2(a)–(e) and 3(a)–(e) represent typical experimental results and their numerical simulations for SiC and $Al_2O_3$ samples, respectively. $P_1$ is the pressure ahead of both the incident and the transmitted shock waves, $P_2$ is the theoretical pressure which should have been reached behind the incident shock wave, and $P_5$ is the theoretical pressure which should have been reached had the incident shock wave reflected head-on from a solid surface. The symbols represent the experimental results, the solid lines are the numerically predicted values when accounting for Forchheimer terms and the dash-dot lines represent the simulations without Forchheimer terms. The pressure histories of the pure gas just ahead of the front edge of the porous sample are shown in Figure 2(a); Figure 2(b) for the SiC and Figure 3(a) for the $Al_2O_3$. The pressure histories of the gas occupying the pores of the porous material along the shock tube side-wall and at its end-wall are shown in Figures 2(c)–(e) for the SiC samples and Figures 3(a)–(e) for the $Al_2O_3$ samples.

Figures 2 and 3, clearly emphasize the advantage of simulations with Forchheimer terms in retrieving the experimental observations. Major differences between the two cases are evident. While the simulations without Forchheimer terms, (dashed doted lines) do not agree with the experimental results, simulations with Forchheimer terms (solid lines) results in predictions which excellently reproduce the experiments. When Forchheimer terms are neglected the shock waves which are reflected, head-on, from the front edge of the porous sample, are weaker than those obtained when the Forchheimer terms are accounted for (compare the second jump in pressure traces

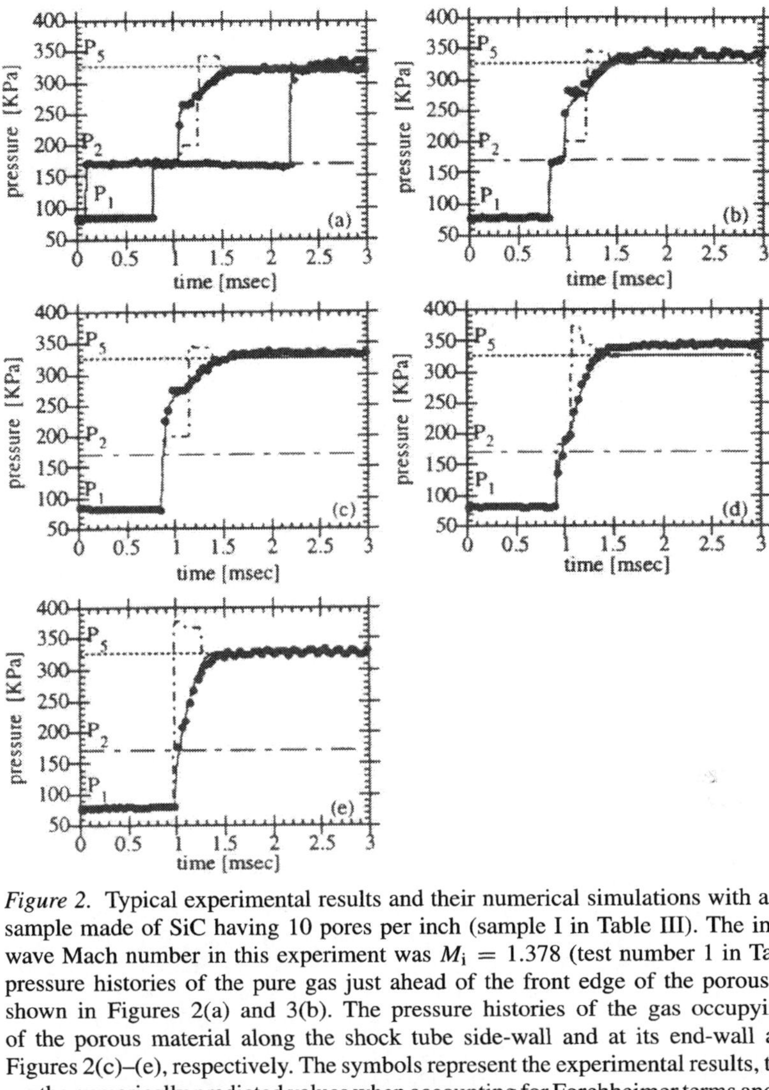


*Figure 2.* Typical experimental results and their numerical simulations with a 40 mm long sample made of SiC having 10 pores per inch (sample I in Table III). The incident shock wave Mach number in this experiment was $M_i = 1.378$ (test number 1 in Table IV). The pressure histories of the pure gas just ahead of the front edge of the porous material are shown in Figures 2(a) and 3(b). The pressure histories of the gas occupying the pores of the porous material along the shock tube side-wall and at its end-wall are shown in Figures 2(c)–(e), respectively. The symbols represent the experimental results, the solid lines are the numerically predicted values when accounting for Forchheimer terms and the dash-dot lines represent the simulations without Forchheimer terms. $P_1$ is the pressure ahead of both the incident and the transmitted shock waves, $P_2$ is the theoretical pressure which should have been reached behind the incident shock wave, and $P_5$ is the theoretical pressure which would have been reached had the incident shock wave reflected head-on from a solid end-wall.

shown in Figures 2(a), 2(b) and 3(a)). These figures also indicate that when the Forchheimer terms are neglected, a sharp shock wave is transmitted at a later time from the porous material into the gaseous phase resulting in a third jump in pressure longer than $P_5$. These high pressures are later decayed by the transmitted expansion waves and reach as expected the value of $P_5$.

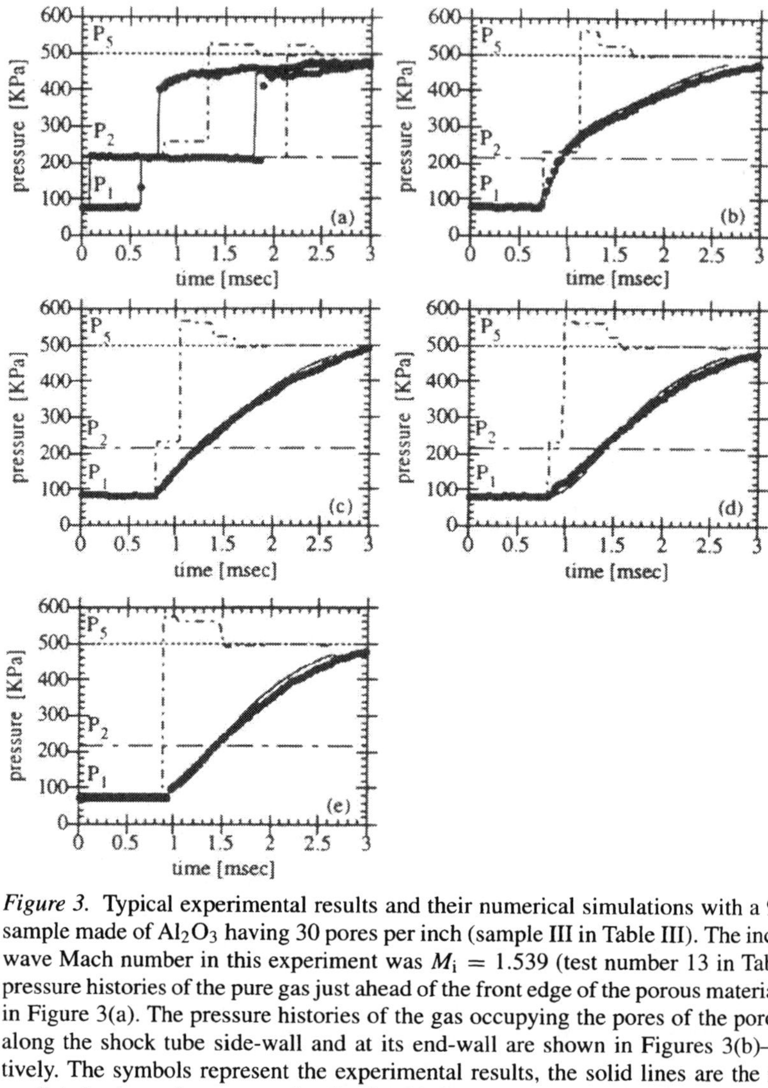


*Figure 3.* Typical experimental results and their numerical simulations with a 93 mm long sample made of $Al_2O_3$ having 30 pores per inch (sample III in Table III). The incident shock wave Mach number in this experiment was $M_i = 1.539$ (test number 13 in Table IV). The pressure histories of the pure gas just ahead of the front edge of the porous material are shown in Figure 3(a). The pressure histories of the gas occupying the pores of the porous material along the shock tube side-wall and at its end-wall are shown in Figures 3(b)–(e), respectively. The symbols represent the experimental results, the solid lines are the numerically predicted values when accounting for Forchheimer terms and the dash-dot lines represent the simulations without Forchheimer terms. $P_1$, $P_2$ and $P_5$ are defined in the caption of Figure 2.

In addition, as a result of the neglection of the Forchheimer terms, the shock waves which are transmitted into the porous material samples do not develop a dispersed structure, as is the situation in reality. Instead, they maintain their sharp fronts. The fact that the transmitted shock waves maintain their sharp structure result in a jump to pressures larger than $P_2$ behind them (see the sharp jumps in the pressures inside the porous samples in Figures 2(c), 2(d), 3(c) and 3(d)). Figures 2(e) and 3(e) in which the pressure histories at the back edge of the porous samples (that is at the shock tube end-wall) are shown again, indicate that when the Forchheimer terms are neglected, in contrast to the experimental results, sharp shock waves reach the end-wall. The reason for the above-mentioned differences lies in the fact that the Forchheimer terms emphasize the friction between the two phases and as a result increase the rate of momentum exchange. This, in turn, results in a situation in which much larger impulses $\int P\,\mathrm{d}t$, are predicted numerically when the Forchheimer terms are neglected.

It is clearly evident from these comparisons that the agreement between the experimental results and the numerical predictions is very good. Note that there are disagreements between the numerical and the experimental positions of the reflected shock wave after it traveled to a relatively long distance from the front edges of the samples (see Figures 2(a) and 3(a)). The agreement regarding the reflected shock wave position was better in the case of low incident shock wave Mach numbers and short samples. This may be caused by the way the porosity gradient terms were treated in the source vector $\mathbf{Q}$.

Although Figures 2(a)–(e) and 3(a)–(e) describe the results of only two typical experiments, one in the SiC sample and one in an alumina ($Al_2O_3$) sample, similar good to excellent agreements were obtained in the comparisons with all the experiments which were conducted by Levy [29].

## 8. Sensitivity Analysis

The developed macroscopic model also accounts for geometrical properties. Hence, for example, the speed of the wave is not only based on properties of the phases but also a function of the porous material structure. Some macroscopic parameters which resulted from the averaging process were the tortuosity factor which represents a tensor associated with the matrix directional cosines, the hydraulic radius of the pore spaces and the porosity which represents the volume fraction of the pores filled by the fluid. As was also shown, a most important factor is the Forchheimer tensor accounting for the microscopic inertia exchange between the phases at the solid/fluid interface. In what follows, we will present a 1-D numerical parametric study of the flow field which develops when a planar shock wave strikes a rigid porous material head-on. This is done in terms of studying the effect of the tortuosity, Forchheimer and Lamé coefficients and the initial porosity in the case of a 81 mm long sample of SiC and an incident shock wave having a Mach number $M_{\mathrm{i}} = 1.377$.

### 8.1. THE TORTUOSITY EFFECT

The dependence of the pressure field on the tortuosity is shown in Figures 4(a)–(c). It is evident from Figure 4(a) that the compression (shock) wave velocity increases as the tortuosity increases. It also indicates that the pressure rise is faster when the tortuosity is larger. It is evident from Figure 4(c) that the reflected shock wave moves faster when the tortuosity is smaller.

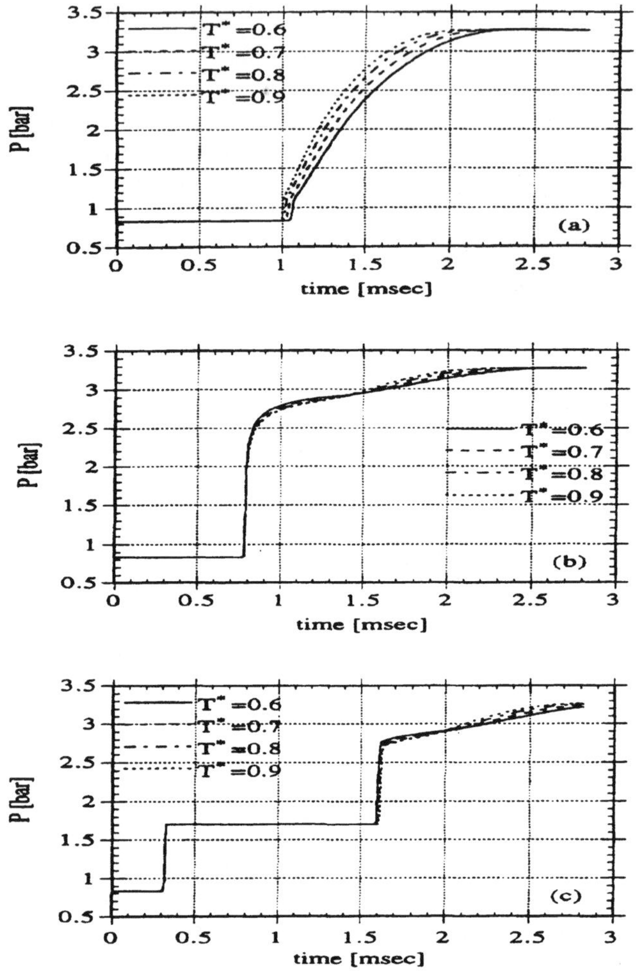


*Figure 4.* The influence of the tortuosity on the pressure: (a) at the back edge of the sample, (b) at the front edge of the sample and (c) 42 cm ahead of the sample. The sample was made of SiC and it was 81 mm long. The incident shock wave Mach number was $M_i = 1.377$.

The influence of the tortuosity on the fluid velocity, at the front edge and ahead of the sample, is shown in Figures 5(a) and 5(b), respectively. We note that the final velocity values become larger for smaller tortuosities.

It is important to note here that two waves are transmitted into the porous material when the incident shock wave strikes it. The first is a very rapid compaction wave which propogates in the solid matrix, and the second is a compression (shock) wave which propagates through the gaseous phase occupying the pores. These waves interact at the solid/fluid interfaces of the porous material. The compaction wave, which is very fast, is responsible for the fast porosity changes.

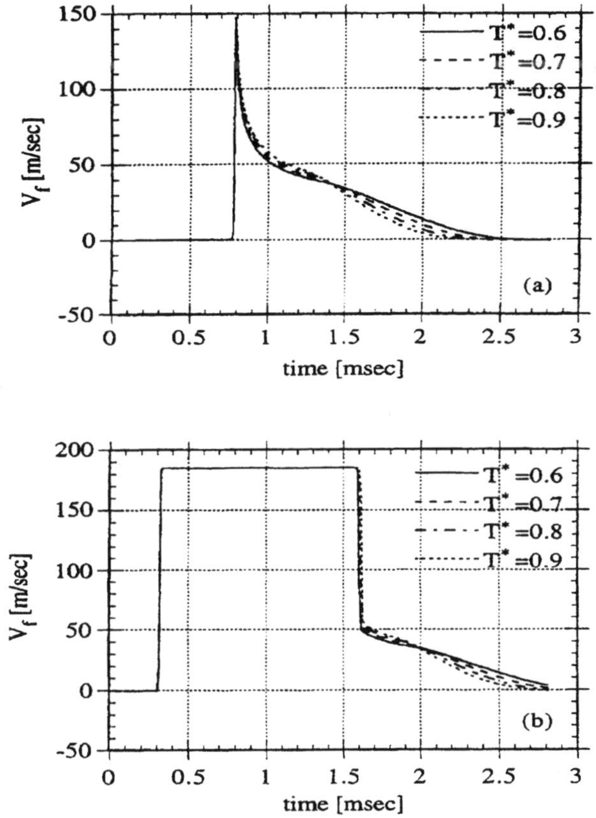


*Figure 5.* The influence of the tortuosity on the fluid velocity: (a) at the front edge of the sample and (b) 42 cm ahead of the sample. The sample was made of SiC and it was 81 mm long. The incident shock wave Mach number was $M_i = 1.377$.

Although not demonstrated, we concluded that practically the porosity could be assumed to be constant and independent of the tortuosity.

Similarly, the solid matrix velocity was found to be practically independent of the tortuosity. Smaller tortuosities result in larger normal stress. However, given enough time after the initiation of the interaction the finally obtained normal stresses becomes independent of the tortuosity.

## 8.2. THE FORCHHEIMER FACTOR EFFECT

The dependence of the pressure field on the Forchheimer factor is depicted in Figures 6(a)–(c). We note from Figure 6(a) that the compression (shock) wave velocity is not influenced by the Forchheimer factor as the profiles for the various Forchheimer factors start to rise at the same time. The finally obtained pressure is also seen to be independent of the Forchheimer factor. However, the pressure rises faster

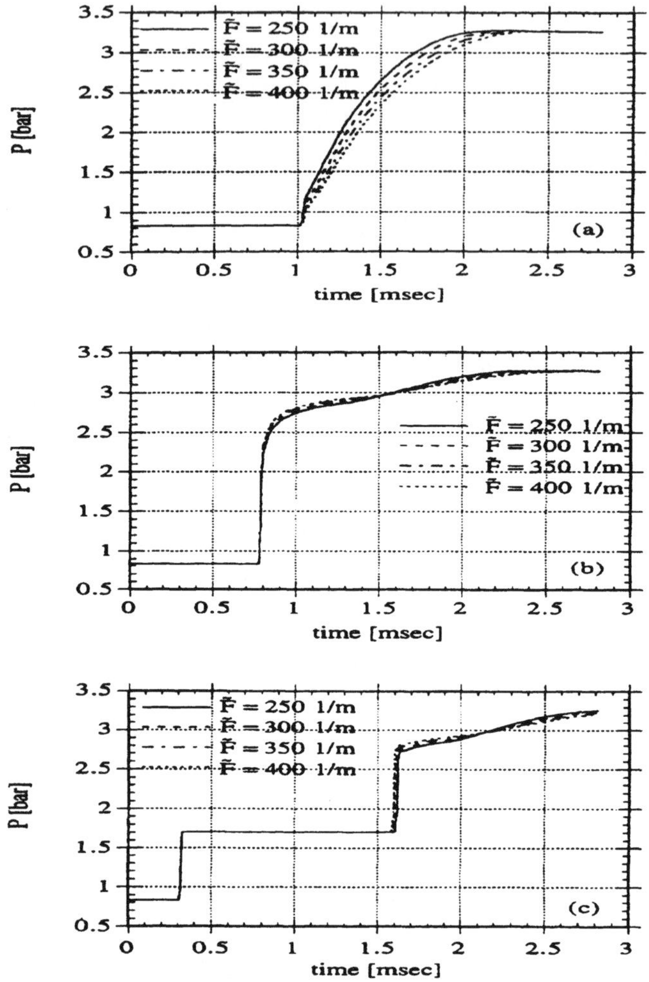


*Figure 6.* The influence of the Forchheimer coefficient on the pressure: (a) at the back edge of the sample, (b) at the front edge of the sample and (c) 42 cm ahead of the sample. The sample was made of SiC and it was 81 mm long. The incident shock wave Mach number was $M_i = 1.377$.

for smaller Forchheimer factors. Figures 6(b) and 6(c) demonstrate, respectively, that the pressures at the front edge of the sample and ahead of it are less influenced by the Forchheimer factor.

In view of Figure 7 we note that different Forchheimer factors can result in fluid velocity differences of up to 7 m/s (for example at about 1 and 2 ms in Figure 7(a)). However, the finally obtained fluid velocities are independent of the Forchheimer factor.

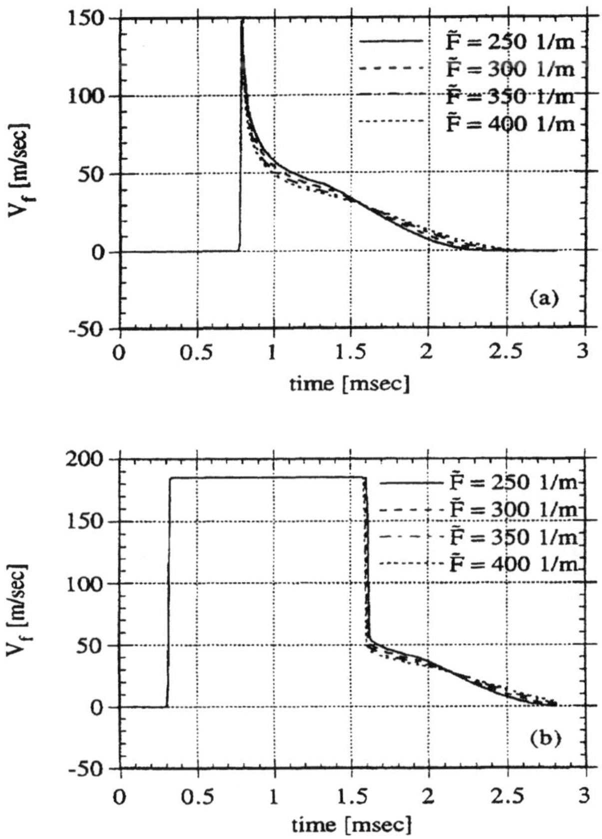


*Figure 7.* The influence of the Forchheimer coefficient on the fluid velocity: (a) at the front edge of the sample and (b) 42 cm ahead of the sample. The sample was made of SiC and it was 81 mm long. The incident shock wave Mach number was $M_{\mathrm{i}} = 1.377$.

### 8.3. THE EFFECT OF THE LAMÉ COEFFICIENTS

The effect of the Lamé $E_\varepsilon(\equiv \lambda_s'' + \mu_s')$ coefficient was most noticeable on the normal stress profiles (Figure 8). It is demonstrated that the normal stresses reach higher values for the smallest value of $E_\varepsilon$. In addition, the finally obtained values of the normal stress are constant and they are independent of $E_\varepsilon$. It should also be noted that the compression (shock) wave reaches the back edge of the sample faster for larger values of $E_\varepsilon$. The difference in the time of arrival between $E_\varepsilon = 380 \times 10^8$ Pa and $E_\varepsilon = 380 \times 10^8$ Pa is about 90 ms.

### 8.4. THE INITIAL POROSITY EFFECT

A significant dependence on the initial porosity was exhibited for the normal stress distribution (Figure 9). Larger porosity values, yield higher normal stresses. For

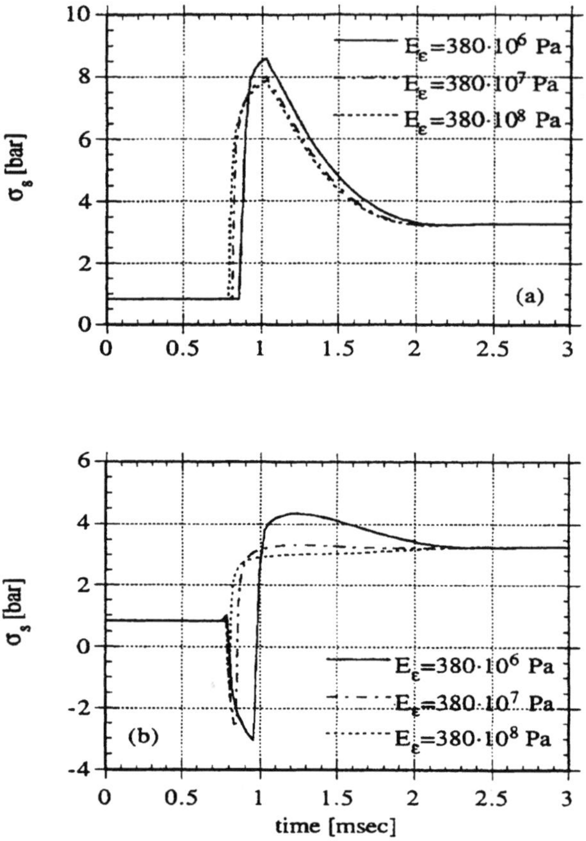


*Figure 8.* The influence of the $E_\varepsilon\ (\equiv \lambda_s'' + \mu_s')$ Lamé coefficient on the normal stress: (a) at the back edge of the sample and (b) at the front edge of the sample. The sample was made of SiC and it was 81 mm long. The incident shock wave Mach number was $M_i = 1.377$.

example, Figure 9(a) indicates that at the rear edge for $\phi_0 = 0.68$ we obtain $(\sigma_s)_{max} \approx 6.9$ KPa while $\phi_0 = 0.78$ yields $(\sigma_s)_{max} \approx 9.5$ KPa.

## 9. Conclusions

Macroscopic balance equations of mass, momentum and energy are described for the case of saturated thermoelastic porous media.

We show that microscopic terms of momentum should be accounted for at the interfaces between the solid and the fluid phases. These microscopic terms lead to the macroscopic Forchheimer terms. The macroscopic terms, represent sources of momentum and energy that are exchanged between the fluid phases and the solid phase, at their solid–fluid interfaces.

In the case of the phases momentum balance equation we present a separate momentum balance equation for the fluid accounting for momentum flux terms,

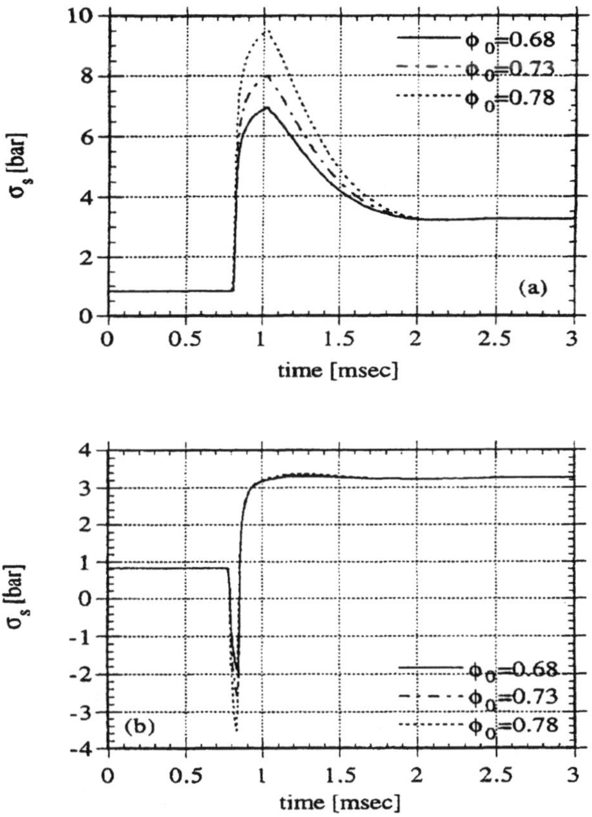


*Figure 9.* The influence of the initial porosity on the normal stress: (a) at the back edge of the sample and (b) at the front edge of the sample. The sample was made of SiC and it was 81 mm long. The incident shock wave Mach number was $M_i = 1.377$.

deviating from the macroscopic ones. These were assumed to be negligible in comparison to the averaged, macroscopic terms. The solution of this second-order equation, subjected to boundary and initial conditions similar to the macroscopic counterpart, will improve the conservation of the momentum balance especially for heterogeneous media and high velocities, which may produce significant variations from the average momentum flux terms.

Nondimensional analysis of the balance equations, after the onset of abrupt change in fluids' pressure and temperature, provide the theoretical basis for obtaining dominant balance equations that conform to wave equations. Four different simplified 1-D analytical solutions are then presented. The first considering post compaction wave for small porosity values, considering incompressible solid phase. The second for a compressible solid phase and a matrix of very small porosity values. Both models are compared with experimental observations of head-on collision of planar shock waves with air saturated rigid porous materials. These prove to yield good results.

The third and the fourth analytical solutions, refer to traveling weak waves in the horizontal direction. In both models, the fluid mass balance equation or its momentum balance equation conform to a pressure wave equation, with gas and matrix quantities that can be evaluated analytically. The third model refers to an isothermal case while the fourth accounts for nonisothermal conditions.

Numerical simulations of compaction waves in a 1-D saturated porous medium, are then compared with shock tube experiments. These simulations were done with and without Forchheimer terms. The case of accounting for Forchheimer terms demonstrates almost no deviation from experimental results and was by far better than the case when we did not include these terms.

## References

1. Attenborough, K.: Acoustical characteristics of porous materials, *Phys. Reports* **82**(3) (1982), 179–227.
2. Bachmat, Y. and Bear, J.: Macroscopic modeling of transport phenomena in porous media: 1. The continuum approach, *Transport in Porous Media* **1** (1986), 213–240.
3. Baer, M. R.: Numerical studies of dynamic compaction of inert and energetic granular materials, *ASME J. Appl. Mech.* **55** (1988), 36–43.
4. Baer, M. R.: A numerical study of shock wave reflections on low density foam, *Shock Waves* **2** (1992), 121–124.
5. Baer, M. R. and Nunziato, J. W.: A two-phase mixture theory for the deflagration to detonation transition (DDT) in reactive granular materials, *Int. J. Multiphase Flow* **12** (1986), 861–889.
6. Bear, J. and Bachmat, Y.: *Introduction to Modeling of Transport Phenomena in Porous Media*, Kluwer Academic Publishers, Dordrecht, The Netherlands, 1990.
7. Bear, J. and Sorek, S.: Evolution of governing mass and momentum balance equations following an abrupt pressure impact in a porous medium, *Transport in Porous Media* **5** (1990), 169–185.
8. Bear, J., Sorek, S., Ben-Dor, G. and Mazor, G.: Displacement waves in saturated thermoelastic porous media: I. Basic equations, *Fluid Dynam. Res.* **9** (1992), 155–164.
9. Ben-Dor, G.: *Shock Wave Reflection Phenomena*, Springer-Verlag, New York, 1991.
10. Ben-Dor, G., Mazor, G., Igra, O., Sorek, S. and Onodera, H.: Shock wave interaction with cellular materials: II. Open cell foams; Experimental and numerical results, *Shock Waves* **3**(3) (1994), 167–179.
11. Ben-Dor, G. and Zaretsky, E. B.: Head-on interaction of planar shock waves with polyurethane foams – A semi-empirical model, *Archivem Appl. Mech.* **64** (1994), 1–8.
12. Biot, M. A.: Theory of propagation of elastic waves in fluid-saturated porous solid, *J. Acoust. Soc. Amer.* **28** (1956), 168–191.
13. Corapcioglu, M. Y.: Wave propagation in porous media – a review, In: J. Bear and M. Y. Corapcioglu (eds), *Transport Processes in Porous Media*, Kluwer Academic Publishers, Dordrecht, The Netherlands, 1991, pp. 373–469.
14. Degrande, G. and de Roeck, G.: FFT-based spectral analysis methodology for one-dimensional wave propagation in poroelastic media, *Transport in Porous Media* **9** (1992), 85–97.
15. Du Plessis, J. P. and Masliyan, J. H.: Mathematical modeling of flow through consolidated isotropic porous media, *Transport in Porous Media* **3** (1988), 145–164.
16. Du Plessis, J. P., Montillet, A., Comiti, J. and Legrand, J.: Pressure drop prediction for flow through high porosity metallic foams, *Chem. Engg. Sci.* **49**(21) (1994), 3545–3553.
17. Gelfand, B. E., Gubanov, A. V. and Timofeev, E. I.: Interaction between shock waves and porous screen, *Sov. Phys.-Fluid Mech.* **4** (1983), 79–84.

18. Gibson, L. J. and Ashby M. F.: *Cellular Solids – Structure & Properties*, Program Press, Headington Hill Hall, Oxford, England, 1988.
19. Glass, I. I. and Hall, G. J.: *Shock Tubes, Handbook of Supersonic Aerodynamics*, Sec. 18, Navond Report 1488, Vol. 6, 1959, p. 66.
20. van der Grinten, J. G. M., Smits, M. A., van der Kogel, H. and van Dongen, M. E. H.: Shock-induced wave propagation in and reflection from porous column partially saturated with water, In: H. Gronig (ed.), *Shock Tubes and Waves*, VCH Verlagsgesellschaft mbH, Weinheim, Germany, 1988, pp. 357–362.
21. Gvozdeva, L. G., Faresov, Yu. M. and Fokeev, V. P.: Interaction between air shock wave and porous compressible material, *Sov. Phys. Appl. Math. Tech. Phys.* **3** (1985), 111–115.
22. Gvozdeva, L. G. and Faresov, Yu. M.: Approximate calculation of steady state shock wave parameters in porous compressible materials, *J. Appl. Mech. Tech. Phys.* (English translation of PMTF, *Zh. Prik. Mekh. Tekh. Fiz.*), **27** (1986), 107–111.
23. Harten, A.: High resolution schemes for hyperbolic conservation laws, *J. Comput. Phys.* **49** (1983), 357–393.
24. Henderson, L. F., Virgona, R. J., Di, J. and Gvozdeva, L. G.: Refraction of a normal shock wave from nitrogen into polyurethane foam, In: Y. W. Kim (ed.), *Current Topics in Shock Waves*, American Institute of Physics, New York, U.S.A., 1990, pp. 814–818.
25. Hsu, C. T. and Cheng, P.: Thermal dispersion in a porous medium, *Int. J. Heat Mass Transfer* **33**(8) (1990), 1587–1597.
26. Igra, O. and Ben-Dor, G.: Dusty shock waves, *ASME Appl. Mech. Rev.* **41** (1988), 379–437.
27. Krylov, A., Sorek, S., Levy, A. and Ben-Dor, G.: Simple waves in saturated porous media: I. The isothermal case, *JSME Int. J. Series B* **39**(2) (1996), 294–298.
28. Landau, L. D. and Lifshitz, E. M.: *Fluid Mechanics*, 2nd edn, Pergamon Press, 1987, pp. 378–385.
29. Levy, A.: Wave Propagation in a Saturated Porous Media, PhD Thesis, Department of Mechanical Engineering, Ben-Gurion University of the Negev, Beer Sheva, Israel, 1996, (in Hebrew).
30. Levy, A., Ben-Dor, G., Skews, B. W. and Sorek, S.: Head-on collision of normal shock waves with rigid porous materials, *Experiments in Fluids* **15** (1993), 183–190.
31. Levy, A., Ben-Dor, G., Sorek, S. and Bear, J.: Jump conditions across strong compaction waves in gas saturated rigid porous media, *Shock Waves* **3**(2) (1993), 105–111.
32. Levy, A., Sorek, S., Ben-Dor, G. and Bear, J.: Evolution of the balance equations in saturated thermoelastic porous media following abrupt simultaneous changes in pressure and temperature, *Transport in Porous Media* **21** (1995), 241–268.
33. Levy, A., Sorek, S., Ben-Dor, G. and Skews B. W.: Waves propagation in saturated rigid porous media: analytical model and comparison with experimental results, *Fluid Dynam. Res.* **17** (1996), 49–65.
34. Levy A., Ben-Dor G., and Sorek, S.: Numerical investigation of the propagation of shock waves in rigid materials: development of the computer code and comparison with experimental results, *J. Fluid Mech.* **324** (1996), 163–179.
35. Li, H. and Ben-Dor, G.: Head-on interaction of weak planar shock waves with flexible porous materials – analytical model, *Int. J. Multiphase Flow* **21**(5) (1995), 941–947.
36. Li, H., Levy, A. and Ben-Dor, G.: Analytical prediction of regular reflection over porous surfaces and comparison with experiments, *J. Fluid Mech.* **282** (1995), 219–232.
37. Liepmann, H. W. and Roshko, A.: *Elements of Gas Dynamics*, Wiley, New York, U.S.A., 1957.
38. Mazor, G., Ben-Dor, G., Igra, O. and Sorek, S.: Shock wave interaction with cellular materials: Part I. Analytical investigation and governing equations, *Shock Waves* **3** (1994), 159–165.
39. Nield, D. A.: The limitations of the Brinkman–Forchheimer equation in modelling flow in a saturated porous medium and at an interface, *Int. J. Heat Fluid Flow* **12** (1991), 269–272.

40. Nield, D. A.: Modelling high speed flow of compressible fluid in a saturated porous medium, *Transport in Porous Media* **14** (1994), 85–88.
41. Nigmatulin, R. I. and Gubaidullin, A. A.: Linear waves in saturated porous media, *Transport in Porous Media* **9** (1992), 135–142.
42. Nikolaevskij, V. M.: *Mechanics of Porous and Fractured Media*, World Scientific, Singapore, 1990.
43. Olim, M., van Dongen, M. E. W., Kitamura, T. and Takayama, K.: Numerical simulation of the propagation of shock waves in compressible open-cell porous foams, *Int. J. Multiphase Flow* **20**(3) (1994), 557–568.
44. Powers, J. M., Stewart, D. S. and Krier, H.: Analysis of steady compaction waves in porous materials, *ASME J. Appl. Mech.* **56** (1989), 15–24.
45. Sandusky, H. W. and Liddiard, T. P.: *Dynamic Compaction of Porous Beds*, NSWC TR 83-256, NAVAL Surface Weapons Center, White Oak, Md., U.S.A., 1985.
46. Skews, B. W.: The reflected pressure field in the interaction of weak shock waves with a compressible porous foams, *Shock Waves* **1** (1991), 205–211.
47. Skews, B. W., Atkins, M. D. and Seitz, M. W.: The impact of shock wave on porous compressible foams, *J. Fluid Mech.* **253** (1993), 245–265.
48. Smeulders, D. M. J., de La Rosette, S. P. M. and Van Dongen, M. E. H.: Waves in partially saturated porous media, *Transport in Porous Media* **9** (1992), 25–37.
49. Sorek, S., Bear, J., Ben-Dor, G. and Mazor, G.: Shock waves in saturated thermoelastic porous media, *Transport in Porous Media* **9** (1992), 3–13.
50. Sorek, S., and Levi-Hevroni, D.: A model for solute transport following an abrupt pressure impact in nondeformable porous media, *2nd Inter. Conf. on Water Pollution Milan* 1993, pp. 77–84.
51. Sorek, S., Krylov, A., Levy, A. and Ben-Dor, G.: Simple waves in saturated porous media: II. The nonisothermal case, *JSME Int. J. Series B* **39**(2) (1996), 299–304.
52. Sorek, S.: A model for solute transport following an abrupt pressure impact in saturated porous media, *Transport in Porous Media* **22** (1996), 271–285.
53. Zaretsky, E. B. and Igra, O.: Head-on collision of a normal shock wave with a polyester foam, *The 19th Int. Symp. Shock Waves*, Marseille, France, 1993.

# Saturated Compressible and Incompressible Porous Solids: Macro- and Micromechanical Approaches

ANJANI KUMAR DIDWANIA[1] and REINT DE BOER[2]
[1]*Department of Applied Mechanics and Engineering Sciences, University of California, San Diego, La Jolla, CA, 92093–0411, U.S.A.*
[2]*Institut für Mechanik, FB 10-Bauwesen, Universität-GH Essen, D-45117 Essen, Germany*

(Received: 22 June 1997; in final form: 20 August 1997)

**Abstract.** In this paper two complementary approaches are used to describe the mechanical behavior of saturated compressible and incompressible porous solids. The macroscopic investigation is based on the mixture theory, restricted by the volume fraction concept. In the micromechanical approach, a hierarchy of conditionally ensemble averaged fluid and solid phase momentum balance equations are derived for a simple model of quasi-static liquid saturated porous media. The ensemble averaged equations for both the phases agree remarkably well with the macroscopic results. A micromechanical basis for Terzhagi's effective stress concept is presented. In addition, an expression for additional partial solid stress modifying the effective stress principle, to account for deformability of solid materials, is also derived.

**Key words:** porous media, macro- and microscopic approaches, compressible and incompressible porous solids, incompressible fluid, effective stress concept.

## 1. Introduction

The macroscopic theory of porous media with *incompressible* constituents has been investigated by several researchers over the years (for a detailed history, see de Boer, 1996). Valuable early contributions by Fillunger (1936) towards a systematic and consistent approach, however, were completely forgotten with time. Later, within the framework of a revival of the porous media theory in the 1970s and 1980s using the mixture theory, Bowen (1980) among others (see Mow *et al.*, 1980) constructed a general mathematical model for incompressible constituents including thermal effects. Early attempts by Biot and co-workers to develop a model of saturated porous solids with *compressible* constituents were not rigorous and limited to linear poroelasticity (Biot, 1941, 1955; Biot and Willis, 1957). Bowen (1982) worked out a consistent mathematical model for compressible constituents using the mixture theory. While introducing volume fractions of constituent phases in the mixture theory, he recognized the need for additional closure relations. His choice of evolution equations for the volume fractions as closure relations however does not contain the effective stress concept and thus contradicts well-known test observations in rock mechanics (see, e.g., Lade and de Boer, 1997). Another closure relation in this context has been discussed by de Boer (1996). In this closure scheme, the saturation condition, that is, the sum of all volume fractions is equal to unity, is a constraint independent of the true material behavior (see also Passman, Nunziato and Walsh, 1984) and must always be considered in the constitutive theory, in the evaluation of the dissipation concept. This new macroscopic approach leads to physically meaningful and mathematically consistent results, as will be shown later in this paper.

Micromechanical investigations of the uplift force in saturated porous solids with incompressible constituents were recently carried out by de Boer and Didwania (1997). Their expression for the uplift force agree remarkably with that obtained from the macroscopic approach. However, no attempts to this date have been made to examine saturated porous solids with compressible solid material from a micromechanical ansatz.

The goal of this paper is to investigate at the macro- as well as at the microlevel a binary model consisting of a compressible or an incompressible porous solid filled with an incompressible fluid. In Section 2, using the saturation condition as a constraint, mechanical behavior of both phases are deduced from the entropy inequality in the macroscopic approach. In Section 3, using a micromechanical approach, a hierarchy of conditionally ensemble averaged fluid and solid phase momentum balance equations are derived for a simple model of quasi-static liquid saturated porous media. The ensemble averaged equations for both the phases agree remarkably well with the macroscopic results. A micromechanical basis for Terzaghi's effective stress concept is presented. In addition, an expression for additional solids phase stress modifying the effective stress concept, to account for deformability of solids material, is also derived.

## 2. The Macroscopic Approach

In this section the volume fraction concept and the kinematics are introduced. Moreover, the entropy inequality is reformulated considering the saturation condition as a constraint.

### 2.1. MATERIAL INDEPENDENT RELATIONS AND ENTROPY INEQUALITY

For the binary model under study we assume the existence of two volume fractions

$$n^{\mathrm{S}} = n^{\mathrm{S}}(\mathbf{x}, t), \qquad n^{\mathrm{F}} = n^{\mathrm{F}}(\mathbf{x}, t) \tag{2.1}$$

for the solid- and fluid phases, where $\mathbf{x}$ denotes the position vector to the spatial point $x$ and $t$ the time. The volume fractions must obey the saturation condition (see de Boer, 1996):

$$n^{\mathrm{S}} + n^{\mathrm{F}} = 1. \tag{2.2}$$

Moreover, with the introduction of the volume fractions $n^{\mathrm{S}}$, $n^{\mathrm{F}}$ two substitute smooth continua with reduced densities $\rho^{\mathrm{S}}$, $\rho^{\mathrm{F}}$

$$\rho^{\mathrm{S}} = n^{\mathrm{S}}\rho^{\mathrm{SR}}, \qquad \rho^{\mathrm{F}} = n^{\mathrm{F}}\rho^{\mathrm{FR}} \tag{2.3}$$

are created, where the realistic densities $\rho^{\mathrm{SR}}$, $\rho^{\mathrm{FR}}$ are the average densities of the porous media constituents. The substitute continua show the same properties as the constituents in a mixture, namely that the particles of both continua simultaneously occupy the spatial point $x$ at any time $t$. Therefore, the mixture theory is an adequate basis for the treatment of the proposed binary model.

The particles $X_{\mathrm{S}}$ and $X_{\mathrm{F}}$ of the solid and fluid phases proceed from different reference positions $\mathbf{X}_{\mathrm{S}}$ and $\mathbf{X}_{\mathrm{F}}$ at time $t = t_0$ to the spatial point $x$. It is assumed that each constituent is assigned its own independent motion function $\boldsymbol{\chi}_{\mathrm{S}}, \boldsymbol{\chi}_{\mathrm{F}}$

$$\mathbf{x} = \boldsymbol{\chi}_{\mathrm{S}}(\mathbf{X}_{\mathrm{S}}, t), \qquad \mathbf{x} = \boldsymbol{\chi}_{\mathrm{F}}(\mathbf{X}_{\mathrm{F}}, t) \tag{2.4}$$

from which the velocities $\mathbf{x}'_{\mathrm{S}}$, $\mathbf{x}'_{\mathrm{F}}$ and the deformation gradients $\mathbf{F}_{\mathrm{S}}$, $\mathbf{F}_{\mathrm{F}}$ can be derived:

$$\mathbf{x}'_{\mathrm{S}} = \frac{\partial \boldsymbol{\chi}_{\mathrm{S}}}{\partial t}, \qquad \mathbf{x}'_{\mathrm{F}} = \frac{\partial \boldsymbol{\chi}_{\mathrm{F}}}{\partial t}, \qquad \mathbf{F}_{\mathrm{S}} = \frac{\partial \boldsymbol{\chi}_{\mathrm{S}}}{\partial \mathbf{X}_{\mathrm{S}}}, \qquad \mathbf{F}_{\mathrm{F}} = \frac{\partial \boldsymbol{\chi}_{\mathrm{F}}}{\partial \mathbf{X}_{\mathrm{F}}}. \tag{2.5}$$

It is convenient for further considerations to multiplicatively decompose the deformation gradients $\mathbf{F}_{\mathrm{S}}, \mathbf{F}_{\mathrm{F}}$ into deformation gradients $\mathbf{F}_{\mathrm{SR}}, \mathbf{F}_{\mathrm{FR}}$ which reflect the microscopic deformations of the true materials at the macroscale and into tensors $\mathbf{F}_{\mathrm{SN}}, \mathbf{F}_{\mathrm{FN}}$ which describe the changing of the volume fractions (for more information see Bluhm and de Boer, 1997; de Boer, 1996; Bluhm, 1997):

$$\mathbf{F}_{S} = \mathbf{F}_{\mathrm{SN}}\mathbf{F}_{\mathrm{SR}}, \qquad \mathbf{F}_{\mathrm{F}} = \mathbf{F}_{\mathrm{FN}}\mathbf{F}_{\mathrm{FR}}. \tag{2.6}$$

In the following some additional deformation relations are needed:

$$\begin{aligned}
&\mathbf{C}_{\mathrm{S}} = \mathbf{F}_{\mathrm{S}}^{T}\mathbf{F}_{\mathrm{S}}, \qquad &&\mathbf{C}_{\mathrm{SR}} = \mathbf{F}_{\mathrm{SR}}^{T}\mathbf{F}_{\mathrm{SR}},\\
&J_{\mathrm{SR}} = \det \mathbf{F}_{\mathrm{SR}}, \qquad &&J_{\mathrm{FR}} = \det \mathbf{F}_{\mathrm{FR}},\\
&J_{\mathrm{SN}} = \det \mathbf{F}_{\mathrm{SN}}, \qquad &&J_{\mathrm{FN}} = \det \mathbf{F}_{\mathrm{FN}}.
\end{aligned} \tag{2.7}$$

Moreover, the symmetric parts of the velocity gradients are introduced:

$$\begin{aligned}
&\mathbf{D}_{\mathrm{S}} = \tfrac{1}{2}(\mathbf{L}_{\mathrm{S}} + \mathbf{L}_{\mathrm{S}}^{T}), \qquad &&\mathbf{D}_{\mathrm{F}} = \tfrac{1}{2}(\mathbf{L}_{\mathrm{F}} + \mathbf{L}_{\mathrm{F}}^{T}),\\
&\mathbf{D}_{\mathrm{SR}} = \tfrac{1}{2}(\mathbf{L}_{\mathrm{SR}} + \mathbf{L}_{\mathrm{SR}}^{T}), \qquad &&\mathbf{D}_{\mathrm{FR}} = \tfrac{1}{2}(\mathbf{L}_{\mathrm{FR}} + \mathbf{L}_{\mathrm{FR}}^{T}).
\end{aligned} \tag{2.8}$$

Here,

$$\begin{aligned}
&\mathbf{L}_{\mathrm{S}} = (\mathbf{F}_{\mathrm{S}})'_{\mathrm{S}}\mathbf{F}_{\mathrm{S}}^{-1}, \qquad &&\mathbf{L}_{\mathrm{F}} = (\mathbf{F}_{\mathrm{F}})'_{\mathrm{F}}\mathbf{F}_{\mathrm{F}}^{-1},\\
&\mathbf{L}_{\mathrm{SR}} = (\mathbf{F}_{\mathrm{SR}})'_{\mathrm{S}}\mathbf{F}_{\mathrm{SR}}^{-1}, \qquad &&\mathbf{L}_{\mathrm{FR}} = (\mathbf{F}_{\mathrm{FR}})'_{\mathrm{F}}\mathbf{F}_{\mathrm{FR}}^{-1}
\end{aligned} \tag{2.9}$$

are the velocity gradients and $(\cdots)'_{\mathrm{S}}$, $(\cdots)'_{\mathrm{F}}$ mark the material time derivatives.

In order to gain restrictions for the constitutive equations the dissipation principle is useful; this means that the entropy inequality must be evaluated. The entropy inequality for a binary model with isothermal deformations, excluding any mass exchange, reads as

$$-\rho^{S}(\psi^{S})'_{S} - \rho^{F}(\psi^{F})'_{F} - \hat{\mathbf{p}}^{F} \cdot (\mathbf{x}'_{F} - \mathbf{x}'_{S}) + \mathbf{T}^{S} \cdot \mathbf{D}^{S} + \mathbf{T}^{F} \cdot \mathbf{D}_{F} \geqslant 0. \qquad (2.10)$$

In (2.10) $\psi^{S}$ and $\psi^{F}$ are the free Helmholtz energy functions, $\hat{\mathbf{p}}^{F}$ denotes the interaction force between both constituents, and $\mathbf{T}^{S}$ and $\mathbf{T}^{F}$ are Cauchy's stress tensors.

The entropy inequality in the form (2.10) is not suitable for an evaluation of the proposed *compressible* porous solids filled with an *incompressible* fluid as the motions of the mixture body are severely restricted by two constraints (see de Boer, 1996): The first one is due to the saturation condition (2.2) which states in the rate formulation:

$$\begin{aligned} &- n^{S}(\mathbf{D}_{S} \cdot \mathbf{I}) + n^{S}(\mathbf{D}_{SR} \cdot \mathbf{I}) - n^{F}(\mathbf{D}_{F} \cdot \mathbf{I}) + n^{F}(\mathbf{D}_{FR} \cdot \mathbf{I}) - \\ &- \operatorname{grad} n^{F} \cdot (\mathbf{x}'_{F} - \mathbf{x}'_{S}) = 0, \end{aligned} \qquad (2.11)$$

the second one results from the incompressibility of the fluid phase

$$n^{F}(\mathbf{D}_{FR} \cdot \mathbf{I}) = 0. \qquad (2.12)$$

Both constraints (2.11) and (2.12) must be taken into account provided with Lagrangian multipliers $\lambda$ and $\kappa^{FR}$ in the thermodynamical treatment and must be added to the entropy inequality (2.10):

$$\begin{aligned} &- \rho^{S}(\psi^{S})'_{S} - \rho^{F}(\psi^{F})'_{F} + \mathbf{D}_{S} \cdot (\mathbf{T}^{S} + n^{S}\lambda\mathbf{I}) + \\ &\quad + \mathbf{D}_{F} \cdot (\mathbf{T}^{F} + n^{F}\lambda\mathbf{I}) - \lambda(\mathbf{D}_{SR} \cdot \mathbf{I})n^{S}\mathbf{I} - \\ &\quad - \mathbf{D}_{FR} \cdot (\lambda - \kappa^{FR})n^{F}\mathbf{I} - (\hat{\mathbf{p}}^{F} - \lambda \operatorname{grad} n^{F}) \cdot (\mathbf{x}'_{F} - \mathbf{x}'_{S}) \geqslant 0. \end{aligned} \qquad (2.13)$$

In the following, only elastic compressible solid materials will be considered. This is sufficient for our purpose, namely to disclose the structure of the constitutive equations for the binary model. The elastic behavior of the true solid material can be described by the right Cauchy Green deformation tensor $\mathbf{C}_{SR}$ $(2.7)_2$. However, as $\mathbf{C}_{SR}$ cannot be determined by a balance equation like $\mathbf{C}_{S}$ (the porous media theory does not contain balance equations at the microscale) a constitutive equation will be assumed for $\mathbf{C}_{SR}$ in order to keep closure:

$$\mathbf{C}_{SR} = \mathbf{C}_{SR}(\mathbf{C}_{S}). \qquad (2.14)$$

From (2.14)

$$\mathbf{D}_{SR} \cdot \mathbf{I} = \mathbf{F}_{S} \left(\frac{\partial \mathbf{C}_{SR}}{\partial \mathbf{C}_{S}}\right)^{T} \mathbf{C}_{SR}^{-1}\mathbf{F}_{S}^{T} \cdot \mathbf{D}_{S} \qquad (2.15)$$

is gained. Considering (2.15) the entropy inequality (2.13) changes slightly to

$$\begin{aligned} &- \rho^{S}(\psi^{S})'_{S} - \rho^{F}(\psi^{F})'_{F} + \mathbf{D}_{S} \cdot \{\mathbf{T}^{S} + n^{S}\lambda(\mathbf{I} - \boldsymbol{\mathcal{F}}_{S})\} + \\ &\quad + \mathbf{D}_{F} \cdot (\mathbf{T}^{F} + n^{F}\lambda\mathbf{I}) - \mathbf{D}_{FR} \cdot (\lambda - \kappa^{FR})n^{F}\mathbf{I} - \\ &\quad - (\hat{\mathbf{p}}^{F} - \lambda \operatorname{grad} n^{F}) \cdot (\mathbf{x}'_{F} - \mathbf{x}'_{S}) \geqslant 0, \end{aligned} \tag{2.16}$$

where

$$\boldsymbol{\mathcal{F}}_{S} = \mathbf{F}_{S} \left( \frac{\partial \mathbf{C}_{SR}}{\partial \mathbf{C}_{S}} \right)^{T} \mathbf{C}_{SR}^{-1} \mathbf{F}_{S}^{T}. \tag{2.17}$$

The entropy inequality (2.16) must be further rearranged for the *incompressible* binary model (incompressible porous solid filled with an incompressible fluid). In this case the incompressibility condition of the solid material must also be taken into account:

$$n^{S}(\mathbf{D}_{SR} \cdot \mathbf{I}) = 0. \tag{2.18}$$

The constraint (2.18) provided with a Lagrangian multiplier $\kappa^{SR}$ must be added to the entropy inequality (2.16). This leads to

$$\begin{aligned} &- \rho^{S}(\psi^{S})'_{S} - \rho^{F}(\psi^{F})'_{F} + \mathbf{D}_{S} \cdot \{\mathbf{T}^{S} + n^{S}\lambda[\mathbf{I} - \breve{\boldsymbol{\mathcal{F}}}_{S}] + n^{S}\kappa^{SR}\breve{\boldsymbol{\mathcal{F}}}_{S}\} + \\ &\quad + \mathbf{D}_{F} \cdot (\mathbf{T}^{F} + n^{F}\lambda\mathbf{I}) - \mathbf{D}_{FR} \cdot (\lambda - \kappa^{FR})n^{F}\mathbf{I} - \\ &\quad - (\hat{\mathbf{p}}^{F} - \lambda \operatorname{grad} n^{F}) \cdot (\mathbf{x}'_{F} - \mathbf{x}'_{S}) \geqslant 0. \end{aligned} \tag{2.19}$$

Here, $\breve{\boldsymbol{\mathcal{F}}}_{S}$ is equal to (2.17) containing, however, the deformation tensor $\breve{\mathbf{C}}_{SR}$ which represents in the proposed incompressible model only the volume-preserving part of $\mathbf{C}_{SR}$ so that

$$\det \breve{\mathbf{C}}_{SR} = 1. \tag{2.20}$$

Neglecting the deformations at the microscale ($\mathbf{C}_{SR}$ equal to zero) inequality (2.19) simplifies to

$$\begin{aligned} &- \rho^{S}(\psi^{S})'_{S} - \rho^{F}(\psi^{F})'_{F} + \mathbf{D}_{S} \cdot (\mathbf{T}^{S} + n^{S}\lambda\mathbf{I}) - \\ &\quad - \mathbf{D}_{SR} \cdot (\lambda - \kappa^{SR})n^{S}\mathbf{I} + \mathbf{D}_{F} \cdot (\mathbf{T}^{F} + n^{F}\lambda\mathbf{I}) - \\ &\quad - \mathbf{D}_{FR} \cdot (\lambda - \kappa^{FR})n^{F}\mathbf{I} - (\hat{\mathbf{p}}^{F} - \lambda \operatorname{grad} n^{F}) \cdot (\mathbf{x}'_{F} - \mathbf{x}'_{S}) \geqslant 0. \end{aligned} \tag{2.21}$$

After these preliminary investigations restrictions for the constitutive equations of the binary model will be developed.

### 2.2. CONSTITUTIVE RELATIONS

The development of constitutive equations for saturated porous media is based on the same principles that govern the derivation of constitutive equations in classical continuum mechanics of one-component materials. These are: *determinism, equipresence, local action, material objectivity,* and *dissipation*. Guided by these principles, constitutive relations for a binary model with a compressible elastic solid skeleton and an incompressible fluid will be derived. Response functions (constitutive relations) are needed for the following set $\mathcal{R}$ of mechanical quantities:

$$\mathcal{R}(s) := \{\psi^{\mathrm{S}}, \psi^{\mathrm{F}}, \mathbf{T}^{\mathrm{S}} + n^{\mathrm{S}}\lambda(\mathbf{I} - \boldsymbol{\mathcal{F}}_{\mathrm{S}}), \mathbf{T}^{\mathrm{F}} + n^{\mathrm{F}}\lambda\mathbf{I}, \hat{\mathbf{p}}^{\mathrm{F}} - \lambda \operatorname{grad} n^{\mathrm{F}}\}. \tag{2.22}$$

The principles of determinism and local action state that the response functions in (2.22) for a material point $X_{\mathrm{S}}$ and $X_{\mathrm{F}}$, at any time $t$ and any place $\mathbf{x}$, must be determined by the history of an arbitrary small neighborhood of this material point. The history of this small neighborhood is laid down by the process variables $s(\mathbf{x}, t)$ which reflect the history of the motions of the two partial constituents solid and fluid. For the elastic model under study, the following process variables are postulated:

$$s := \{\mathbf{C}_{\mathrm{S}}, \mathbf{x}'_{\mathrm{F}} - \mathbf{x}'_{\mathrm{S}}\}. \tag{2.23}$$

The elastic deformations of the partial solid are described by $\mathbf{C}_{\mathrm{S}}$ and velocity difference $\mathbf{x}'_{\mathrm{F}} - \mathbf{x}'_{\mathrm{S}}$ govern dissipative effects.

For the present model, thermodynamical restrictions result from the entropy inequality – principle of dissipation – together with the constitutive assumptions (2.23). Using standard arguments, the investigations yield the models to be governed by constitutive equations which are relatively simple. In the following, only the main results are listed:

$$\begin{aligned}\mathbf{T}^{\mathrm{S}} &= -n^{\mathrm{S}}\lambda(\mathbf{I} - \boldsymbol{\mathcal{F}}_{\mathrm{S}}) + 2\rho^{\mathrm{S}}\mathbf{F}_{\mathrm{S}}\frac{\partial\psi^{\mathrm{S}}}{\partial\mathbf{C}_{\mathrm{S}}}\mathbf{F}_{\mathrm{S}}^{T} + \\ &\quad + \rho^{\mathrm{S}}\frac{\partial\psi^{\mathrm{S}}}{\partial(\mathbf{x}'_{\mathrm{F}} - \mathbf{x}'_{\mathrm{S}})} \otimes (\mathbf{x}'_{\mathrm{F}} - \mathbf{x}'_{\mathrm{S}}),\end{aligned} \tag{2.24}$$

$$\mathbf{T}^{\mathrm{F}} = -n^{\mathrm{F}}\lambda\mathbf{I} + \rho^{\mathrm{F}}\frac{\partial\psi^{\mathrm{F}}}{\partial(\mathbf{x}'_{\mathrm{F}} - \mathbf{x}'_{\mathrm{S}})} \otimes (\mathbf{x}'_{\mathrm{F}} - \mathbf{x}'_{\mathrm{S}}), \tag{2.25}$$

$$\lambda = \kappa^{\mathrm{FR}}. \tag{2.26}$$

Moreover, a dissipation inequality remains

$$-(\hat{\mathbf{p}}^{\mathrm{F}} - \lambda \operatorname{grad} n^{\mathrm{F}}) \cdot (\mathbf{x}'_{\mathrm{F}} - \mathbf{x}'_{\mathrm{S}}) \geqslant 0. \tag{2.27}$$

From (2.26) it is recognized that the Lagrangian multiplier $\lambda$ connected with the saturation condition is equal to the Lagrangian multiplier $\kappa^{\mathrm{FR}}$ connected with the incompressibility condition of the fluid. The Lagrangian multiplier $\kappa^{\mathrm{FR}}$ is in hydraulics usually denoted by $p$, the porewater pressure. Thus,

$$\lambda = \kappa^{\mathrm{FR}} = p. \tag{2.28}$$

In order to be able to compare the results of this macroscopic approach with those of micromechanical investigations to follow only mechanical states characterized by small velocity differences close to the mixture equilibrium are considered. In this case considering (2.28) the aforementioned constitutive restrictions (2.24), (2.25) and (2.27) simplify to

$$\mathbf{T}^{\mathrm{S}} = -n^{\mathrm{S}} p(\mathbf{I} - \mathcal{F}_{\mathrm{S}}) + \mathbf{T}^{\mathrm{S}}_{\mathrm{E}}. \tag{2.29}$$

Here,

$$\mathbf{T}^{\mathrm{S}}_{\mathrm{E}} = 2\rho^{\mathrm{S}} \mathbf{F}_{\mathrm{S}} \frac{\partial \psi^{\mathrm{S}}}{\partial \mathbf{C}_{\mathrm{S}}} \mathbf{F}^{T}_{\mathrm{S}} \tag{2.30}$$

is the effective stress. Moreover,

$$\mathbf{T}^{\mathrm{F}} = -n^{\mathrm{F}} p \mathbf{I}, \tag{2.31}$$

$$\hat{\mathbf{p}}^{\mathrm{F}} = p \operatorname{grad} n^{\mathrm{F}} - \beta(\mathbf{x}'_{\mathrm{F}} - \mathbf{x}'_{\mathrm{S}}), \tag{2.32}$$

where $\beta$ is a material dependent response parameter, always positive.

As has already been mentioned in the introduction, Equation (2.29) contains the effective stress concept of Suklje (1969). For a purely hydrostatic stress state considering an appropriate assumption for $\mathcal{F}_{\mathrm{S}}$ of the mixture body, it turns out that for the total pressure $\bar{p}$ (see de Boer, 1996):

$$\begin{aligned} \bar{p} &= p^{\mathrm{S}} + p^{\mathrm{F}}, \\ \bar{p} &= p\left(1 - n^{\mathrm{S}} \frac{K_{\mathrm{S}}}{K_{\mathrm{SR}}}\right) + p^{\mathrm{S}}_{\mathrm{E}}. \end{aligned} \tag{2.33}$$

Thus, Suklje's effective stress concept is derived, where $K_{\mathrm{S}}$ and $K_{\mathrm{SR}}$ represent the compressibility moduli of the partial solid body and of the real solid material, respectively. If the solid material is incompressible the modulus $K_{\mathrm{SR}}$ is much larger than $K_{\mathrm{S}}$; thus, $K_{\mathrm{S}}/K_{\mathrm{SR}}$ can be neglected and von Terzaghi's effective stress concept is gained.

For a model with incompressible porous solid and incompressible fluid taking into account the volume-preserving parts of the deformations at the microscale the following constitutive equations are obtained from the evaluation of the entropy inequality (2.19) close to the mixture equilibrium state

$$\mathbf{T}^{\mathrm{S}} = -n^{\mathrm{S}} \lambda (\mathbf{I} - \breve{\mathcal{F}}_{\mathrm{S}}) - n^{\mathrm{S}} \kappa^{\mathrm{SR}} \breve{\mathcal{F}}_{\mathrm{S}} + 2\rho^{\mathrm{S}} \mathbf{F}_{\mathrm{S}} \frac{\partial \psi^{\mathrm{S}}}{\partial \mathbf{C}_{\mathrm{S}}} \mathbf{F}^{T}_{\mathrm{S}}, \tag{2.34}$$

and

$$\mathbf{T}^{\mathrm{F}} = -n^{\mathrm{F}} p \mathbf{I}, \qquad \hat{\mathbf{p}}^{\mathrm{F}} = p \operatorname{grad} n^{\mathrm{F}} - \beta(\mathbf{x}'_{\mathrm{F}} - \mathbf{x}'_{\mathrm{S}}). \tag{2.35}$$

If the deformations at the microscale are neglected, the constitutive equation for Cauchy's stress tensor $\mathbf{T}^{\mathrm{S}}$ of the incompressible porous solid simplifies to

$$\mathbf{T}^{\mathrm{S}} = -n^{\mathrm{S}} \lambda \mathbf{I} + 2\rho^{\mathrm{S}} \mathbf{F}_{\mathrm{S}} \frac{\partial \psi^{\mathrm{S}}}{\partial \mathbf{C}_{\mathrm{S}}} \mathbf{F}^{T}_{\mathrm{S}}, \tag{2.36}$$

$$\lambda = \kappa^{\mathrm{SR}} = p. \tag{2.37}$$

The constitutive equations for the fluid phase remain unchanged.

In the next section it will be shown that the above stated constitutive relations correspond to those developed by micromechanical investigations

## 3. The Micromechanical Approach

### 3.1. SYSTEM MODEL AND SOME DEFINITIONS

In this section we present a continuum level description of a homogeneous liquid-saturated porous media by ensemble averaging of the micro-mechanical realizations. We adopt a particular model of the liquid-saturated porous media in which the solid skeleton is assumed to consist of $N$ identical grains of a specific configuration $C^N$, specified by a set of position vectors $\mathbf{r}^\alpha$ for $\alpha = 1, 2, \ldots, N$. For simplicity we restrict our discussion here to spherical grains immersed in an incompressible liquid phase. We have further assumed there is no gravitational field, no relative velocity between the two phases and both phases are isothermal. The gravitational terms can be easily included in a manner outlined earlier by de Boer and Didwania (1997).

Before proceeding further, we recall some definitions and manipulations commonly employed in the ensemble averaging approach (Irving and Kirkwood, 1950, Sangani and Didwania, 1993). We consider an ensemble of realizations and denote by $f(C^N)$ the probability of a specific configuration $C^N$. In view of the identity of the grains, the appropriate normalization is

$$\int \mathrm{d}C^N\ f(C^N) = N!. \tag{3.1}$$

A reduced probability distribution $f(C^M)$ in which the positions of $M$ grains are specified, can be obtained from $f(C^N)$ by integration

$$f(C^M) = \frac{1}{(N-M)!}\int \mathrm{d}C^{N-M}\ f(C^N),$$

with

$$\int \mathrm{d}C^M\ f(C^M) = \frac{N!}{(N-M)!}. \tag{3.2}$$

Furthermore, a conditional probability can also be defined as

$$f(C^{N-M}) = \frac{f(C^N)}{f(C^M)} \quad \text{with} \quad \int \mathrm{d}C^{N-M}\ f(C^{N-M}) = (N-M)!. \tag{3.3}$$

We define an indicator function $\boldsymbol{\chi}^{\mathrm{S}}(\mathbf{x}; C^N)$, such that $\boldsymbol{\chi}^{\mathrm{S}}(\mathbf{x}; C^N) = 1$ if the point $\mathbf{x}$ is in the solid phase (that is, inside a grain), otherwise $\boldsymbol{\chi}^{\mathrm{S}}(\mathbf{x}; C^N) = 0$. For spherical grains of radii $R$, the indicator function is of the form

$$\boldsymbol{\chi}^{\mathrm{S}}(\mathbf{x}; C^N) = \sum_{\alpha=1}^{N} H(R - |\mathbf{x} - \mathbf{r}^\alpha|), \tag{3.4}$$

with $H$ the Heaviside distribution. The gradient of the indicator function is then

$$\nabla \chi^{\mathrm{S}}(\mathbf{x}; C^N) = -\sum_{\alpha=1}^{N} \delta(R - |\mathbf{x} - \mathbf{r}^\alpha|) \frac{\mathbf{x} - \mathbf{r}^\alpha}{|\mathbf{x} - \mathbf{r}^\alpha|}, \tag{3.5}$$

where $\delta$ refers to the Dirac delta function. Since the boundary of the grains has a measure of zero, the indicator function of the liquid phase $\chi^{\mathrm{F}}$, is $1 - \chi^{\mathrm{S}}$ in case of liquid-saturated porous solids. The liquid and solid phase volume fractions are then defined by

$$n^{\mathrm{F,S}}(\mathbf{x}) = \frac{1}{N!} \int \mathrm{d}C^N \, f(C^N) \, \chi^{\mathrm{F,S}}(\mathbf{x}; C^N), \tag{3.6}$$

from which, by Equation (3.1) we obtain the saturation condition,

$$n^{\mathrm{F}} + n^{\mathrm{S}} = 1. \tag{3.7}$$

By substituting here the expression (3.3) for $\chi^{\mathrm{S}}$ one readily finds

$$n^{\mathrm{S}}(\mathbf{x}) = \int_{|\mathbf{x} - \mathbf{r}^1| \leqslant R} \mathrm{d}^3 r^1 \, f(C^1 = \mathbf{r}^1), \tag{3.8}$$

where $\mathrm{d}^3 r^1$ denotes the differential volume element within the grain centred at $\mathbf{r}^1$.

We further define a conditional volume fraction $n_M^{\mathrm{F,S}}$ for a specified configuration of $M$ grains

$$\begin{aligned} n_M^{\mathrm{F,S}}(\mathbf{x}) &= n^{\mathrm{F,S}}(\mathbf{x}; C^M) \\ &= \frac{1}{(N-M)! f(C^M)} \int \mathrm{d}C^{N-M} \, f(C^N) \, \chi^{\mathrm{F,S}}(\mathbf{x}; C^N) \\ &= \frac{1}{(N-M)!} \int \mathrm{d}C^{N-M} \, f(C^{N-M}) \, \chi^{\mathrm{F,S}}(\mathbf{x}; C^N). \end{aligned} \tag{3.9}$$

We note that the conditional volume fraction also satisfies the saturation condition $n_M^{\mathrm{F}} + n_M^{\mathrm{S}} = 1$, see (3.7).

Let $\theta^{\mathrm{F,S}}(\mathbf{x}; C^N)$ be any flow quantity pertaining to the liquid or solid phase at position $\mathbf{x}$ in the presence of the configuration $C^N$ of the grains. The phase-ensemble average of $\theta^{\mathrm{F,S}}$ is defined by averaging over all the configurations such that the point $\mathbf{x}$ is in the appropriate phase,

$$\langle \theta^{\mathrm{F,S}} \rangle(\mathbf{x}) = \frac{1}{n^{\mathrm{F,S}} N!} \int \mathrm{d}C^N \, \theta^{\mathrm{F,S}}(\mathbf{x}; C^N) \, \chi^{\mathrm{F,S}}(\mathbf{x}; C^N) \, f(C^N). \tag{3.10}$$

The conditional average of $\theta^{\mathrm{F,S}}(\mathbf{x}; C^N)$ on the specified configuration $C^M$ of $M$ grains can be similarly defined as

$$\begin{aligned} \langle \theta^{\mathrm{F,S}} \rangle_M(\mathbf{x}) &= \langle \theta^{\mathrm{F,S}} \rangle(\mathbf{x}; C^M) = \frac{1}{n_M^{\mathrm{F,S}} (N-M)!} \times \\ &\quad \times \int \mathrm{d}C^{N-M} \, \theta^{\mathrm{F,S}}(\mathbf{x}; C^N) \chi^{\mathrm{F,S}}(\mathbf{x}; C^N) \, f(C^{N-M}). \end{aligned} \tag{3.11}$$

It can be easily shown that

$$\begin{aligned}\nabla\langle\theta^{\mathrm{F}}\rangle &= \langle\nabla\theta^{\mathrm{F}}\rangle - \frac{\nabla n^{\mathrm{F}}}{n^{\mathrm{F}}}\langle\theta^{\mathrm{F}}\rangle + \\ &+ \frac{1}{n^{\mathrm{F}}}\int_{|\mathbf{x}-\mathbf{r}^1|=R} \mathrm{d}S_{r^1}\,\mathbf{n}_{r^1} f(C^1=\mathbf{r}^1)\,\langle\theta^{\mathrm{F}}\rangle_1(\mathbf{x};C^1),\end{aligned} \tag{3.12}$$

where $\langle\theta^{\mathrm{F}}\rangle_1$ is the average of $\theta^{\mathrm{F}}$ conditional to the presence of a grain centred at $\mathbf{r}^1$, $\mathrm{d}S_{r^1}$ is the differential surface element of this grain and $\mathbf{n}_{r^1}$ the unit normal directed out of the grain. The integration is over all the grains touching the point $\mathbf{x}$. Similarly,

$$\begin{aligned}\nabla\langle\theta^{\mathrm{F}}\rangle_M &= \nabla\langle\theta^{\mathrm{F}}\rangle(\mathbf{x};C^M) \\ &= \langle\nabla\theta^{\mathrm{F}}\rangle_M - \frac{\nabla n^{\mathrm{F}}_M}{n^{\mathrm{F}}_M}\langle\theta^{\mathrm{F}}\rangle_M + \frac{1}{n^{\mathrm{F}}_M}\times \\ &\times\int_{|\mathbf{x}-\mathbf{r}^{M+1}|=R} \mathrm{d}S_{r^{M+1}}\,\hat{\mathbf{n}}_{r^{M+1}} f(C^{M+1})\langle\theta^{F}\rangle_{M+1}(\mathbf{x};C^{M+1}).\end{aligned} \tag{3.13}$$

Expressions (3.12) and (3.13) can also be rearranged in another convenient form as

$$\begin{aligned}\nabla\langle\theta^{\mathrm{F}}\rangle_M &= \langle\nabla\theta^{\mathrm{F}}\rangle_M + \frac{1}{n^{\mathrm{F}}_M}\int_{|\mathbf{x}-\mathbf{r}^{M+1}|=R} \mathrm{d}S_{r^{M+1}}\,\hat{\mathbf{n}}_{r^{M+1}}\, f(C^{M+1})\times \\ &\times[\langle\theta^{\mathrm{F}}\rangle_{M+1}(\mathbf{x};C^{M+1}) - \langle\theta^{\mathrm{F}}\rangle_M(\mathbf{x};C^M)],\end{aligned} \tag{3.14}$$

which yields as a special case

$$\begin{aligned}\nabla\langle\theta^{\mathrm{F}}\rangle &= \langle\nabla\theta^{\mathrm{F}}\rangle + \frac{1}{n^{\mathrm{F}}}\int_{|\mathbf{x}-\mathbf{r}^1|=R} \mathrm{d}S_{r^1}\,\hat{\mathbf{n}}_{r^1}\, f(C^1=\mathbf{r}^1)\times \\ &\times[\langle\theta^{\mathrm{F}}\rangle_1(\mathbf{x};C^1) - \langle\theta^{\mathrm{F}}\rangle(\mathbf{x})].\end{aligned} \tag{3.15}$$

The average of $\theta^{\mathrm{S}}(\mathbf{x};C^N)$ pertaining to the solid phase is given by (3.10) and by virtue of the identity of the grains reduces to

$$\langle\theta^{\mathrm{S}}\rangle(\mathbf{x}) = \frac{1}{n^{\mathrm{S}}(\mathbf{x})}\int_{|\mathbf{x}-\mathbf{q}|\leqslant R} \mathrm{d}^3q\; f(C^1=\mathbf{x})\langle\theta^{\mathrm{S}}\rangle_1(\mathbf{q};\mathbf{x}), \tag{3.16}$$

where $\mathbf{q}$ is a point within the grain. We note that $f(C^1=\mathbf{x})$ is also the local number density of particles by definition.

### 3.2. ENSEMBLE-AVERAGED FLUID-PHASE MOMENTUM BALANCE

Next we derive continuum momentum balance equations for the fluid phase using these concepts that we have developed. For the case of quasi-static limit with no relative velocity between the two phases, as considered here, we ignore the fluid viscosity and inertia contributions. The fluid pressure, at any point $\mathbf{q}$ in the fluid, is then given as

$$\nabla p^{\mathrm{F}}(\mathbf{q};C^N) = \mathbf{0}. \tag{3.17}$$

By conditionally ensemble averaging over (3.17), we have by (3.14) a hierarchy of equations with $N$ members given as

$$\nabla\langle p^{\mathrm{F}}\rangle - \frac{1}{n^{\mathrm{F}}}\int_{|\mathbf{x}-\mathbf{r}^1|=R} \mathrm{d}S_{r^1}\,\hat{\mathbf{n}}_{r^1} f(C^1=\mathbf{r}^1)[\langle p^{\mathrm{F}}\rangle_1(\mathbf{x};\,C^1) - \langle p^{\mathrm{F}}\rangle(\mathbf{x})] = \mathbf{0}. \tag{3.18}$$

$$\nabla\langle p^{\mathrm{F}}\rangle_M - \frac{1}{n^{\mathrm{F}}_M}\int_{|\mathbf{x}-\mathbf{r}^{M+1}|=R} \mathrm{d}S_{r^{M+1}}\,\hat{\mathbf{n}}_{r^{M+1}}\, f(C^{M+1}) \times \\ \times [\langle p^{\mathrm{F}}\rangle_{M+1}(\mathbf{x};\,C^{M+1}) - \langle p^{\mathrm{F}}\rangle_M(\mathbf{x};\,C^M)] = \mathbf{0}, \tag{3.19}$$

with $M = 0, \ldots, N-2$.

To obtain the continuum level description, in principle, we have to solve this hierarchy sequentially to obtain $\langle p^{\mathrm{F}}\rangle_1$. Once $\langle p^{\mathrm{F}}\rangle_1$ is obtained, we have an ensemble-averaged balance of momentum, Equation (3.18), for the fluid phase.

We rearrange this hierarchy of equations in a more desirable form with the first equation given as

$$-\nabla[n^{\mathrm{F}}\langle p^{\mathrm{F}}\rangle] + \langle p^{\mathrm{F}}\rangle\nabla n^{\mathrm{F}} + \\ + \int_{|\mathbf{x}-\mathbf{r}^1|=R} \mathrm{d}S_{r^1}\hat{\mathbf{n}}_{r^1} f(C^1=\mathbf{r}^1)[\langle p^{\mathrm{F}}\rangle_1(\mathbf{x};\,C^1) - \langle p^{\mathrm{F}}\rangle(\mathbf{x})] = \mathbf{0}. \tag{3.20}$$

### 3.3. ENSEMBLE-AVERAGED SOLID-PHASE MOMENTUM BALANCE

Next we direct our attention to the solid phase. The force balance on a single grain located at position $\mathbf{x}$ is

$$\mathbf{0} = -\int_{|\mathbf{x}-\mathbf{q}|=R} \mathrm{d}S_q \mathbf{n}_q\, p^{\mathrm{F}}(\mathbf{q};\,C^N) + \mathbf{F}_{\mathrm{c}}(\mathbf{x};\,C^N), \tag{3.21}$$

where $p^{\mathrm{F}}(\mathbf{q};\,C^N)$ is the fluid pressure at a point $\mathbf{q}$ in the presence of a specific grain configuration $C^N$ with one of the grains located at $\mathbf{x}$ and $\mathbf{F}_{\mathrm{c}}$ is the sum of all contact forces due to other grains in contact with this particular grain.

Upon averaging this force balance according to (3.14), we obtain

$$\mathbf{0} = -f(C^1=\mathbf{x})\int_{|\mathbf{x}-\mathbf{q}|=R} \mathrm{d}S_q\, \mathbf{n}_q\langle p^{\mathrm{F}}\rangle_1(\mathbf{q};\,\mathbf{x}) + \nabla\cdot\mathbf{T}^{\mathrm{S}}_{\mathrm{E}}, \tag{3.22}$$

with $\mathbf{T}^{\mathrm{S}}_{\mathrm{E}}$ denoting the contact stress among the grains and can be computed exactly from the knowledge of $\mathbf{F}_{\mathrm{c}}$ (See Zhuang, Didwania and Goddard 1995, Goddard and Didwania 1997 for rigid grains).

We also note that once we know $\langle p^{\mathrm{F}}\rangle_1$ from the previous section, in principle, we can evaluate the first integral assuming that it does not diverge. However, we take an alternative approach and try to cast this term in a more familiar form.

For further simplicity, we assume that $f(\mathbf{x})$ and $\langle p^{\mathrm{F}}\rangle$ vary slowly on the scale of a single grain. Then, by Taylor series expansion around the grain center,

$$n^{\mathrm{S}}(\mathbf{x}) = \int_{|\mathbf{x}-\mathbf{r}^1| \leqslant R} \mathrm{d}^3 r^1 f(C^1 = \mathbf{r}^1) = v\left[f(\mathbf{x}) + \frac{R^2}{5}\nabla^2 f(\mathbf{x}) + \mathrm{O}(R^4)\right], \tag{3.23a}$$

$$\int_{|\mathbf{x}-\mathbf{q}|=R} \mathrm{d}S_q \mathbf{n}_q \langle p^{\mathrm{F}}\rangle(\mathbf{q}) = v\left[\nabla\langle p^{\mathrm{F}}\rangle(\mathbf{x}) + \frac{\pi R^2}{10}\nabla\nabla^2\langle p^{\mathrm{F}}\rangle(\mathbf{x}) + \mathrm{O}(R^4)\right], \tag{3.23b}$$

where $v$ denotes the volume of a single grain. For small grains, neglecting the higher order terms, we obtain

$$n^{\mathrm{S}}(\mathbf{x}) = v f(\mathbf{x}), \tag{3.24a}$$

and

$$\int_{|\mathbf{x}-\mathbf{q}|=R} \mathrm{d}S_q\, \mathbf{n}_q \langle p^{\mathrm{F}}\rangle(\mathbf{q}) = v\nabla\langle p^{\mathrm{F}}\rangle(\mathbf{x}). \tag{3.24b}$$

With these two results, we write Equation (3.22) as

$$\begin{aligned}&- f(C^1 = \mathbf{x}) \int_{|\mathbf{x}-\mathbf{q}|=R} \mathrm{d}S_q\, \mathbf{n}_q[\langle p^{\mathrm{F}}\rangle_1(\mathbf{q}; \mathbf{x}), -\langle p^{\mathrm{F}}\rangle(\mathbf{q})] \\ &- f(C^1 = \mathbf{x}) \int_{|\mathbf{x}-\mathbf{q}|=R} \mathrm{d}S_q\, \mathbf{n}_q \langle p^{\mathrm{F}}\rangle(\mathbf{q}) + \nabla \cdot \mathbf{T}^{\mathrm{S}}_{\mathrm{E}} = \mathbf{0}.\end{aligned} \tag{3.25}$$

Using Equation (3.24b) and then (3.24a), we find Equation (3.25) reduces to

$$\begin{aligned}&- f(C^1 = \mathbf{x}) \int_{|\mathbf{x}-\mathbf{q}|=R} \mathrm{d}S_q\, \mathbf{n}_q[\langle p^{\mathrm{F}}\rangle_1(\mathbf{q}; \mathbf{x}), -\langle p^{\mathrm{F}}\rangle(\mathbf{q})] \\ &- n^{\mathrm{S}}\nabla\langle p^{\mathrm{F}}\rangle(\mathbf{x}) + \nabla \cdot \mathbf{T}^{\mathrm{S}}_{\mathrm{E}} = \mathbf{0}.\end{aligned} \tag{3.26}$$

The terms under the integral sign in Equations (3.20) and (3.26) look similar and are of opposite sign but do not cancel each other. In Equation (3.26) the integration is carried out over the surface of a particle centred at $\mathbf{x}$, while in Equation (3.20) the integration is over all the particles that touch the point $\mathbf{x}$. The integrand of the term in Equation (3.20) depends strongly on the distance from the particle centre but only weakly on the position $\mathbf{r}^1$ of the centre itself and therefore a Taylor series expansion can be carried out in this variable. For brevity, denote the bracketed quantity in the integrand of Equation (3.20) by $F(\mathbf{x}; \mathbf{r}^1)$ and let $\mathbf{x} = \mathbf{r}^1 + \mathbf{s}$. Then we have (Hinch, 1977)

$$f(\mathbf{r}^1)F(\mathbf{r}^1 + \mathbf{s}; \mathbf{r}^1) = f(\mathbf{x})F(\mathbf{x} + \mathbf{s}; \mathbf{x}) - \mathbf{s} \cdot \nabla_{\mathbf{x}}[f(\mathbf{x})\, F(\mathbf{x} + \mathbf{s}; \mathbf{x})] + \cdots . \tag{3.27}$$

By changing the integration variable $\mathbf{x}+\mathbf{s}$ to $\mathbf{q}$ we find that the first term in the expansion is equal to the integral term in Equation (3.26). The second term in the expansion corresponds to an extra stress term.

Making the appropriate substitution in expression (3.26) by using (3.27), we obtain the two overall momentum balance equations as

*Fluid Phase*:

$$\nabla \cdot \mathbf{T}^{\mathrm{F}} + \mathbf{I}^{\mathrm{F}} = \mathbf{0}. \tag{3.28}$$

*Solid Phase*:

$$\nabla \cdot \mathbf{T}^{\mathrm{S}} + \mathbf{I}^{\mathrm{S}} = \mathbf{0}, \tag{3.29}$$

where

$$\mathbf{T}^{\mathrm{F}} = -n^{\mathrm{F}} \langle p^{\mathrm{F}} \rangle \mathbf{I}, \tag{3.30a}$$

$$\mathbf{I}^{\mathrm{F}} = \langle p^{\mathrm{F}} \rangle \nabla n^{\mathrm{F}} + \int_{|\mathbf{x}-\mathbf{r}^1|=R} \mathrm{d}S_{r^1}\, \hat{\mathbf{n}}_{r^1} f(C^1 = \mathbf{r}^1)[\langle p^{\mathrm{F}} \rangle_1(\mathbf{x}; C^1) - \langle p^{\mathrm{F}} \rangle(\mathbf{x})], \tag{3.30b}$$

$$\mathbf{T}^{\mathrm{S}} = -n^{\mathrm{S}} \langle p^{\mathrm{F}} \rangle [\mathbf{I} - \mathbf{T}^{\mathrm{SA}}] + \mathbf{T}^{\mathrm{S}}_{\mathrm{E}}, \tag{3.30c}$$

$$\begin{aligned} \mathbf{I}^{\mathrm{S}} &= -\mathbf{I}^{\mathrm{F}} = \langle p^{\mathrm{F}} \rangle \nabla n^{\mathrm{S}} \\ &\quad - \int_{|\mathbf{x}-\mathbf{r}^1|=R} \mathrm{d}S_{r^1} \hat{\mathbf{n}}_{r^1}\, f(C^1 = \mathbf{r}^1)[\langle p^{\mathrm{F}} \rangle_1(\mathbf{x}; C^1) - \langle p^{\mathrm{F}} \rangle(\mathbf{x})], \end{aligned} \tag{3.30d}$$

and

$$\mathbf{T}^{\mathrm{SA}} = \frac{R}{v\langle p^{\mathrm{F}} \rangle} \int_{|\mathbf{x}-\mathbf{q}|=R} \mathrm{d}S_q\, \hat{\mathbf{n}}_q\, \hat{\mathbf{n}}_q [\langle p^{\mathrm{F}} \rangle(\mathbf{q}) - \langle p^{\mathrm{F}} \rangle_1(\mathbf{q}; \mathbf{x})]. \tag{3.30e}$$

As mentioned earlier, the integral term in Equations (3.30b) and (3.30d) is over all the particles that touch the point $\mathbf{x}$. In an isotropic porous media this integral term, that is the contribution $[\langle p^{\mathrm{F}} \rangle_1(\mathbf{x}; C^1) - \langle p^{\mathrm{F}} \rangle(\mathbf{x})]$ averaged over all possible particle orientations at point $\mathbf{x}$, is identically zero. Hence, the set of Equations (3.30a)–(3.30e) simplify to

$$\mathbf{T}^{\mathrm{F}} = -n^{\mathrm{F}} \langle p^{\mathrm{F}} \rangle \mathbf{I}, \tag{3.31a}$$

$$\mathbf{I}^{\mathrm{F}} = \langle p^{\mathrm{F}} \rangle \nabla n^{\mathrm{F}}, \tag{3.31b}$$

$$\mathbf{T}^{\mathrm{S}} = -n^{\mathrm{S}} \langle p^{\mathrm{F}} \rangle [\mathbf{I} - \mathbf{T}^{\mathrm{SA}}] + \mathbf{T}^{\mathrm{S}}_{\mathrm{E}}, \tag{3.31c}$$

$$\mathbf{I}^{\mathrm{S}} = -\mathbf{I}^{\mathrm{F}} = \langle p^{\mathrm{F}} \rangle \nabla n^{\mathrm{S}} \tag{3.31d}$$

and

$$\mathbf{T}^{\mathrm{SA}} = \frac{R}{v\langle p^{\mathrm{F}}\rangle} \int_{|\mathbf{x}-\mathbf{q}|=R} \mathrm{d}S_q \, \hat{\mathbf{n}}_q \, \hat{\mathbf{n}}_q [\langle p^{\mathrm{F}}\rangle(\mathbf{q}) - \langle p^{\mathrm{F}}\rangle_1(\mathbf{q}; \mathbf{x})]. \tag{3.31e}$$

In a quasi-static porous media $\langle p^{\mathrm{F}}\rangle$ is the same as the pore pressure $p$ of the macroscopic porous media theory. On comparison with Equations (2.29)–(2.32) as derived from the entropy inequality considerations in the macroscopic approach, we find that the expressions for interaction force and the fluid phase stress, as derived from both approaches are identical. The solids phase stress is also the same from both the approaches when we identify $\mathcal{F}_{\mathrm{S}}$ with $\mathbf{T}^{\mathrm{SA}}$ for the elastic compressible materials. In fact, we have found a micromechanical expression for $\mathcal{F}_{\mathrm{S}}$ and furthermore, we note that our micromechanical stress $\mathbf{T}^{\mathrm{SA}}$ holds for all types of partial solid materials, that is no assumption of any kind, for example, elastic compressible etc.

## 4. Concluding Remarks

In this paper, we have derived a micromechanical basis for Terzhagi's effective stress concept for nondeformable incompressible constituents (Equations (3.31a)–(3.31e) $\mathbf{T}^{\mathrm{SA}} = 0$ for this case). We also derive results to account for compressibility or deformability of the solids phase, leading to non-zero $\mathcal{F}_{\mathrm{S}}$ from the macroscopic and $\mathbf{T}^{\mathrm{SA}}$ from the micromechanical ansatz.

## Acknowledgements

The first author (AKD) is grateful to the University of Essen for offering him a University Gast-Professorship during which most of the work was completed. AKD also thanks the members of the Institut für Mechanik for their kind hospitality during this period and acknowledges the partial support of U.S. National Science Foundation (Grants INT-9605036, CTS-9510121) and U. S. Air Force Office of Sponsored Research (Grant F49620-96-1-02446) towards completion of the present work. The financial support of the Deutsche Forschungsgemeinschaft (DFG) is gratefully acknowledged by the second author.

## References

Biot, M. A.: 1941, General theory of three-dimensional consolidation, *J. Appl. Phys.* **12**, 155–164.

Biot, M. A.: 1955, Theory of elasticity and consolidation for a porous anisotropic solid, *J. Appl. Phys.* **26**, 182–185.

Biot, M. A. and Willis, D. G.: 1957, The elastic coefficients of the theory of consolidation, *J. Appl. Mech.* **24**, 594–601.

Bluhm, J.: 1997, A consistent model for saturated and empty porous media, Habilitation thesis, Universität Essen.

Bluhm, J. and de Boer, R.: 1997, The volume fraction concept in the porous media theory, *Z. Angew. Math. Mech. (ZAMM)* **77**, 563–577.

de Boer, R.: 1996, Highlights in the historical developement of the theory of porous media-towards a consistent macroscopic theory, *Appl. Mech. Rev.* **49**, 201–262.
de Boer, R. and Didwania, A. K.: 1997, The effect of uplift in liquid-saturated porous solids – Karl Terzaghi's contributions and recent findings, *Geotechnique* **47**, 289–298.
Bowen, R. M.: 1980, Incompressible porous media models by use of the theory of mixtures, *Int. J. Engng. Sci.* **20**, 1129–1148.
Bowen, R. M.: 1982, Compressible porous media models by use of the theory of mixtures, *Int. J. Engng. Sci.* **20**, 697–735.
Goddard, J. D. and Didwania, A. K.: 1998, Computations of dilatancy and yieldsurfaces for assemblies of rigid frictional spheres, *Quart. J. Mech. Appl. Math.* **51**(1), 15–43.
Fillunger, P.: 1936, Erdbaumechanik?, Selbstverlag des Verfassers, Wien.
Hinch, E. J.: 1977, An averaged-equation approach to particle interactions in a fluid suspension, *J. Fluid Mech.* **83**, 695–720.
Irving, J. H. and Kirkwood, J. G.: 1950, The statistical mechanical theory of transport processes: IV. The equations of hydrodynamics, *J. Chem. Phys.* **18**, 817–829.
Lade, P. and de Boer, R.: 1997, The concept of effected stress for soil, concrete, and rock, *Geotechnique* **47**, 61–78.
Mow, V. C., Kuei, S. C., Lai, W. M. and Armstrong, C. G.: 1980, Biphasic creep and stress relaxation of articular cartilage in compression: theory and experiments, *J. Biomech. Engng.* **102**, 73–84.
Passman, S. L., Nunziato, J. W. and Walsh, E. K.: 1984, A theory of multiphase mixtures, In: C. Truesdell (ed.), *Rational Thermodynamics*, McGraw-Hill, New York, pp. 286–325.
Sangani, A. S. and Didwania, A. K.: 1993, A dispersed-phase stress tensor in flows of bubbly liquids at large Reynolds number, *J. Fluid Mech.* **248**, 27–54.
Suklje, L.: 1969, *Rheological Aspects of Soil Mechanics*, Wiley Interscience, New York.
Zhuang, X., Didwania, A. K. and Goddard, J. D.: 1995, Simulation of the quasi-static mechanics and scalar transport properties of granular assemblies, *J. Comp. Phys.* **121**, 331–346.

# Wave Dynamics of Saturated Porous Media and Evolutionary Equations

INNA YA. EDELMAN
*United Institute of Physics of the Earth, Russian Academy of Sciences, B. Gruzinskaya 10, Moscow, 123810, Russia*

(Received: 3 July 1997)

**Abstract.** Nonlinear wave dynamics of an elastically deformed saturated porous media is investigated following the Biot approach. Mathematical models under research are the Biot model and its generalization by consideration of viscous stresses inside liquids. Using two-scales and linear WKB methods, the classical Biot system is transformed to a first-order wave equation. To construct the solution of the other system, an asymptotic modified two-scales method is developed. Initial system of equations is transformed to a nonlinear generalized Korteweg–de Vries–Burgers equation for quick elastic wave. Distinctions of wave propagation in the context of the Biot model and its generalization are shown.

**Key words:** porous media, Biot's theory, nonlinear waves, asymptotic methods, evolutionary equations.

## 1. Introduction

Practical geomechanical phenomena as well as numerous experimental data, which show essential nonlinearity of wave propagation in real media, call for a theoretical description of wave processes in saturated porous media [1, 2]. We study nonlinear wave dynamics of an elastically deformed saturated porous media following the classical Biot theory concerning the propagation of elastic waves [3, 4]. Thus, it is assumed that a medium is mixture of interpenetrating solid, liquid or/and gas continua. The mathematical models proposed are formulated as differential conservation laws and closing relations. The first one is the classical Biot model for an isothermal porous medium saturated by liquid, where momentum balances include interphase interaction force with Biot's corrections. The Biot correction for interphase interaction force in momentum balances denotes the deviations from the Darcy law for the case of ultrasound wave propagation. The system is closed by the modified Hook law and the equations of phases state. The other mathematical model is a generalization of the Biot model by consideration of viscous stresses inside a liquid phase [5]. It is formulated as mass, momentum, and energy differential conservation laws, and closing rheological and thermodynamic relations. The model describes thermo-visco-elastically deformed porous medium, consisting of an elastic skeleton, a viscous liquid film which is bound by skeleton, and viscous liquid phase. It is assumed that the liquid phase as well as the solid phase is weakly compressed. Nonlinear effects in saturated porous medium can be stipulated both by the dynamics of the medium and inherent rheological and thermodynamic properties of solid and fluid phases. Indeed, it is considered a nonlinearity due to interphase interaction and due to viscous friction inside the fluid phase.

To construct the solution of the Biot model, slow and quick variables are introduced and linear WKB method is applied [7–9]. For one-dimensional motion, the system is transformed to a first-order wave equation.

To construct the solution for the other proposed system an asymptotic modified two-scale method is developed [5, 6]. It is based on the standard two-scales method and nonlinear WKB method [7–9]. The separation of slow and quick variables leads to the distinguishing of two types of motion: slow background motion and quick oscillating motion. For problems concerning wave processes in porous media this approach is new in principle, but it corresponds to a physical pattern of different motion types in real media [10]. The substitution of the asymptotic solution into initial equation system allows one to obtain the dispersion equation for the definition of the frequencies of the waves. In the case under research, the dispersion equation has nonzero roots, which correspond to the frequencies of the longitudinal waves of the first and second kinds, and frequencies of the transverse waves of different polarization [5, 14]. Using an asymptotic modified two-scales method, a Cauchy problem for the initial system of equations is transformed into a Cauchy problem for a nonlinear evolutionary equation for quick wave amplitudes and an inhomogeneous set of equations for slow background motion. The derived evolutionary equation is a generalized Korteweg–de Vries–Burgers equation for the amplitudes of modulated waves.

## 2. The Biot Model

### 2.1. PROBLEM STATEMENT

Consider a liquid saturated isothermal elastically deformed porous medium. The mass and momentum balances for liquid and solid phases, suggested by Biot [3] can be written in a dimensionless form ($i, j = 1, 2, 3$).

*Mass conservation equations:*

$$\begin{aligned} &\frac{\partial}{\partial t} m\rho_l + \frac{\partial}{\partial x_i} m\rho_l v_{li} = 0, \\ &\frac{\partial}{\partial t}(1-m)\rho_s + \frac{\partial}{\partial x_i}(1-m)\rho_s v_{si} = 0. \end{aligned} \tag{1}$$

Here, $m$ is the porosity, $\rho$ the density and $v$ the velocity vector. Indices l and s are symbols of liquid and solid phases, respectively.

*Momentum conservation equations:*

$$m\rho_l\left[\frac{\partial}{\partial t}+\left(v_{lj},\frac{\partial}{\partial x_j}\right)\right]v_{li}+m\frac{\partial}{\partial x_j}p\delta_{ij}+m^2L(v_{li}-v_{si})=0,$$

$$(1-m)\rho_s\left[\frac{\partial}{\partial t}+\left(v_{sj},\frac{\partial}{\partial x_j}\right)\right]v_{si}-\frac{\partial}{\partial x_j}\sigma_{ij}+ \qquad (2)$$

$$+(1-m)\frac{\partial}{\partial x_j}p\delta_{ij}-m^2L(v_{li}-v_{si})=0.$$

Here, $p$ is the pressure and $\sigma_{ij}$ the effective stress tensor. $L$ is the differential operator:

$$L=1-a\frac{\partial}{\partial t}+b\frac{\partial^2}{\partial t^2}, \qquad (3)$$

which reflects the Biot correction to the Darcy law for the case of ultrasound wave propagation [11]. The constants $a$ and $b$ can be estimated experimentally [11].

*The equations of phases state:*

$$\rho_l=\rho_{l0}[1+\beta_l(p_l-p_0)],$$
$$\rho_s=\rho_{s0}[1-\beta_s(\tfrac{1}{3}\sigma^s_{kk}-\sigma_0)], \qquad (4)$$

where

$$\sigma^s_{ij}=\frac{\sigma_{ij}}{1-m}-p\delta_{ij},$$

$\beta$ is the compressibility coefficient, $\sigma^s_{ij}$ is the true stress tensor inside the solid phase. It is assumed that the liquid as well as the solid phase is weakly compressed.

*The equations for stress and strain tensors:*

$$\sigma_{ij}=Ke_{kk}\delta_{ij}+2G(e_{ij}-\tfrac{1}{3}e_{kk}\delta_{ij})+\beta_sKp\delta_{ij},$$
$$\frac{\partial e_{ij}}{\partial t}=\frac{1}{2}\left(\frac{\partial v_{si}}{\partial x_j}+\frac{\partial v_{sj}}{\partial x_i}\right). \qquad (5)$$

Here, $K$ is the volume elasticity modulus, $G$ the shear modulus and $e_{ij}$ the strain tensor.

The system of Equations (1)–(5) is closed with respect to unknown scalar $m$, $p$, vector $v_{li}$, $v_{si}$, and tensor $\sigma_{ij}$ functions. We consider the following Cauchy problem with initial conditions:

$$m|_{t=0}=m^0, \qquad p\,|_{t=0}=p^0,$$
$$v_{li}|_{t=0}=v^0_{li}, \qquad v_{si}|_{t=0}=v^0_{si}, \qquad \sigma_{ij}|_{t=0}=\sigma^0_{ij}. \qquad (6)$$

### 2.2. ASYMPTOTIC SOLUTION

Let us introduce slow $T$, $X$ and quick $\tau$ variables:

$$T = \epsilon t, \quad X = \epsilon x, \quad \tau = \frac{S(T, X)}{\epsilon}. \tag{7}$$

Here, $S(X, T) = kX - \omega T$, $k$ is a wave vector, $\omega$ a frequency and $\epsilon$ a small parameter. The solution of the Cauchy problem (1)–(6) is sought as the formal series:

$$U(\tau, T, X) = U^{(0)} + \epsilon U^{(1)}(\tau, T, X) + \epsilon^2 U^{(2)}(\tau, T, X) + \cdots, \tag{8}$$

where

$$U(\tau, T, X) = (m, v_{li}, v_{si}, p, \sigma_{ij}), \quad i, j = 1, 2, 3$$

is the vector-function of the desired variables, $U^{(0)} = (m_0, 0, 0, 0, 0)$. Function $U(\tau, T, X)$ is smooth and $2\pi$-periodic with respect to $\tau$.

Below we consider one-dimensional motion ($i, j = 1$). Making the change of variables (7) and substituting (8) into (1)–(5), one can obtain that

$$\hat{L}U^{(1)} = 0, \tag{9}$$

$$\hat{L}U^{(2)} = \Pi. \tag{10}$$

Here, the differential operator $\hat{L} = \hat{L}(\partial/\partial\tau, \partial^2/\partial\tau^2)$ and $\Pi$ depends on $U^{(1)}$ and its derivatives (their lengthy expressions are not reported here).

Integrating (9), one finds that

$$\begin{aligned}
v^{(1)} &\equiv v_{s1}^{(1)} - v_{l1}^{(1)} = \varphi_0(T, X)\cos(\Omega\tau) + \varphi_1(T, X)\sin(\Omega\tau),\\
m^{(1)} &= f_1(T, X) + c_1(\varphi_0(T, X)\cos(\Omega\tau) + \varphi_1(T, X)\sin(\Omega\tau)),\\
v_{l1}^{(1)} &= f_2(T, X) + c_2(\varphi_0(T, X)\cos(\Omega\tau) + \varphi_1(T, X)\sin(\Omega\tau)),\\
p^{(1)} &= f_3(T, X) + c_3(\varphi_0(T, X)\cos(\Omega\tau) + \varphi_1(T, X)\sin(\Omega\tau)),\\
\sigma_{11}^{(1)} &= f_4(T, X) + c_4(\varphi_0(T, X)\cos(\Omega\tau) + \varphi_1(T, X)\sin(\Omega\tau)),
\end{aligned} \tag{11}$$

where the constants $c_i (i = \overline{1, 4})$ are defined during the construction of asymptotic solution. The functions $\varphi_0(T, X)$, $\varphi_1(T, X)$ – the amplitudes, and $f_i(T, X)$ $(i = \overline{1, 4})$ – the background, are unknown.

$2\pi$-periodicity condition for $U(\tau, T, X)$ with respect to $\tau$ has the form $\Omega = 1$ and allows one to obtain a dispersion equation for the definition of wave frequencies $\omega(k)$. For one-dimensional motion (the case under research) a dispersion equation has four nonzero roots which correspond to the frequencies of the longitudinal waves of the first and second kinds of forward and backward directions.

Necessary and sufficient condition for solvability of the system (10) within the class of smooth, periodic with respect to $\tau$ functions is orthogonality condition in $L_2(0, 2\pi)$ of the right-hand side $\Pi$ to the periodic solutions of the equation

$$\hat{L}^* U^* = 0, \tag{12}$$

where $\hat{L}^*$ is the adjoint operator. The orthogonality condition leads to the following system of equations with respect to unknown functions $\varphi_0(T, X)$, $\varphi_1(T, X)$, $f_i(T, X)$:

$$\frac{\partial f}{\partial T} + A_0 \frac{\partial f}{\partial X} = 0, \tag{13}$$

$$\frac{\partial \varphi}{\partial T} + A_1 \frac{\partial \varphi}{\partial X} + A_2(f)\varphi = 0. \tag{14}$$

Here, $f = (f_1, f_2, f_3, f_4)^{\mathrm{T}}$, $\varphi = (\varphi_0, \varphi_1)^{\mathrm{T}}$. The matrices $A_0, A_1, A_2$ depend on the parameters of solid and liquid phases and the frequency and wave vector of the propagating wave. The matrix $A_2$ depends additionally on the functions $f_i(T, X)$. Equation (13) can be written as

$$\frac{\partial f}{\partial T} + P^{-1}\Lambda P \frac{\partial f}{\partial X} = 0, \tag{15}$$

where $\Lambda$ is the diagonal matrix consisting of the eigenvalues $\lambda_i$ $(i = \overline{1, 4})$ of the matrix $A_0$ and $P$ a matrix of similarity transformation.

After the change $F = Pf$, (15) takes the form $(i = \overline{1, 4})$:

$$\frac{\partial F_i}{\partial T} + \lambda_i \frac{\partial F_i}{\partial X} = 0, \tag{16}$$

Equation (16) has the solutions as

$$F_i = F_{0i}(X - \lambda_i T), \tag{17}$$

which correspond to a plane wave propagating with the velocity $\lambda_i$ without shape modification. The functions $F_{0i}$ are defined by the initial conditions (6).

The evolutionary equation (14) describes the amplitude variation of the propagating waves. It is obvious that the solution of Equation (14) attenuates exponentially due to the term $A_2(f)\varphi$.

## 3. The Biot Generalized Model

### 3.1. PROBLEM STATEMENT

Consider the system of equations for liquid and solid phases in the dimensionless form $(i, j = 1, 2, 3)$ [5].

*Mass conservation equations:*

$$\begin{aligned} &\frac{\partial}{\partial t}(1-\alpha)m\rho_1 + \frac{\partial}{\partial x_i}(1-\alpha)m\rho_1 v_{1i} = 0, \\ &\frac{\partial}{\partial t}(\alpha m\rho_\alpha + (1-m)\rho_s) + \frac{\partial}{\partial x_i}(\alpha m\rho_\alpha + (1-m)\rho_s)v_{si} = 0. \end{aligned} \tag{18}$$

Here, $m$ is the porosity, $\rho$ the density, $v$ the velocity vector and $\alpha$ the volume fraction of liquid film. Indices l, s, $\alpha$ are symbols of liquid, solid phases, and liquid film, respectively. The porous medium skeleton and the bound liquid form an effective visco-elastic solid phase which shows the elastic properties of the skeleton and the viscous properties of the liquid film. The skeleton and bound liquid have the same velocity and pressure.

*Momentum conservation equations:*

$$\begin{aligned}
&(1-\alpha)m\rho_{\mathrm{l}}\left[\frac{\partial}{\partial t}+\left(v_{\mathrm{l}j},\frac{\partial}{\partial x_j}\right)\right]v_{\mathrm{l}i} - \\
&\quad -(1-\alpha)m\frac{\partial}{\partial x_j}P^{\mathrm{l}}_{ij}+m^2(1-\alpha)^2(v_{\mathrm{l}i}-v_{\mathrm{s}i})=0, \\
&[\alpha m\rho_\alpha+(1-m)\rho_{\mathrm{s}}]\left[\frac{\partial}{\partial t}+\left(v_{\mathrm{s}j},\frac{\partial}{\partial x_j}\right)\right]v_{\mathrm{s}i}-\frac{\partial}{\partial x_j}\sigma_{ij} - \\
&\quad -[1-(1-\alpha)m]\frac{\partial}{\partial x_j}P^{\mathrm{l}}_{ij}-m^2(1-\alpha)^2(v_{\mathrm{l}i}-v_{\mathrm{s}i})=0.
\end{aligned} \tag{19}$$

Here, $P^{\mathrm{l}}_{ij}$ is the stress tensor inside a liquid phase:

$$P^{\mathrm{l}}_{ij}=-p\delta_{ij}+\epsilon^2\nu_{\mathrm{l}}\left[\frac{\partial v_{\mathrm{l}i}}{\partial x_j}+\frac{\partial v_{\mathrm{l}j}}{\partial x_i}-\frac{2}{3}\frac{\partial v_{\mathrm{l}k}}{\partial x_k}\delta_{ij}\right], \tag{20}$$

where $p$ is the pressure, $\nu$ the viscosity and $\sigma_{ij}$ the effective stress tensor.

*Energy conservation equation (it is assumed that the temperatures of solid and liquid phases are equal):*

$$\begin{aligned}
&(1-\alpha)m\rho_{\mathrm{l}}\left[\frac{\partial}{\partial t}+\left(v_{\mathrm{l}i},\frac{\partial}{\partial x_i}\right)\right]E_{\mathrm{l}}+(1-m)\rho_{\mathrm{s}}\left[\frac{\partial}{\partial t}+\left(v_{\mathrm{s}i},\frac{\partial}{\partial x_i}\right)\right]E_{\mathrm{s}} + \\
&\quad +\alpha m\rho_\alpha\left[\frac{\partial}{\partial t}+\left(v_{\mathrm{s}i},\frac{\partial}{\partial x_i}\right)\right]E_\alpha-(1-\alpha)mP^{\mathrm{l}}_{ij}\frac{\partial v_{\mathrm{l}i}}{\partial x_j} - \\
&\quad -[\sigma_{ij}+(1-(1-\alpha)m)P^{\mathrm{l}}_{ij}]\frac{\partial v_{\mathrm{s}i}}{\partial x_j}-m^2(1-\alpha)^2|v_{\mathrm{l}}-v_{\mathrm{s}}|^2 - \\
&\quad -\epsilon^2\frac{\partial}{\partial x_i}[(1-\alpha)m\lambda_{\mathrm{l}}+(1-m)\lambda_{\mathrm{s}}+\alpha m\lambda_\alpha]\frac{\partial}{\partial x_i}T=0.
\end{aligned} \tag{21}$$

Here, $T$ is the temperature, $E$ the internal energy, and $\lambda$ the heat conductivity coefficient.

*The equations of phases state:*

$$\begin{aligned}
\rho_{\mathrm{l}} &= \rho_{\mathrm{l}0}[1+\beta_{\mathrm{l}}(p_{\mathrm{l}}-p_0)-\varphi_{\mathrm{l}}(T-T_0)], \\
\rho_\alpha &= \rho_{\alpha 0}[1-\beta_\alpha(\tfrac{1}{3}\sigma^{\mathrm{s}}_{kk}-\sigma_0)-\varphi_\alpha(T-T_0)], \\
\rho_{\mathrm{s}} &= \rho_{\mathrm{s}0}[1-\beta_{\mathrm{s}}(\tfrac{1}{3}\sigma^{\mathrm{s}}_{kk}-\sigma_0)-\varphi_{\mathrm{s}}(T-T_0)],
\end{aligned} \tag{22}$$

where

$$\sigma_{ij}^{\mathrm{s}} = \frac{\sigma_{ij}}{1-(1-\alpha)m} + P_{ij}^{\mathrm{l}},$$

$\beta$ is the compressibility coefficient, $\varphi$ the heat expansion coefficient and $\sigma_{ij}^{\mathrm{s}}$ the true stress tensor inside a solid phase. It is assumed that the liquid as well as the solid phase is weakly compressed.

*The equations for stress and strain tensors and displacement vector:*

$$\begin{aligned}\frac{\partial \sigma_{ij}}{\partial t} &= K\delta_{ij}\frac{\partial e_{kk}}{\partial t} + 2G\frac{\partial}{\partial t}\left[e_{ij} - \frac{e_{kk}\delta_{ij}}{3}\right] + \beta_{\mathrm{s}}\delta_{ij}K\frac{\partial p}{\partial t} - \varphi_{\mathrm{s}}\delta_{ij}K\frac{\partial T}{\partial t} + \\ &\quad + \epsilon^2 \alpha m \nu_\alpha \frac{\partial}{\partial t}\left[\frac{\partial v_{\mathrm{s}i}}{\partial x_j} + \frac{\partial v_{\mathrm{s}j}}{\partial x_i} - \frac{2}{3}\frac{\partial v_{\mathrm{s}k}}{\partial x_k}\delta_{ij}\right],\end{aligned} \tag{23}$$

$$\frac{\partial u_i}{\partial t} - v_{\mathrm{s}i} = 0, \tag{24}$$

$$\frac{\partial e_{ij}}{\partial t} - \frac{1}{2}\left(\frac{\partial v_{\mathrm{s}i}}{\partial x_j} + \frac{\partial v_{\mathrm{s}j}}{\partial x_i}\right) = 0, \tag{25}$$

where $K$ is the volume elasticity modulus, $G$ the shear modulus, $e_{ij}$ the strain tensor and $u$ the displacement vector. Equation (23) is the modified Hook law for saturated thermoelastic porous media [4].

*Thermodynamic relations:*

$$\begin{aligned}\rho_{\mathrm{l}}\mathrm{d}E_{\mathrm{l}} &= \rho_{\mathrm{l}}C_{\mathrm{l}}\,\mathrm{d}T + \frac{p}{\rho_{\mathrm{l}}}\,\mathrm{d}\rho_{\mathrm{l}} - \varphi_{\mathrm{l}}T\,\mathrm{d}p,\\ \rho_{\mathrm{s}}\,\mathrm{d}E_{\mathrm{s}} &= \rho_{\mathrm{s}}C_{\mathrm{s}}\,\mathrm{d}T + \sigma_{ij}^{\mathrm{s}}\,\mathrm{d}e_{ij} + \tfrac{1}{3}\varphi_{\mathrm{s}}T\,\mathrm{d}\sigma_{kk}^{\mathrm{s}},\\ \rho_\alpha\,\mathrm{d}E_\alpha &= \rho_\alpha C_\alpha\,\mathrm{d}T - \frac{\sigma_{kk}^{\mathrm{s}}}{3\rho_\alpha}\,\mathrm{d}\rho_\alpha + \tfrac{1}{3}\varphi_\alpha T\,\mathrm{d}\sigma_{kk}^{\mathrm{s}}.\end{aligned} \tag{26}$$

Here, $C$ is the heat capacity.

Small parameter $\epsilon$ is obtained as a result of dimensional analysis done according to the standard procedure. It is defined by characteristic values of liquid viscosity, modulus of elasticity, density and medium permeability: $\epsilon = \nu_l/\sqrt{K_0\rho_0\kappa}$ [5]. It is easy to show that $\epsilon^2 = \mathrm{Re}_{\mathrm{a}}^{-1}$($\mathrm{Re}_{\mathrm{a}}$ is the acoustic Reynolds number). This small parameter accounts for the relationship between viscous and inertial effects and defines the scale of an appearance of the dispersion properties.

The principal part of the system under research is hyperbolic. The system of equations is closed with respect to unknown scalar $m$, $p$, $T$, vector $v_{\mathrm{l}i}$, $v_{\mathrm{s}i}$, $u_i$, and tensor $\sigma_{ij}$, $e_{ij}$ functions. For the system (18)–(26) we consider the following Cauchy problem with initial conditions:

$$u_i|_{t=0} = u_i^0, \quad m|_{t=0} = m^0, \quad p|_{t=0} = p^0, \quad v_{si}|_{t=0} = v_{si}^0, \quad v_{li}|_{t=0} = v_{li}^0,$$

$$T|_{t=0} = T^0, \quad \sigma_{ij}|_{t=0} = \sigma_{ij}^0, \quad e_{ij}|_{t=0} = \frac{1}{2}\left[\frac{\partial u_i^0}{\partial x_j} + \frac{\partial u_j^0}{\partial x_i}\right]. \tag{27}$$

### 3.2. ASYMPTOTIC SOLUTION

The construction of the solution of the Cauchy problem (18)–(27) is carried out by an asymptotic modified two-scales method. It is based on the classical two-scales method and nonlinear WKB method [7–9]. The method developed assumes a separation of slow and quick variables and leads to a distinguishing of two types of motion of different scales. Such an approach corresponds to physical facts concerning the existence of two types of motion in saturated porous media: slow macromotion and quick oscillating motion (10). An asymptotic solution is sought as a sum of two functions:

$$U(\tau, x, t) = U_b(x, t) + W(\tau, x, t), \tag{28}$$

where

$$U(\tau, x, t) = (m, v_{li}, v_{si}, p, \sigma_{ij}, e_{ij}, T, u_i)$$

is the vector-function of the desired variables ($i, j = 1, 2, 3$).

Function $U_b(x, t)$ is the slow background. It depends on slow variables $x, t$ and characterizes the nonperiodic medium motion.

Function $W(\tau, x, t)$ describes quick periodic motions and depends on $x$, $t$, and, in addition on the quick variable $\tau = S(x, t)/\epsilon$, where $S$ is a phase. Function $W(\tau, x, t)$ is smooth in variable totality and $2\pi$-periodic, with zero mean with respect to $\tau$. Thus, slow background $U_b(x, t)$ is the mean of the solution $U$ of the initial equation system.

First, consider slow medium motion. Function $U_b(x, t)$ is sought in the form of

$$U_b(x, t) = U_b^{(0)}(x, t) + \epsilon U_b^{(1)}(x, t) + \epsilon^2 U_b^{(2)}(x, t) + O(\epsilon^3). \tag{29}$$

The construction of the background solution is carried out by the substitution of (28) into (18)–(26). It is easy to see that any constant may be zero approximation $U_b^{(0)}$ of the background solution. For the first approximation of the background solution (mod $O(\epsilon)$), one obtains a homogeneous hyperbolic system of equations. In the problem under research, Cauchy data (27) mod $O(\epsilon)$ are equal to zero. Therefore, this hyperbolic system has a unique zero solution. Further, for $U_b^{(2)}(x, t)$ (the solution mod $O(\epsilon^2)$) one obtains hyperbolic system of equations with 'sources'. These 'sources' are integral terms which are the mean values of nonlinear terms in the initial system. For a linear problem, the condition of a zero mean would lead to a homogeneous system of equations which has a trivial solution only. For the nonlinear case under consideration, the system for $U_{\mathrm{b}}^{(2)}(x,t)$ has a nontrivial solution even for zero Cauchy data (these Cauchy data are the consequence of the initial conditions (27)). The solution of the derived inhomogeneous system of equations explains an appearance of slow nonperiodic background motion as a result of averaged contribution of quick elastic waves propagating through porous medium [6].

Now consider a quick medium motion. Function $W(\tau,x,t)$ is sought in the form of

$$W(\tau,x,t)=\epsilon H W^{(1)}(\tau,x,t)+\epsilon^2(W^{(2)}(\tau,x,t)+Q(x,t,\epsilon))+\mathrm{O}(\epsilon^3). \quad (30)$$

Here, $H(x,t)$ is a null-vector of matrix $A$:

$$A(t,x,S_t,S_x)\cdot H(x,t)=0, \quad (31)$$

where $A$ is the symbol of linearized operator $A(t,x,\partial/\partial t,\nabla_x)$ of the initial system of equations on the background $U_{\mathrm{b}}^{(0)}$. $Q$ is an infinitely differentiated bounded function [12].

Upon substituting (28) into the initial system (18)–(26), one obtains the dispersion equation, which governs the frequencies of the waves under research and defines the form of function $W^{(2)}(\tau,x,t)$. The dispersion equation has the form:

$$\operatorname{Det} A(t,x,-\omega,k)=0, \quad (32)$$

where the derivatives $S_t$ and $S_x$ were denoted by $-\omega$ and $k$, respectively; $\omega$ is a frequency and $k$ a wave vector. A null-vector $H(x,t)$ corresponds to the frequency $\omega(x,t,k)$,see Equation (31), and its coordinates refer to the desired functions.

The Hamilton–Jacobi equation for a wave phase $S(x,t)$ follows from (32):

$$\begin{aligned}&\frac{\partial S}{\partial t}+\omega(x,t,\nabla_x S)=0,\\&S|_{t=0}=S_0(x).\end{aligned} \quad (33)$$

Hereafter, we will consider for simplicity the case of zero background velocities of solid and liquid motions: $v_{\mathrm{s}i}^{(0)}=v_{\mathrm{l}i}^{(0)}=0$. In this case, the dispersion equation (32) has eight nonzero roots $\omega_i(x,t,k)$, which correspond to the frequencies of longitudinal waves of the first and second kinds, and multiple frequencies of the transverse waves (of forward and backward directions) [14].

As a result of the construction of the asymptotic solution, Cauchy problem (18)–(27) for the initial system of equations is transformed to a Cauchy problem for a nonlinear evolutionary equation for quick waves at second approximation. The derived equation is the generalized Korteweg–de Vries–Burgers equation. It describes a propagation of the modulated waves:

$$\frac{dW^{(1)}}{dt_A} + a_1\frac{\partial^2 W^{(1)}}{\partial\tau^2} + a_2 W^{(1)}\frac{\partial W^{(1)}}{\partial\tau} + a_3 W^{(1)} +$$
$$+\epsilon\left[a_4\frac{\partial^3 W^{(1)}}{\partial\tau^3} + a_5 W^{(1)}\frac{\partial^2 W^{(1)}}{\partial\tau^2} + a_6\left(W^{(1)}\right)^2\frac{\partial W^{(1)}}{\partial\tau} +\right.$$
$$\left. + a_7\left(W^{(1)}\right)^2 + a_8\left(\frac{\partial W^{(1)}}{\partial\tau}\right)^2 + a_9\frac{\partial W^{(1)}}{\partial\tau}\right] = 0, \tag{34}$$

$$W^{(1)}|_{t=0} = W_0^{(1)}.$$

Here, the operator $d/dt_A$ is the total derivative along the characteristics of Equation (33), function $W^{(1)}(\tau, x, t)$ is an amplitude. The coefficients $a_i (i = \overline{1, 9})$ are defined during the construction of the asymptotic solution. They are complicated combinations of medium parameters, the frequency and wave vector of the propagating wave, and parameters of equilibrium background medium state $U_b^{(0)}$. Solving Equation (34) and multiplying function $W^{(1)}$ by the corresponding coordinate of null-vector $H$, one can obtain the distribution of the amplitudes of the desired functions: porosity, solid and liquid phase motion velocities, pressure, and others. The solutions of Equation (34), that have been constructed [14], show the existence of a few regimes of wave propagation, the realization of which depends mainly on the relation between the coefficients of the second and third derivatives in the equation: the diffusive regime with monotonically decreasing amplitude and oscillating regime.

It is clear that there is coupling of different types of motion: the propagation of elastic waves is defined significantly by the parameters of slow background motion, and an appearance of background motion is stipulated by the consideration of averaged contribution of quick nonlinear waves.

*Remark.* The model (18)–(26) can be easily extended to the case when a porous medium is saturated by liquid and gas [6]. Consideration of the pressure difference in the liquid and gas phases, i.e., the capillary pressure as

$$(P_{ij}^{l} - P_{ij}^{g})n_j = \gamma J(\theta)n_i, \quad \gamma = \eta\sqrt{\frac{\kappa}{m}},$$

where $P_{ij}^{g}$ is the stress tensor inside a gas phase, $\eta$ the interphase tension coefficient, $\kappa$ the medium permeability, $J(\theta)$ the dimensionless Leverett function [4], $\mathbf{n}$ the unit normal vector leads to two additional nonzero roots of Equation (32). These roots correspond to the frequencies of the longitudinal wave (forwardly and backwardly directed), which we will call the wave of interphase tension [6]. If one assumes that there is no pressure difference in the liquid and gas then the frequency of these waves becomes zero. The waves of interphase tension are very slow (the frequency is two to three orders lower than the frequencies of the longitudinal waves of the first and second kinds, respectively). The experimental data [13] clearly confirms the existence of this new slow mode. It should be noted that consideration of two-phase saturation of a porous medium does not result in the modification of the structure of evolutionary equation (34) but leads to the change of the coefficients of the Equation (34). Indeed, in contrast to one-phase saturation of porous medium, here the shock wave (kink) regimes of propagation for the longitudinal waves of the first and second kinds and interphase tension waves proved possible. These soliton-like solutions occur either in transition from a diffusive regime or as a mixed solition-oscillating regime [6].

## 4. Conclusions

In this paper the results concerning a new approach for the investigation of wave processes in saturated porous media are presented. This approach assumes the distinguishing of quick and slow medium motions and corresponds to a physical pattern of different motion types in real media [10].

1. In the context of both the Biot model and its generalization, equation system of nonlinear wave dynamics is reduced to an evolutionary equation for the wave amplitudes. It has been possible due to the application of linear WKB method (for the Biot classical model) and modified two-scales method (for the Biot generalized model). The solutions of the derived evolutionary equations completely govern possible regimes of wave propagation in saturated porous media.
2. An evolutionary equation for the Biot classical model is a first-order wave equation as (14). Its solutions are exponentially attenuating waves, which propagate on slow background. The background moves at a constant velocity without shape modification.
3. An evolutionary equation for the Biot modified model is a generalized nonlinear Korteweg–de Vries–Burgers equation as (34). This equation has a few types of solutions, namely diffusive and oscillating solution [14].
4. The Biot classical model would not have been sufficient for the description of nonlinear wave propagation in real media. On the other hand, experimental facts of waves frequency transformation and an appearance of oscillating waves of the envelopes [1, 2] are explained qualitatively on the basis of the Biot generalized model and nonlinear Korteweg–de Vries–Burgers equation.
5. In the context of the Biot generalized model, the structure of the derived Korteweg–de Vries–Burgers equation does not change depending on consideration of non-isothermal effects or/and two-phase saturation of porous media by liquid and gas. The modification of a generalized model by taking into account these effects leads to the change of the coefficients of the equation. It should be emphasized that the type of equation is defined by the presence of high space derivatives in the initial system. Indeed, in the case investigated, the type of the derived evolutionary equation is defined by the consideration of viscous stresses in the liquid phase.

## References


1. Vilchinska, N. and Dzilna, A.: Resonance frequencies in sands, *Geotec. Ecol. UNICONE Proc.* **1**(1) (1992), 31–35.
2. Plona, T. J.: Observation of a second bulk compressional wave in a porous medium at ultrasonic frequencies, *Appl. Phys. Lett.* **36**(4) (1980), 259–261.
3. Biot, M. A.: Theory of propagation of elastic waves in a fluid-saturated porous solid: I. Low frequency range, and II. High frequency range, *J. Acoust. Soc. Amer.* **28** (1956), 168–186.
4. Nikolaevskiy, V. N.: *Mechanics of Porous and Fractured Media*, World Scientific, Singapore, 1990.
5. Maksimov, A. M. and Radkevich, E. V.: On modulated waves in Biot–Nikolaevskiy model, *Doklady Akademii Nauk (Physics Doklady)* **38**(10) (1993), 416–418 (translated from Russian).
6. Edelman, I. Ya.: Asymptotic research of nonlinear wave processes in saturated porous media, *Int. J. Nonlinear Dynam.* **13** (1997), 83–98.
7. Dobrokhotov, S. Yu. and Maslov V. P.: Multiphase asymptotics of nonlinear partial differential equations with a small parameter, *Sov. Sci. Rev.* **3** (1982), 221–311.
8. Nayfeh, A. H.: *Perturbations Methods*, Wiley, New York, 1973.
9. Whitham, G. B.: *Linear and Nonlinear Waves*, Wiley, New York, 1974.
10. Kuznetsov, O. L. and Efimova, S. A.: *An Application of Ultrasound in Oil Industry*, Nedra, Moscow, 1983 (in Russian).
11. Biot, M. A.: Generalized theory of acoustic propagation in porous dissipative media, *J. Acoust. Soc. Amer.* **34**(9) (1962), 1254–1264.
12. Maslov, V. P.: *Resonance Processes in the Wave Theory and Self-Focalization*, MIEM, Moscow, 1983.
13. Nagy, P. B.: Observation of a new surface mode on a fluid-saturated permeable solid, *Appl. Phys. Lett.* **60**(22) (1992), 2735–2737.
14. Maksimov, A. M., Radkevich, E. V. and Edelman, I. Ya.: Generation of oscillating envelopes for waves propagating in saturated porous rock media, *Doklady Akademii nauk (Earth Science Sections)* **345** (1995), 108–114 (translated from Russian).

# Thermo-Chemo-Electro-Mechanical Formulation of Saturated Charged Porous Solids

J. M. HUYGHE and J. D. JANSSEN
*Department of Mechanical Engineering, Eindhoven University of Technology, Eindhoven, The Netherlands*

(Received: 4 July 1997; in final form: 10 December 1997)

**Abstract.** A thermo-chemo-electro-mechanical formulation of quasi-static finite deformation of swelling incompressible porous media is derived from a mixture theory including the volume fraction concept. The model consists of an electrically charged porous solid saturated with an ionic solution. Incompressible deformation is assumed. The mixture as a whole is assumed locally electroneutral. Different constituents following different kinematic paths are defined: solid, fluid, anions, cations and neutral solutes. Balance laws are derived for each constituent and for the mixture as a whole. A Lagrangian form of the second law of thermodynamics for incompressible porous media is used to derive the constitutive restrictions of the medium. The material properties are shown to be contained in one strain energy function and a matrix of frictional tensors. A principle of reversibility results from the constitutive restrictions. Existing theories of swelling media should be evaluated with respect to this principle.

**Key words:** mixture theory, biology, hydrogel, shale, polymer.

## Nomenclature

| | |
|---|---|
| $a, A$ | scalar. |
| $\mathbf{a}$ | vector. |
| $\mathbf{A}$ | tensor |
| $\underline{a}$ | column. |
| $\underline{\mathbf{A}}$ | matrix. |
| $\mathbf{b}^\alpha$ | body force of constituent $\alpha$ per unit volume constituent. |
| $\mathbf{b}$ | body force of the mixture per unit mixture volume. |
| $\mathbf{B}^{\beta\gamma}$ | frictional tensor between constituents $\beta$ and $\gamma$. |
| $C^\alpha$ | concentration of constituent $\alpha$ per unit current mixture volume. |
| $\mathbf{E}$ | Green strain tensor of the solid. |
| F | Faraday's constant. |
| $\mathbf{F}$ | deformation gradient tensor of the solid. |
| $J$ | ratio of current mixture volume over initial mixture volume. |
| $n^\alpha$ | ratio of current volume of constituent $\alpha$ over current mixture volume. |
| $N^\alpha$ | ratio of current volume of constituent $\alpha$ over initial mixture volume. |
| $p$ | pressure. |
| $\mathbf{q}^\alpha$ | heat flux of constituent $\alpha$. |
| $\mathbf{q}$ | an energy flux of the mixture (23). |
| $r^\alpha$ | heat supply of constituent $\alpha$ per unit volume constituent. |
| $r$ | heat supply of the mixture per unit mixture volume. |

| | |
|---|---|
| $\mathbf{v}^{\alpha}$ | velocity of constituent $\alpha$. |
| $\mathbf{v}^{\alpha s}$ | Lagrangian velocity of constituent $\alpha$ relative to the solid. |
| $\mathbf{v}^{qs}$ | Lagrangian form of energy flux $\mathbf{q}$. |
| $\overline{V}^{\alpha}$ | molar volume of constituent $\alpha$. |
| $W$ | Helmholtz free energy per unit initial mixture volume. |
| $z^{\alpha}$ | valence of constituent $\alpha$. |
| $\epsilon^{\alpha}$ | internal energy of constituent $\alpha$ per unit volume constituent. |
| $\epsilon$ | internal energy of the mixture per unit volume. |
| $\eta^{\alpha}$ | entropy of constituent $\alpha$ per unit volume constituent. |
| $\eta$ | entropy of the mixture per unit volume. |
| $\eta_0$ | entropy of the mixture per unit initial mixture volume. |
| $\mu^{\alpha}$ | electrochemical potential of constituent $\alpha$ per unit volume constituent. |
| $\pi^{\alpha}$ | momentum interaction of constituent $\alpha$. |
| $\Psi^{\alpha}$ | Helmholtz free energy of constituent $\alpha$ per unit volume constituent. |
| $\psi^{\alpha}$ | Helmholtz free energy of constituent $\alpha$ per unit mixture volume. |
| $\sigma^{\alpha}$ | partial Cauchy stress of constituent $\alpha$. |
| $\theta^{\alpha}$ | temperature of constituent $\alpha$. |
| $\theta$ | temperature of the mixture. |
| $\nabla$ | gradient operator in current configuration. |
| $\nabla_0$ | gradient tensor in initial configuration. |

## 1. Introduction

Many biological, mineral and synthetic porous media exhibit swelling or shrinking when in contact with changing salt concentrations. This phenomenon, observed in clays, shales, cartilage and gels, is caused by electric charges fixed to the solid, counteracted by corresponding charges in the fluid. These charges result in a variety of features, including swelling, electro-osmosis, streaming potentials and streaming currents. In physiology, quantitative understanding of both intracellular and extracellular events requires correct modelling of these phenomena in deforming tissue. At least four components are involved in the swelling mechanics: a solid; a fluid; anions; and cations. Lai *et al.* [7] have realized a major breakthrough by developing a triphasic theory for soft hydrated tissue. The three phases involved are the solid, the fluid and the ionic phase. They have applied the theory to cartilage while neglecting geometric nonlinearities. They have verified the theory for one-dimensional equilibrium results of swelling. Sherwood [8], working in the field of petroleum engineering, has presented a triphasic model of shale poroelasticity which, unlike the theory of Lai *et al.* does not include electrical charges. Snijders *et al.* [10] have developed a family of finite elements for two-dimensional, three-dimensional and axisymmetric analysis of finite deformation of triphasic media. This novel finite element software features fixed charge density as one of its inputs. However, they use it only to model Donnan osmotic swelling and ignore both electrical current and electric potential fields within the medium. Oomens *et al.* [14] have performed swelling pressure experiments and confined swelling and compression experiments on an acrylamide acrylic acid copolymer gel and have found the gel properties analoguous to properties of soft charged hydrated tissue as measured by Drost *et al.* [13]. The model of Snijders

*Table I.* Direct and coupled flow phenomena in charged porous media

| Flow/force | Pressure | Chemical potential | Electric potential | Temperature |
|---|---|---|---|---|
| Fluid | filtration | osmosis | electro-osmosis | thermo-osmosis |
| Solute | ultrafiltration | diffusion | electroforesis | thermal diffusion |
| Current | streaming current | diffusion current | electric conduction | thermo-electricity |
| Heat | isothermal heat transfer | Dufour effect | Peltier effect | thermal conduction |

*et al.* has been unable to fit the experimental data of either of the two materials. Huyghe and Janssen [23] have developed a finite deformation quadriphasic theory which includes both electric current and electric potential, and which has turned out to fit experimental data of confined swelling and compression reasonably well, using material parameters consistent with other experimental data [24].

The key role of swelling in the behaviour of shales [17] and the associated problems of bore hole instability in shale drilling force petroleum engineers to develop continuum models of swelling shales. Numerous difficulties are associated with *in situ* measurement of relevant quantities around the borehole and with taking adequate samples of shale for subsequent testing in the laboratory. Therefore, there is a need for advanced models able to translate the experimental data available into predictions of the *in situ* situation. Unlike biological tissues, the shale is exposed to high overburden loads – in the order of hundreds of atmospheres – causing swelling forces other than osmosis to play a dominant role. Heidug and Wong [15] have developed a triphasic model which includes hydration forces in addition to osmosis, and have applied it to shales.

Table I gives an overview of direct and coupled flow phenomena induced by pressure, chemical potential, electrical potential and temperature gradients in charged porous media [16].

In this paper, we present a theory of swelling incompressible charged porous media. In order to simplify the mathematics as much as possible, a Lagrangian form of the entropy inequality [18] is used which leads to equations consistent with Biot's porous media theories in a more straightforward way than the more familiar Eulerian approach of Bowen [5]. The saturation and electroneutrality conditions are introduced by means of two Lagrange multipliers; the latter is physically interpreted as an electrical potential, the former as a pressure.

## 2. Theory

We assume all constituents $\alpha$ intrinsically incompressible, that is if one adds a mass $m$ of constituent $\alpha$ to the mixture, the added volume of the mixture is always proportional to the mass $m$, irrespective of the value $m$, of the initial composition of the mixture or of the state of deformation. The proportionality constant, the intrinsic density

$$\rho_i^\alpha = \frac{\rho^\alpha}{n^\alpha}, \tag{1}$$

is constant. $\rho^\alpha$ represents the apparent density of the $\alpha$th constituent. $n^\alpha$ is the volume fraction of the $\alpha$th constituent. We use the word incompressible in the sense used by Bowen [5]: deformation of the constituents does not result in changes of intrinsic densities, nor do changes in mixing ratios result in changes of intrinsic densities. Although it is more common to express solute quantities in molar or mass values than in volume values, we will use volume fractions throughout this paper, to reduce the number of symbols introduced. Because of incompressibility, the ratio between mass and volume is constant, and the two quantities are interchangeable in the equations. Excluding mass transfer between constituents, the mass balance of each constituent is then written as

$$\frac{\partial n^\alpha}{\partial t} + \nabla \cdot (n^\alpha \mathbf{v}^\alpha) = 0, \tag{2}$$

in which $\mathbf{v}^\alpha$ is the velocity of constituent $\alpha$ and $\nabla$ the Eulerian gradient operator. As we assume saturation, we find

$$\sum_\alpha n^\alpha = 1. \tag{3}$$

We will agree that summation over $\alpha$ is taken over all constituents and summation over $\beta$ or $\gamma$ is taken over all constituents but the solid. Differentiation of Equation (3) with respect to $t$ and substitution of the mass balance equations (2) yields the mass balance of the mixture

$$\nabla \cdot \mathbf{v}^{\mathrm{s}} + \sum_\beta \nabla \cdot [n^\beta (\mathbf{v}^\beta - \mathbf{v}^{\mathrm{s}})] = 0. \tag{4}$$

The deformation gradient tensor $\mathbf{F}$ maps an infinitesimal material line segment in the initial state of the solid onto the corresponding line segment in the current state. The relative volume change from the initial to the current state is the determinant of the deformation gradient tensor $J = \det \mathbf{F}$. If we introduce volume fractions

$$N^\alpha = J n^\alpha \tag{5}$$

per unit initial volume, we can rewrite the mass balance equation (2) as follows:

$$\frac{\mathrm{D}^{\mathrm{s}} N^\alpha}{\mathrm{D}t} + J \nabla \cdot [n^\alpha (\mathbf{v}^\alpha - \mathbf{v}^{\mathrm{s}})] = 0 \tag{6}$$

when using the identity

$$\frac{\mathrm{D}^{\mathrm{s}}}{\mathrm{D}t} J = J \nabla \cdot \mathbf{v}^{\mathrm{s}}. \tag{7}$$

In Equations (6) and (7), $D^s/Dt = \partial/\partial t + \mathbf{v}^s \cdot \nabla$ is the time derivative for an observer moving with the solid. The electrostatic interactions between ions and the charged solid are accounted for by means of an electroneutrality condition. Disturbances of charge neutrality are corrected in such media in a time lapse in the order of the quotient of the dielectric constant over the electrical conductivity of the medium. The order of magnitude of this time constant is 1 ns [11]. As we are not concerned with phenomena faster than 1 μs, we assume electroneutrality to be valid at all times. Magnetic fields generated by the electric fluxes through the medium are considered negligible. The electroneutrality condition requires that

$$C^{fc} + \sum_{\beta} z^{\beta} C^{\beta} = 0, \tag{8}$$

in which $z^{\beta}$ are the valences, $C^{fc}$ the fixed charge density per unit initial mixture volume and $C^{\beta}$ the current molar concentration of mobile ions in the solvent per unit initial mixture volume

$$C^{\beta} = \frac{N^{\beta}}{\overline{V}^{\beta}}, \tag{9}$$

in which $\overline{V}^{\beta}$ are the partial molar volumes of the solvent and solutes. Equation (8) is rewritten in a more convenient differentiated form, using the fixity of the fixed charges ($D^s C^{fc}/Dt = 0$) and Equation (9) [23]

$$\sum_{\beta} \frac{z^{\beta}}{\overline{V}^{\beta}} \nabla \cdot [n^{\beta}(\mathbf{v}^{\beta} - \mathbf{v}^{s})] = 0. \tag{10}$$

Neglecting inertia, the momentum balance takes the form

$$\nabla \cdot (\boldsymbol{\sigma}^{\alpha})^{C} + \boldsymbol{\pi}^{\alpha} = n^{\alpha}\mathbf{b}^{\alpha}, \tag{11}$$

which, after summation over all constituents, yields

$$\nabla \cdot (\boldsymbol{\sigma})^{C} = \sum_{\alpha} \nabla \cdot (\boldsymbol{\sigma}^{\alpha})^{C} = \mathbf{b}, \tag{12}$$

if use is made of the balance condition:

$$\sum_{\alpha} \boldsymbol{\pi}^{\alpha} = \mathbf{0}, \tag{13}$$

$\boldsymbol{\sigma}^{\alpha}$ is the partial Cauchy stress tensor of constituent $\alpha$, $\boldsymbol{\pi}^{\alpha}$ the momentum interaction with constituents other than $\alpha$, $\mathbf{b}^{\alpha}$ the body force of constituent $\alpha$ per unit volume and $\mathbf{b} = \sum_{\alpha} n^{\alpha}\mathbf{b}^{\alpha}$ the body force of the mixture per unit volume. The partial Cauchy stress is defined for all constituents in the same way in accordance with standard literature, for example [12]. One should expect, for example, that electrostatic interactions cause significant forces between mobile ions. Balance of moment of momentum requires that the stress tensor $\boldsymbol{\sigma}$ be symmetric. If no moment of momentum interaction between components occurs, the partial stresses $\boldsymbol{\sigma}^\alpha$ are also symmetric. In this paper we assume all partial stresses to be symmetric. The balance of energy for the $\alpha$th constituent reads

$$n^\alpha \frac{\mathrm{D}^\alpha \epsilon^\alpha}{\mathrm{D}t} = \boldsymbol{\sigma}^\alpha : \nabla \mathbf{v}^\alpha - \nabla \cdot \mathbf{q}^\alpha + n^\alpha r^\alpha + \hat{\epsilon}^\alpha, \tag{14}$$

in which $\epsilon^\alpha$ is the partial internal energy of the $\alpha$th constituent per unit volume, $\mathbf{q}^\alpha$ the heat flux of constituent $\alpha$, $r^\alpha$ the partial heat supply per unit volume constituent and $\hat{\epsilon}^\alpha$ the energy interaction with other constituents per unit volume. Total energy balance requires

$$\sum_\alpha (\hat{\epsilon}^\alpha + \mathbf{v}^\alpha \cdot \boldsymbol{\pi}^\alpha) = 0. \tag{15}$$

Substitution of Equation (14) into Equation (15) yields an equivalent form of the total energy balance

$$\frac{\mathrm{D}^s \epsilon}{\mathrm{D}t} = \sum_\alpha \{\boldsymbol{\sigma}^\alpha : \nabla \mathbf{v}^\alpha - \nabla \cdot [\mathbf{q}^\alpha + n^\alpha \epsilon^\alpha (\mathbf{v}^\alpha - \mathbf{v}^s)] - \mathbf{v}^\alpha \cdot \boldsymbol{\pi}^\alpha\} - \epsilon \nabla \cdot \mathbf{v}^s + r, \tag{16}$$

in which

$$\epsilon = \sum_\alpha n^\alpha \epsilon^\alpha \tag{17}$$

is the internal energy of the mixture per unit volume and

$$r = \sum_\alpha n^\alpha r^\alpha \tag{18}$$

is the total heat supply of the mixture per unit volume. Under incompressible conditions, the entropy inequality for a unit volume of mixture reads [5]

$$\frac{\mathrm{D}^s \eta}{\mathrm{D}t} - \eta \nabla \cdot \mathbf{v}^s + \sum_\alpha \left[ \nabla \cdot \left( \frac{\mathbf{q}^\alpha}{\theta^\alpha} + n^\alpha \eta^\alpha (\mathbf{v}^\alpha - \mathbf{v}^s) \right) - \frac{n^\alpha r^\alpha}{\theta^\alpha} \right] \geqslant 0, \tag{19}$$

in which $\eta^\alpha$ is the entropy of constituent $\alpha$ per unit volume constituent, $\eta = \sum_\alpha n^\alpha \eta^\alpha$ is the entropy of the mixture per unit mixture volume and $\theta^\alpha$ the temperature of the constituent $\alpha$. If the temperature field is assumed to be the same for all constituents, we find

$$\frac{\mathrm{D}^s \eta}{\mathrm{D}t} - \eta \nabla \cdot \mathbf{v}^s + \nabla \cdot \left[ \frac{1}{\theta} \sum_\alpha (\mathbf{q}^\alpha + n^\alpha \eta^\alpha \theta (\mathbf{v}^\alpha - \mathbf{v}^s)) \right] - \frac{r}{\theta} \geqslant 0, \tag{20}$$

in which $\theta$ is the temperature of the mixture. Substitution of the energy balance (16) into the entropy inequality eliminates the heat supply $r$ from the inequality

$$-\eta\frac{\mathrm{D}^{s}\theta}{\mathrm{D}t}-\frac{\mathbf{q}\cdot\nabla\theta}{\theta}+\sum_{\alpha}\left(-n^{\alpha}\frac{\mathrm{D}^{\alpha}\Psi^{\alpha}}{\mathrm{D}t}+\boldsymbol{\sigma}^{\alpha}:\nabla\cdot\mathbf{v}^{\alpha}-\boldsymbol{\pi}^{\alpha}\cdot\mathbf{v}^{\alpha}\right)\geqslant 0, \tag{21}$$

in which

$$\Psi^{\alpha}=\epsilon^{\alpha}-\theta\eta^{\alpha} \tag{22}$$

is the Helmholtz free energy of constituent $\alpha$ per unit volume constituent and

$$\mathbf{q}=\sum_{\alpha}[\mathbf{q}^{\alpha}+n^{\alpha}\eta^{\alpha}\theta(\mathbf{v}^{\alpha}-\mathbf{v}^{s})] \tag{23}$$

is an energy flux vector of the mixture. We introduce the strain energy function

$$W=J\sum_{\alpha}n^{\alpha}\Psi^{\alpha}=J\sum_{\alpha}\psi^{\alpha} \tag{24}$$

as the Helmholtz free energy of a mixture volume which in the *initial* state of the solid equals unity. $\psi^{\alpha}$ is the Helmholz free energy of constituent $\alpha$ per unit mixture volume. Rewriting the inequality (21) for the entropy production per initial mixture volume – that is, we multiply inequality (21) by the relative volume change $J$ – we find

$$-J\eta\frac{\mathrm{D}^{s}\theta}{\mathrm{D}t}-J\frac{\mathbf{q}\cdot\nabla\theta}{\theta}-\frac{\mathrm{D}^{s}}{\mathrm{D}t}W+J\boldsymbol{\sigma}:\nabla\mathbf{v}^{s}-J\nabla\cdot\sum_{\beta}(\mathbf{v}^{\beta}-\mathbf{v}^{s})\psi^{\beta}+$$
$$+J\sum_{\beta}[\nabla(\mathbf{v}^{\beta}-\mathbf{v}^{s}):\boldsymbol{\sigma}^{\beta}-(\mathbf{v}^{\beta}-\mathbf{v}^{s})\cdot\boldsymbol{\pi}^{\beta}]\geqslant 0. \tag{25}$$

The entropy inequality should hold for an arbitrary state of the mixture, complying with the balance laws, incompressibility, saturation and electroneutrality. There are two ways to comply with these restrictions. One is substitution of the restriction into the inequality, resulting in the elimination of a field variable. The other is by introduction of a Lagrange multiplier. The balance laws and the incompressibility conditions (1) are accounted for by means of substitution. The differentiated forms of the saturation condition (4) and the electroneutrality (10) are accounted for by means of a Lagrange multiplier. The inequality (25) shows that the apparent densities $\rho^{\alpha}$, the body forces $\mathbf{b}^{\alpha}$ and the heat supplies $r^{\alpha}$ are already eliminated from the inequality. In other words the conditions of incompressibility (1) are fulfilled by choosing $\rho^{\alpha}=\rho_{i}^{\alpha}n^{\alpha}$, the momentum balances (11) are fulfilled by choosing $\mathbf{b}^{\alpha}$ accordingly and the energy balance is fulfilled by choosing the heat supply $r$ accordingly. Therefore, restrictions still to be fulfilled are the mass balances, saturation and the electroneutrality. The differentiated form of the saturation condition (4) is substituted by means of a Lagrange multiplier $p$

$$-\eta_0 \frac{\mathrm{D}^{\mathrm{s}}\theta}{\mathrm{D}t} - \frac{J\mathbf{v}^{q\mathrm{s}} \cdot \nabla_0\theta}{\theta} - \frac{\mathrm{D}^{\mathrm{s}}}{\mathrm{D}t}W + J\boldsymbol{\sigma}^{\mathrm{eff}} : \nabla\mathbf{v}^{\mathrm{s}} + \\ + J\sum_{\beta}[\boldsymbol{\sigma}^{\beta} + (pn^{\beta} - \psi^{\beta})\mathbf{I}] : \nabla(\mathbf{v}^{\beta} - \mathbf{v}^{\mathrm{s}}) + \\ + J\sum_{\beta}(\mathbf{v}^{\beta} - \mathbf{v}^{\mathrm{s}}) \cdot (-\nabla\psi^{\beta} + p\nabla n^{\beta} - \boldsymbol{\pi}^{\beta}) \geqslant 0, \tag{26}$$

in which the effective stress $\boldsymbol{\sigma}^{\mathrm{eff}}$ is defined as

$$\boldsymbol{\sigma}^{\mathrm{eff}} = \boldsymbol{\sigma} + p\mathbf{I}, \tag{27}$$

the heat flux $\mathbf{v}^{q\mathrm{s}}$ through an area equal to unity in the initial state of the solid as

$$\mathbf{v}^{q\mathrm{s}} = \mathbf{F}^{-1} \cdot \mathbf{q} \tag{28}$$

and the entropy per unit initial volume as

$$\eta_0 = J\eta. \tag{29}$$

Introducing the restriction (10) into inequality (26) by means of a Lagrange multiplier $\lambda$, yields

$$-\eta_0 \frac{\mathrm{D}^{\mathrm{s}}\theta}{\mathrm{D}t} - \frac{J\mathbf{v}^{q\mathrm{s}} \cdot \nabla_0\theta}{\theta} - \frac{\mathrm{D}^{\mathrm{s}}}{\mathrm{D}t}W + J\boldsymbol{\sigma}^{\mathrm{eff}} : \nabla\mathbf{v}^{\mathrm{s}} + \\ + J\sum_{\beta}\left[\boldsymbol{\sigma}^{\beta} + \left(\left(p + \frac{z^{\beta}\lambda}{\overline{V}^{\beta}}\right)n^{\beta} - \psi^{\beta}\right)\mathbf{I}\right] : \nabla(\mathbf{v}^{\beta} - \mathbf{v}^{\mathrm{s}}) + \\ + J\sum_{\beta}(\mathbf{v}^{\beta} - \mathbf{v}^{\mathrm{s}}) \cdot \left[-\nabla\psi^{\beta} + \left(p + \frac{z^{\beta}\lambda}{\overline{V}^{\beta}}\right)\nabla n^{\beta} - \boldsymbol{\pi}^{\beta}\right] \geqslant 0, \tag{30}$$

in which $\nabla_0 = \mathbf{F}^{C} \cdot \nabla$ is the gradient operator with respect to the initial configuration. We choose as independent variables the Green strain $\mathbf{E}$, the temperature $\theta$, the Lagrangian form of the volume fractions of the fluid and the ions $N^{\beta}$, and of the relative velocities $\mathbf{v}^{\beta\mathrm{s}} = \mathbf{F}^{-1} \cdot (\mathbf{v}^{\beta} - \mathbf{v}^{\mathrm{s}})$ and $\mathbf{v}^{q\mathrm{s}}$. The dependent variables are the Helmholtz free energy per unit initial mixture volume $W$, the entropy per unit initial mixture volume $\eta_0$, the effective stress $\boldsymbol{\sigma}^{\mathrm{eff}}$ of the mixture, the Helmholtz free energy $\psi^{\gamma}$ of constituent $\gamma$, the effective partial stresses $\boldsymbol{\sigma}^{\gamma} + (p + z^{\gamma}\lambda/\overline{V}^{\gamma})n^{\gamma}\mathbf{I}$, the effective momentum interactions $\boldsymbol{\pi}^{\gamma} - (p + z^{\gamma}\lambda/\overline{V}^{\gamma})\nabla n^{\gamma}$ and the temperature gradient $\nabla_0\theta$. We apply the principle of equipresence, that is, all dependent variables depend on all independent variables, unless the entropy inequality requires otherwise:

$$W = W(\mathbf{E}, \theta, N^{\beta}, \mathbf{v}^{\beta\mathrm{s}}, \mathbf{v}^{q\mathrm{s}}), \tag{31}$$

$$\eta_0 = \eta_0(\mathbf{E}, \theta, N^{\beta}, \mathbf{v}^{\beta\mathrm{s}}, \mathbf{v}^{q\mathrm{s}}), \tag{32}$$

$$\boldsymbol{\sigma}^{\text{eff}} = \mathbf{F} \cdot \mathbf{S}^{\text{eff}}(\mathbf{E}, \theta, N^\beta, \mathbf{v}^{\beta s}, \mathbf{v}^{qs}) \cdot \mathbf{F}^C, \tag{33}$$

$$\psi^\gamma = \psi^\gamma(\mathbf{E}, \theta, N^\beta, \mathbf{v}^{\beta s}, \mathbf{v}^{qs}), \tag{34}$$

$$\boldsymbol{\sigma}^\gamma + \left(p + \frac{z^\gamma \lambda}{\overline{V}^\gamma}\right) n^\gamma \mathbf{I} = \mathbf{F} \cdot \mathbf{S}^\gamma(\mathbf{E}, \theta, N^\beta, \mathbf{v}^{\beta s}, \mathbf{v}^{qs}) \cdot \mathbf{F}^C, \tag{35}$$

$$\boldsymbol{\pi}^\gamma - \left(p + \frac{z^\gamma \lambda}{\overline{V}^\gamma}\right) \boldsymbol{\nabla} n^\gamma = \mathbf{F} \cdot \mathbf{p}^\gamma(\mathbf{E}, \theta, N^\beta, \mathbf{v}^{\beta s}, \mathbf{v}^{qs}), \tag{36}$$

$$\boldsymbol{\nabla}_0 \theta = \mathbf{g}_0(\mathbf{E}, \theta, N^\beta, \mathbf{v}^{\beta s}, \mathbf{v}^{qs}). \tag{37}$$

We apply the chain rule for time differentiation of $W$, which implies:

$$\begin{aligned} &- \left(\eta_0 + \frac{\partial W}{\partial \theta}\right) \frac{\mathrm{D}^s \theta}{\mathrm{D}t} - \frac{\mathbf{v}^{qs} \cdot \boldsymbol{\nabla}_0 \theta}{\theta} \left(J \boldsymbol{\sigma}^{\text{eff}} - \mathbf{F} \cdot \frac{\partial W}{\partial \mathbf{E}} \cdot \mathbf{F}^C\right) : \boldsymbol{\nabla} \mathbf{v}^s - \\ &- \frac{\partial W}{\partial \mathbf{v}^{qs}} \cdot \frac{\mathrm{D}^s}{\mathrm{D}t} \mathbf{v}^{qs} + \sum_\beta \left\{ - \frac{\partial W}{\partial \mathbf{v}^{\beta s}} \cdot \frac{\mathrm{D}^s}{\mathrm{D}t} \mathbf{v}^{\beta s} + \right. \\ &+ J[\boldsymbol{\sigma}^\beta + (\mu^\beta n^\beta - \psi^\beta)\mathbf{I}] : \boldsymbol{\nabla}(\mathbf{v}^\beta - \mathbf{v}^s) + \\ &\left. + J(\mathbf{v}^\beta - \mathbf{v}^s) \cdot (-\boldsymbol{\nabla} \psi^\beta + \mu^\beta \boldsymbol{\nabla} n^\beta - \boldsymbol{\pi}^\beta) \right\} \geqslant 0, \end{aligned} \tag{38}$$

in which $\mu^\beta$ are the electrochemical potentials of constituents other than the porous solid

$$\mu^\beta = \frac{\partial W}{\partial N^\beta} + p + \frac{z^\beta \lambda}{\overline{V}^\beta}. \tag{39}$$

Comparison of the above equation to the classical equations of electrochemistry indicates that the Lagrange multiplier $p$ can be interpreted as the fluid pressure and $\lambda$ as the electrical potential of the medium multiplied by the constant of Faraday. Inequation (38) should be true for any value of the state variables. In view of the set of independent variables, the first term of (38) is linear in the time derivative of the temperature $\mathrm{D}^s\theta/\mathrm{D}t$, the third term is linear in the solid velocity gradient $\boldsymbol{\nabla}\mathbf{v}^s$, the fourth term linear in $(\mathrm{D}^s/\mathrm{D}t)\mathbf{v}^{qs}$, the fifth term linear in $(\mathrm{D}^s/\mathrm{D}t)\mathbf{v}^{\beta s}$ and the sixth term linear in the relative velocity gradients $\boldsymbol{\nabla}(\mathbf{v}^\beta - \mathbf{v}^s)$. Therefore, by a standard argument [3], we find

$$\eta_0 = -\frac{\partial W}{\partial \theta}, \tag{40}$$

$$\boldsymbol{\sigma}^{\text{eff}} = \frac{1}{J} \mathbf{F} \cdot \frac{\partial W}{\partial \mathbf{E}} \cdot \mathbf{F}^C, \tag{41}$$

$$\frac{\partial W}{\partial \mathbf{v}^{qs}} = 0, \tag{42}$$

$$\frac{\partial W}{\partial \mathbf{v}^{\beta s}} = 0, \tag{43}$$

$$\boldsymbol{\sigma}^{\beta} = (\psi^{\beta} - \mu^{\beta} n^{\beta})\mathbf{I}, \tag{44}$$

leaving as inequality

$$-\frac{J\mathbf{v}^{qs}\cdot\boldsymbol{\nabla}_0\theta}{\theta} + \sum_{\beta} J(\mathbf{v}^{\beta} - \mathbf{v}^{s})\cdot(-\boldsymbol{\nabla}\psi^{\beta} + \mu^{\beta}\boldsymbol{\nabla} n^{\beta} - \boldsymbol{\pi}^{\beta}) \geqslant 0. \tag{45}$$

Equation (41) indicates that the effective stress of the mixture can be derived from a strain energy function $W$ which represents the free energy of the mixture. Equation (43) shows that the strain energy function cannot depend on the relative velocities. Thus, the effective stress of a charged porous medium can be derived from a regular strain energy function, which physically has the same meaning as in single phase or biphasic media, but which can depend on both strain and solute concentrations in the medium. According to Equation (44), the partial stress of the fluid and the solutes are scalars. Transforming the relative velocities to their Lagrangian equivalents, we find, instead of (45),

$$-\frac{\mathbf{v}^{qs}\cdot\boldsymbol{\nabla}_0\theta}{\theta} + \sum_{\beta} \mathbf{v}^{\beta s}\cdot[-\boldsymbol{\nabla}_0\psi^{\beta} + \mu^{\beta}\boldsymbol{\nabla}_0 n^{\beta} - \mathbf{F}^{C}\cdot\boldsymbol{\pi}^{\beta}] \geqslant 0. \tag{46}$$

If we assume that the system is not too far from equilibrium, we can express the dissipation (46) associated with relative flow of fluid, solutes and heat as a quadratic function of the relative velocities

$$-\boldsymbol{\nabla}_0\psi^{\beta} + \mu^{\beta}\boldsymbol{\nabla}_0 n^{\beta} - \mathbf{F}^{C}\cdot\boldsymbol{\pi}^{\beta} = \sum_{\gamma} \mathbf{B}^{\beta\gamma}\cdot\mathbf{v}^{\gamma s} + \mathbf{B}^{\beta q}\cdot\mathbf{v}^{qs}, \tag{47}$$

$$-\frac{\boldsymbol{\nabla}_0\theta}{\theta} = \sum_{\gamma} \mathbf{B}^{q\gamma}\cdot\mathbf{v}^{\gamma s} + \mathbf{B}^{qq}\cdot\mathbf{v}^{qs}, \tag{48}$$

in which

$$\begin{bmatrix} \mathbf{B}^{\beta\gamma} & \mathbf{B}^{\beta q} \\ \mathbf{B}^{q\gamma} & \mathbf{B}^{qq} \end{bmatrix} \tag{49}$$

is a positive definite matrix of frictional tensors. Substituting Equations (11) and (44) into Equation (47) yields Lagrangian forms of the classical equations of irreversible thermodynamics

$$-n^{\beta}\boldsymbol{\nabla}_0\mu^{\beta} - \mathbf{F}^{C}\cdot n^{\beta}\mathbf{b}^{\beta} = \sum_{\gamma} \mathbf{B}^{\beta\gamma}\cdot\mathbf{v}^{\gamma s} + \mathbf{B}^{\beta q}\cdot\mathbf{v}^{qs}, \tag{50}$$

$$-\frac{\boldsymbol{\nabla}_0\theta}{\theta} = \sum_{\gamma} \mathbf{B}^{q\gamma}\cdot\mathbf{v}^{\gamma s} + \mathbf{B}^{qq}\cdot\mathbf{v}^{qs}. \tag{51}$$

The boundary conditions are given by a no-jump condition of the temperature, the electrochemical potential of the ions and the fluid across the boundary and the momentum balance of the boundary.

## 3. Discussion

The analysis presented in this paper differs in a few aspects from related work in the literature. We demonstrate that the constitutive behaviour needed for analysis of the deformation of swelling material (Equations (39)–(41), (50) and (51)) is entirely contained in the strain energy function $W$ and the frictional matrix $\mathbf{B}^{\beta\gamma}$. This is a more compact description than the more classical Eulerian analysis requiring separate free energy functions to be evaluated for each constituent (e.g. [5, 7]) and might save a great deal of tedious work for experimentalists responsible for quantifying these constitutive functions. The definition of the function $W$ as the Helmholtz free energy of a mixture volume equal to unity in the initial state of the solid is consistent with the classical definition of a strain energy function of a single phase material or of a saturated porous medium by Biot [4]. It is a straightforward extension of the work of Wilmansky on Lagrangian porous media models [20].

We apply the method of Coleman and Noll [3] to derive constitutive relationships from the second law of thermodynamics. This method has been criticized [21] and shown to be more restrictive than Mueller's method [19]. However, the criticisms are not shared by all experts in the field [19, 20], and for our purposes the use of either method is satisfactory.

Another feature distinguishes this thermodynamic analysis of swelling materials from the pioneering work of Lai *et al.* [7]. Lai *et al.* neglect to introduce the electroneutrality condition as a restriction on the second law of thermodynamics, although this condition mutually links the ion velocity gradients according to Equation (10). The electrical potential, obtained in our analysis as the Lagrange multiplier associated with the electroneutrality condition, is introduced in their analysis as an external volume force, which it is not. Our Lagrange multiplier $\lambda$ is an internally generated electric field adding up to external electric fields contained in the body force vector $\mathbf{b}^{\alpha}$.

Equations (39)–(41) are rewritten in their differentiated form

$$\begin{bmatrix} \mathrm{d}\boldsymbol{\sigma}^{\mathrm{eff}} \\ \mathrm{d}\left(\mu^{\beta} - p - \dfrac{z^{\beta}\lambda}{\overline{V}^{\beta}}\right) \\ -\,\mathrm{d}\eta_0 \end{bmatrix} = \begin{bmatrix} \dfrac{\partial^2 W}{\partial \mathbf{E}^2} & \dfrac{\partial^2 W}{\partial N^{\gamma}\partial \mathbf{E}} & \dfrac{\partial^2 W}{\partial \theta \partial \mathbf{E}} \\ \dfrac{\partial^2 W}{\partial \mathbf{E}\partial N^{\beta}} & \dfrac{\partial^2 W}{\partial N^{\gamma}\partial N^{\beta}} & \dfrac{\partial^2 W}{\partial \theta \partial N^{\beta}} \\ \dfrac{\partial^2 W}{\partial \mathbf{E}\partial \theta} & \dfrac{\partial^2 W}{\partial N^{\gamma}\partial \theta} & \dfrac{\partial^2 W}{\partial \theta^2} \end{bmatrix} \begin{bmatrix} \mathrm{d}\mathbf{E} \\ \mathrm{d}N^{\gamma} \\ \mathrm{d}\theta \end{bmatrix} \tag{52}$$

or

$$\mathrm{d}\underline{y} = \underline{\mathbf{C}}\,\mathrm{d}\underline{x}. \tag{53}$$

Assuming smoothness of the function $W$, the matrix $\underline{\mathbf{C}}$ should be symmetric. The existence of the function $W$ and the symmetry of $\underline{\mathbf{C}}$ is closely associated with the principle of reversibility. This principle means that if the changes occur at an infinitely slow rate, the work done to bring the material from the initial condition to its final state of strain and composition is independent of the path by which the final state is reached. In the absence of such a condition, an infinite amount of energy could be drawn from the material by loading and unloading it along a closed cycle. The above derivation generalises the approach of Huyghe and Janssen [23] to nonisothermal conditions in which temperatures are the same for all components. From the above equations several well-known physical theories can be derived [23], such as the Nernst equation for diffusion potentials, streaming potentials or the finite strain poroelasticity theory of Biot [4]. A field of application where temperature gradients may play an important role in swelling materials is for example human skin.

The reasons for choosing the quadriphasic theory [23] as a basis for the development are both experimental and conceptual. Alternative theories are available in the literature [7, 8, 10, 22]. Sherwood and Heidug have developed, triphasic theories in which fixed charges are absent altogether and, hence, in which electrical effects are not present. Although this is a valid approach for specific applications, it is not sufficient for our purposes, involving materials with high fixed charge densities. Snijders *et al.* [10] have handled somewhat less stringent assumptions. Lai *et al.* [7] have developed a triphasic theory in which fixed charges are included. Although at the outset the theory is general when applying it to articular cartilage, they assume that deformations are infinitesimal. Similar assumptions are handled by [22]. Gu *et al.* [9] have used the theory of Lai *et al.* in the context of one-dimensional permeation experiments. Those three papers use constitutive assumptions which lead to asymmetrical forms of the matrix $\mathbf{C}^{\beta\gamma}$. The chemical expansion stress $\mathbf{T}_c$ as introduced by Lai *et al.* [7], for example, contributes to $\mathbf{C}^{s\beta}$, while corresponding lower-diagonal terms $\mathbf{C}^{\beta s}$ vanish. Although such an assumption is acceptable for infinitesimal theory, caution should be taken in extending such constitutive modelling to the finite strain domain. Moreover, recent comparison of experimental and numerical data on transient swelling [24] has not shown evidence of the existence of a chemical expansion stress.

## Acknowledgements

The research of Dr J. M. Huyghe has been made possible through a fellowship of the Royal Netherlands Academy of Arts and Sciences. Part of this research was supported by the Technology Foundation (STW). The authors thank M. M. Molenaar for his valuable comments to this manuscript.

## References

1. Biot, M. A.: General theory of three-dimensional consolidation, *J. Appl. Phys.* **12** (1941), 155–164.
2. Staverman, A. J.: Non-equilibrium thermodynamics of membrane processes, *Trans. Faraday Soc.* **48** (1952), 176–185.
3. Coleman, B. D. and Noll, W.: The thermodynamics of elastic materials with heat conduction and viscosity, *Arch. Rational Mech. Anal.* **13** (1963), 167–178.

4. Biot, M.: Theory of finite deformation of porous solids, *Indiana University Math. J.* **21** (1972), 597–620.
5. Bowen, R. M.: Incompressible porous media models by use of the theory of mixtures, *Int. J. Engng. Sci.* **18** (1980), 1129–1148.
6. Richards, E. G.: *An Introduction to Physical Properties of Large Molecules in Solution*, Cambridge University Press, Cambridge, 1980.
7. Lai, W. M., Hou, J. S. and Mow, V. C.: A triphasic theory for the swelling and deformation behaviours of articular cartilage, *J. Biomech. Engng.* **113** (1991), 245–258.
8. Sherwood, J. D.: Biot poroelasticity of a chemically active shale, *Proc. Roy. Soc. London A* **440** (1993), 367–377.
9. Gu, W. Y., Lai, W. M. and Mow, V. C.: Transport of fluid and ions through a porous-permeable charged-hydrated tissue, and streaming potential data on normal bovine articular cartilage, *J. Biomech.* **26** (1993), 709–723.
10. Snijders, H., Huyghe, J. M. and Janssen, J. D.: Triphasic finite element model for swelling porous media, *Int. J. Num. Meth. Fluids* **20** (1995), 1039–1046.
11. Grodzinsky, A. J., Roth, V., Meyers, E., Grossman, W. D. and Mow, V. C.: The significance of electromechanical and osmotic forces in the non-equilibrium swelling behaviour of articular cartilage in tension, *J. Biomech. Engng.* **103** (1981), 221–231.
12. Bluhm, J. and de Boer R.: Effective stresses – a clarification, *Arch. Appl. Mech.* **66** (1996), 479–492.
13. Drost, M. R., Willems P., Snijders, H., Huyghe, J. M., Janssen, J. D. and Huson, A.: Confined compression of canine annulus fibrosus under chemical and mechanical loading, *J. Biomech. Engng.* **117** (1995), 390–396.
14. Oomens, C. W. J., Heus, H. de, Huyghe, J. M., Nelissen, L. and Janssen, J. D.: Validation of the triphasic mixture theory for a mimic of intervertebral disk tissue, *Biomimetics* **3** (1994), 171–185.
15. Heidug, W. K. and Wong, S.-W.: Hydration swelling of water absorbing rocks: a constitutive model, *Int. J. Num. Anal. Meth. Geomech.* **20** (1996), 403–430.
16. Yeung, A. T. and Mitchell, J. K.: Coupled fluid, electrical and chemical flows in soil, *Geotechnique* **43** (1993), 121.
17. Oort, E. van, Hale, A. H., Mody, F. K. and Sanjit, R.: Critical parameters in modelling the chemical aspects of borehole stability in shales and in designing improved water-based shale drilling fluids, Society of Petroleum Engineers, paper 28309, presented at the *SPE 69th Annual Technical Conference and Exhibition*, New Orleans, La., September, 1994.
18. Vankan, W. J., Huyghe, J. M., Janssen, J. D. and Huson, A.: Poroelasticity of saturated solids with an application to blood perfusion, *Int. J. Engng. Sci.* **34** (1996), 1019–1031.
19. Mueller, I.: *Thermodynamics*, Pitman, Boston, 1985.
20. Wilmansky, K.: Lagrangian model of two-phase porous material, *J. Non-Equil. Thermodyn.* **20** (1995), 50–77.
21. Woods, L. C.: Thermodynamic inequalities in continuum mechanics, *IMA J. Appl. Math.* **29** (1982), 221–246.
22. Simon, B. R., Liable, J. P., Pflaster, D., Yuan, Y. and Krag, M. H.: A poroelastic finite element formulation including transport and swelling in soft tissue structures, *J. Biomech. Engng.* **118** (1996), 1–9.
23. Huyghe, J. M. and Janssen, J. D.: Quadriphasic theory of swelling incompressible porous media, *Int. J. Engng. Sci.* **35** (1997), 793–802.
24. Frijns, A. J. H., Huyghe, J. M. and Janssen, J. D.: A validation of the quadriphasic mixture theory for intervertebral disc tissue, *Int. J. Engng. Sci.* **35** (1997), 1419–1429.

# Transport of Multi-Electrolytes in Charged Hydrated Biological Soft Tissues

W. Y. GU*, W. M. LAI and V. C. MOW
*Orthopaedic Research Laboratory, Departments of Mechanical Engineering and Orthopaedic Surgery, Columbia University, New York, NY, U.S.A.*

(Received: 11 July 1997; in final form: 10 December 1997)

**Abstract.** A mechano-electrochemical theory for charged hydrated soft tissues with multi-electrolytes was developed based on the continuum mixture theory. The momentum equations for water and ions were derived in terms of a mechanochemical force (gradient of water chemical potential), electrochemical forces (gradient of Nernst potentials) and an electrical force (gradient of electrical potential). The theory was shown to be consistent with all existing specialized theories. Using this theory, some mechano-electrokinetic properties of charged isotropic tissues were studied. The well-known Hodgkin–Huxley equation for resting cell membrane potential was derived and the phenomenon of electro-osmotic flow in charged hydrated soft tissues was investigated. Analyses show that the tissue fixed charge density plays an important role in controlling the transport of water and ions in charged hydrated soft tissues.

**Key words:** transport, mixture theory, electro-osmosis, charged hydrated soft tissues.

## 1. Introduction

Many biological tissues exhibit mechano-electrochemical coupling phenomena such as ion-induced swelling, streaming potential or current, anomalous osmosis, and electro-osmosis. The existence of the fixed charges on the bio-macromolecules (e.g. proteoglycans) within a biological tissue is responsible for the mechano-electrochemical coupling behavior of the tissue. Knowledge of transport of electrolyte solutions within such biological tissues under mechanical, osmotic, or electrical loading is important for understanding cell/tissue physiology (e.g. signal transduction for modulation of cellular metabolism) and medical treatment (e.g. drug delivery into tissues).

A number of theoretical models have been proposed for describing the transport of electrolyte solutions in charged porous biological tissues and ion-exchange membranes (e.g. Teorell, 1953; Kedem and Katchalsky, 1961; Helfferich, 1962; Fair and Osterle, 1971; Silberberg, 1982; Salzstein *et al.*, 1987; Frank and Grodzinsky, 1987b; Comper *et al.*, 1990; Lai *et al.*, 1991; Coelho *et al.*, 1996; Huyghe and Janssen, 1997). Some of them are based on the continuum mixture theory (e.g. Lai *et al.*, 1991;

* *Current Address:* Department of Biomedical Engineering, University of Miami, P.O. Box 248294, Coral Gables, FL 33124, U.S.A., e-mail: wgu@exchange.ir.miami.edu

Achanta *et al.*, 1994; Heidug and Wong, 1996; Huyghe and Janssen, 1997) and others are based on the continuum micromechanics approach (e.g. Fair and Osterle, 1971; Coelho *et al.*, 1996), or continuum mixture models coupled with a phenomenological theory (e.g. Salzstein *et al.*, 1987; Frank and Grodzinsky, 1987b). It has been shown that the triphasic mechano-electrochemical model (Lai *et al.*, 1991), based on the nonlinear continuum mixture theory (see Truesdell and Noll, 1965; Mow *et al.*, 1980; Bowen, 1980) and the physicochemical theory for electrolyte solution (e.g. Donnan, 1924; Katchalsky and Curran, 1975; de Groot and Mazur, 1984), is capable of describing many aspects of deformational and electrokinetic behaviors of charged hydrated soft tissues with 1:1 electrolyte solutions (Lai *et al.*, 1991, Gu *et al.*, 1993a, b; Lai *et al.*, 1994; Gu *et al.*, 1997). Recently, we have extended the triphasic theory to model charged hydrated soft tissues with multi-electrolyte solutions (Gu *et al.*, 1994; Gu *et al.*, 1998). Our hypothesis is that the fixed charges on the bio-macromolecules within a tissue play an important role in regulating the transport of water and ions in biological tissues and cells. In this paper, we will briefly review our multiphasic (component) mixture theory, and present applications for it to investigate the role of the fixed charges in mechano-electrokinetic transduction effects.

## 2. Theory

Following the approach developed by Lai *et al.* (1991), we model a charged-hydrated soft tissue as a continuum mixture consisting of: solid phase, interstitial water phase, and $n$ different species of mono- or multi-valent ions. Thus, there are a total of $n+2$ constituents in the tissue. Any components other than water and mobile ions within a tissue are considered as the solid phase in this model. The solid phase of the tissue is in general charged and its charged nature derives from the ionic groups (e.g. $SO_3^-$ and $COO^-$) on the bio-macromolecules (e.g. proteoglycan) of a tissue. Relative to the mobile ions dissolved in the water, these charges are 'fixed' to the solid phase. The amount of fixed charges per unit of volume or weight of interstitial water or tissue is called the fixed charge density (FCD). In the model, each phase (component) is assumed to be intrinsically incompressible. We further assume the tissue to be saturated (e.g., see de Boer, 1996), thus, the summation of the volume fractions of each constituent is unity, that is

$$\sum_{\beta} \phi^{\beta} = 1 \quad \text{(summation over all constituents)}, \tag{1}$$

where $\beta(= \text{s, w}, 1, 2, \ldots, n)$ denotes the constituents (solid, water, ion-species 1, ion-species 2, …, etc.) within the tissue.

### 2.1. CONTINUITY EQUATIONS

The continuity equations for the constituent of the mixture are

$$\frac{\partial \rho^\beta}{\partial t} + \operatorname{div}(\rho^\beta \mathbf{v}^\beta) = 0, \quad \beta = \mathrm{s}, \mathrm{w}, 1, 2, \dots, n, \tag{2}$$

where the apparent mass density $\rho^\beta$ is related to its true density, $\rho_T^\beta$, by $\rho^\beta = \phi^\beta \rho_T^\beta$. The apparent densities of ion species can also be written as $\rho^\alpha = \phi^{\mathrm{w}} c_\alpha M_\alpha (\alpha = 1, 2, \dots, n)$ where $c_\alpha$ is the ion concentration, defined as moles per unit of *water volume* within the mixture and $M_\alpha$ is the molar weight of ions. Equation (2) reduces to

$$\frac{\partial \phi^\beta}{\partial t} + \operatorname{div}(\phi^\beta \mathbf{v}^\beta) = 0, \quad \beta = \mathrm{s}, \mathrm{w}, 1, 2, \dots, n, \tag{3}$$

when the true density ($\rho_T^\beta$) of each constituent is constant (incompressible).

### 2.2. ELECTRONEUTRALITY CONDITION

On the macroscopic continuum level, it is assumed that each point (elemental volume) is electrically neutral. That is, the net charge must be zero at any point within the tissue

$$\sum_\alpha z_\alpha c_\alpha + \omega c^{\mathrm{F}} = 0 \quad (\text{summation over}\, \alpha = 1, 2, \dots, n), \tag{4}$$

where $c^{\mathrm{F}}$ is the value of the FCD on the solid phase, defined as the *equivalent moles of mono-valent ions per unit of water volume in the mixture*; $z_\alpha$ is the valency of $\alpha$-species of ions, and $\omega = +1$ represents for positively charged tissues while $\omega = -1$ for negatively charged tissues. In the present theory, the fixed charges on the solid phase are assumed to be conserved, thus, the FCD is a function of deformation and water volume fraction (porosity) only:

$$c^{\mathrm{F}} = \frac{c_r^{\mathrm{F}}}{1 + (J - 1)/\phi_r^{\mathrm{w}}}, \tag{5}$$

where $c^{\mathrm{F}}$ and $\phi_r^{\mathrm{w}}$ are the FCD and porosity at the reference configuration, respectively, and $J$ $(= \mathrm{d}V/\mathrm{d}V_0)$ is the ratio of current volume to the undeformed volume.

### 2.3. CONSTITUTIVE RELATIONS

Let $A^\beta (\beta = \mathrm{s}, \mathrm{w}, 1, 2, \dots, n)$ be the Helmholtz energy density of $\beta$-constituent per unit of mass and

$$A^\beta = A^\beta\,(\mathbf{C}, \rho^{\mathrm{w}}, \rho^1, \rho^2, \dots, \rho^n), \tag{6}$$

where $\mathbf{C}$ is the right Cauchy–Green tensor of the solid phase and $\rho^\alpha$ ($\alpha =$ w, $1, 2, \ldots, n$) are the mass densities of water and ions per unit of current mixture volume (i.e. apparent mass densities). Making use of the principles of conservation of mass, momentum, and energy, and the second law of thermodynamics, we can obtain the following constitutive relations for the mixture subjected to the incompressibility and electrical neutrality constraints (3, 4) (Bowen, 1980; Huyghe and Janssen, 1997):

$$\sigma^{\mathrm{s}} = -\lambda_1\phi^{\mathrm{s}}\mathbf{I} + 2\mathbf{F}\left(\frac{\partial\Psi}{\partial\mathbf{C}}\right)\mathbf{F}^{\mathrm{T}} + \rho^{\mathrm{s}}A^{\mathrm{s}}\mathbf{I} + \lambda_2\omega c^{\mathrm{F}}\phi^{\mathrm{w}}\mathbf{I}, \tag{7}$$

$$\sigma^\alpha = -\lambda_1\phi^\alpha\mathbf{I} - \rho^\alpha\left(\frac{\partial\Psi}{\partial\rho^\alpha}\right)\mathbf{I} + \rho^\alpha A^\alpha\mathbf{I} + \lambda_2 z_\alpha c_\alpha\phi^{\mathrm{w}}\mathbf{I},$$
$$\alpha = 1, 2, \ldots, n, \tag{8}$$

where $\sigma^\beta$ ($\beta = $ s, w, $1, 2, \ldots, n$) is the apparent Cauchy stress tensor of $\beta$-constituent, $\lambda_1$ and $\lambda_2$ are the Lagrange multipliers, $\mathbf{F}$ is the deformation gradient tensor and $\Psi$ ($= \Sigma\rho^\beta A^\beta$) is the inner part of the Helmholtz energy density of mixture per unit of deformed volume. By introducing the (electro) chemical potential for water and ions defined as $\tilde{\mu}^\alpha\mathbf{I} = A^\alpha\mathbf{I} - \sigma^\alpha/\rho^\alpha$, we have

$$\tilde{\mu}^\alpha = \frac{p}{\rho_{\mathrm{T}}^\alpha} + \frac{\partial\Psi}{\partial\rho^\alpha} + \frac{\mathrm{F_c}\psi z_\alpha}{M_\alpha}, \quad \alpha = 1, 2, \ldots, n. \tag{9}$$

In arriving at Equation (9) the Lagrange multiplier $\lambda_1$ has been identified as the fluid pressure ($p$) and $\lambda_1$ as $-\mathrm{F_c}\psi$ where $\mathrm{F_c}$ is the Faraday constant and $\psi$ is the electrical potential (Huyghe and Janssen, 1997).

### 2.4. MOMENTUM EQUATIONS

Neglecting the effect of magnetic field on the movement of each constituent, it can be shown that the gradient of electrochemical potential is the sole driving force for the movement of water and dissolved ions (Lai *et al.*, 1991). Under a quasi-static condition, the momentum equations for water and ions are

$$-\rho^\alpha\nabla\tilde{\mu}^\alpha + \sum_\beta f_{\alpha\beta}(\mathbf{v}^\beta - \mathbf{v}^\alpha) = 0$$
$$\text{(summation over } \beta = \text{s, w, } 1, 2, \ldots, n), \tag{10}$$

where $f_{\alpha\beta}$ is the frictional coefficient between $\alpha$ and $\beta$ phases (or constituents) with $f_{\alpha\beta}$ assumed to be equal to $f_{\alpha\beta}$(Lai *et al.*, 1991).*

In order to develop the equations for ion and water transport through the tissue, we need to solve for $[\mathbf{v}^\alpha - \mathbf{v}^{\mathrm{s}}]$ from the momentum Equations (10). By inverting

* Note that $f_{\alpha\beta} = 0$ if $\alpha = \beta$.

Equations (10), we obtain

$$[\mathbf{v}^\alpha - \mathbf{v}^s] = [A]^{-1}[\rho^\alpha \nabla \tilde{\mu}^\alpha] = [B][\rho^\alpha \nabla \tilde{\mu}^\alpha] = \left[\sum_\beta B_{\alpha\beta}\rho^\beta \nabla \tilde{\mu}^\beta\right]$$

$$\text{(summation over } \beta = \mathrm{w}, 1, 2, \ldots n), \qquad (11)$$

where $[\mathbf{v}^\alpha - \mathbf{v}^s]$ and $[\rho^\alpha \nabla \tilde{\mu}^\alpha]$ are $n+1$ column vectors ($\alpha = \mathrm{w}, 1, 2, \ldots, n$), and $[A]$ and $[B]$ are $n+1$ by $n+1$ matrices with $[B] = [A]^{-1}$. The components of matrix $[A]$ in Equation (11) are related to the frictional coefficients $f_{\alpha\beta}$ in Equation (10) by

$$A_{\alpha\beta} = f_{\alpha\beta} - \mathrm{d}_\alpha \delta_{\alpha\beta} \quad \text{(no summation)} \quad \alpha, \beta = \mathrm{w}, 1, 2, \ldots, n, \qquad (12)$$

where $\delta_{\alpha\beta}$ is the Kronecker delta and $d_\alpha = \sum_\beta f_{\alpha\beta}$ (summation over $\beta = \mathrm{s}, \mathrm{w}, 1, \ldots, n$). Both $[A]$ and $[B]$ in Equation (11) are symmetric since $f_{\alpha\beta} = f_{\beta\alpha}$. For hydrated tissues, the frictional coefficients between ions and solid, and the frictional coefficients among the ions may be assumed to be negligible. Under this situation, the coefficients of matrix $[B]$ can be found by (Gu *et al.*, 1998)

$$\begin{aligned} B_{\mathrm{w}\alpha} &= -(f_{\mathrm{ws}})^{-1} \quad \text{for } \alpha = \mathrm{w}, 1, 2, \ldots, n \\ B_{\alpha\beta} &= -(f_{\mathrm{ws}})^{-1} - \delta_{\alpha\beta}(f_{\alpha\mathrm{w}})^{-1} \quad \text{for } \alpha, \beta = 1, 2, \ldots, n. \end{aligned} \qquad (13)$$

Finally, the momentum equation for the entire mixture (i.e. the tissue) is (Mow *et al.*, 1980; Lai *et al.*, 1991)

$$\operatorname{div} \sigma = \mathbf{0}, \qquad (14)$$

where $\sigma = \Sigma\sigma^\beta$ (summation over all constituents).

## 3. Transport of Water and Ions in Isotropic Tissues

From this general formulation, as a special case, the constitutive equations for an isotropic mixture with infinitesimal deformation may be derived:

$$\sigma = -p\mathbf{I} + \lambda_s \operatorname{tr}(\mathbf{E})\,\mathbf{I} + 2\,\mu_s\,\mathbf{E}, \qquad (15)$$

where $p$ is the fluid pressure, $\lambda_s$ and $\mu_s$ are the Lame coefficients for the mixture, $\mathbf{E}$ is the infinitesimal strain tensor for the solid phase measured relative to the hypertonic reference configuration (Lai *et al.*, 1991), and tr$\mathbf{E}$ is the dilatation. The constitutive relation for water chemical potential is

$$\mu^{\mathrm{w}} = \mu_0^{\mathrm{w}} + \left[p - RT\sum_\alpha(\Phi_\alpha c_\alpha) + B_{\mathrm{w}}\operatorname{tr}(\mathbf{E})\right] / \rho_{\mathrm{T}}^{\mathrm{w}},$$

$$\text{(summation over } \alpha = 1, 2, \ldots, n), \qquad (16)$$

where $\mu_0^{\mathrm{w}}$ is the reference chemical potential of water, $R$ is the universal gas constant, $T$ is the absolute temperature, $\Phi_\alpha$ are the osmotic coefficients, and $B_{\mathrm{w}}$ is a coupling coefficient. The constitutive equation for the electrochemical potential of the $\alpha$th ionic species ($\alpha = 1, 2, \ldots, n$) is

$$\tilde{\mu}^{\alpha} = \mu_0^{\alpha} + \left(\frac{RT}{M_{\alpha}}\right)\ln(\gamma_{\alpha}c_{\alpha}) + \frac{z_{\alpha}\mathrm{F_c}\psi}{M_{\alpha}}, \tag{17}$$

where $\mu_0^{\alpha}$ are the reference chemical potentials and $\gamma_{\alpha}$ are the activity coefficients (Katchalsky and Curran, 1975; Gu *et al.*, 1993a). We note that the electrochemical potentials of ions generally depend on the fluid pressure (see Equation (9); however, this dependency is neglected in the present model.

### 3.1. VOLUME AND IONIC FLUXES

Neglecting small values of ion volume fractions, the relative fluid volume flux ($\mathbf{J}_{\mathrm{v}}$) is equal to the water volume flux, given by

$$\mathbf{J}_{\mathrm{v}} = \phi^{\mathrm{w}}(\mathbf{v}^{\mathrm{w}} - \mathbf{v}^{\mathrm{s}}). \tag{18}$$

The ionic molar flux, $\mathbf{J}_{\alpha}$, is defined as

$$\mathbf{J}_{\alpha} = \phi^{\mathrm{w}}c_{\alpha}(\mathbf{v}^{\alpha} - \mathbf{v}^{\mathrm{s}}), \quad \alpha = 1, 2, \ldots, n. \tag{19}$$

The ionic current, $(\mathbf{I}_{\mathrm{e}})_{\alpha}$, carried by $\alpha$-species is given as

$$(\mathbf{I}_{\mathrm{e}})_{\alpha} = \mathrm{F_c}z_{\alpha}\mathbf{J}_{\alpha}, \quad \alpha = 1, 2, \ldots, n, \tag{20}$$

and the electrical current density, $\mathbf{I}_{\mathrm{e}}$, carried by the mobile ions and fixed charges is given by*

$$\mathbf{I}_{\mathrm{e}} = \mathrm{F_c}\phi^{\mathrm{w}}\left(\sum_{\alpha} z_{\alpha}c_{\alpha}\mathbf{v}_{\alpha} + \omega c^{\mathrm{F}}\mathbf{v}^{\mathrm{s}}\right). \tag{21}$$

Finally, by making use of Equations (4), (9) and (20), the electrical current density can be expressed as

$$\mathbf{I}_{\mathrm{e}} = \mathrm{F_c}\sum_{\alpha} z_{\alpha}\mathbf{J}_{\alpha} = \sum_{\alpha}(\mathbf{I}_{\mathrm{e}})_{\alpha}. \tag{22}$$

### 3.2. FORCE-FLUX RELATIONS

From Equations (11), (16)–(22), we can obtain the force-flux relations for the volume flux of water, $\mathbf{J}_{\mathrm{v}}$, ionic current of each species, $(\mathbf{I}_{\mathrm{e}})_{\alpha}$, and electrical current density ($\mathbf{I}_{\mathrm{e}}$) carried by all mobile ions and fixed charges:

$$\mathbf{J}_{\mathrm{v}} = -k_0\nabla P + \sum_{\alpha} b_{\alpha}\nabla N_{\alpha} - g_0\nabla\psi, \tag{23}$$

* In the following, summations run over all species of ions.

$$(\mathbf{I}_e)_\alpha = -b_\alpha \nabla P + \sum_\beta h_{\alpha\beta} \nabla N_\beta - g_\alpha \nabla \psi, \quad \alpha = 1, 2, \ldots, n, \tag{24}$$

$$\mathbf{I}_e \equiv \sum_\alpha (\mathbf{I}_e)_\alpha = -g_0 \nabla P + \sum_\alpha g_\alpha \nabla N_\alpha - \chi_0 \nabla \psi, \tag{25}$$

where

$$\nabla P \equiv \rho_T^w \nabla \mu^w, \tag{26}$$

$$\nabla N_\alpha = -\frac{RT\nabla[\ln(\gamma_\alpha c_\alpha)]}{z_\alpha F_c}, \quad \alpha = 1, 2, \ldots, n, \tag{27}$$

and $k_0$, $b_\alpha$, $h_{\alpha\beta}$, $g_0\left(=\sum_\alpha b_\alpha\right)$, $g_\alpha\left(=\sum_\beta h_{\alpha\beta}\right)$, and $\chi_0\left(=\sum_\alpha g_\alpha\right)$ are material parameters which are functions of the ion concentrations and the coefficients $B_{\alpha\beta}$ in Equations (11) and (13) (Gu *et al.*, 1998). In Equations (23)–(25) $k_0$ is the hydraulic permeability (or conductivity), $b_\alpha$ is the mechano-electrochemical coupling coefficient, $g_\alpha$ is the ionic conductivity, and $\chi_0$ is the tissue conductivity. Further, $b_\alpha$ and $h_{\alpha\beta}$ are given by (Gu *et al.*, 1998)

$$b_\alpha = -(\phi^w)^2 F_c z_\alpha c_\alpha B_{w\alpha} \quad \alpha = 1, 2, \ldots, n, \tag{28}$$

$$h_{\alpha\beta} = -(\phi^w F_c)^2 z_\alpha c_\alpha B_{\alpha\beta} z_\beta c_\beta \quad \alpha, \beta = 1, 2, \ldots, n. \tag{29}$$

Equations (23) and (24) are the momentum equations for water and ions in terms of the mechanochemical force ($\nabla P$), electrochemical forces ($\nabla N_\alpha$) and electrical force ($\nabla\psi$). It should be pointed out that $\nabla P$ includes the osmotic pressure effect and $\nabla N_\alpha$ is the gradient of the Nernst potential (Nernst, 1888).

Equations (23)–(25) can be used to predict and interpret many mechano-electrochemical coupling phenomena associated with the transport of water and ions through charged hydrated soft tissues, such as streaming potential and current (e.g. Maroudas, 1968; Frank and Grodzinsky, 1987a, b; Gu *et al.*, 1993a, b), anomalous osmosis (Gu *et al.*, 1997), and electro-osmosis (e.g. Grodzinsky, 1990; Gu *et al.*, 1993b) in articular cartilage. In Equation (23), the last term on the right hand side represents the fluid flow (electro-osmosis) due to the electrical force ($\nabla\psi$). The coefficient ($g_0$) for the electro-osmosis in Equation (23) is identical to the coefficient for the electrical current due to the mechanochemical force ($\nabla P$) in Equation (25). Thus, the Onsager reciprocity holds for the relations between the fluxes ($\mathbf{J}_v$, $\mathbf{I}_e$) and the forces ($\nabla P$, $\nabla\psi$) (Onsager, 1931a, b). This result is related to the assumption that $f_{\alpha\beta} = f_{\beta\alpha}$ in Equation (10) (Lai *et al.*, 1991). Therefore, Equations (23) and (25) are consistent with the phenomenological theory (Katchalsky and Curran, 1975; de Groot and Mazur, 1984).

## 4. Derivation of the Hodgkin–Huxley Equation

It is of interest to investigate the transport of water and ions across a cell membrane and the associated electrical phenomena due to ion concentration gradients (diffusion problem). To do this, let us consider a flat membrane (thickness $h$) separating two (intracellular and extracellular) media. The ion concentrations in the intracellular medium are denoted by $(c^*_\alpha)_\mathrm{i}$ and those in the extracellular medium by $(c^*_\alpha)_\mathrm{o}$. Here, '$*$' refers to the quantities *external to the membrane*. The intracellular ion concentrations are in general not equal to the extracellular ion concentrations. Let $J_\mathrm{v}$ and $I_\mathrm{e}$ denote the water volume flux and electrical current density from the intracellular to the extracellular media, respectively; and $\Delta(\,) = (\,)_\mathrm{i} - (\,)_\mathrm{o}$. Using the boundary condition that water and ion (electro)chemical potentials are continuous at the surfaces of the tissue (membrane) (Katchalsky and Curran, 1975; Lai *et al.*, 1991; Gu *et al.*, 1993a), we can obtain the following equations similar to Equations (23) and (25):

$$\mathbf{J}_\mathrm{v} = \frac{k_0 \Delta P^*}{h} - \sum_\alpha \frac{b_\alpha \Delta N^*_\alpha}{h} + \left(\sum_\alpha b_\alpha\right) \frac{\Delta \psi^*_\alpha}{h}, \tag{30}$$

$$\mathbf{I}_\mathrm{e} = \frac{g_0 \Delta P^*}{h} - \sum_\alpha \frac{g_\alpha \Delta N^*_\alpha}{h} + \left(\sum_\alpha g_\alpha\right) \frac{\Delta \psi^*_\alpha}{h}. \tag{31}$$

In the above equations, $\Delta P^* = \rho^\mathrm{w}_\mathrm{T} \Delta \mu^{\mathrm{w}*}$ and $\Delta N^*_\alpha$ is the Nernst potential of the $\alpha$-species (Nernst, 1888), given as

$$\Delta N^*_\alpha = \left[\frac{RT}{z_\alpha \mathrm{F_c}}\right] \ln \left[\frac{(\gamma^*_\alpha c^*_\alpha)_\mathrm{o}}{(\gamma^*_\alpha c^*_\alpha)_\mathrm{i}}\right]. \tag{32}$$

In arriving at Equations (30)–(32), we have assumed that the membrane is very thin and the activity coefficients of ions are constant. The material parameters, $k_0$, $g_0$, $b_\alpha$, and $g_\alpha$, in Equations (30) and (31) are the mean values for the membrane. If both electrical current and volume flux are zero, Equations (30) and (31) yield the following well-known equation for membrane potential:

$$\Delta \psi^* = \frac{\sum_\alpha (\hat{g}_\alpha \Delta N^*_\alpha)}{\sum_\alpha \hat{g}_\alpha}, \quad (\text{for } \mathbf{I}_\mathrm{e} = \mathbf{J}_\mathrm{v} = 0), \tag{33}$$

where

$$\hat{g}_\alpha = g_\alpha - \frac{g_0 b_\alpha}{k_0}. \tag{34}$$

Equation (33), presented first by Hodgkin and Huxley (1952a, b), states that the cell membrane potential under $J_\mathrm{v} = 0$ and $I_\mathrm{e} = 0$ is the mean value of weighted Nernst potentials of all ion species. The weights, $\hat{g}_\alpha$, are the ionic conductivities of $\alpha$-species at zero-volume flux.

## 5. Electro-Osmotic Flow Through Charged Hydrated Tissues

To further demonstrate the application of this theory, let us consider a case where fluid flow and ion transport through a charged hydrated soft tissue under an electrical

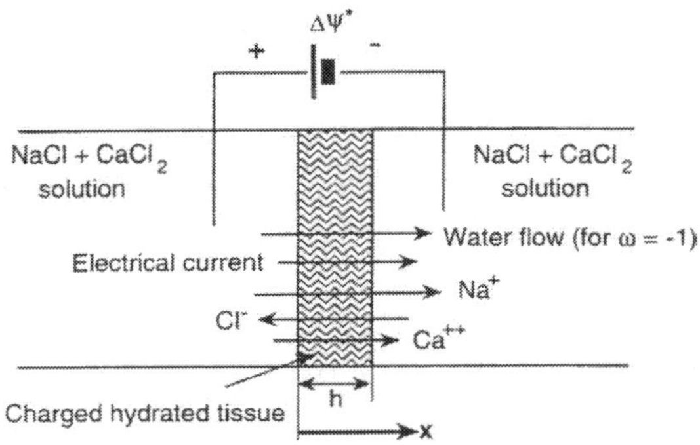


*Figure 1.* Schematic of the electro-osmosis of a multi-electrolyte solution through a charged hydrated tissue.

potential difference, see Figure 1. The tissue is initially equilibrated with an external bathing solution of NaCl and $CaCl_2$. Let $c_1^*$ denote the $Na^+$ concentration, $c_2^*$ the $Ca^{++}$ concentration, and $c_3^*$ the $Cl^-$ concentration in the bath. According to the electroneutrality condition, $c_3^* = c_1^* + 2c_2^*$. At equilibrium, there is a swelling strain of the tissue relative to the hypertonic reference configuration (Lai *et al.*, 1991). The initial swelling strain, fluid pressure, and ion concentrations within the tissue (without applied $\Delta\psi^*$) can be calculated by the following boundary conditions (Katchalsky and Curran, 1975; Lai *et al.*, 1991; Gu et al, 1993a):

$$\mu^{\mathrm{w}} = \mu^{\mathrm{w}*} \quad \text{and} \quad \tilde{\mu}^{\alpha} = \mu^{\alpha*}, \quad \alpha = 1, 2, 3, \tag{35}$$

$$\sigma \mathbf{n} = \mathbf{t}, \tag{36}$$

where

$$\mu^{\mathrm{w}*} = \mu_0^{\mathrm{w}*} + \left[ p^* - RT \sum_{\alpha} (\Phi_{\alpha}^* c_{\alpha}^*) \right] / \rho_{\mathrm{T}}^{\mathrm{w}} \quad \text{(summation over } \alpha = 1, 2, 3), \tag{37}$$

$$\tilde{\mu}^{\alpha*} = \mu_0^{\alpha*} + \left( \frac{RT}{M_{\alpha}} \right) \ln (\gamma_{\alpha}^* c_{\alpha}^*) + \frac{z_{\alpha} \mathrm{F_c} \psi^*}{M_{\alpha}}, \quad \alpha = 1, 2, 3, \tag{38}$$

and $\mathbf{t}$ is the traction vector which is set equal to zero for this case. For hydrated tissues the water content is high (e.g. $\phi^{\mathrm{w}} \sim 80\%$ for articular cartilage), thus, in Equation (10) we may assume the frictional coefficients between solid and ions, as well as those among the ions, to be negligible when compared with the frictional coefficients between water and solid and those between water and ions (Lai *et al.*, 1991; Gu *et al.*, 1993a; 1997). The electrolyte solution on the left side (L) of the tissue is assumed to be identical to that on the right side (R) and the changes in their concentrations are negligible during the electro-osmosis under $\Delta\psi^* = \psi_{\mathrm{L}}^* - \psi_{\mathrm{R}}^*$.

From Equations (11)–(19) and one-dimensional assumption, at steady state ($v^s = 0$), we can obtain the following governing equations:

$$c_1 + 2c_2 - c_3 + \omega c^F = 0, \tag{39}$$

$$c^F = c_0^F \left(1 - \frac{e_{xx}}{\phi_0^w}\right) \quad (\text{for } e_{xx} \ll 1). \tag{40}$$

$$\frac{dJ_v}{dx} = 0, \quad \frac{dJ_1}{dx} = 0, \quad \frac{dJ_2}{dx} = 0, \quad \text{and} \quad \frac{dJ_3}{dx} = 0, \tag{41}$$

$$\begin{aligned} J_v = & -k_0 \left[\frac{dp}{dx} - RT\left[(1-\Phi_1)\frac{dc_1}{dx} + (1-\Phi_2)\frac{dc_2}{dx} + (1-\Phi_3)\frac{dc_3}{dx}\right] + \right. \\ & \left. + B_w \frac{de_{xx}}{dx} - \omega c^F F_c \frac{d\psi}{dx}\right], \end{aligned} \tag{42}$$

$$J_1 = c_1 J_v - D_1 \phi^w \frac{dc_1}{dx} - D_1 \phi^w c_1 \left(\frac{F_c}{RT}\right) \frac{d\psi}{dx}, \tag{43}$$

$$J_2 = c_2 J_v - D_2 \phi^w \frac{dc_2}{dx} - 2D_2 \phi^w c_2 \left(\frac{F_c}{RT}\right) \frac{d\psi}{dx}, \tag{44}$$

$$J_3 = c_3 J_v - D_3 \phi^w \frac{dc_3}{dx} + D_3 \phi^w c_3 \left(\frac{F_c}{RT}\right) \frac{d\psi}{dx}, \tag{45}$$

$$\frac{d\sigma_{xx}}{dx} = 0, \tag{46}$$

where $\sigma_{xx} = -p + (\lambda_s + 2\mu_s)e_{xx}$.

In Equations (39)–(46), $c_1, c_2$, and $c_3$ are $Na^+$, $Ca^{++}$, and $Cl^-$ concentrations within the tissue, respectively; $c^F$ and $c_0^F$ are the values of the FCD at current state and hypertonic reference state, respectively; $J_1$, $J_2$ and $J_3$ are the $Na^+$, $Ca^{++}$ and $Cl^-$ molar fluxes, respectively; $J_v$ is the water volume flux; $D_1$, $D_2$ and $D_3$ are the ionic diffusivities for $Na^+$, $Ca^{++}$ and $Cl^-$, respectively, given by $D_\alpha = RT\phi^w c_\alpha / f_{\alpha w}$; $\sigma_{xx}$ is the $x$-component of stress tensor of the mixure, and $e_{xx}$ the $x$-component of strain tensor. In arriving at Equations (42)–(46), we have made the assumptions that all material properties are constant. Of course, the water volume fraction $\phi^w$ is not constant and is related to the deformation by (e.g. Gu *et al.*, 1993a)

$$\phi^w = \phi_0^w + (1 - \phi_0^w)e_{xx} \quad (\text{for } e_{xx} \ll 1), \tag{47}$$

where $\phi_0^w$ is the water volume fraction at the reference configuration where $e_{xx} = 0$.

There is an analytic solution to the governing Equations (39)–(46) and the boundary conditions (35) and (36):

$$e_{xx} = \text{const.}, \quad p = \text{const.}, \quad c^F = \text{const.}, \tag{48a}$$

$$c_1 = \text{const.}, \quad c_2 = \text{const.}, \quad c_3 = \text{const.}, \tag{48b}$$

$$J_v = -\omega k_0 c^F F_c \frac{\Delta\psi^*}{h}, \tag{49}$$

$$J_1 = c_1 J_v + D_1 \phi^w c_1 \left(\frac{F_c}{RT}\right) \frac{\Delta\psi^*}{h}, \tag{50a}$$

$$J_2 = c_2 J_v + 2D_2 \phi^w c_2 \left(\frac{F_c}{RT}\right) \frac{\Delta\psi^*}{h}, \tag{50b}$$

$$J_3 = c_3 J_v - D_3 \phi^w c_3 \left(\frac{F_c}{RT}\right) \frac{\Delta\psi^*}{h}, \tag{50c}$$

$$I_e = \frac{\chi_0 \Delta\psi^*}{h}, \tag{51}$$

where $\chi_0$ is the electrical conductivity of the tissue bathed in NaCl and $CaCl_2$:

$$\chi_0 = [RT(c^F)^2 k_0 + \phi^w c_1 D_1 + 4\phi^w c_2 D_2 + \phi^w c_3 D_3] \left(\frac{F_c^2}{RT}\right). \tag{52}$$

It can be seen from Equation (49) that within the theory assumptions, the electro-osmotic flow ($J_v$) in charged hydrated tissues is directly related to the values of hydraulic permeability ($k_0$) and the FCD ($c^F$), and independent of ion diffusivities ($D_1$, $D_2$, and $D_3$). The ion molar fluxes, see Equations (50), have two terms: the first term on the right hand side of Equations (50) is the convection term (due to the electro-osmotic flow, $J_v$) and the second term is the electrophoresis term. If the tissue is negatively charged, the electro-osmotic flow occurs in the same direction of the electrophoresis for cations and in the opposite direction of the electrophoresis for anions.

To highlight the nonlinear effect of the FCD ($c^F$) on ion fluxes and electrical conductivity, we calculate ion concentrations ($c_1, c_2, c_3$), water and ion fluxes ($J_v, J_1, J_2, J_3$) and electrical current as a function of $c^F$ for the following parameters: $\omega = -1$ (negatively charged), $k_0 = 10^{-15}$ m$^4$/Ns, $\phi^w = 0.8$, $D_1 = 7 \times 10^{-10}$ m$^2$/s, $D_2 = 2 \times 10^{-10}$ m$^2$/s, $D_3 = 10 \times 10^{-10}$ m$^2$/s, $T = 300°$ K, $h = 1.0$ mm, $\Delta\psi^* = 10$ mV, $c_1^* = 0.15$ M NaCl, $c_2^* = 0.015$ M $CaCl_2$, and all activity coefficients are assumed to be unity.

Figure 2 shows that for negatively charged tissues the electro-osmotic flow ($J_v$) occurs in the same direction of the electrical current ($I_e$). It also shows the electrical current depends nonlinearly on the value of the FCD, indicating the electrical conductivity is a nonlinear function of the FCD. This is because the electrical conductivity is a nonlinear function of $c^F$ (see Equation (52)) and the fact that the ion concentrations within the tissue are nonlinearly dependent on the FCD, see Figure 3. As the FCD increases, the concentrations of cations or counter-ions (with the opposite sign of

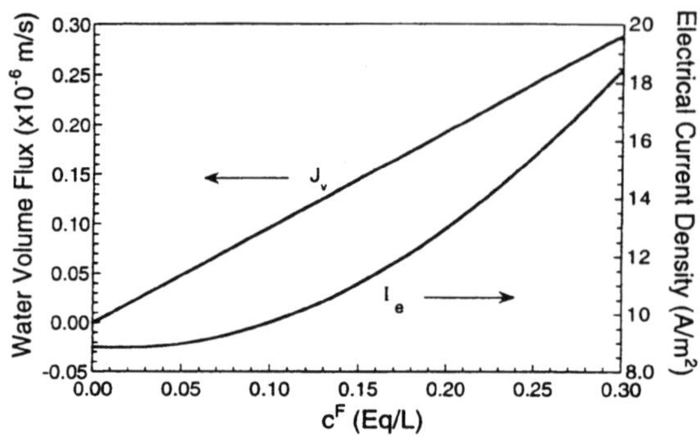


*Figure 2.* Electro-osmotic flow ($J_v$) and electrical current density ($I_e$) as a function of the fixed charge density ($c^F$). The electro-osmosis occurs in the same direction of the electrical current for the negatively charged tissues.

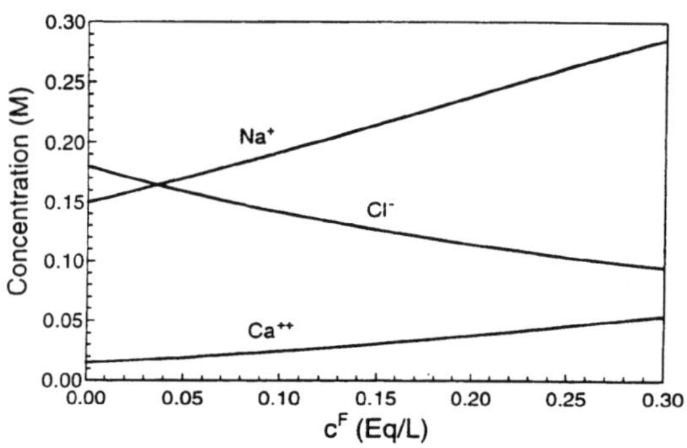


*Figure 3.* Ion concentrations within the tissues as a function of the fixed charge density.

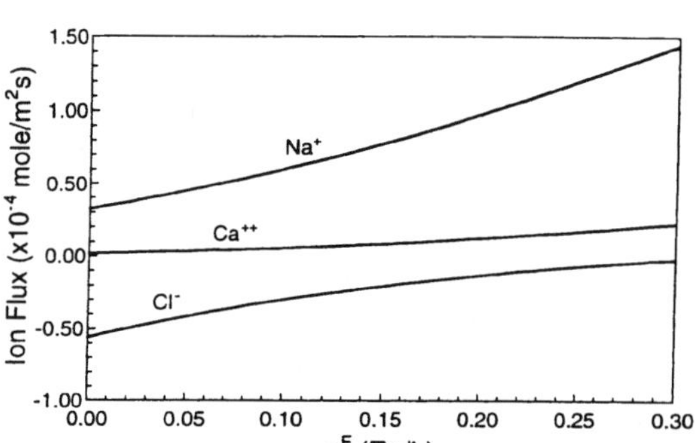


*Figure 4.* Nonlinear dependence of ion fluxes on the fixed charge density. The $Na^+$ and $Ca^{++}$ ions move in the positive $x$-direction while $Cl^-$ ions move in the negative $x$-direction.

fixed charges) increase while the concentration of anions or co-ions (with the same sign of fixed charges) decreases. This effect is known as the Donnan exclusion effect (Helfferich, 1962).

Figure 4 shows the nonlinear dependence of ion fluxes on the FCD. Note that the negative value of $Cl^-$ flux indicates the $Cl^-$ ions move in the negative $x$-direction. The magnitudes of cation fluxes increase while the magnitude of anion flux decreases with increasing FCD. This nonlinear dependence is due to the fact that ion concentrations are nonlinearly dependent on the FCD (Figure 3).

## 6. Conclusions

The transport of interstitial water and ions in charged hydrated soft tissues has been studied using a continuum mixture model in which the electrical effect has been taken into consideration. The momentum equations for water and ions have been derived in terms of the mechanochemical force (gradient of water chemical potential), the electrochemical forces (gradient of Nernst potentials) and the electrical force (gradient of electrical potential). The results for the force-flux relations for water and ions are consistent with the phenomenological theory. Moreover, with this multiphasic (component) theory, the phenomenological coefficients are now defined by the fundamental coefficients (e.g. $k_0$, $b_\alpha$, and $h_{\alpha\beta}$) used in the model. In this paper the well-known Hodgkin–Huxley equation for resting cell membrane potential has been derived and the electro-osmosis in charged hydrated soft tissues has been studied. Analyses show that the tissue fixed charge density plays a dominant role in controlling rate of transport of water and ions in charged hydrated tissues. This study demonstrates that the present continuum mixture model is capable of describing the mechano-electrokinetic behavior of charged hydrated tissues containing multi-electrolytes.

## Acknowledgement

This work was supported in part by the grants from the NIH (AR 41913) and the OREF (96-043).

## References

Achanta, S., Cushman, J. H. and Okos, M. R.: 1994, On multicomponent, multiphasic thermomechanics with interfaces, *Int. J. Engng. Sci.* **32**, 1717–1738.

Bowen, R. M.: 1980, Incompressible porous media models by use of the theory of mixtures, *Int. J. Engng. Sci.* **18**, 1129–1148.

Coelho, D., Shapiro, M., Thovert, J. F. and Adler, P. M.: 1996, Electroosmotic phenomena in porous media, *J. Colloid Interf. Sci.* **181**, 169–190.

Donnan, F. G.: 1924, The theory of membrane equilibria, *Chem. Rev.* **1**, 73–90.

de Boer, R.: 1996, Highlights in the historical development of the porous media theory: toward a consistent macroscopic theory, *Appl. Mech. Rev.* **49**, 201–262.

de Groot, S. R. and Mazur, P.: 1984, *Non-Equilibrium Theormodynamics*, Dover, New York.

Fair, J. C. and Osterle, J. F.: 1971, Reverse electrodialysis in charged capillary membranes, *J. Chem. Phys.* **54**, 3307–3316.
Frank, E. H. and Grodzinsky, A. J.: 1987a, Cartilage electromechanics – I. Electrokinetic transduction and the effects of electrolyte ph and ionic strength, *J. Biomech.* **20**, 615–627.
Frank, E. H. and Grodzinsky, A. J.: 1987b, Cartilage electromechanics – II. A continuum model of cartilage electrokinetics and correlation with experiments, *J. Biomech.* **20**, 629–639.
Frank, E. H., Grodzinsky, A. J., Phillips, S. L. and Grimshaw, P. E.: 1990, Physicochemical and bioelectrical determinants of cartilage material properties, In: V. C. Mow, A. Ratcliffe and S. L.-Y. Woo (eds), *Biomechanics of Diarthrodial Joints*, Vol. 1, Springer-Verlag, New York, pp. 363–390.
Grodzinsky, A. J.: 1990, Mechanical and electrical properties and their relevance to physiological processes, In: A. Maroudas and K. Kuettner (eds), *Methods in Cartilage Research*, Academic Press, San Diego, pp. 275–281.
Gu, W. Y., Lai, W. M. and Mow, V. C.: 1993a, Transport of fluid and ions through a porous-permeable charged-hydrated tissue, and streaming potential data on normal bovine articular cartilage, *J. Biomech.* **26**, 709–723.
Gu, W. Y., Lai, W. M. and Mow, V. C.: 1993b, Theoretical basis for measurements of cartilage fixed-charge density using streaming current and electro-osmosis effects, In: J. M. Tarbell (ed.), *Adv. Bioeng. ASME*, Vol. 26, BED, New York, pp. 55–58.
Gu, W. Y., Lai, W. M. and Mow, V. C.: 1994, A generalized triphasic theory for multi-electrolyte transport in charged hydrated soft tissues, In: M. J. Askew (ed.), *Adv. Bioeng. ASME*, Vol. 28, BED, New York, pp. 217–218.
Gu, W. Y., Lai, W. M. and Mow, V. C.: 1997, A triphasic analysis of negative osmotic flow through charged-hydrated soft tissues, *J. Biomech.* **30**, 71–78.
Gu, W. Y., Lai, W. M. and Mow, V. C.: 1998, A mixture theory for charged-hydrated soft tissues containing multi-electrolytes: passive transport and swelling behaviors, *J. Biomech. Engng.* **120**, 169–180.
Heodug, W. K. and Wong, S-W.: 1996, Hydration swelling of water-absorbing rocks: a constitutive model, *Int. J. Num. Anal. Methods Geomech.* **20**, 403–430.
Helfferich, F.: 1962, *Ion Exchange*, McGraw-Hill, New York.
Hodgkin, A. L. and Huxley, A. F.: 1952a, The components of membrane conductance in the giant axon of *Loligo*, *J. Physiol.* (London) **116**, 473–496.
Hodgkin, A. L. and Huxley, A. F.: 1952b, A quantitative description of membrane current and its application to conduction and excitation in nerve, *J. Physiol.* (London) **117**, 500–544.
Huyghe, J. M. and Janssen, J. D.: 1997, Quadriphasic mechanics of swelling incompressible porous media, *Int. J. Engng. Sci.* **35**, 793–802.
Katchalsky, A. and Curran, P. F.: 1975, *Nonequilibrium Thermodynamics in Biophysics*, 4th edn, Harvard University Press, Cambridge, Mass.
Kedem, O. and Katchalsky, A.: 1961, A physical interpretation of the phenomenological coefficients of membrane permeability, *J. Gen. Physiol.* **45**, 143–179.
Lai, W. M., Hou, J. S. and Mow, V. C.: 1991, A triphasic theory for the swelling and deformation behaviors of articular cartilage, *J. Biomech. Engng.* **113**, 245–258.
Lai, W. M., Gu, W. and Mow, V. C.: 1994, Flows of electrolytes through charged hydrated biologic tissue, *Appl. Mech. Rev.* **47** (part 2), 277–281.
Maroudas, A.: 1968, Physicochemical properties of cartilage in the light of ion exchange theory, *Biophy. J.* **8**, 575–595.
Mow, V. C., Kuei, S. C., Lai, W. M. and Armstrong, C. G.: 1980, Biphasic creep and stress relaxation of articular cartilage in compression: theory and experiments, *J. Biomech. Engng.* **102**, 73–84.
Nernst, W.: 1888, Zur Kinetik der Lösung Befindlichen Körper: Theorie der Diffusion, *Z. Phys. Chem.* **3**, 613–637.
Onsager, L.: 1931a, Reciprocal relations in irreversible processes. I, *Phys. Rev.* **37**, 405–426.
Onsager, L.: 1931b, Reciprocal relations in irreversible processes. II, *Phys. Rev.* **38**, 2265–2279.

Salzstein, R. A., Pollack, S. R. and Mak, A. F. T.: 1987, Electromechanical potentials in cortical bone – I. A continuum approach, *J. Biomech.* **20**, 261–270.

Silberberg, A.: 1982, The mechanics and thermodynamics of separation flow through porous, molecularly disperse, solid media, *Biorheology* **19**, 111–127.

Teorell, T.: 1953, Transport processes and electrical phenomena in ionic membranes, *Prog. Biophys. Physicochem.* **3**, 305–369.

# Localization Phenomena in Liquid-Saturated and Empty Porous Solids

W. EHLERS and W. VOLK
*Institut für Mechanik (Bauwesen), Universität Stuttgart, D-70569 Stuttgart, Germany.*
*e-mail: {ehlers, volk}@mechbau.uni-stuttgart.de*

(Received: 4 August 1997; in final form: 7 November 1997)

**Abstract.** Localization phenomena occur as a result of local concentrations of plastic deformations in small bands of finite width (shear bands). Porous materials, as, for instance, soil, rock, concrete and sinter materials as well as polymeric and metallic foams exhibit a strong tendency towards shear banding caused by plastic dilatation in the brittle deformation range. This kind of behaviour is of great practical importance in engineering design, for example in the study and computation of failure mechanisms in soil mechanics (base failure, slope failure, etc.). From the mathematical point of view, the computation of localization phenomena, for example within the framework of the finite element method (FEM), yields an ill-posed problem, since each mesh refinement leads to smaller shear bands until one obtains (ideally) a singular surface. Following this, regularization mechanisms should be introduced to obtain reliable and robust results.

In the present article, two natural regularization mechanisms for liquid-saturated and empty granular porous materials are discussed. These mechanisms are (1) the inclusion of additional independent degrees of freedom in the sense of the Cosserat brothers for the granular porous solid and (2) the inclusion of pore-fluid viscosity in the saturated case.

**Key words:** saturated and empty materials, elasto-plastic skeleton, micropolarity, viscous pore-fluid, shear band localization.

## 1. Introduction

The continuum mechanical description of saturated and empty porous solid materials as for instance soil, rock, concrete, sinter materials, polymeric and metallic foams, temperate ice, biological tissues, etc., naturally falls in the category of the well-founded Theory of Porous Media (TPM), which is defined by the classical theory of mixtures combined with the concept of volume fractions. The reader who is interested in the various details of this approach, is referred to the work by Truesdell and Toupin (1960), Bowen (1976, 1980), de Boer and Ehlers (1986), Ehlers (1989, 1993a, b, 1996) and de Boer (1996). Basically, porous media theories are restricted to the description of seepage and infiltration flow problems in rigid and nondominant deformable porous solid materials. However, there are a lot of problems, where the elasto-plastic deformation behaviour of the skeleton matrix is dominant and, furthermore, leads to the possibility of localization phenomena, such as shear banding. The mathematical formulation and computation of the localization problem via the finite element method (FEM) usually leads to an ill-posed problem, when single phase solid materials and empty porous materials are concerned. In particular, it is known that each mesh refinement leads to a decrease of the shear band width until one obtains (ideally) a singular surface. In reality, this behaviour is unphysical, since the shear band width can be found by biaxial tests to yield around 20–30 average diameters of the granular material under study (Mühlhaus and Vardoulakis, 1987; Desrues, 1990). The second problem arises from the fact that the generalized Jacobian of the numerical FEM system tends to become singular when localization phenomena and, as a result, material instabilities occur. Thus, systems of this type can only be solved by numerical regularization techniques, for example, line search methods, as far as no physical changes are included into the model equations.

In the present contribution, two natural physical regularization techniques are discussed. The first method consists of the fact that the individual grain rotations of granular materials are taken into consideration by introducing an independent kinematical rotation vector, which can be understood as the local average grain rotation of the porous material. Naturally, this method is related to micropolar theories in the sense of the Cosserat brothers, compare, for example, the work by Cosserat and Cosserat (1909), Günther (1958) or Eringen and Kafadar (1976). The application of micropolar theories to single phase materials has been studied by de Borst (1991, 1993), Steinmann (1994, 1995), Dietsche *et al.* (1993) and others. Applications to saturated and empty porous materials can be found in the publications by Schrefler (e.g. Schrefler *et al.*, 1995, 1996) or Ehlers and Volk (1997a, b). The second possibility is restricted to saturated media, thus yielding a natural regularization mechanism by the inclusion of pore-fluid viscosity. However, both methods can also be applied together.

In the present contribution, use is made of a binary model of an elasto-plastic micropolar porous solid skeleton and a viscous pore-liquid. Mass exchanges between the solid and the liquid materials are excluded and both constituents are assumed to exhibit the property of material incompressibility. Following this, the non-symmetric effective stress of the skeleton material, in the geometrical linear deformation range, is governed by an extension of the Hookean elasticity law, whereas the couple stress is given as a function of the elastic curvature tensor. In frame of the ideal-plastic approximation, the plastic yield limit is governed by an extension of the single-surface yield criterion by Ehlers (1993a, 1995). Based on the concept of the non-associated plasticity, an additional plastic potential function is introduced to govern the plastic strain rate. The plastic rate of curvature tensor, however, is directly computed from the plastic strain rate by use of the micropolar compatibility condition. In the framework of elastoplasticity, this condition has been used for the first time by the authors (Ehlers and Volk, 1997b).

## 2. Governing Basic Equations

By use of the TPM, fluid-saturated porous solid skeleton materials are considered as a mixture of immiscible constituents $\varphi^\alpha$ with particles $X^\alpha$ ($\alpha$ = S: solid skeleton;

$\alpha$ = F: pore-fluid). In this framework, empty skeleton materials are treated as a special case of saturated media, where the mechanical effect of the pore content is neglected. In general, at any time $t$, each spatial point $\mathbf{x}$ of the current configuration is simultaneously occupied by material points $X^\alpha$ of all constituents $\varphi^\alpha$ (superimposed continua). These particles proceed from different reference positions $\mathbf{X}_\alpha$ at time $t_0$. Thus, each constituent is assigned an individual motion function

$$\mathbf{x} = \boldsymbol{\chi}_\alpha(\mathbf{X}_\alpha, t). \tag{1}$$

The volume fractions

$$n^\alpha = n^\alpha(\mathbf{x}, t) \tag{2}$$

are defined as the local ratios of the constituent volumes $v^\alpha$ with respect to the bulk volume $v$. Thus, the saturation condition yields

$$n^{\mathrm{S}} + n^{\mathrm{F}} = 1. \tag{3}$$

Associated with each constituent $\varphi^\alpha$ is an effective or material density $\rho^{\alpha R}$ and a partial or bulk density $\rho^\alpha$. The effective density $\rho^{\alpha R}$ is defined as the local mass of $\varphi^\alpha$ per unit of $v^\alpha$, whereas the bulk density $\rho^\alpha$ exhibits the same mass per unit of $v$. Thus,

$$\rho^\alpha = n^\alpha \rho^{\alpha R}. \tag{4}$$

From (4), it is seen that the property of material incompressibility (defined by $\rho^{\alpha R}$ = const.) is not equivalent to global incompressibility, since the partial density functions can still change through changes in the volume fractions $n^\alpha$.

It follows from (1) that each constituent is assigned its own velocity field (Lagrangian description):

$$\overset{\prime}{\mathbf{x}}_\alpha = \frac{\partial \boldsymbol{\chi}_\alpha(\mathbf{X}_\alpha, t)}{\partial t}. \tag{5}$$

Thus, with the aid of the inverse motion function $\boldsymbol{\chi}_\alpha^{-1}$, the corresponding Eulerian description reads

$$\overset{\prime}{\mathbf{x}}_\alpha = \overset{\prime}{\mathbf{x}}_\alpha\,(\mathbf{x}, t). \tag{6}$$

Assuming that $\Gamma = \Gamma(\mathbf{x}, t)$ is an arbitrary, continuous and continuously differentiable function. Then, the material time derivatives of $\Gamma$ corresponding to the individual motion functions of $\varphi^\alpha$ yield

$$(\Gamma)'_\alpha := \frac{\mathrm{d}_\alpha}{\mathrm{d}t}\Gamma = \frac{\partial \Gamma}{\partial t} + \operatorname{grad}\Gamma \cdot \overset{\prime}{\mathbf{x}}_\alpha\,. \tag{7}$$

Therein, the operator 'grad ( · )' is defined as the partial derivative of ( · ) with respect to the position vector $\mathbf{x}$ of the actual configuration.

In the framework of saturated or empty porous skeleton materials, the solid motion is considered by the Lagrangian formulation, where

$$\mathbf{u}_S = \mathbf{x} - \mathbf{X}_S, \tag{8}$$

is the displacement vector, while the pore-fluid is characterized by a modified Eulerian description, where the seepage velocity $\mathbf{w}_F$ is taken as the kinematic variable:

$$\mathbf{w}_F = \overset{\prime}{\mathbf{x}}_F - \overset{\prime}{\mathbf{x}}_S. \tag{9}$$

From (1) and (8), the material deformation gradient and the displacement gradient of the solid skeleton are

$$\mathbf{F}_S = \text{Grad}_S\, \mathbf{x}, \qquad \mathbf{H}_S = \text{Grad}_S\, \mathbf{u}_S, \tag{10}$$

where 'Grad$_S$ $(\cdot)$' defines the partial derivative of $(\cdot)$ with respect to the reference position $\mathbf{X}_S$ of $\varphi^S$. Given the displacement gradient, one easily concludes to the classical measures of the geometrically linear theory:

$$\begin{aligned} \mathbf{H}_{S\,\text{sym}} &= \tfrac{1}{2}(\mathbf{H}_S + \mathbf{H}_S^T) =: \boldsymbol{\varepsilon}_S, \\ \mathbf{H}_{S\,\text{skw}} &= \tfrac{1}{2}(\mathbf{H}_S - \mathbf{H}_S^T) =: \boldsymbol{\varphi}_S \times \mathbf{I}. \end{aligned} \tag{11}$$

Therein, $\boldsymbol{\varepsilon}_S$ is the linear Lagrangian solid strain tensor, $\boldsymbol{\varphi}_S$ is the so-called continuum rotation vector, $\mathbf{I}$ is the second-order identity (fundamental tensor of second order), and $(\cdot)^T$ is the transpose of $(\cdot)$. The external tensor product between vectors and tensors (cf. de Boer, 1982) is defined by

$$\boldsymbol{\varphi}_S \times \mathbf{I} = -\overset{3}{\mathbf{E}}\boldsymbol{\varphi}_S; \tag{12}$$

$\overset{3}{\mathbf{E}}$ is the Ricci permutation or the fundamental tensor of the third order, respectively. Given $(11)_2$, it is straightforward to obtain $\boldsymbol{\varphi}_S$ as a function of $\mathbf{H}_S$:

$$\boldsymbol{\varphi}_S = -\tfrac{1}{2}\overset{3}{\mathbf{E}}(\mathbf{H}_{S\,\text{skw}}). \tag{13}$$

Extending the continuum mechanical description of the skeleton material, additional micropolar degrees of freedom are introduced via the independent rotation $\overset{*}{\boldsymbol{\varphi}}_S$, thus defining the total average grain rotation $\bar{\boldsymbol{\varphi}}_S$, in the framework of geometrically linear theories, as the sum of the continuum rotation and the additional micropolar rotation, that is to say

$$\bar{\boldsymbol{\varphi}}_S = \boldsymbol{\varphi}_S + \overset{*}{\boldsymbol{\varphi}}_S\,. \tag{14}$$

Based on the additional micropolar degrees of freedom, the linear Cosserat strain tensor $\boldsymbol{\varepsilon}_{Sc}$ and the linear curvature tensor $\bar{\boldsymbol{\kappa}}_S$ yield

$$\boldsymbol{\varepsilon}_{Sc} = \mathbf{H}_S + \overset{3}{\mathbf{E}}\bar{\boldsymbol{\varphi}}_S, \qquad \bar{\boldsymbol{\kappa}}_S = \text{Grad}_S\, \bar{\boldsymbol{\varphi}}_S. \tag{15}$$

The symmetric and skew symmetric parts of the Cosserat strain are

$$\begin{aligned} \boldsymbol{\varepsilon}_{\mathrm{Sc\,sym}} &= \tfrac{1}{2}\,(\mathbf{H}_{\mathrm{S}} + \mathbf{H}_{\mathrm{S}}^{\mathrm{T}})\,, \\ \boldsymbol{\varepsilon}_{\mathrm{Sc\,skw}} &= \tfrac{1}{2}\,(\mathbf{H}_{\mathrm{S}} - \mathbf{H}_{\mathrm{S}}^{\mathrm{T}}) + \overset{3}{\mathbf{E}}\,\bar{\boldsymbol{\varphi}}_{\mathrm{S}}. \end{aligned} \tag{16}$$

Therein, $\boldsymbol{\varepsilon}_{\mathrm{Sc\,sym}}$ equals the linearized Lagrangian strain $\boldsymbol{\varepsilon}_{\mathrm{S}}$ of nonpolar materials, whereas

$$\boldsymbol{\varepsilon}_{\mathrm{Sc\,skw}} = \overset{3}{\mathbf{E}}(\bar{\boldsymbol{\varphi}}_{\mathrm{S}} - \boldsymbol{\varphi}_{\mathrm{S}}) = \overset{3}{\mathbf{E}}\,\overset{*}{\boldsymbol{\varphi}}_{\mathrm{S}} \tag{17}$$

is a tensorial measure for the additional micropolar rotation $\overset{*}{\boldsymbol{\varphi}}_{\mathrm{S}}$. Finally, with the aid of (11), $(16)_1$ and (17), the Cosserat strain $\boldsymbol{\varepsilon}_{\mathrm{Sc}}$ can also be expressed by

$$\begin{aligned} \boldsymbol{\varepsilon}_{\mathrm{Sc}} &= \tfrac{1}{2}\,(\mathbf{H}_{\mathrm{S}} + \mathbf{H}_{\mathrm{S}}^{\mathrm{T}}) + \overset{3}{\mathbf{E}}\,\overset{*}{\boldsymbol{\varphi}}_{\mathrm{S}} \\ &= \boldsymbol{\varepsilon}_{\mathrm{S}} + \overset{3}{\mathbf{E}}\,\overset{*}{\boldsymbol{\varphi}}_{\mathrm{S}}\,. \end{aligned} \tag{18}$$

It is important to recognize that the Cosserat strain and the linear curvature tensor are related to each other by the micropolar compatibility condition, which can be obtained from $(10)_2$ and (15) by use of the Schwarzian exchangeability rule of partial derivatives:

$$\mathrm{Grad}_{\mathrm{S}}\,\boldsymbol{\varepsilon}_{\mathrm{Sc}} - \mathrm{Grad}_{\mathrm{S}}^{\overset{23}{\mathrm{T}}}\,\boldsymbol{\varepsilon}_{\mathrm{Sc}} = (\overset{3}{\mathbf{E}}\,\bar{\boldsymbol{\kappa}}_{\mathrm{S}})^{\underline{3}} - (\overset{3}{\mathbf{E}}\,\bar{\boldsymbol{\kappa}}_{\mathrm{S}})^{\underline{3}\overset{23}{\mathrm{T}}}. \tag{19}$$

Therein, the transpositions $(\,\cdot\,)^{\overset{ik}{\mathrm{T}}}$ indicate an exchange of the $i$th and $k$th basis systems included into the tensor basis of higher order tensors. The additional topscript $(\,\cdot\,)^{\underline{n}}$ defines the included contraction $(\,\cdot\,)$ to yield a tensor of $n$th order (de Boer, 1982). In a simple index notation, the above compatibility condition has firstly been proposed by Nowacki in 1969 (cf. Nowacki, 1986). In general, (19) consists of 27 scalar equations. However, since there are only 9 independent equations incorporated into (19), this relation can be solved with respect to $\bar{\boldsymbol{\kappa}}_{\mathrm{S}}$ by use of the methods of the general tensor calculus. Thus,

$$\bar{\boldsymbol{\kappa}}_{\mathrm{S}} = \tfrac{1}{2}\overset{3}{\mathbf{E}}(\mathrm{Grad}_{\mathrm{S}}\,\boldsymbol{\varepsilon}_{\mathrm{Sc}} + \mathrm{Grad}_{\mathrm{S}}^{\overset{13}{\mathrm{T}}}\,\boldsymbol{\varepsilon}_{\mathrm{Sc}} - \mathrm{Grad}_{\mathrm{S}}^{\overset{23}{\mathrm{T}}}\,\boldsymbol{\varepsilon}_{\mathrm{Sc}})^{\underline{2}}. \tag{20}$$

This representation of the micropolar compatibility condition is a novelty and has firstly been published, as far as the authors are aware, by Ehlers and Volk (1997b). As a result of (20), it follows immediately that once $\boldsymbol{\varepsilon}_{\mathrm{Sc}}$ is given, it is straightforward to compute $\bar{\boldsymbol{\kappa}}_{\mathrm{S}}$.

Proceeding from the general formulation of the TPM, the governing balance equations of the problem under study can be taken from the literature, for example, Bowen (1980), de Boer and Ehlers (1986) or Ehlers (1996). However, under the present circumstances, the balance relations of the standard formulation of the TPM must be extended by the introduction of elements of multiphase micropolar theories (cf. Diebels and Ehlers, 1996). Considering the special case of quasi-static processes of a liquid-saturated porous skeleton material within a purely mechanical description, where inertia terms are neglected and mass exchanges between the solid and the fluid constituents are excluded, the balance equations of mass, momentum and moment of momentum (m.o.m.) yield

*Balance of mass:*

$$0 = (\rho^\alpha)'_\alpha + \rho^\alpha \operatorname{div} \overset{\prime}{\mathbf{x}}_\alpha . \tag{21}$$

*Balance of momentum:*

$$\begin{aligned} \mathbf{0} &= \operatorname{div} \mathbf{T}^\alpha + \rho^\alpha \mathbf{b}^\alpha + \hat{\mathbf{p}}^\alpha, \\ \hat{\mathbf{p}}^S &+ \hat{\mathbf{p}}^F = \mathbf{0}. \end{aligned} \tag{22}$$

*Balance of m.o.m.:*

$$\begin{aligned} [2\,\rho^\alpha (\bar{\mathbf{W}}_\alpha\, \mathbf{\Theta}^\alpha)_{\text{sym}}]\, \bar{\boldsymbol{\omega}}_\alpha &= \mathbf{I} \times \mathbf{T}^\alpha + \operatorname{div} \mathbf{M}^\alpha + \rho^\alpha \mathbf{c}^\alpha + \hat{\mathbf{m}}^\alpha, \\ \hat{\mathbf{m}}^S + \hat{\mathbf{m}}^F &= \mathbf{0}. \end{aligned} \tag{23}$$

In these relations, $\mathbf{T}^\alpha$ and $\mathbf{M}^\alpha$ are the non-symmetric Cauchy stress tensor and the couple stress tensor of micropolar materials, $\mathbf{b}^\alpha$ and $\mathbf{c}^\alpha$ are the external body force and body couple per unit of mass, $\mathbf{\Theta}^\alpha$ is the tensor of microinertia, $\bar{\boldsymbol{\omega}}_\alpha = (\bar{\boldsymbol{\varphi}}_\alpha)'_\alpha$ is the total rotational velocity, and $\hat{\mathbf{p}}^\alpha$ and $\hat{\mathbf{m}}^\alpha$ represent the momentum production and the moment of momentum production terms, respectively. Furthermore, 'div $(\cdot)$' is the divergence operator corresponding to 'grad $(\cdot)$', whereas

$$\bar{\mathbf{W}}_\alpha = \bar{\boldsymbol{\omega}}_\alpha \times \mathbf{I} \tag{24}$$

denotes the skew-symmetric micropolar gyration tensor (Diebels and Ehlers, 1996). Finally, $\mathbf{I} \times \mathbf{T}^\alpha$ defines twice the axial vector corresponding to the skew-symmetric part of $\mathbf{T}^\alpha$ (de Boer, 1982):

$$\mathbf{I} \times \mathbf{T}^\alpha = -\overset{3}{\mathbf{E}}\, \mathbf{T}^\alpha = 2\overset{A}{\mathbf{t}}{}^\alpha . \tag{25}$$

Considering the case of a binary model of a micropolar skeleton and a nonpolar pore-fluid, where $\hat{\mathbf{m}}^F = 0$ (de Boer *et al.*, 1991), and assuming the well-known assumption of spherical microparticles, where

$$\mathbf{\Theta}^\alpha = \Theta^\alpha \mathbf{I} \longrightarrow (\bar{\mathbf{W}}_\alpha\, \mathbf{\Theta}^\alpha)_{\text{sym}} = \mathbf{0}, \tag{26}$$

the m.o.m. balance (23) is substituted by

$$\begin{aligned} \mathbf{0} &= \mathbf{I} \times \mathbf{T}^S + \operatorname{div} \mathbf{M}^S + \rho^S \mathbf{c}^S, \\ \mathbf{0} &= \mathbf{I} \times \mathbf{T}^F \longrightarrow \mathbf{T}^F = (\mathbf{T}^F)^T. \end{aligned} \tag{27}$$

Based on the property of material incompressibility of both constituents ($\rho^{\alpha R}$ = const.), the sum of the individual mass balance equations (21) of solid and the fluid materials, the saturation condition (3) and the density relation (4) combine to yield

$$\operatorname{div}[n^{\mathrm{F}}\mathbf{w}_{\mathrm{F}} + (\mathbf{u}_{\mathrm{S}})'_{\mathrm{S}}] = 0. \tag{28}$$

This equation represents the volume balance relation or the incompressibility constraint of the binary material under study.

## 3. Constitutive Setting

Once the basic relations of Section 2 are given, the model is properly defined except for the fact that constitutive equations must be added for the solid and the fluid stress tensors $\mathbf{T}^{\alpha}(\alpha \in \{\mathrm{S}, \mathrm{F}\})$, the solid couple stress tensor $\mathbf{M}^{\mathrm{S}}$ and for the interaction force $\hat{\mathbf{p}}^{\mathrm{F}} = -\hat{\mathbf{p}}^{\mathrm{S}}$. As a result of the incompressibility assumption of both materials, the expressions for the stress tensors and for the interaction force consist of two terms each:

$$\begin{aligned} \mathbf{T}^{\alpha} &= -n^{\alpha} p\,\mathbf{I} + \mathbf{T}^{\alpha}_{\mathrm{E}}, \\ \hat{\mathbf{p}}^{\mathrm{F}} &= p \operatorname{grad} n^{\mathrm{F}} + \hat{\mathbf{p}}^{\mathrm{F}}_{\mathrm{E}} . \end{aligned} \tag{29}$$

Therein, $p$ is a Lagrangian multiplier, which can be identified as the effective liquid pressure, while the so-called extra terms, index $(\cdot)_{\mathrm{E}}$, are governed by the solid deformation and the pore-fluid flow (Ehlers, 1989). Furthermore, it is usual to neglect the liquid extra stress (the frictional stress) in comparison to the other terms included in the momentum balance relation of $\varphi^{\mathrm{F}}$. Thus, the a priori assumption $\mathbf{T}^{\mathrm{F}}_{\mathrm{E}} = \mathbf{0}$ is applied. The internal friction between the skeleton and the viscous pore-liquid is taken into account by the effective drag force or the interaction force, respectively:

$$\hat{\mathbf{p}}^{\mathrm{F}}_{\mathrm{E}} = -\frac{(n^{\mathrm{F}})^2\,\gamma^{\mathrm{F}R}}{k^{\mathrm{F}}}\,\mathbf{w}_{\mathrm{F}} . \tag{30}$$

Therein, $\gamma^{\mathrm{F}R} = \rho^{\mathrm{F}R}|\mathbf{b}|$ with the gravity $\mathbf{b}$ is the effective (true) specific weight of the pore-liquid and $k^{\mathrm{F}}$ is the Darcy permeability coefficient, which is related to the intrinsic (or physical) permeability parameter $K^{\mathrm{S}}$ of the skeleton material via

$$k^{\mathrm{F}} = \frac{\gamma^{\mathrm{F}R}}{\mu^{\mathrm{F}R}}\,K^{\mathrm{S}}, \tag{31}$$

$\mu^{\mathrm{F}R}$ is the effective shear viscosity of $\varphi^{\mathrm{F}}$. It follows from (31) that the permeability parameter $k^{\mathrm{F}}$, in the sense of a macroscopic quantity, provides information on both the local size and isotropic structure of the pore-space (included in $K^{\mathrm{S}}$) and the fluid viscosity. Furthermore, the permeability can be given as a function of the porosity or, alternatively, of the solid deformation. In the present article, it is assumed that

$$k^{\mathrm{F}} = k^{\mathrm{F}}_0[1 + m^{\mathrm{F}}\,n^{\mathrm{S}}_{0\mathrm{S}}(\boldsymbol{\varepsilon}_{\mathrm{Sc}} \cdot \mathbf{I})]. \tag{32}$$

Therein, $m^{\mathrm{F}}$ is a material number, and $n^{\mathrm{S}}_{0\mathrm{S}}$ is the solid volume fraction in the reference configuration of $\varphi^{\mathrm{S}}$.

Proceeding from a geometrically linear description of elasto-plastic frictional micropolar skeleton materials, the following considerations are based on the assumption of an additive split of the Cosserat strain $\boldsymbol{\varepsilon}_{\mathrm{Sc}}$ into elastic and plastic parts. Furthermore, by use of the micropolar compatibility condition (20), it is straightforward to conclude that an additive split of $\boldsymbol{\varepsilon}_{\mathrm{Sc}}$ includes an additive split of the curvature tensor $\bar{\boldsymbol{\kappa}}_{\mathrm{S}}$ (Ehlers and Volk, 1997b). Thus,

$$
\begin{aligned}
\boldsymbol{\varepsilon}_{\mathrm{Sc}} &= \boldsymbol{\varepsilon}_{\mathrm{Sce}} + \boldsymbol{\varepsilon}_{\mathrm{Scp}}, \\
\bar{\boldsymbol{\kappa}}_{\mathrm{S}} &= \bar{\boldsymbol{\kappa}}_{\mathrm{Se}} + \bar{\boldsymbol{\kappa}}_{\mathrm{Sp}}.
\end{aligned} \tag{33}
$$

Concerning the constitutive equation for the solid extra stress, or the effective stress, respectively, an extension of the Hookean elasticity model towards nonsymmetric stresses yields

$$
\mathbf{T}^{\mathrm{S}}_{\mathrm{E}} = [2\mu^{\mathrm{S}} \overset{4}{\mathbf{I}}_{\mathrm{sym}} + 2\mu^{\mathrm{S}}_{\mathrm{c}} \overset{4}{\mathbf{I}}_{\mathrm{skw}} + \lambda^{\mathrm{S}}(\mathbf{I} \otimes \mathbf{I})]\boldsymbol{\varepsilon}_{\mathrm{Sce}}, \tag{34}
$$

where

$$
\begin{aligned}
\overset{4}{\mathbf{I}}_{\mathrm{sym}} &= \tfrac{1}{2}[(\mathbf{I} \otimes \mathbf{I})^{\overset{23}{\mathrm{T}}} + (\mathbf{I} \otimes \mathbf{I})^{\overset{13}{\mathrm{T}}}], \\
\overset{4}{\mathbf{I}}_{\mathrm{skw}} &= \tfrac{1}{2}[(\mathbf{I} \otimes \mathbf{I})^{\overset{23}{\mathrm{T}}} - (\mathbf{I} \otimes \mathbf{I})^{\overset{13}{\mathrm{T}}}].
\end{aligned} \tag{35}
$$

Therein, $\overset{4}{\mathbf{I}}_{\mathrm{sym}}$ and $\overset{4}{\mathbf{I}}_{\mathrm{skw}}$ are fundamental tensors of the fourth order. The application of these tensors to any arbitrary tensor of the second order yields its symmetric and its skew-symmetric parts. The material parameters $\mu^{\mathrm{S}}$ and $\lambda^{\mathrm{S}}$ are the Lamé constants of the porous skeleton material, whereas $\mu^{\mathrm{S}}_{\mathrm{c}}$ is an additional parameter governing the influence of the skew-symmetric part of the elastic Cosserat strain $\boldsymbol{\varepsilon}_{\mathrm{Sce}}$ on the effective stress of the skeleton material. In analogy to the strain relation (18), the symmetric part of $\mathbf{T}^{\mathrm{S}}_{\mathrm{E}}$ represents the solid stress for non-polar skeleton materials, whereas the skew-symmetric part of micropolar skeleton materials is directly related to the independent micropolar rotation $\overset{*}{\boldsymbol{\varphi}}_{\mathrm{S}}$. Finally, it should be noted that in case of materially incompressible skeletons, $\lambda^{\mathrm{S}}$ includes the structural compression modulus (Ehlers, 1993 a, b), thus governing the influence of elastic volume strains based on local porosity variations on the macroscopic stress behaviour.

Concerning a constitutive equation for the couple stress tensor, the simplified set-up

$$
\mathbf{M}^{\mathrm{S}} = 2\mu^{\mathrm{S}}(l^{\mathrm{S}}_{\mathrm{c}})^2\,\bar{\boldsymbol{\kappa}}_{\mathrm{Se}} \tag{36}
$$

by de Borst (1991) is applied. Therein, $l^{\mathrm{S}}_{\mathrm{c}}$ can be interpreted as an internal length scale parameter. This parameter implicitly offers the possibility to include the shear band width into the model; however, $l^{\mathrm{S}}_{\mathrm{c}}$ is not the shear band width itself.

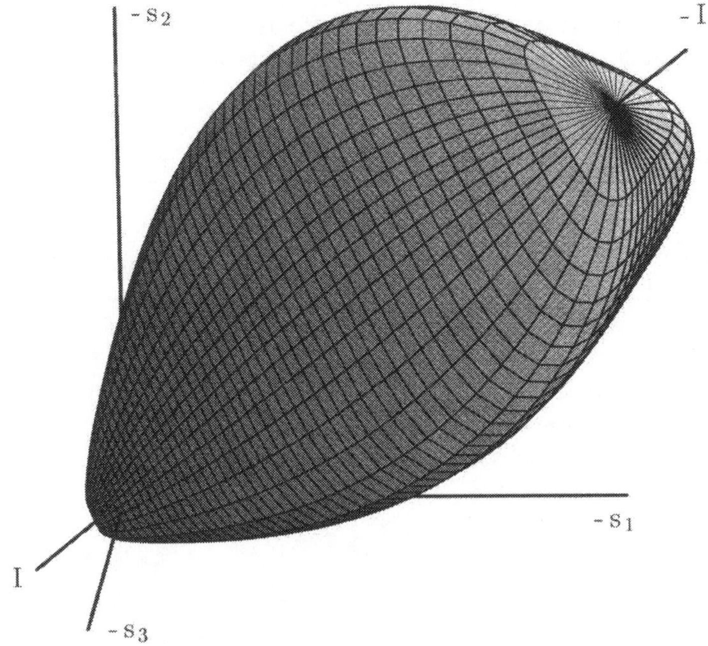


*Figure 1.* Single-surface yield criterion for nonpolar cohesive-frictional materials; $s_1$, $s_2$, $s_3$: principal stresses of $\mathbf{T}^S_{E\,sym}$ (tension positive).

In order to describe the plastic material properties of both nonpolar and polar skeleton materials, one has to introduce a convenient yield function to bound the elastic domain, where, in the present article, use is made of the ideal plasticity concept. Furthermore, evolution equations for the nonsymmetric plastic strain rate $(\boldsymbol{\varepsilon}_{Scp})'_S$ and the plastic rate of curvature tensor $(\bar{\boldsymbol{\kappa}}_{Sp})'_S$ must be formulated.

Concerning the introduction of a yield function for frictional porous materials, an extension of the single-surface yield criterion by Ehlers (1993a, 1995) is applied, compare Figure 1.

By extension of the original yield function towards micropolar cohesive-frictional materials, one obtains (Ehlers and Volk, 1997b)

$$
\begin{aligned}
F^c &= \Phi^{1/2} + \beta\,\mathrm{I} + \epsilon\,\mathrm{I}^2 - \kappa = 0,\\
\Phi &= \mathrm{II}_{D\,sym}(1+\gamma\vartheta)^m + \tfrac{1}{2}\alpha\mathrm{I}^2 + \delta^2\,\mathrm{I}^4 + k_1\,\mathbf{M}^S\cdot\mathbf{M}^S + k_2\,\mathrm{II}_{skw},\\
\vartheta &= \frac{\mathrm{III}_{D\,sym}}{(\mathrm{II}_{D\,sym})^{3/2}}.
\end{aligned}
\tag{37}
$$

In this representation, $\{\alpha,\ \beta,\ \gamma,\ \delta,\ \epsilon,\ m,\ \kappa,\ k_1,\ k_2\}$ is a set of nine material parameters; I, $\mathrm{II}_{D\,sym}$ and $\mathrm{III}_{D\,sym}$ are the first and deviatoric (negative) second and third principal invariants of the symmetric part of the effective stress $\mathbf{T}_E^S$, whereas $\mathrm{II}_{skw}$ defines the second principal invariant of the skew-symmetric part of $\mathbf{T}_E^S$:

$$\begin{aligned} &\mathrm{I} = \mathbf{T}_E^S \cdot \mathbf{I}, \qquad \mathrm{II}_{D\,sym} = \tfrac{1}{2}\,\mathbf{T}_{E\,sym}^{SD} \cdot \mathbf{T}_{E\,sym}^{SD}, \\ &\mathrm{II}_{skw} = \tfrac{1}{2}\,\mathbf{T}_{E\,skw}^{S} \cdot \mathbf{T}_{E\,skw}^{S}, \qquad \mathrm{III}_{D\,sym} = \tfrac{1}{3}\,\mathbf{T}_{E\,sym}^{SD} \cdot (\mathbf{T}_{E\,sym}^{SD}\,\mathbf{T}_{E\,sym}^{SD}). \end{aligned} \tag{38}$$

The index $(\cdot)^D$ characterizes the deviatoric part of the respective tensorial object. For an extended discussion of the single-surface criterion (37) and its application to nonpolar cohesive-frictional materials, the reader is referred to Ehlers (1995) and Ehlers and Volk (1997b). By use of the yield criterion (37), softening effects occurring in the localization zones of dilatant frictional materials can be described even in the framework of the ideal plasticity concept. Furthermore, the ideal plasticity concept implies that all material parameters of the yield criterion are constants, where the following relations between $\beta$ and $\kappa$ and the angle of internal friction $\phi$ and the cohesion $c$ hold

$$\beta = \tfrac{1}{3}\sin\phi, \qquad \kappa = c\cos\phi. \tag{39}$$

In addition to the single-surface yield function defined above, evolution equations (flow rules) for the plastic Cosserat strain rate $(\boldsymbol{\varepsilon}_{Scp})'_S$ and the plastic rate of curvature tensor $(\bar{\boldsymbol{\kappa}}_{Sp})'_S$ must be considered. Concerning the representation of these equations, one has to take into account that the usual associated flow rule concept cannot be applied to frictional materials (Lade and Duncan, 1973). In particular, the use of an associated flow rule for $(\boldsymbol{\varepsilon}_{Scp})'_S$ would predict wrong results in both the description of the deviatoric part and of the volumetric part (dilatancy and contractancy) of the plastic strain rate. Thus, in the framework of the nonassociated plasticity, it is convenient to introduce an additional plastic potential function:

$$\begin{aligned} &G^c = \Gamma^{1/2} + \beta\,\mathrm{I} + \epsilon\,\mathrm{I}^2 - g(\mathrm{I}) = 0, \\ &\Gamma = \mathrm{II}_{D\,sym} + \tfrac{1}{2}\alpha\,\mathrm{I}^2 + \delta^2\,\mathrm{I}^4 + k_2\,\mathrm{II}_{skw}. \end{aligned} \tag{40}$$

The difference between the yield function (37) and the plastic potential function (40) lies in the fact that (40) exhibits, in the principal stress space of nonpolar materials, a circular shape in the deviatoric plane, whereas (37) is characterized by a triangular shape with rounded corners. The function $g(\mathrm{I})$ serves to adjust the dilatation angle $\nu_p$ to test results, compare Figure 2.

Once the plastic potential is given, it is straightforward to determine the plastic strain rate by differentiation of $G^c$ with respect to the effective solid stress:

$$(\boldsymbol{\varepsilon}_{Scp})'_S = \Lambda\,\frac{\partial G^c}{\partial \mathbf{T}_E^S}\,. \tag{41}$$

Therein, $\Lambda$ is the usual plastic multiplier. In good accordance with experimental data, this relation predicts co-axial behaviour of the deviatoric part of $(\boldsymbol{\varepsilon}_{Scp})'_S$. Concerning the volumetric part, the function $g(\mathrm{I})$ is chosen in such a way that the direction of

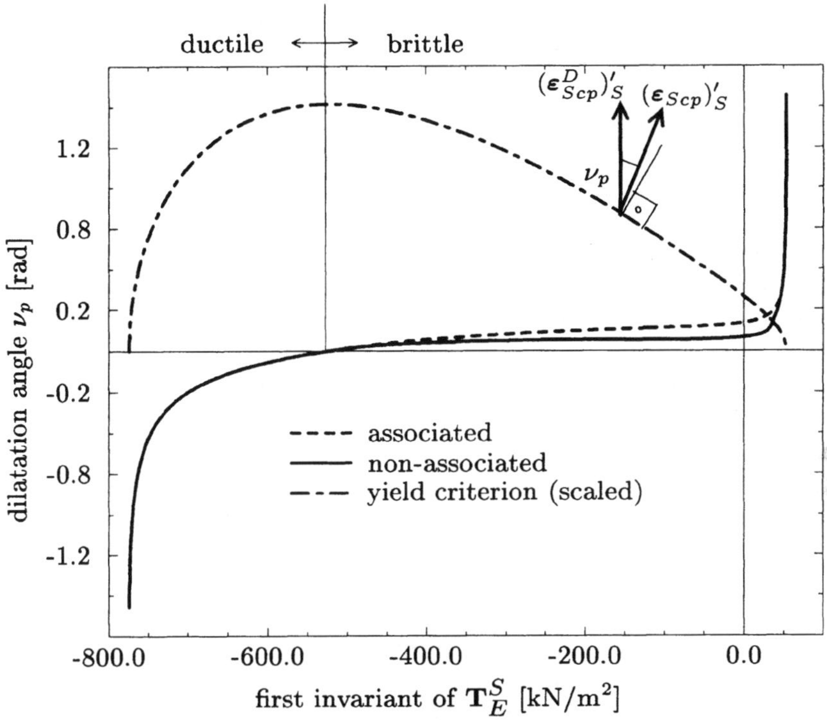


*Figure 2.* Dilatation angle of the nonassociated flow rule.

$(\boldsymbol{\varepsilon}_{\mathrm{Scp}})'_{\mathrm{S}}$, in the brittle range, approximately lies in the middle between the associated and the deviatoric direction. In the ductile range and at the tip of the yield surface, $g(\mathrm{I})$ predicts associated behaviour, compare Figure 2.

In the literature on micropolar elasto-plasticity, it is usually assumed that the flow rule for $(\bar{\boldsymbol{\kappa}}_{\mathrm{Sp}})'_{\mathrm{S}}$ is independent from the flow rule for $(\boldsymbol{\varepsilon}_{\mathrm{Scp}})'_{\mathrm{S}}$. Consequently, an additional evolution equation is proposed. However, it was found that one obtains a very inconvenient convergence behaviour in the framework of finite element computations, when the micropolar degrees of freedom are active, for example in the localization zone. This inconvenience immediately vanishes, when the micropolar compatibility condition is used, that is when the independent flow rule for $(\bar{\boldsymbol{\kappa}}_{\mathrm{Sp}})'_{\mathrm{S}}$ is substituted by an evolution equation directly related to (20) and (41) (Ehlers and Volk, 1997b). Thus,

$$(\bar{\boldsymbol{\kappa}}_{\mathrm{Sp}})'_{\mathrm{S}} = \tfrac{1}{2}\overset{3}{\mathbf{E}}\left(\mathrm{Grad}_{\mathrm{S}}(\boldsymbol{\varepsilon}_{\mathrm{Scp}})'_{\mathrm{S}} + \mathrm{Grad}_{S}^{\overset{13}{\mathrm{T}}}(\boldsymbol{\varepsilon}_{\mathrm{Scp}})'_{\mathrm{S}} - \mathrm{Grad}_{S}^{\overset{23}{\mathrm{T}}}(\boldsymbol{\varepsilon}_{\mathrm{Scp}})'_{\mathrm{S}}\right)^{\underline{2}} . \tag{42}$$

## 4. Weak Formulation and Numerical Tools

Concerning the quasi-static problem under consideration, the finite element discretization is based on the displacement–rotation–pressure formulation of the volumetrically coupled solid–fluid problem. This formulation proceeds from the fact that the seepage velocity $\mathbf{w}_F$ can be eliminated by use of the momentum balance for the pore-fluid $(22)_1$ together with the constitutive Equations (29) and (30) and the density relation (4). Thus, one obtains the Darcy Equation

$$n^F \mathbf{w}_F = -\frac{k^F}{\gamma^{FR}} (\operatorname{grad} p - \rho^{FR} \mathbf{b}). \tag{43}$$

As a result, the independent fields are given by the solid displacement $\mathbf{u}_S$, the effective liquid pressure $p$ (pore pressure) and the total average grain rotation $\bar{\boldsymbol{\varphi}}_S$. The corresponding three equations of the weak formulation can now be obtained from the kinematics, the balance relations and the constitutive equations of the model under study. In particular, the first equation is given by the sum of the solid and fluid momentum balance Equations $(22)_1$ multiplied by the test function $\delta\mathbf{u}_S$:

$$\begin{aligned} &\int_V (\mathbf{T}_E^S - p\mathbf{I}) \cdot \operatorname{grad} \delta\mathbf{u}_S \, \mathrm{d}v - \\ &\quad - \int_V (n^F \rho^{FR} + n^S \rho^{SR}) \, \mathbf{b} \cdot \delta\mathbf{u}_S \, \mathrm{d}v = \int_S \mathbf{t} \cdot \delta\mathbf{u}_S \, \mathrm{d}a. \end{aligned} \tag{44}$$

Therein, $\mathbf{t}$ is the external load vector acting upon both constituents. In addition, the assumption $\mathbf{b}^\alpha = \mathbf{b} \; \forall \, \varphi^\alpha$ has been applied. The second equation results from the incompressibility constraint of the model given by the volume balance (28) of the whole aggregate multiplied by the test function $\delta p$:

$$\int_V \operatorname{div} [n^F \mathbf{w}_F + (\mathbf{u}_S)'_S] \, \delta p \, \mathrm{d}v = 0. \tag{45}$$

Insertion of (43) into (45) yields

$$\begin{aligned} &\int_V \operatorname{div} (\mathbf{u}_S)'_S \delta p \, \mathrm{d}v + \int_V \frac{k^F}{\gamma^{FR}} (\operatorname{grad} p - \rho^{FR} \mathbf{b}) \cdot \operatorname{grad} \delta p \, \mathrm{d}v \\ &\quad = - \int_S n^F \, \mathbf{w}_F \cdot \mathbf{n} \, \delta p \, \mathrm{d}a. \end{aligned} \tag{46}$$

In order to complete the set of equations, multiplication of the moment of momentum balance $(27)_1$ by the test function $\delta\bar{\boldsymbol{\varphi}}_S$ results in

$$\int_V (\mathbf{I} \times \mathbf{T}^S + \operatorname{div} \mathbf{M}^S + \rho^S \mathbf{c}^S) \cdot \delta\bar{\boldsymbol{\varphi}}_S \, \mathrm{d}v = 0. \tag{47}$$

Proceeding from the assumption that there is no external loading by couple stress vectors $\mathbf{c}^S$ and by surface couples $\mathbf{m}^S = \mathbf{M}^S \, \mathbf{n}$ ($\mathbf{n}$ is the outward oriented unit surface normal), one obtains instead of (47)

$$\int_V (\mathbf{I} \times \mathbf{T}^S) \cdot \delta\bar{\boldsymbol{\varphi}}_S \, \mathrm{d}v - \int_V \mathbf{M}^S \cdot \operatorname{Grad}_S \delta\bar{\boldsymbol{\varphi}}_S \, \mathrm{d}v = 0. \tag{48}$$

Equations (44), (46) and (48) represent the weak formulation of the problem, which has basically been used throughout the numerical computations. In case of nonpolar skeleton materials, these equations reduce to the well-known displacement–pressure formulation, where (48) is dropped. Furthermore, the empty skeleton material is either computed from (44) and (48) or from (44) alone, where (44) is reduced to a pure skeleton equation by dropping the influence of $p$ and $\rho^{FR}$.

In the framework of the finite element discretization, use is made of triangular elements with six nodes, where the solid displacement $\mathbf{u}_S$ is approximated by quadratic shape functions, while the liquid pressure $p$ and the solid rotation vector $\bar{\boldsymbol{\varphi}}_S$ are represented by linear shape functions. Finally, by use of the standard Galerkin procedure, the included test functions $\delta\mathbf{u}_S$, $\delta p$ and $\delta\bar{\boldsymbol{\varphi}}_S$ and the related shape functions coincide and correspond to the respective field quantities.

The spatial discretization of the governing weak equations yields a matrix differential equation of first order in time:

$$\mathcal{G}\,(\mathbf{u}) = \mathbf{M}\frac{\mathrm{d}\mathbf{u}}{\mathrm{d}t} + \mathbf{k}(\mathbf{u}) - \mathbf{f} = 0. \tag{49}$$

Therein, $\mathbf{M}$ is the generalized mass matrix, $\mathbf{k}(\mathbf{u})$ is the global stiffness vector, $\mathbf{u}$ is the vector of the unknown field functions, and $\mathbf{f}$ is the generalized force vector. As a consequence of the material incompressibility constraint of both constituents, there is no evolution equation for the effective liquid pressure $p$. Thus, one obtains a system of differential-algebraic equations (DAE) in time, that is $\mathbf{M}$ included into (49) does not have the full rank.

In the following computations, the time integration of (49) is carried out by an implicit Euler scheme, where $n$ is the time step index. Thus,

$$\mathcal{G}\,(\mathbf{u}^{n+1}) = \frac{1}{\Delta t}\mathbf{M}(\mathbf{u}^{n+1} - \mathbf{u}^{n}) + \mathbf{k}(\mathbf{u}^{n+1}) - \mathbf{f} = 0. \tag{50}$$

In order to solve this nonlinear equation, a Newton linearization implies

$$\mathcal{G}\,(\mathbf{u}^{n+1,\,i+1}) = \mathcal{G}\,(\mathbf{u}^{n+1,\,i}) + \left(\frac{\mathrm{d}\mathcal{G}\,(\mathbf{u}^{n+1})}{\mathrm{d}\,\mathbf{u}^{n+1}}\right)\Bigg|_i \Delta\mathbf{u}^{n+1,\,i}, \tag{51}$$

where $i$ is the iteration index. With the aid of (50) and (51) one obtains

$$\left(\frac{1}{\Delta t}\mathbf{M} + \mathbf{K}\right)^{i} \Delta\mathbf{u}^{n+1,\,i} = -\mathcal{G}\,(\mathbf{u}^{n+1,\,i}), \tag{52}$$

where

$$\mathbf{K} = \left(\frac{\mathrm{d}\mathbf{k}\,(\mathbf{u}^{n+1})}{\mathrm{d}\,\mathbf{u}^{n+1}}\right)\Bigg|_i \tag{53}$$

is the stiffness matrix. The governing Equation (52) must be solved in any time step until the tolerance is reached.

## 5. Numerical Computations

The numerical results presented in this paper are computed by use of the noncommercial code PANDAS that has especially been developed for the computation of strongly coupled problems and DAE systems.

In particular, the following example concerns the classical base failure problem induced by a consolidation process, compare Figure 3. From the geotechnical literature (e.g. Terzaghi and Jelinek, 1954), it is well-known that, once a critical loading is reached, the plastic deformations concentrate in small bands of finite width finally leading to a failure situation. The present computations are carried out on an irregular mesh of triangular elements (Figure 4), where a mesh refinement has been applied in the localization zone. The solid material included into the model is fitted in comparison with Schad's model (Schad, 1979), compare Table I.

The following figures show the results of the numerical computations. Three different models have been applied to solve the localization problem. By use of the basic material parameters, an empty micropolar solid skeleton (Figure 5), a liquid-saturated micropolar solid skeleton (Figure 6) and a liquid-saturated non-polar solid skeleton (Figure 7) have been investigated. In particular, the plots are taken directly (Figure 5) or 20 days (Figures 6 and 7) after having applied the total critical external load of $f = 150\,\mathrm{kN/m^2}$. This load has been linearly increased priorly over a period of 15 days. Furthermore, the regularization of the localization problem by the inclusion of micropolar degrees of freedom, Figures 5 and 6, exhibit a mesh-independent symmetric behaviour, although the computations have been carried out on a non-symmetric irregular mesh. Furthermore, in comparison to the static behaviour of the empty material and the time-dependent behaviour of the saturated material (Figures 5 and 6), only very little differences can be observed with respect to the shear band

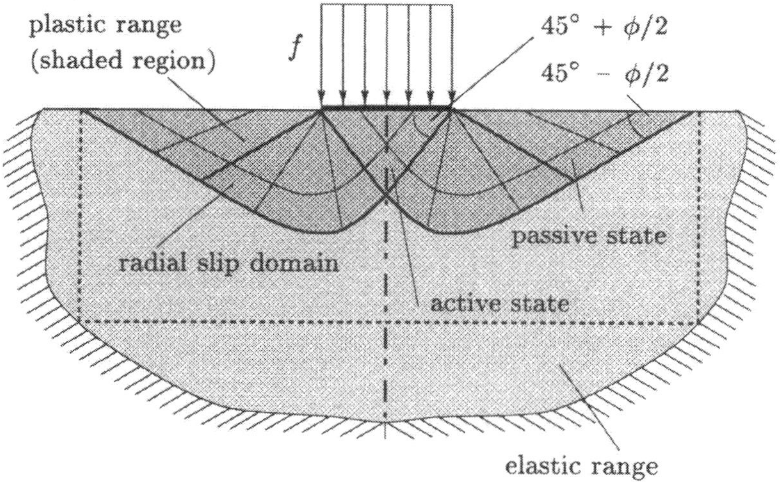


*Figure 3.* The classical base failure problem ($\phi$ is the angle of internal friction).

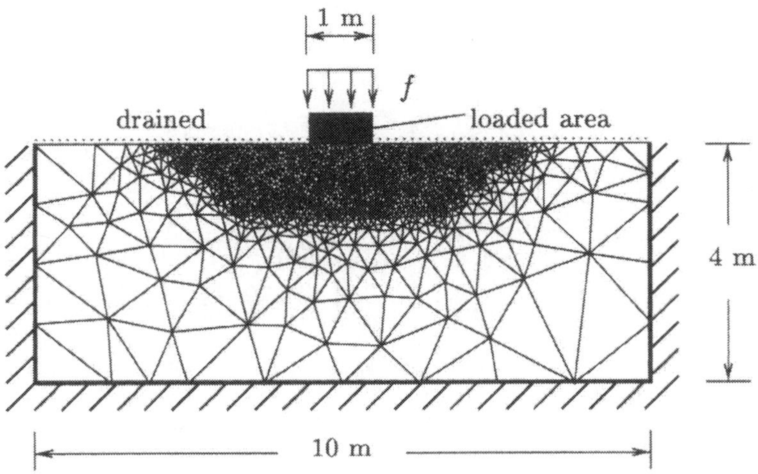


*Figure 4.* Irregular mesh of the FE discretization.

*Table I.* Basic material parameters

| Parameter | Symbol | Value |
|---|---|---|
| Lamé constants | $\mu^{\mathrm{S}}$ | $5583\,\mathrm{kN/m^2}$ |
| | $\lambda^{\mathrm{S}}$ | $8375\,\mathrm{kN/m^2}$ |
| Effective densities | $\rho^{SR}$ | $2600\,\mathrm{kg/m^3}$ |
| | $\rho^{FR}$ | $1000\,\mathrm{kg/m^3}$ |
| Volume fractions | $n^{\mathrm{S}}_{0\mathrm{S}}$ | 0.67 |
| | $n^{\mathrm{F}}_{0\mathrm{S}}$ | 0.33 |
| Effective liquid weight | $\gamma^{FR}$ | $10\,\mathrm{kN/m^3}$ |
| Permeability | $k^{\mathrm{F}}_0$ | $1 \times 10^{-8}\,\mathrm{m/s}$ |
| | $m^{\mathrm{F}}$ | 100 |
| Parameters of the | $\alpha$ | $1.0740 \times 10^{-2}$ |
| single-surface | $\beta$ | 0,1195 |
| yield criterion | $\gamma$ | 1,555 |
| | $\delta$ | $1.377 \times 10^{-4}\,\mathrm{m^2/kN}$ |
| | $\varepsilon$ | $4.330 \times 10^{-6}\,\mathrm{m^2/kN}$ |
| | $\kappa$ | $10.27\,\mathrm{kN/m^2}$ |
| | $m$ | 0.5935 |
| Cosserat parameters | $l^{\mathrm{S}}_{\mathrm{c}}$ | $5 \times 10^{-2}\,\mathrm{m}$ |
| | $\mu^{\mathrm{S}}_{\mathrm{c}}$ | $2 \times 10^{3}\,\mathrm{kN/m^2}$ |
| | $k_1$ | 0 |
| | $k_2$ | 1 |

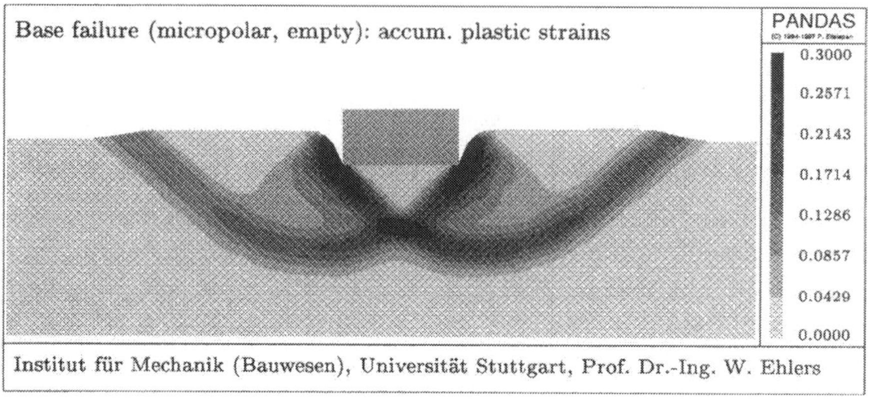


*Figure 5.* Accumulated plastic strains of an empty skeleton material (micropolar formulation) after having applied the external load.

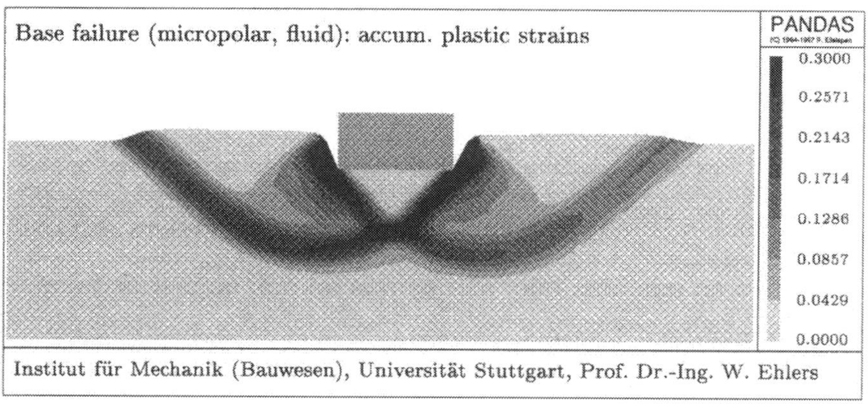


*Figure 6.* Accumulated plastic strains of a liquid-saturated skeleton material (micropolar formulation) 20 days after having applied the external load.

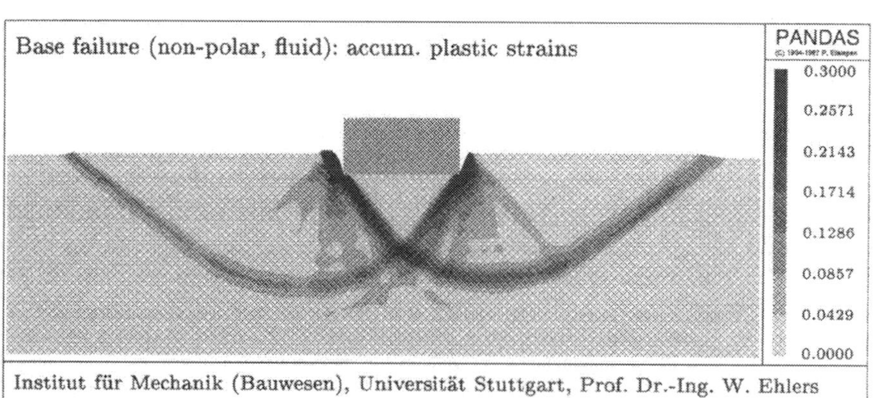


*Figure 7.* Accumulated plastic strains of a liquid-saturated skeleton material (nonpolar formulation) 20 days after having applied the external load.

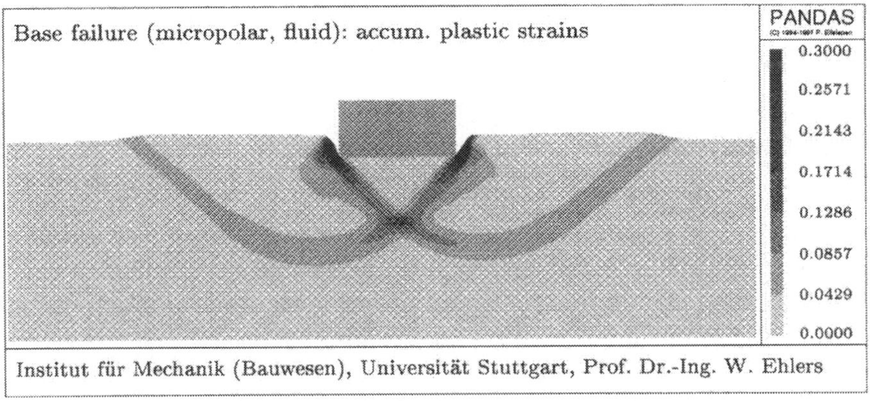


*Figure 8.* Accumulated plastic strains of a liquid-saturated skeleton material (micropolar formulation) 20 days after having applied the external load.

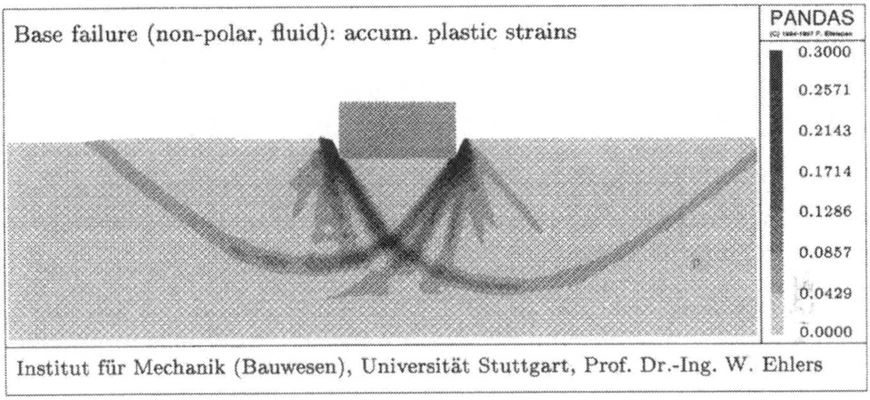


*Figure 9.* Accumulated plastic strains of a liquid-saturated skeleton material (nonpolar formulation) 10 days after having applied the external load.

width. As a result of the internal length scale $l_c$ and the Cosserat parameter $\mu_c^S$ included in the basic set of material parameters, the shear band width proves to be rather strong. On the other hand, a variation of these parameters towards $l_c = 10^{-4}$ m and $\mu_c^S = 10^3\ \mathrm{kN/m^2}$, compare Figure 8, reveals the expected decrease of the shear band width. In addition, Figures 5, 6 and 8 show that the fluid viscosity included into $k_0^F$ plays a subordinate role. This effect can also be seen from Figure 7, where the fluid viscosity alone has been used to regularize the problem. In particular, Figure 7 only shows a nearly symmetric shear band evolution on the nonsymmetric mesh. However, it is seen from Figure 9, that an increase of $k_0^F$ towards $k_0^F = 10^{-7}$ m/s immediately yields a worse regularization, since in this case, the shear band evolution is neither symmetric nor mesh-independent. Note in passing that an increase of $k_0^F$ by the factor 10 means a decrease of the effective liquid viscosity by the same factor.

Finally, it should be noted that, without taking into consideration the fluid viscosity, no solution can be obtained for the nonpolar skeleton.

## 6. Concluding Remarks

In the present contribution, nonpolar and micropolar elasto-plastic frictional porous solid materials have been studied. In particular, the continuum mechanical results have been applied to localization phenomena, where the regularization of a basically ill-posed problem, as for instance an empty nonpolar skeleton material, could be carried out by the inclusion of additional degrees of freedom in the sense of the Cosserat brothers and/or by the inclusion of a viscous pore-fluid. However, the regularization effect obtained by the micropolar formulation is dominant in comparison to the inclusion of fluid viscosity.

The numerical examples demonstrate the efficiency of the proposed method. In particular, the micropolar formulation opens the possibility to implicitly include the shear band width into the problem by proposing convenient numbers for the internal length scale parameter. However, since the inclusion of a viscous pore-fluid concerns a different mechanical problem than the description of an empty porous material, regularization by fluid viscosity is clearly restricted to volumetrically coupled saturated problems.

The numerical procedure carried out in the present contribution offers the possibility of describing saturated and empty skeleton materials in the framework of both nonpolar and polar formulations.

## References

Bowen, R. M.: 1976, Theory of mixtures, In: A. C. Eringen (ed.), *Continuum Physics*, Vol. III, Academic Press, New York, pp. 1–127.

Bowen, R. M.: 1980, Incompressible porous media models by use of the theory of mixtures, *Int. J. Engng. Sci.* **18**, 1129–1148.

Cosserat, E. and Cosserat, F.: 1909, *Théorie des Corps Déformables*, A. Hermann et Fils, Paris.

de Boer, R.: 1982, *Vektor- und Tensorrechnung für Ingenieure*, Springer-Verlag, Berlin.

de Boer, R.: 1996, Highlights in the historical development of the porous media theory: Toward a consistent macroscopic theory, *Appl. Mech. Rev.* **49**, 201–262.

de Boer, R. and Ehlers, W.: 1986, Theorie der Mehrkomponentenkontinua mit Anwendung auf bodenmechanische Probleme, Forschungsberichte aus dem Fachbereich Bauwesen 40, Universität-GH-Essen.

de Boer, R., Ehlers, W., Kowalski, S. and Plischka, J.: 1991, Porous media – A survey of different approaches, Forschungsberichte aus dem Fachbereich Bauwesen 54, Universität-GH-Essen.

de Borst, R.: 1991, Numerical modelling of bifurcation and localisation in cohesive-frictional materials, *Pageoph.* **137**, 368–390.

de Borst, R.: 1993, A generalisation of $J_2$-flow theory for polar continua, *Comput. Methods Appl. Mech. Engng.* **103**, 347–362.

Desrues, J.: 1990, Shear band initiation in granular materials: Experimentation and theory, In: F. Darve (ed.), *Geomaterials: Constitutive Equations and Modelling*, Elsevier Applied Science, London, pp. 283–310.

Diebels, S. and Ehlers, W.: 1996, On basic equations of multiphase micropolar materials, *Technische Mechanik* **16**, 77–88.
Dietsche, A., Steinmann, P. and Willam, K.: 1993, Micropolar elastoplasticity and its role in localisation, *Int. J. Plasticity* **9**, 813–831.
Ehlers, W.: 1989, Poröse Medien – ein kontinuumsmechanisches Modell auf der Basis der Mischungstheorie, Forschungsberichte aus dem Fachbereich Bauwesen 47, Universität-GH-Essen.
Ehlers, W.: 1993a, Constitutive equations for granular materials in geomechanical context, In: K. Hutter (ed.), *Continuum Mechanics in Environmental Sciences and Geophysics*, CISM Courses and Lecture Notes No. 337, Springer-Verlag, Berlin, pp. 313–402.
Ehlers, W.: 1993b, Compressible, incompressible and hybrid two-phase models in porous media theories, *ASME: AMD* **158**, 25–38.
Ehlers, W.: 1995, A single surface yield function for geomaterials, *Arch. Appl. Mech.* **65**, 63–76.
Ehlers, W.: 1996, Grundlegende Konzepte in der Theorie Poröser Medien, *Technische Mechanik* **16**, 63–76.
Ehlers, W. and Volk, W.: 1997a, On shear band localization phenomena of liquid-saturated granular elasto-plastic porous solid materials accounting for fluid viscosity and micropolar solid rotations, *Mech. Cohesive-Frictional Mater.* **2**, 301–320.
Ehlers, W. and Volk, W.: 1997b, On shear band localization phenomena induced by elasto-plastic consolidation of fluid-saturated soils, In: D. J. R. Owen, E. Oñate and E. Hinton (eds), *Computational Plasticity – Fundamentals and Applications*, CIMNE, Barcelona, pp. 1657–1664.
Eringen, A. C. and Kafadar, C. B.: 1976, Polar field theories, In: A. C. Eringen (ed.), *Continuum Physics*, Vol. IV, Academic Press, New York, pp. 1–73.
Lade, P. V. and Duncan, J. M.: 1973, Cubical triaxial tests on cohesionless soils, *ASCE: J. Soil Mech. Found. Div.* **99**, 793–812.
Mühlhaus, H.-B. and Vardoulakis, I.: 1987, The thickness of shear bands in granular materials, *Géotechnique* **37**, 271–283.
Günther, W.: 1958, Zur Statik und Kinematik des Cosseratschen Kontinuums, Abhandlungen der Braunschweig, *Wiss. Gesellschaft* **10**, 195–213.
Nowacki, W.: 1986, *Theory of Asymmetric Elasticity*, Pergamon Press, Oxford.
Plischka, J.: 1992, Die Bedeutung der Durchschnittsbildungstheorie für die Theorie poröser Medien, Dissertation, Fachbereich Bauwesen, Universität-GH-Essen.
Schad, H.: 1979, Nichtlineare Stoffgleichungen für Böden und ihre Verwendung bei der numerischen Analyse von Grundbauaufgaben, Mitteilungen des Baugrundinstituts Stuttgart 10, Universität Stuttgart.
Schrefler, B. A., Majorna, C. E. and Sanavia, L.: 1995, Shear band localization in saturated porous media, *Arch. Mech.* **47**, 577–599.
Schrefler, B. A., Sanavia, L. and Majorna, C.: 1996, A multiphase media model for localisation and post-localisation simulation in geomaterials, *Mech. Cohesive-Frictional Mater.* **1**, 95–114.
Steinmann, P.: 1994, A micropolar theory of finite deformation and finite rotation multiplicative elastoplasticity, *Int. J. Solids Structures* **31**, 1063–1084.
Steinmann, P.: 1995, Theory and numerics of ductile micropolar elastoplastic damage, *Int. J. Numer. Methods Engng.* **38**, 583–606.
Terzaghi, K. and Jelinek, R.: 1954, *Theoretische Bodenmechanik*, Springer-Verlag, Berlin.
Truesdell, C. and Toupin, R. A.: 1960, The classical field theories, In: S. Flügge (ed.), *Handbuch der Physik*, Band III/1, Springer-Verlag, Berlin, pp. 226–902.

# Finite Elastic Deformations in Liquid-Saturated and Empty Porous Solids

*(Dedicated to Franz Ziegler on the occasion of his 60th birthday)*

W. EHLERS and G. EIPPER
*Institut für Mechanik (Bauwesen), Universität Stuttgart, Pfaffenwaldring 7, D-70569 Stuttgart, Germany*

(Received: 4 August 1997)

**Abstract.** Based on the Theory of Porous Media (TPM), a formulation of a fluid-saturated porous solid is presented where both constituents, the solid and the fluid, are assumed to be materially incompressible. Therefore, the so-called point of compaction exists. This deformation state is reached when all pores are closed and any further volume compression is impossible due to the incompressibility constraint of the solid skeleton material. To describe this effect, a new finite elasticity law is developed on the basis of a hyperelastic strain energy function, thus governing the constraint of material incompressibility for the solid material. Furthermore, a power function to describe deformation dependent permeability effects is introduced.

After the spatial discretization of the governing field equations within the finite element method, a differential algebraic system in time arises due to the incompressibility constraint of both constituents. For the efficient numerical treatment of the strongly coupled nonlinear solid-fluid problem, a consistent linearization of the weak forms of the governing equations with respect to the unknowns must be used.

**Key words:** deformation dependent permeability, point of compaction, finite hyperelasticity, elasticity tensor, differential algebraic equations (DAE), consistent linearization.

## 1. Introduction

Fluid-saturated porous solids, for example polymer foams, metal foams, sponges or biological tissues can undergo large elastic deformations in the macroscopic range. These foams consist of a porous solid skeleton which is filled with a fluid. Based on the Theory of Porous Media (TPM), these types of materials can be described by a two-phase model. The Theory of Porous Media is defined by the classical theory of mixtures combined with the concept of volume fractions. The reader who is interested in a broad review of fundamentals of this theory is referred to Bowen (1976, 1980), de Boer and Ehlers (1986) or Ehlers (1989, 1993).

In this work, a fluid-saturated solid with materially incompressible constituents is described. Incompressibility of a constituent (material incompressibility) implies incompressibility in the material's micro range, which does not lead to macroscopic incompressibility. Even if the realistic density of a constituent is constant, the partial density can still change through changes in the volume fraction. Therefore, the so-called point of compaction exists. This deformation state is reached, when all pores are closed and further volume deformations are impossible due to the incompressibility constraint of the solid skeleton material. To describe this effect, a new finite elasticity law is developed in the framework of the hyperelasticity concept. Fluid viscosity is taken into account by the drag force, where the permeability is deformation dependent and assumed to be a function of the actual porosity of the solid skeleton.

Concerning the numerical treatment of the volumetrically coupled solid–fluid problem, incompressibility of both constituents leads to a system of differential-algebraic equations (DAE) in the time domain. This DAE system must be handled by an appropriate time integration scheme. The spatial discretization is carried out by the finite element method (FEM). Finally, a consistent linearization of the nonlinear equations guarantees quadratic convergence of the solution within the Newton procedure.

## 2. Preliminaries

### 2.1. KINEMATICS

In the TPM, a fluid-saturated porous medium can be treated as an immiscible mixture of constituents $\varphi^\alpha$ ($\alpha$ = S: solid phase; $\alpha$ = F: fluid phase). A macroscopic description implies a model of superimposed continua. Thus, at any time $t$ each spatial point of the current configuration is simultaneously occupied by particles of both constituents, solid and fluid. These particles proceed from different reference positions $\mathbf{X}_\alpha$ at the time $t_0$. Thus, each constituent has its own motion function

$$\mathbf{x} = \chi_\alpha(\mathbf{X}_\alpha, t) \tag{1}$$

and its own velocity field

$$\overset{\prime}{\mathbf{x}}_\alpha = \frac{\partial \chi_\alpha}{\partial t}. \tag{2}$$

With the introduction of the volume fractions

$$n^\alpha = n^\alpha(\mathbf{x}, t), \tag{3}$$

which are defined as the local ratios between the current constituent volumes $\mathrm{d}v^\alpha$ and a current bulk volume element $\mathrm{d}v$, the saturation condition reads

$$n^{\mathrm{S}} + n^{\mathrm{F}} = 1. \tag{4}$$

Associated with each constituent $\varphi^\alpha$ is a realistic density $\rho^{\alpha\mathrm{R}}$, defined as the mass of $\varphi^\alpha$ per unit of $\mathrm{d}v^\alpha$, and a partial density $\rho^\alpha$, defined as the mass of $\varphi^\alpha$ per unit of $\mathrm{d}v$. The density functions are related by the volume fractions

$$\rho^\alpha = n^\alpha\, \rho^{\alpha\mathrm{R}}. \tag{5}$$

Thus, it is obvious that incompressibility of a constituent ($\rho^{\alpha R} = \text{const.}$) does not lead to macroscopic incompressibility, because the partial density $\rho^\alpha$ can still change through changes in the volume fractions $n^\alpha$.

For any given differentiable field function $\Gamma = \Gamma(\mathbf{x}, t)$, the material time derivatives of $\Gamma$ corresponding to the individual motion functions of the respective constituent $\varphi^\alpha$ are defined by

$$\frac{\mathrm{d}_\alpha}{\mathrm{d}t}\Gamma = (\Gamma)'_\alpha = \frac{\partial \Gamma}{\partial t} + \operatorname{grad}\Gamma \cdot \overset{\prime}{\mathbf{x}}_\alpha, \tag{6}$$

where 'grad' denotes the partial differentiation with respect to $\mathbf{x}$.

The solid motion can be described in a Lagrangian setting by introducing the displacement vector of the solid phase

$$\mathbf{u}_\mathrm{S} = \mathbf{x} - \mathbf{X}_\mathrm{S}. \tag{7}$$

Thus, the solid deformation gradient is defined by

$$\mathbf{F}_\mathrm{S} = \frac{\partial \mathbf{x}}{\partial \mathbf{X}_\mathrm{S}} =: \operatorname{Grad}_\mathrm{S} \mathbf{x}. \tag{8}$$

The Jacobian determinant of the deformation gradient

$$J_\mathrm{S} = \det \mathbf{F}_\mathrm{S} \tag{9}$$

describes the volume deformation of a volume element of the mixture in the actual configuration by $\mathrm{d}v = J_\mathrm{S}\, \mathrm{d}V_{0\mathrm{S}}$, where $\mathrm{d}V_{0\mathrm{S}}$ is the volume element of the mixture in the reference configuration of the solid.

The fluid motion is described in an Eulerian setting with the seepage velocity, relating the fluid motion to the deforming solid skeleton matrix:

$$\mathbf{w}_\mathrm{F} = \overset{\prime}{\mathbf{x}}_\mathrm{F} - \overset{\prime}{\mathbf{x}}_\mathrm{S}\,. \tag{10}$$

## 2.2. BALANCE RELATIONS

Excluding mass exchanges between the solid skeleton and the pore-fluid and neglecting acceleration terms, the mechanical balance equations for each constituent $\varphi^\alpha$ read, see, e.g., Ehlers (1993):

*Balance of mass:*

$$(\rho^\alpha)'_\alpha + \rho^\alpha \operatorname{div} \overset{\prime}{\mathbf{x}}_\alpha = 0. \tag{11}$$

*Balance of linear momentum (quasi-static):*

$$\operatorname{div} \mathbf{T}^\alpha + \rho^\alpha\, \mathbf{b} + \hat{\mathbf{p}}^\alpha = \mathbf{0}. \tag{12}$$

In the above relations, 'div' is the divergence operator corresponding to 'grad', $\mathbf{T}^\alpha$ is the partial Cauchy stress tensor, $\mathbf{b}$ is the body force density acting upon both constituents, and $\hat{\mathbf{p}}^\alpha$ represents the momentum production term. For two-phase materials, this term can be interpreted as the interaction force per unit of bulk volume. Thus, the constraint

$$\hat{\mathbf{p}}^{\mathrm{S}} + \hat{\mathbf{p}}^{\mathrm{F}} = \mathbf{0} \tag{13}$$

must hold, due to the fact that the sum of the constituent balance equations yields the same form as the balance equation of a single-phase material. Assuming microscopically nonpolar constituents, the partial Cauchy stress tensor $\mathbf{T}^\alpha$ is symmetric.

With the use of the incompressibility condition ($\rho^{\alpha\mathrm{R}} = \text{const.}$) and the definition of the partial densities (5), the mass balance equations lead to volume balance equations

$$(n^\alpha)'_\alpha + n^\alpha \operatorname{div} \overset{\prime}{\mathbf{x}}_\alpha = 0. \tag{14}$$

By integration of the solid volume balance relation and by use of the Jacobian determinant (9) of the deformation gradient, the solid volume fraction reads

$$n^{\mathrm{S}} = \det \mathbf{F}_{\mathrm{S}}^{-1}\, n_{0\mathrm{S}}^{\mathrm{S}} = \frac{1}{J_{\mathrm{S}}}\, n_{0\mathrm{S}}^{\mathrm{S}}, \tag{15}$$

where $n_{0\mathrm{S}}^{\mathrm{S}}$ is the solid volume fraction referring to the solid reference configuration. The incompressibility constraint for the whole binary medium (the mixture) is found by combining the volume balance equations (14) for the solid and the fluid phases together with the saturation constraint (4):

$$\operatorname{div} (n^{\mathrm{F}}\, \mathbf{w}_{\mathrm{F}} + (\mathbf{u}_{\mathrm{S}})'_{\mathrm{S}}) = 0. \tag{16}$$

### 2.3. BASIC CONSTITUTIVE EQUATIONS

The binary model with incompressible constituents has to be completed by constitutive assumptions for the partial solid and fluid stress tensors, $\mathbf{T}^{\mathrm{S}}$ and $\mathbf{T}^{\mathrm{F}}$, as well as for the interaction term (the drag force) $\hat{\mathbf{p}}^{\mathrm{F}} = -\hat{\mathbf{p}}^{\mathrm{S}}$. In the present section, we refer to results given by Ehlers (1993). Due to the incompressibility constraints of the constituents, each of the constitutive equations consists of two terms. The first term includes the effective fluid pressure $p$, which could be interpreted as a Lagrangian multiplier to govern the incompressibility requirement. The second term represents the so-called extra quantity, characterized by the additional index $(..)_{\mathrm{E}}$. Therefore, the stress tensors are given by

$$\mathbf{T}^\alpha = -n^\alpha p\, \mathbf{I} + \mathbf{T}^\alpha_{\mathrm{E}}, \tag{17}$$

where $\mathbf{I}$ is the second order identity tensor. The interaction force is given by

$$\hat{\mathbf{p}}^{\mathrm{F}} = p \operatorname{grad} n^{\mathrm{F}} + \hat{\mathbf{p}}^{\mathrm{F}}_{\mathrm{E}}. \tag{18}$$

As is usual in hydraulics, it is assumed that the friction stress is negligible:

$$\mathbf{T}_{\mathrm{E}}^{\mathrm{F}} \approx \mathbf{0}. \tag{19}$$

The internal friction between the solid and the fluid constituents is then taken into account by the extra interaction force, which, in the case of isotropic permeability, is given by

$$\hat{\mathbf{p}}_{\mathrm{E}}^{\mathrm{F}} = -\frac{(n^{\mathrm{F}})^2\,\gamma^{\mathrm{FR}}}{k^{\mathrm{F}}}\;\mathbf{w}_{\mathrm{F}}. \tag{20}$$

Therein, $\gamma^{\mathrm{FR}}$ is the effective weight of the fluid and $k^{\mathrm{F}}$ is the Darcy permeability parameter, which is a macroscopic quantity that depends both on the viscosity of the fluid and on the size and structure of the pore space. From the balance of momentum (12) for the fluid phase in combination with the constitutive assumptions for the fluid, (19) and (20), the relation

$$n^{\mathrm{F}}\,\mathbf{w}_{\mathrm{F}} = -\frac{k^{\mathrm{F}}}{\gamma^{\mathrm{FR}}}\left(\operatorname{grad} p - \rho^{\mathrm{FR}}\,\mathbf{b}\right) \tag{21}$$

holds, which is known as Darcy's law.

The constitutive assumptions for the solid extra stress tensor $\mathbf{T}_{\mathrm{E}}^{\mathrm{S}}$ will be discussed in greater detail in Section 3.

### 2.4. DEFORMATION DEPENDENT DARCY PERMEABILITY PARAMETER

In general, the Darcy permeability parameter $k^{\mathrm{F}}$, which relates the filter velocity $(n^{\mathrm{F}}\,\mathbf{w}_{\mathrm{F}})$ to the pressure gradient grad $p$, is deformation dependent. The value of $k^{\mathrm{F}}$ decreases when the pore volume decreases and must be zero when all pores are closed. In this case, no fluid flow is possible. It is assumed that the permeability depends on the volume fraction of the fluid, which represents the porosity of the solid skeleton in the saturated two-phase model. The fluid volume fraction is given by the saturation constraint (4) and with the aid of the Jacobian determinant:

$$n^{\mathrm{F}} = 1 - n^{\mathrm{S}} = 1 - \frac{1}{J_{\mathrm{S}}}\,n_{0\mathrm{S}}^{\mathrm{S}}. \tag{22}$$

The deformation dependency of the Darcy parameter is described by a power function

$$k^{\mathrm{F}}(n^{\mathrm{F}}) = k_0^{\mathrm{F}}\left(\frac{n^{\mathrm{F}}}{n_{0\mathrm{S}}^{\mathrm{F}}}\right)^{\kappa}, \tag{23}$$

where $k_0^{\mathrm{F}}$ is the initial permeability at the initial porosity $n_{0\mathrm{S}}^{\mathrm{F}} = 1 - n_{0\mathrm{S}}^{\mathrm{S}}$. The parameter $\kappa$ governs the strength of the deformation dependency. For $\kappa = 0$, $k^{\mathrm{F}}$ is constant.

## 3. A New Finite Elasticity Law

We proceed from a hyperelastic strain energy function $W^{\mathrm{S}} = \rho_{0\mathrm{S}}^{\mathrm{S}}\,\psi^{\mathrm{S}}$, where $\psi^{\mathrm{S}}$ is the partial free energy density of the solid. For isotropic materials, the strain energy function must be a function of the principal invariants or eigenvalues of the right or left Cauchy–Green deformation tensors

$$\mathbf{C}_S = \mathbf{F}_S^T \, \mathbf{F}_S, \qquad \mathbf{B}_S = \mathbf{F}_S \, \mathbf{F}_S^T, \tag{24}$$

see for example Ogden (1984). The elasticity law describes the deformation of the porous solid skeleton structure. Thus, the use of a compressible elasticity law is necessary, even if material incompressibility of the solid skeleton is assumed, since large volumetric strains can occur as a result of porosity changes (changes of the volume fractions $n^S$ and $n^F$). Therefore, the material parameters included in the following elasticity law, in this work the Lamé constants $\mu^S$ and $\Lambda^S$, are macroscopical parameters of the solid skeleton structure and not of the microscopic skeleton material itself.

In what follows, a strain energy function of the Neo-Hookean type (Rivlin, 1948) with an extension term for the description of compressibility effects is used:

$$W^S = \tfrac{1}{2}\,\mu^S\,(I_1 - 3) - \mu^S \ln J_S + U^S(J_S). \tag{25}$$

Therein, $I_1$ is the first principal invariant of $\mathbf{C}_S$ or $\mathbf{B}_S$. The second term in (25) guarantees that the reference configuration is stress free. The extension function $U^S(J_S)$ describes the finite volume changing domain including the concept of the compaction point.

Recall that in the case of an incompressible matrix material, the point of compaction is reached when all pores are closed and further volume deformations are impossible due to the incompressibility constraint of the solid material. On the other hand, the pores can be widened infinitely, allowing large volume extensions. Thus, with (15) and $0 \leqslant n^S \leqslant 1$, the Jacobian determinant of $\mathbf{F}_S$ is limited by

$$n_{0S}^S < J_S < \infty. \tag{26}$$

This effect must be expressible by the elasticity law. Therefore, the constitutive formulation of $U^S$ is restricted to certain physical requirements. Under hydrostatic loading, the point of compaction can only be reached with an infinite strain energy $U^S$ and an infinite pressure resulting from the hydrostatic Cauchy stress $\partial U^S/\partial J_S$. Thus, the conditions

$$\left.\begin{aligned} U^S &\to +\infty, \\ \frac{\partial U^S}{\partial J_S} &\to -\infty, \end{aligned}\right\} \quad \text{for} \quad J_S \to n_{0S}^S \tag{27}$$

must hold. Furthermore, the usual definition of the Lamé parameter

$$\Lambda^S = \left.\frac{\partial^2 U^S}{\partial J_S^2}\right|_{J_S=1} \tag{28}$$

must be satisfied. This guarantees a transition to an elasticity law of the Hookean type in the case of a linearization procedure. If the initial volume fraction $n_{0S}^S$ declines, the reference solid volume decreases, with an increasing pore space. Thus, if $n_{0S}^S \to 0$, the deformation state associated with the point of compaction reduces to a singular point. In this case, the elasticity law should formally coincide with an elasticity law of a compressible nonporous solid, which allows for this deformation state. Furthermore, for $n_{0S}^S \to 1$ a 'one-sided' incompressible material law must result. In this case, a volume compression is not possible, whereas a volume expansion by increasing of the pores can occur. A further requirement is that $U^S$ is a convex function of $J_S$,

$$\frac{\partial^2 U^S}{\partial J_S^2} > 0, \tag{29}$$

thus guaranteeing uniqueness of the material model.

Based on the function $U^S$ found by Ogden (1972) in the context of finite elasticity for non-porous solids, a modification of this term is presented meeting the above requirements. For the description of a porous solid skeleton consisting of an incompressible matrix material, the modified function can be derived by

$$U^S(J_S) = \Lambda^S\,(1 - n_{0S}^S)^2 \left( \frac{J_S - 1}{1 - n_{0S}^S} - \ln \frac{J_S - n_{0S}^S}{1 - n_{0S}^S} \right), \tag{30}$$

see Ehlers and Eipper (1997). However, the function $U^S$ from (30) is not only valid in combination with the Neo-Hookean model (25). It is possible as well to combine the function $U^S$ with other known strain energy functions, for example, the Ogden model (Ogden, 1972) formulated in eigenvalues of the deformation tensors.

In the case of hyperelasticity, the Kirchhoff extra stress tensor, which is a weighted Cauchy stress tensor ($\boldsymbol{\tau}_E^S = J_S\,\mathbf{T}_E^S$), can be obtained by

$$\boldsymbol{\tau}_E^S = \mathbf{F}_S \left( 2\,\frac{\partial W^S}{\partial \mathbf{C}_S} \right) \mathbf{F}_S^T \tag{31}$$

$$= 2\,\frac{\partial W^S}{\partial \mathbf{B}_S}\,\mathbf{B}_S. \tag{32}$$

For the Neo-Hookean model (25) in combination with $U^S$ from Equation (30), the extra stress tensor reads

$$\boldsymbol{\tau}_E^S = \mu^S\,(\mathbf{B}_S - \mathbf{I}) + \Lambda^S(1 - n_{0S}^S)^2 \left( \frac{J_S}{1 - n_{0S}^S} - \frac{J_S}{J_S - n_{0S}^S} \right) \mathbf{I}. \tag{33}$$

The Oldroyd stress rate (lower Lie derivative of $\boldsymbol{\tau}_E^S$) is defined by

$$(\boldsymbol{\tau}_E^S)^\triangledown = (\boldsymbol{\tau}_E^S)'_S - \mathbf{L}_S\,\boldsymbol{\tau}_E^S - \boldsymbol{\tau}_E^S\,\mathbf{L}_S^T, \tag{34}$$

where $\mathbf{L}_S = \operatorname{grad} \overset{\prime}{\mathbf{x}}_S$ is the spatial velocity gradient. The elasticity tensor $\overset{4}{\mathbf{C}}$ of the actual configuration relates the Oldroyd stress rate to the solid deformation rate $\mathbf{D}_S = 1/2(\mathbf{L}_S + \mathbf{L}_S^T)$:

$$(\boldsymbol{\tau}_E^S)^\triangledown =: \overset{4}{\mathbf{C}}\,\mathbf{D}_S. \tag{35}$$

Therein, $\overset{4}{\mathbf{C}}$ can be derived from the elastic tangent of the reference configuration

$$\overset{4}{\mathbf{B}} = 4\,\frac{\partial^2 W^{\mathrm{S}}}{\partial \mathbf{C}_{\mathrm{S}} \otimes \partial \mathbf{C}_{\mathrm{S}}} \tag{36}$$

by the fourth-order push-forward transport mechanism

$$\overset{4}{\mathbf{C}} = (\mathbf{F}_{\mathrm{S}} \otimes \mathbf{F}_{\mathrm{S}})^{\overset{23}{\mathrm{T}}}\ \overset{4}{\mathbf{B}}\ (\mathbf{F}_{\mathrm{S}}^{\mathrm{T}} \otimes \mathbf{F}_{\mathrm{S}}^{\mathrm{T}})^{\overset{23}{\mathrm{T}}}. \tag{37}$$

Therein, the transposition $(..)^{\overset{ik}{\mathrm{T}}}$ indicates an exchange of the $i$th and $k$th basis system of higher order tensors. For the model under consideration, the elasticity tensor is given by the following isotropic representation:

$$\overset{4}{\mathbf{C}} = 2\alpha(\mathbf{I} \otimes \mathbf{I})^{\overset{23}{\mathrm{T}}} + \beta(\mathbf{I} \otimes \mathbf{I}), \tag{38}$$

where

$$\begin{aligned}
\alpha(J_{\mathrm{S}}) &= \left(\mu^{\mathrm{S}} - J_{\mathrm{S}}\frac{\partial U^{\mathrm{S}}}{\partial J_{\mathrm{S}}}\right)\\
&= \left[\mu^{\mathrm{S}} - \Lambda^{\mathrm{S}}(1 - n_{0S}^{\mathrm{S}}) J_{\mathrm{S}}\ \frac{J_{\mathrm{S}} - 1}{J_{\mathrm{S}} - n_{0S}^{\mathrm{S}}}\right],\\
\beta(J_{\mathrm{S}}) &= \left(J_{\mathrm{S}}\,\frac{\partial U^{\mathrm{S}}}{\partial J_{\mathrm{S}}} + J_{\mathrm{S}}^2\,\frac{\partial^2 U^{\mathrm{S}}}{\partial J_{\mathrm{S}}^2}\right)\\
&= \Lambda^{\mathrm{S}}\left[(1 - n_{0\mathrm{S}}^{\mathrm{S}})\ J_{\mathrm{S}}\ \frac{J_{\mathrm{S}}^2 - 2\,n_{0\mathrm{S}}^{\mathrm{S}}\,J_{\mathrm{S}} + n_{0\mathrm{S}}^{\mathrm{S}}}{(J_{\mathrm{S}} - n_{0\mathrm{S}}^{\mathrm{S}})^2}\right];
\end{aligned} \tag{39}$$

$(\mathbf{I} \otimes \mathbf{I})^{\overset{23}{\mathrm{T}}}$ represents the fourth order identity tensor.

The described elasticity law is also suitable for the computation of empty porous solid skeleton matrices, that is, porous materials, where the mechanical description of the fluid constituent can be neglected.

## 4. Weak Form and FEM-Equations

For the numerical treatment of the problem within the finite element method (FEM), the weak forms of the field equations are used. Therefore, the governing equations are multiplied by independent test functions and are then integrated over the spatial domain $\mathcal{B}$ with boundary $\partial\mathcal{B}$.

The balance equation of momentum (12) for the binary medium under consideration together with the constitutive Equations (17) and (19), are transformed into the weak formulation

$$\int_{\mathcal{B}} (\mathbf{T}_{\mathrm{E}}^{\mathrm{S}} - p\,\mathbf{I}) \cdot \operatorname{grad}\delta\mathbf{u}_{\mathrm{S}}\ \mathrm{d}v - \int_{\mathcal{B}} (\rho^{\mathrm{S}} + \rho^{\mathrm{F}})\,\mathbf{b} \cdot \delta\mathbf{u}_{\mathrm{S}}\ \mathrm{d}v = \int_{\partial\mathcal{B}} \mathbf{t} \cdot \delta\mathbf{u}_{\mathrm{S}}\ \mathrm{d}a, \tag{40}$$

where $\delta \mathbf{u}_S$ is the test function corresponding to the solid displacement and $\mathbf{t}$ is the surface tension. The solid extra stress tensor $\mathbf{T}_E^S$ is computed from the elasticity law presented in the previous section.

Considering quasi-static problems, the seepage velocity $\mathbf{w}_F$ can be eliminated from the field equations. Inserting Darcy's law (21) into the volume balance of the mixture (16), the weak form of the volume balance reads

$$\int_{\mathcal{B}} \operatorname{div} (\mathbf{u}_S)'_S \, \delta p \, dv + \int_{\mathcal{B}} \frac{k^F}{\gamma^{FR}} \left(\operatorname{grad} p - \rho^{FR} \, \mathbf{b}\right) \cdot \operatorname{grad} \delta p \, dv$$
$$= - \int_{\partial \mathcal{B}} q \, \delta p \, da, \tag{41}$$

where $\delta p$ is the test function corresponding to $p$, $(\mathbf{u}_S)'_S$ is the displacement velocity of the solid and $q = n^F \, \mathbf{w}_F \cdot \mathbf{n}$ is the filter volume flow of the fluid draining through the surface $\partial \mathcal{B}$.

In a standard Galerkin procedure (see e.g. Zienkiewicz and Taylor, 1984), the unknown fields $\mathbf{u}_S$ and $p$ and the corresponding test functions $\delta \mathbf{u}_S$ and $\delta p$ are approximated by the same shape functions. In the present investigation, quadratic shape functions for the approximation of the solid displacement and linear shape functions for the approximation of the pore pressure are used.

Applying a spatial discretization, the global system of equations reads

$$\mathbf{G} = \mathbf{M} \frac{d\mathbf{y}}{dt} + \mathbf{k}(\mathbf{y}) - \mathbf{f}(t) = \mathbf{0}. \tag{42}$$

Therein, the vector $\mathbf{y}$ of the unknowns contains the values of the solid displacement and the pore pressure at the FEM-nodes. $\mathbf{M}$ is the generalized mass matrix, $\mathbf{k}$ the generalized stiffness vector, and $\mathbf{f}$ is the vector of the generalized forces. Due to the incompressibility constraint, the pore pressure has no evolution equation and the matrix $\mathbf{M}$ is singular. Therefore, (42) is an index one system of DAE of first order in time. For the time integration of the DAE, an implicit Euler integration scheme can be used, see Brenan *et al.* (1989).

Time discretization of Equation (42) leads to a system of nonlinear equations. For the solution, a Newton procedure is applied in each time step. For an efficient treatment, it is necessary to apply a consistent linearization of the weak forms with respect to the unknowns, see Marsden and Hughes (1983), Wriggers (1988). This guarantees quadratic convergence within a Newton procedure. A consistent linearization of the weak forms (40) and (41) with respect to the unknown field functions $\mathbf{u}_S$ and $p$ leads, after a spatial discretization, to a global tangent operator $\mathbf{DG}$ of (42), where the elasticity tensor (38) plays an important role. Thus, in each time step $n$, a Newton iteration (iteration step $k$)

$$\mathbf{DG}_{n+1}^k \, \Delta \mathbf{y}_{n+1}^k = -\mathbf{G}_{n+1}^k \tag{43}$$

must be computed, until convergence is reached. The solution vector is updated by the incremental solution vector $\Delta \mathbf{y}$:

$$\mathbf{y}_{n+1}^{k+1} = \mathbf{y}_{n+1}^k + \Delta \mathbf{y}_{n+1}^k. \tag{44}$$

## 5. Examples

### 5.1. COMPRESSION OF A POROUS TUBE: POINT OF COMPACTION

The first example handles a fluid-saturated porous tube, where the upper boundary is perfectly drained and loaded by a time dependent external force. The other boundaries are rigid and undrained (see Figure 1). In this one-dimensional deformation problem, the point of compaction can be calculated analytically. The height of the tube is 1 m and the volume fraction of the solid in the reference configuration is $n^S_{0S} = 0.67$. Therefore, the point of compaction is reached when the subsidence of the surface has the amount of 0.33 m. The other material parameters were chosen as follows: $\mu^S = 5583$ kN/m$^2$, $\Lambda^S = 8375$ kN/m$^2$ and $\gamma^{FR} = 10^4$ N/m$^3$. In this example, the Darcy permeability parameter is assumed to be deformation independent and is chosen to $k^F_0 = 10^{-4}$ m/s. The body force density **b** is neglected. Figure 1 shows the typical time-deformation history of a consolidation process, where a step loading of $f = 7500$ kN/m$^2$ is applied (solid line). The viscosity of the fluid leads to a time dependent deformation under the step loading. While the consolidation process is running, fluid is drained out through the upper surface. The maximum deformation is reached, when the consolidation process is finished.

For the validation of the finite hyperelastic law presented in Section 3, a second load case is applied. After the same step loading at time $t_0 = 0$, as in the first load case, the external load is increased in time until it has reached 20 times the initial step loading. The deformation of the surface asymptotically approaches the point of compaction (dashed line). In this deformation state, the total pore-fluid is drained

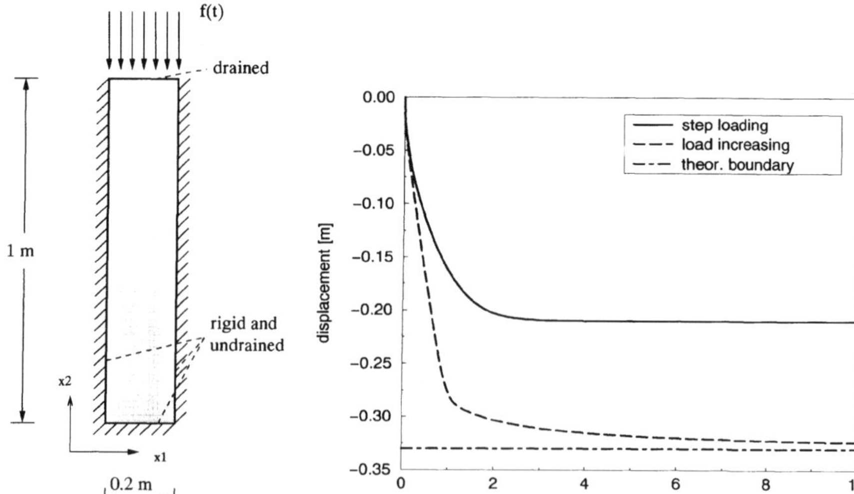


*Figure 1.* Boundary conditions and displacement of the surface under step loading and under load increasing.

out, and the remaining solid body consists of an incompressible material. Thus, no further volume deformation is possible, and the tube locks.

This result cannot be obtained with finite elasticity laws of usual nonporous solid materials. In this case, the computation of the porous tube would lead to physically impossible results. The point of compaction would be surpassed and the saturation constraint would only be fulfilled with negative volume fractions of the fluid, which is obviously physically impossible.

## 5.2. COMPRESSION OF A POROUS TUBE: DEFORMATION DEPENDENT PERMEABILITY

Now, the influence of the deformation dependent Darcy permeability to the time scale of the time-deformation history in the consolidation problem should be examined. Therefore, the same boundary value problem as in the first example, the compression of a porous cube under a step loading, is used. The material parameters are also the same as in the first example, except that the permeability $k^{\mathrm{F}}$ is not constant. Here, the power function (23) for the Darcy parameter is used, where the exponent $\kappa$ is varied.

Figure 2 shows that the time scale in the consolidation problem changes significantly, when the parameter $\kappa$ is changed. The higher the exponent $\kappa$ is, the more the permeability in the volume compression range decreases, and therefore it takes longer until the consolidation process is finished.

## 5.3. CONSOLIDATION OF A CUBE

In this example, a quarter of the upper surface of a fluid-saturated porous cube is left undrained and loaded by a ramp load. The remainder of the upper surface is unloaded and perfectly drained. The spatial discretization of the problem is based on

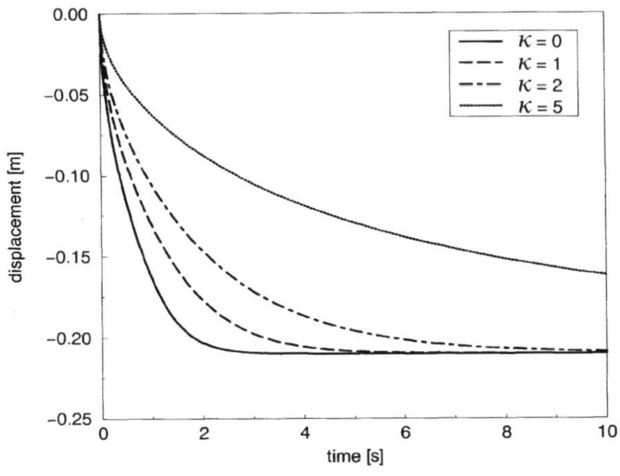


*Figure 2.* Displacement of the tube surface under step loading, with different parameters $\kappa$ in the deformation dependent Darcy permeability parameter.

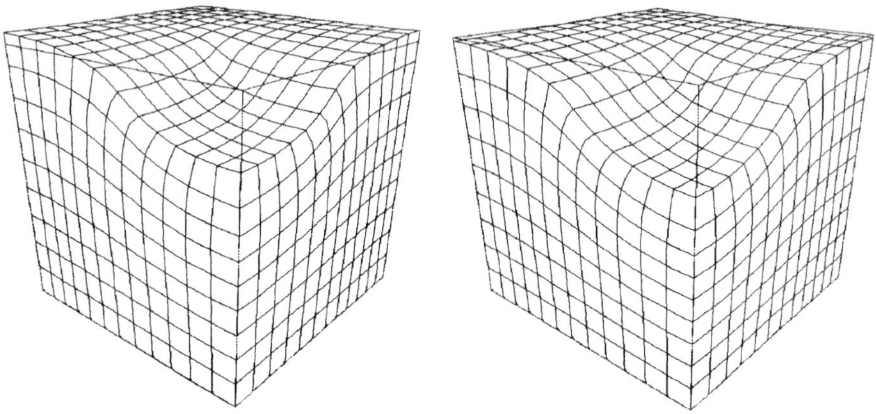

*Figure 3.* Deformation of the cube after applying external load (left) and when consolidation process is finished (right).

20-node 3-d brick elements. Due to the viscosity of the fluid, the deformation process is time dependent. Directly after having applied the full external load, the unloaded corners of the cube vault up. While the consolidation process is running, the fluid is drained out through the (drained) upper surface of the cube. Therefore, a subsidence of the unloaded corners under the original form of the cube is only reached when the consolidation process is finished, see Figure 3.

## 6. Conclusion

Proceeding from the TPM, the model equations for a fluid-saturated solid skeleton have been derived. Even if material incompressibility of the solid phase is assumed, large volumetric strains may occur as a result of porosity changes. Therefore, the point of compaction cannot be passed. This deformation state is reached when all pores are closed and further volume compression is impossible. To describe this effect, an extension of a hyperelastic strain energy function has been developed, in which all important physical requirements are satisfied. Combining this extension term with the well known Neo-Hookean model, the extra stress tensor and the consistent elastic tangent have been derived for possible finite deformations.

In the case of finite deformations, the influence of the deformation dependent permeability on the solution of a boundary value problem is significant. Therefore, a power function describing the dependency of the Darcy permeability parameter from the porosity of the solid skeleton has been introduced. Further investigations must be done for the description of anisotropic permeability effects, induced by isochoric deformations.

As a result of the assumed incompressibility of the solid and fluid phases, spatial discretization of the weak forms of the governing equations leads to a system of DAE in time. After a time discretization, a system of nonlinear equations arises.

For the efficient numerical treatment of the strongly coupled solid–fluid problem, a consistent linearization of the governing equations has been applied.

In the numerical examples it has been shown that the viscosity of the fluid leads to a time dependent deformation under a step loading, which is known as the consolidation problem. The effectiveness of the developed elasticity law has been shown in numerical examples. A description of the point of compaction is impossible with known material laws for nonporous solids. Their application leads to physically impossible results.

## References

Bowen, R. M.: 1976, Theory of mixtures, In: A. C. Eringen (ed.), *Continuum Physics*, Vol. III, Academic Press, New York, pp. 1–127.

Bowen, R. M.: 1980, Incompressible porous media models by use of the theory of mixtures, *Int. J. Engng. Sci.* **18**, 1129–1148.

de Boer, R. and Ehlers, W.: 1986, Theorie der Mehrkomponentenkontinua mit Anwendung auf bodenmechanische Probleme, Teil I, Forschungsberichte aus dem Fachbereich Bauwesen der Universität-GH-Essen 40, Essen.

Brenan, K. E., Campbell, S. L. and Petzold, L. R.: 1989, *Numerical Solution of Initial-Value Problems in Differential-Algebraic Equations*, Elsevier Science, New York.

Ehlers, W.: 1989, Poröse Medien – ein kontinuumsmechanisches Modell auf der Basis der Mischungstheorie, Forschungsberichte aus dem Fachbereich Bauwesen der Universität-GH-Essen 47, Essen.

Ehlers, W.: 1993, Constitutive equations for granular materials in geomechanical context, In: K. Hutter (ed.), *Continuum Mechanics in Environmental Sciencies and Geophysics*, CISM Courses and Lecture Notes No. 337, Springer-Verlag, Wien, pp. 313–402.

Ehlers, W. and Eipper, G.: 1997, Finite Elastizität bei fluidgesättigten hochporösen Festkörpern, *ZAMM* **77**, S 79–S 80.

Marsden, J. E. and Hughes, T. J. R.: 1983, *Mathematical Foundations of Elasticity*, Prentice Hall, Englewood Cliffs, N.J.

Ogden, R. W.: 1972, Large deformation isotropic elasticity: On the correlation of theory and experiment for compressible rubberlike solids, *Proc. Royal Soc. London, Series A* **328**, 567–583.

Ogden, R. W.: 1984, *Non-Linear Elastic Deformations*, Ellis Horwood, Chichester, U.K.

Rivlin, R. S.: 1948, Large elastic deformations of isotropic materials: I. Fundamental concepts, *Proc. Royal Soc. London, Series A* **240**, 459–490.

Zienkiewicz, O. C. and Taylor, R. L.: 1984, *The Finite Element Method*, Vol. 1, 4th edn, McGraw-Hill, London.

Wriggers, P.: 1989, Konsistente Linearisierung in der Kontinuumsmechanik und ihre Anwendung auf die Finite-Element-Methode, Bericht F88/4, Institut für Baumechanik und Numerische Mechanik, Universität Hannover.

# A Micropolar Theory of Porous Media: Constitutive Modelling

S. DIEBELS
*Institut für Mechanik (Bauwesen), Lehrstuhl II, Universität Stuttgart, Pfaffenwaldring 7, 70569 Stuttgart, Germany. e-mail: diebels@mechbau.uni-stuttgart.de*

(Received: 4 August 1997; in final form: 22 October 1997)

**Abstract.** The extension of the classical mixture theory by the concept of volume fractions leads to the theory of porous media. In this article, the theory of porous media is generalised to micropolar constituents. The kinematic relations and the balance equations for a porous medium are developed without restricting the number of constituents. Based on the entropy inequality, the general form of the constitutive equations are derived for a binary medium consisting of a porous elastic skeleton saturated by a viscous pore-fluid. Both constituents are assumed to be compressible. Handling the saturation constraint by a Lagrangian multiplier leads to a compatibility of the proposed model to so-called hybrid and incompressible models.

**Key words:** micropolar theory, thermodynamically consistent constitutive modelling.

## 1. Introduction

In several branches of engineering, the behaviour of granular materials is of high interest, for example in geotechnique, mining engineering, and silo technique. Therefore, it becomes necessary to model such materials theoretically and to allow for the computation of their behaviour. One possibility to model granular materials is a microscopic description, where the motion of each particle is computed, for example Hemmingsson *et al.* (1996). The disadvantage of this method is the restriction to several thousands of particles due to the big computational effort. Therefore, a microscopic theory based on the motion of individual particles is not applicable in real engineering problems. The alternative is a macroscopic theory such as the Theory of Porous Media, which describes the behaviour of mixtures with immiscible constituents and internal structure, for example Bowen (1976, 1980, 1982), de Boer and Ehlers (1986), Ehlers (1989, 1993), and others. Such a theory is based on the behaviour of a representative elementary volume (REV), which is set up by several particles of each constituent. This leads to the idea of superimposed continua, where material points of different constituents occupy the same spatial position in the actual configuration. The internal structure of a mixture is described by scalar internal variables, the volume fractions, relating the volume occupied by the individual constituents to the bulk volume of the whole mixture. Macroscopic theories may be used to obtain a better understanding of the behaviour of granular materials when rotational degrees of freedom are taken into account. The additional degrees of freedom represent the average rotation of the particles within the REV. The idea to include rotational degrees of freedom into continuum descriptions was firstly proposed by the Cosserats (1909) at the beginning of the century. During the sixties and seventies, the interest in so-called Cosserat or micropolar theories increased, for example Eringen (1964), Eringen and Kafadar (1976), and Besdo (1974), and it was found in the last few years, that such theories lead to a regularisation of localisation phenomena such as shear banding (de Borst, 1991; Dietsche *et al.*, 1993; Tejchman and Wu, 1993). The combination of the theory of porous media with the micropolar theory allows a detailed description of fluid-saturated granular materials.

It is the aim of this article to summarise the kinematic relations governing the behaviour of micropolar mixtures. Furthermore, the balance equations of mass, momentum, moment of momentum, energy, and entropy are discussed. The so-called balance of microinertia introduced by Eringen (1964) and Eringen and Kafadar (1976) is interpreted as a kinematic constraint, which describes the material points as rigid particles on the microscale. The kinematic relations and the balance equations are independent of the material under study. To obtain a closed theory, constitutive equations are required. Based on the entropy inequality, a general framing for the constitutive equations is developed.

As an example, the presented results are applied to a binary mixture consisting of a porous elastic skeleton and a viscous pore-fluid. The saturation constraint (Svendsen and Hutter, 1995) is taken into account by a Lagrangian multiplier, which may be identified as a thermodynamic pressure of the pore-fluid.

## 2. Kinematics

### 2.1. CONCEPT OF VOLUME FRACTIONS

In mixture theories, material points of each constituent $\varphi^\alpha$ ($\alpha = 1, \dots, n$) occupy the same spatial point in the actual configuration (superimposed continua). The volume element $\mathrm{d}v$ is given by the sum of the partial volume elements $\mathrm{d}v^\alpha$, which, on the microscale, are occupied by $\varphi^\alpha$ only:

$$\mathrm{d}v = \sum_{\alpha=1}^{n} \mathrm{d}v^\alpha. \tag{1}$$

The scalar structural variables to describe the microstructure on the macroscopic scale are the volume fractions, defined as the local ratio of the partial volume element $\mathrm{d}v^\alpha$ to the bulk volume element $\mathrm{d}v$

$$n^\alpha = \frac{\mathrm{d}v^\alpha}{\mathrm{d}v}. \tag{2}$$

The combination of Equations (1) and (2) leads to the saturation constraint

$$\sum_\alpha n^\alpha = 1. \tag{3}$$

As pointed out by Svendsen and Hutter (1995), the saturation condition is a true constraint that must be taken into account if constitutive equations are developed.

### 2.2. MOTION AND MICROMOTION

Starting from different reference positions $\mathbf{X}_\alpha$, particles of each constituent $\varphi^\alpha$ follow their own motion

$$\mathbf{x} = \boldsymbol{\chi}_\alpha(\mathbf{X}_\alpha, t). \tag{4}$$

In addition to the motion (4), each point undergoes a micromotion, which describes the total rotation of the material points. In the case of micropolar media, the material points are assumed to be rigid particles on the microscale. Therefore, each point has attached directors $\boldsymbol{\Xi}_\alpha$, which are rotated into the directors of the actual configuration called $\boldsymbol{\xi}_\alpha$. The micromotion $\bar{\mathbf{R}}_\alpha$ is represented by an orthogonal tensor with the following properties:

$$\boldsymbol{\xi}_\alpha = \bar{\mathbf{R}}_\alpha \boldsymbol{\Xi}_\alpha, \qquad \bar{\mathbf{R}}_\alpha \bar{\mathbf{R}}_\alpha^T = \mathbf{I}, \qquad \det \bar{\mathbf{R}}_\alpha = 1. \tag{5}$$

While the deformation of classical continua is described by the deformation gradient

$$\mathbf{F}_\alpha = \mathrm{G{\scriptsize RAD}}_\alpha \boldsymbol{\chi}_\alpha(\mathbf{X}_\alpha, t), \tag{6}$$

where the operator 'GRAD$_\alpha$' means differentiation with respect to the reference position $\mathbf{X}_\alpha$, the deformation of micropolar continua requires two deformation tensors, the first Cosserat deformation tensor $\bar{\mathbf{U}}_\alpha$ and the curvature tensor $\bar{\mathbf{K}}_\alpha$. They are defined by

$$\bar{\mathbf{U}}_\alpha = \bar{\mathbf{R}}_\alpha^T \mathbf{F}_\alpha, \tag{7}$$

$$\bar{\mathbf{K}}_\alpha = -\frac{1}{2}\left[\overset{3}{\mathbf{E}}\left(\bar{\mathbf{R}}_\alpha^T \mathrm{G{\scriptsize RAD}}_\alpha \bar{\mathbf{R}}_\alpha\right)^{\underline{3}}\right]^{\underline{2}}. \tag{8}$$

In this expression the tensor $\overset{3}{\mathbf{E}}$ is the Ricci or permutation tensor, and the underlined superscript indicates the order of the resulting tensor (de Boer, 1982). As discussed in detail by Diebels and Ehlers (1996), the first Cosserat deformation tensor $\bar{\mathbf{U}}_\alpha$ describes the scalar product of line elements $\mathrm{d}\mathbf{x}_\alpha$ and directors $\boldsymbol{\xi}_\alpha$ of the actual configuration with respect to the corresponding scalar product of vectors of the reference configuration:

$$\boldsymbol{\xi}_\alpha \cdot \mathrm{d}\mathbf{x}_\alpha = \boldsymbol{\Xi}_\alpha \cdot \bar{\mathbf{U}}_\alpha \, \mathrm{d}\mathbf{X}_\alpha. \tag{9}$$

The curvature tensor $\bar{\mathbf{K}}_\alpha$ may be interpreted as a generalised Christoffel symbol (Besdo, 1974; Diebels and Ehlers, 1996).

2.3. DEFORMATION VELOCITIES

The velocity gradient $\mathbf{L}_\alpha$ describes the change in time of line elements $\mathrm{d}\mathbf{x}_\alpha$ associated with the constituent $\varphi^\alpha$:

$$(\mathrm{d}\mathbf{x}_\alpha)'_\alpha = \mathbf{L}_\alpha \mathrm{d}\mathbf{x}_\alpha, \qquad \mathbf{L}_\alpha = (\mathbf{F}_\alpha)'_\alpha \mathbf{F}_\alpha^{-1}. \tag{10}$$

The symbol $(\cdots)'_\alpha$ represents the material time derivative following the motion of $\varphi^\alpha$. The time rate of the directors $\boldsymbol{\xi}_\alpha$ is described by the gyration tensor $\bar{\mathbf{W}}_\alpha$:

$$(\boldsymbol{\xi}_\alpha)'_\alpha = \bar{\mathbf{W}}_\alpha \boldsymbol{\xi}_\alpha, \qquad \bar{\mathbf{W}}_\alpha = (\bar{\mathbf{R}}_\alpha)'_\alpha \bar{\mathbf{R}}_\alpha^T. \tag{11}$$

If the skew symmetry of $\bar{\mathbf{W}}_\alpha$ is taken into account, the angular velocity $\bar{\boldsymbol{\omega}}_\alpha$ is given by the axial vector of the gyration tensor:

$$\bar{\boldsymbol{\omega}}_\alpha = \tfrac{1}{2}\overset{3}{\mathbf{E}}\bar{\mathbf{W}}_\alpha^T. \tag{12}$$

The push-forward transformation of the deformation rates $(\bar{\mathbf{U}}_\alpha)'_\alpha$ and $(\bar{\mathbf{K}}_\alpha)'_\alpha$ from the reference configuration to the actual configuration is found as

$$(\bar{\mathbf{U}}_\alpha)'_\alpha = \bar{\mathbf{R}}_\alpha^T (\mathbf{L}_\alpha - \bar{\mathbf{W}}_\alpha)\mathbf{F}_\alpha \;:=\; \bar{\mathbf{R}}_\alpha^T \bar{\mathbf{L}}_\alpha \mathbf{F}_\alpha, \tag{13}$$

$$(\bar{\mathbf{K}}_\alpha)'_\alpha = \bar{\mathbf{R}}_\alpha^T \operatorname{grad} \bar{\boldsymbol{\omega}}_\alpha \mathbf{F}_\alpha. \tag{14}$$

The operator 'grad' denotes the differentiation with respect to the actual position $\mathbf{x}$.

## 3. Balance Equations

As discussed by Haupt (1993) and several others, the balance equations of mechanics may be summarised by a so-called master balance, which represents the general structure of the balance equations. Specifying the physical quantities $\Psi^\alpha$, their fluxes $\boldsymbol{\Phi}^\alpha$, supplies $\sigma^\alpha$, and productions $\hat{\Psi}^\alpha$, the well-known balances of mass, momentum, moment of momentum, energy and entropy may be derived. The concept of the master balance was successfully applied to mixtures (Ehlers, 1996). Here, the results obtained by Diebels and Ehlers (1996) for mixtures with micropolar constituents are summarised without presenting details.

The structure of the local form of a balance equation is given by

$$(\Psi^\alpha)'_\alpha + \Psi^\alpha \operatorname{div} \mathbf{x}'_\alpha = \operatorname{div} \boldsymbol{\Phi}^\alpha + \sigma^\alpha + \hat{\Psi}^\alpha, \tag{15}$$

where the left hand side describes the change of the physical quantity $\Psi^\alpha$ and the right hand side gives the reason for this change, cf. Haupt (1993). According to Truesdell's metaphysical principles (Truesdell, 1984), the same structure is found for the mixture as a whole:

$$\dot{\Psi} + \Psi \operatorname{div} \dot{\mathbf{x}} = \operatorname{div} \boldsymbol{\Phi} + \sigma + \hat{\Psi}, \tag{16}$$

where $(\cdots)^{\cdot}$ is the material time derivative according to the barycentric motion.

*Table I.* Physical quantities $\Psi^\alpha$, fluxes $\mathbf{\Phi}^\alpha$, supplies $\sigma^\alpha$, and productions $\hat{\Psi}^\alpha$ for the balance equations of the constituents $\varphi^\alpha$

| | $\Psi^\alpha$ | $\mathbf{\Phi}^\alpha$ | $\sigma^\alpha$ | $\hat{\Psi}^\alpha$ |
|---|---|---|---|---|
| Mass | $\rho^\alpha$ | — | — | $\hat{\rho}^\alpha$ |
| Microinertia | $\rho^\alpha \mathbf{\Theta}^\alpha$ | — | $2\,\mathrm{sym}(\rho^\alpha \bar{\mathbf{W}}_\alpha \mathbf{\Theta}^\alpha)$ | $\hat{\rho}^\alpha \mathbf{\Theta}^\alpha$ |
| Momentum | $\rho^\alpha \mathbf{x}'_\alpha$ | $\mathbf{T}^\alpha$ | $\rho^\alpha \mathbf{b}^\alpha$ | $\hat{\mathbf{s}}^\alpha$ |
| Moment of momentum | $\mathbf{x} \times \rho^\alpha \mathbf{x}'_\alpha + \rho^\alpha \mathbf{\Theta}^\alpha \bar{\boldsymbol{\omega}}_\alpha$ | $\mathbf{x} \times \mathbf{T}^\alpha + \mathbf{M}^\alpha$ | $\mathbf{x} \times \rho^\alpha \mathbf{b}^\alpha + \rho^\alpha \mathbf{c}^\alpha$ | $\hat{\mathbf{h}}^\alpha$ |
| Energy | $\rho^\alpha \varepsilon^\alpha + \frac{1}{2}\rho^\alpha \mathbf{x}'_\alpha \cdot \mathbf{x}'_\alpha + \frac{1}{2}\rho^\alpha \mathbf{\Theta}^\alpha \bar{\boldsymbol{\omega}}_\alpha \cdot \bar{\boldsymbol{\omega}}_\alpha$ | $\mathbf{T}^{\alpha T} \mathbf{x}'_\alpha + \mathbf{M}^{\alpha T} \bar{\boldsymbol{\omega}}_\alpha - \mathbf{q}_\alpha$ | $\rho^\alpha \mathbf{b}^\alpha \cdot \mathbf{x}'_\alpha + \rho^\alpha \mathbf{c}^\alpha \cdot \bar{\boldsymbol{\omega}}_\alpha + \rho^\alpha r^\alpha$ | $\hat{e}^\alpha$ |
| Entropy | $\rho^\alpha \eta^\alpha$ | $-\dfrac{\mathbf{q}^\alpha}{\vartheta^\alpha}$ | $\dfrac{\rho^\alpha r^\alpha}{\vartheta^\alpha}$ | $\hat{\eta}^\alpha$ |

Following Truesdell (1984), the mixture must behave like the sum of its parts. Therefore, the following restrictions hold (Ehlers, 1996):

$$\begin{aligned} \Psi &= \sum_\alpha \Psi^\alpha, \qquad & \mathbf{\Phi} &= \sum_\alpha (\mathbf{\Phi}^\alpha - \Psi^\alpha \mathbf{d}_\alpha), \\ \sigma &= \sum_\alpha \sigma^\alpha, \qquad & \hat{\Psi} &= \sum_\alpha \hat{\Psi}^\alpha, \end{aligned} \tag{17}$$

with the diffusion velocity $\mathbf{d}_\alpha = \mathbf{x}'_\alpha - \dot{\mathbf{x}}$. Note in passing that the production terms $\hat{\Psi}$ for the mixture are zero in general, except for the entropy balance. Furthermore, the production terms $\hat{\Psi}^\alpha$ for each constituent $\varphi^\alpha$ are required in every balance equation to model the interaction between the different constituents. Following Diebels and Ehlers (1996) the physical quantities, their fluxes, supplies and productions are listed in Table I.

The used symbols have the following meaning: partial density $\rho^\alpha$, density production $\hat{\rho}^\alpha$, tensor of microinertia $\rho^\alpha\,\mathbf{\Theta}^\alpha$, Cauchy stress tensor $\mathbf{T}^\alpha$, volume force $\rho^\alpha\,\mathbf{b}^\alpha$, momentum production $\hat{\mathbf{s}}^\alpha$, couple stress $\mathbf{M}^\alpha$, body moment $\rho^\alpha \mathbf{c}^\alpha$, supply of moment of momentum $\hat{\mathbf{h}}^\alpha$, internal energy $\varepsilon^\alpha$, heat influx $\mathbf{q}^\alpha$, radiation $\rho^\alpha r^\alpha$, energy production $\hat{e}^\alpha$, entropy $\eta^\alpha$, Kelvin's temperature $\vartheta^\alpha$, entropy production $\hat{\eta}^\alpha$. The operator 'sym' renders the symmetric part of a second-order tensor.

Making use of the quantities listed in Table I, the following balance equations for the constituents $\varphi^\alpha$ may be obtained (Diebels and Ehlers, 1996):

*Balance of mass:*

$$(\rho^\alpha)'_\alpha + \rho^\alpha \operatorname{div} \mathbf{x}'_\alpha = \hat{\rho}^\alpha, \tag{18}$$

Restrictions:

$$\rho = \sum_\alpha \rho^\alpha, \qquad \mathbf{0} = \sum_\alpha \rho^\alpha \mathbf{d}_\alpha, \qquad 0 = \sum_\alpha \hat{\rho}^\alpha. \tag{19}$$

*Balance of microinertia:*
The balance of microinertia follows from the assumption, that the material points as carrier of the physical properties are rigid bodies on the microscale (Eringen, 1964). Thus, the inertia associated with the special shape of the rigid particles keeps constant with respect to a frame rotating with the particle. This may be expressed in terms of the microinertia rotated back to the reference configuration:

$$(\bar{\mathbf{R}}_\alpha^T\, \mathbf{\Theta}^\alpha\, \bar{\mathbf{R}}_\alpha)'_\alpha = \mathbf{0}. \tag{20}$$

From this equation, the balance of microinertia is found for each constituent (Diebels and Ehlers, 1996):

$$(\rho^\alpha \mathbf{\Theta}^\alpha)'_\alpha + \rho^\alpha \mathbf{\Theta}^\alpha \operatorname{div} \mathbf{x}'_\alpha = 2\,\mathrm{sym}\big(\bar{\mathbf{W}}_\alpha \rho^\alpha \mathbf{\Theta}^\alpha\big) + \hat{\rho}^\alpha \mathbf{\Theta}^\alpha, \tag{21}$$

From the restrictions (17) the corresponding quantities for the mixture as a whole may be computed.

Restrictions:

$$\begin{aligned} \rho\,\mathbf{\Theta} &= \sum_\alpha \rho^\alpha\, \mathbf{\Theta}^\alpha, & \overset{3}{\mathbf{\Phi}}_\Theta &= -\sum_\alpha \rho^\alpha\, \mathbf{\Theta}^\alpha \otimes \mathbf{d}_\alpha, \\ \overset{2}{\boldsymbol{\sigma}}_\Theta &= 2\sum_\alpha\ \mathrm{sym}\left[\rho^\alpha\, \bar{\mathbf{W}}_\alpha\, \mathbf{\Theta}^\alpha\right], & \hat{\mathbf{\Theta}} &= \sum_\alpha \hat{\rho}^\alpha\, \mathbf{\Theta}^\alpha. \end{aligned} \tag{22}$$

*Balance of momentum:*

$$\rho^\alpha \mathbf{x}''_\alpha = \operatorname{div} \mathbf{T}^\alpha + \rho^\alpha \mathbf{b}^\alpha + \hat{\mathbf{p}}^\alpha \tag{23}$$

with

$$\hat{\mathbf{p}}^\alpha = \hat{\mathbf{s}}^\alpha - \hat{\rho}^\alpha \mathbf{x}'_\alpha. \tag{24}$$

Restrictions:

$$\begin{aligned} \rho\dot{\mathbf{x}} &= \sum_\alpha \rho^\alpha \mathbf{x}'_\alpha, & \mathbf{T} &= \sum_\alpha \big(\mathbf{T}^\alpha - \rho^\alpha \mathbf{x}'_\alpha \otimes \mathbf{d}_\alpha\big), \\ \rho\mathbf{b} &= \sum_\alpha \rho^\alpha \mathbf{b}^\alpha, & \mathbf{0} &= \sum_\alpha \big(\hat{\mathbf{p}}^\alpha + \hat{\rho}^\alpha \mathbf{x}'_\alpha\big). \end{aligned} \tag{25}$$

*Balance of moment of momentum:*

$$\rho^\alpha \mathbf{\Theta}^\alpha (\bar{\boldsymbol{\omega}}_\alpha)'_\alpha \;=\; \mathbf{I} \times \mathbf{T}^\alpha + \operatorname{div} \mathbf{M}^\alpha + \rho^\alpha \mathbf{c}^\alpha + \hat{\mathbf{m}}^\alpha - 2\,\mathrm{sym}\big(\bar{\mathbf{W}}_\alpha \rho^\alpha \mathbf{\Theta}^\alpha\big)\bar{\boldsymbol{\omega}}_\alpha \tag{26}$$

with

$$\hat{\mathbf{m}}^\alpha = \hat{\mathbf{h}}^\alpha - \mathbf{x} \times \hat{\rho}^\alpha \mathbf{x}'_\alpha - \hat{\rho}^\alpha \mathbf{\Theta}^\alpha \bar{\boldsymbol{\omega}}_\alpha - \mathbf{x} \times \hat{\mathbf{p}}^\alpha. \tag{27}$$

Restrictions:

$$\rho\boldsymbol{\Theta}\bar{\boldsymbol{\omega}} = \sum_\alpha \rho^\alpha \boldsymbol{\Theta}^\alpha \bar{\boldsymbol{\omega}}_\alpha, \qquad \mathbf{M} = \sum_\alpha \big(\mathbf{M}^\alpha - \rho^\alpha \boldsymbol{\Theta}^\alpha \bar{\boldsymbol{\omega}}_\alpha \otimes \mathbf{d}_\alpha\big),$$
$$\rho\mathbf{c} = \sum_\alpha \rho^\alpha \mathbf{c}^\alpha, \qquad \mathbf{0} = \sum_\alpha \big(\hat{\mathbf{m}}^\alpha + \hat{\rho}^\alpha \boldsymbol{\Theta}^\alpha \bar{\boldsymbol{\omega}}_\alpha\big). \tag{28}$$

*Balance of energy:*

$$\begin{aligned}\rho^\alpha(\varepsilon^\alpha)'_\alpha &= \mathbf{T}^\alpha \cdot \bar{\mathbf{L}}_\alpha + \mathbf{M}^\alpha \cdot \operatorname{grad}\bar{\boldsymbol{\omega}}_\alpha - \operatorname{div}\mathbf{q}^\alpha + \\ &\quad + \rho^\alpha r^\alpha - \bar{\boldsymbol{\omega}}_\alpha \cdot \operatorname{sym}\big(\bar{\mathbf{W}}_\alpha \rho^\alpha \boldsymbol{\Theta}^\alpha\big)\bar{\boldsymbol{\omega}}_\alpha + \hat{\varepsilon}^\alpha\end{aligned} \tag{29}$$

with

$$\hat{\varepsilon}^\alpha = \hat{e}^\alpha - \hat{\mathbf{p}}^\alpha \cdot \mathbf{x}'_\alpha - \hat{\mathbf{m}}^\alpha \cdot \bar{\boldsymbol{\omega}}_\alpha - \hat{\rho}^\alpha\left(\varepsilon^\alpha + \tfrac{1}{2}\mathbf{x}'_\alpha \cdot \mathbf{x}'_\alpha + \tfrac{1}{2}\bar{\boldsymbol{\omega}}_\alpha \cdot \rho^\alpha \boldsymbol{\Theta}_\alpha \bar{\boldsymbol{\omega}}^\alpha\right) \tag{30}$$

Restrictions:

$$\rho\varepsilon = \sum_\alpha \left(\rho^\alpha\varepsilon^\alpha + \frac{1}{2}\rho^\alpha \mathbf{d}_\alpha \cdot \mathbf{d}_\alpha + \frac{1}{2}\bar{\boldsymbol{\delta}}_\alpha \cdot \rho^\alpha \boldsymbol{\Theta}^\alpha \bar{\boldsymbol{\delta}}_\alpha\right),$$

$$\begin{aligned}\mathbf{q} + \frac{1}{2}\left(\overset{3}{\boldsymbol{\Phi}}{}^{\overset{13}{T}}_{\Theta}\bar{\boldsymbol{\omega}}\right)\bar{\boldsymbol{\omega}} &= \sum_\alpha \Big[\mathbf{q}^\alpha - \mathbf{T}^{\alpha T}\mathbf{d}_\alpha - \mathbf{M}^{\alpha T}\bar{\boldsymbol{\delta}}_\alpha + \rho^\alpha\varepsilon^\alpha \mathbf{d}_\alpha + \\ &\quad + \frac{1}{2}\rho^\alpha(\mathbf{d}_\alpha \cdot \mathbf{d}_\alpha)\mathbf{d}_\alpha + \frac{1}{2}(\bar{\boldsymbol{\delta}}_\alpha \cdot \rho^\alpha \boldsymbol{\Theta}^\alpha \bar{\boldsymbol{\delta}}_\alpha)\mathbf{d}_\alpha\Big]\end{aligned}$$

$$\rho r = \sum_\alpha [\rho^\alpha r^\alpha + \rho^\alpha \mathbf{b}^\alpha \cdot \mathbf{d}_\alpha + \rho^\alpha \mathbf{c}^\alpha \cdot \bar{\boldsymbol{\delta}}_\alpha] \tag{31}$$

$$0 = \sum_\alpha \left[\hat{\varepsilon}^\alpha + \hat{\rho}^\alpha\left(\varepsilon^\alpha + \frac{1}{2}\mathbf{x}'_\alpha \cdot \mathbf{x}'_\alpha + \frac{1}{2}\bar{\boldsymbol{\omega}}_\alpha \cdot \boldsymbol{\Theta}^\alpha \bar{\boldsymbol{\omega}}_\alpha\right) + \hat{\mathbf{p}}^\alpha \cdot \mathbf{x}'_\alpha + \hat{\mathbf{m}}^\alpha \cdot \bar{\boldsymbol{\omega}}_\alpha\right],$$

where $\overset{3}{\boldsymbol{\Phi}}_\Theta$ is the flux of the microinertia of the whole mixture computed according to $(22)_2$; $\bar{\boldsymbol{\delta}}_\alpha = \bar{\boldsymbol{\omega}}_\alpha - \bar{\boldsymbol{\omega}}$ is the diffusion angular velocity.

*Balance of entropy:*

$$(\rho^\alpha\eta^\alpha)'_\alpha + \rho^\alpha\eta^\alpha \operatorname{div}\mathbf{x}'_\alpha = -\operatorname{div}\left(\frac{\mathbf{q}^\alpha}{\vartheta^\alpha}\right) + \frac{\rho^\alpha r^\alpha}{\vartheta^\alpha} + \hat{\eta}^\alpha. \tag{32}$$

Restrictions:

$$\rho\eta = \sum_\alpha \rho^\alpha\eta^\alpha, \qquad \boldsymbol{\Phi}_\eta = \sum_\alpha\left(\frac{\mathbf{q}^\alpha}{\vartheta^\alpha} + \rho^\alpha\eta^\alpha\mathbf{d}_\alpha\right),$$
$$\sigma_\eta = \sum_\alpha \sigma^\alpha_\eta, \qquad \hat{\eta} = \sum_\alpha \hat{\eta}^\alpha. \tag{33}$$

The second law of thermodynamics may only be applied to the mixture as a whole and not to each constituent for its own. It requires that the total production of entropy is nonnegative:

$$\hat{\eta} = \sum_{\alpha} \hat{\eta}^{\alpha} \geqslant 0. \tag{34}$$

## 4. Constitutive Modelling for a Two-Phase Model

On the basis of the kinematic relations and the balance equations presented in the preceding sections, a two phase model is developed in this section. It is assumed that the porous medium under study consists of a porous elastic solid skeleton $\varphi^{S}$ and a viscous pore-fluid $\varphi^{F}$. In addition, further assumptions are made at the beginning:

1. The binary porous medium under study is saturated:

$$n^{S} + n^{F} = 1. \tag{35}$$

The incremental version of the saturation condition, which is required in the entropy principle in combination with a Lagrangian multiplier $P$ to guarantee saturation, is given by

$$(n^{S})'_{S} + (n^{F})'_{F} - \operatorname{grad} n^{F} \cdot \mathbf{w}_{F} = 0. \tag{36}$$

Herein, the seepage velocity $\mathbf{w}_{F} = \mathbf{x}'_{F} - \mathbf{x}'_{S}$ is introduced.

2. There is no mass exchange between both constituents:

$$\hat{\rho}^{\alpha} = 0. \tag{37}$$

3. It is assumed that the material points are represented by spherical particles on their microscale. In this case, the microinertia is given by one scalar value $\Theta^{\alpha}$:

$$\boldsymbol{\Theta}^{\alpha} = \Theta^{\alpha}\,\mathbf{I}. \tag{38}$$

Therefore, the microinertia becomes constant,

$$(\Theta^{\alpha})'_{\alpha} = 0, \tag{39}$$

according to $(5)_2$ and (20).

4. The temperature of both constituents is assumed to be the same:

$$\vartheta^{S} = \vartheta^{F} = \vartheta. \tag{40}$$

In general, this is the case, if the diffusion motion is slow enough to allow for a compensation of temperature differences between both constituents. Keeping in mind that the model should be applicable to geomechanical problems, this assumption is usually fulfilled.

Note in passing that in this case only the energy equation for the mixture as a whole must be taken into account.

5. For both constituents the free energy functions $\psi^\alpha$ are introduced by the definition

$$\psi^\alpha := \varepsilon^\alpha - \vartheta\eta^\alpha. \tag{41}$$

By use of these assumptions, the entropy inequality for the mixture reads

$$\begin{aligned}
&-\rho^S(\psi^S)'_S - \rho^F(\psi^F)'_F - \rho\eta\dot{\vartheta} - \frac{1}{\vartheta}\mathbf{q}\cdot\operatorname{grad}\vartheta + \\
&\quad + \mathbf{T}^S\cdot\bar{\mathbf{L}}_S + \mathbf{T}^F\cdot\bar{\mathbf{L}}_F + \mathbf{M}^S\cdot\operatorname{grad}\bar{\boldsymbol{\omega}}_S + \mathbf{M}^F\cdot\operatorname{grad}\bar{\boldsymbol{\omega}}_F - \\
&\quad - \hat{\mathbf{p}}^F\cdot\mathbf{w}_F - \hat{\mathbf{m}}^F\cdot\bar{\boldsymbol{\nu}}_F - \\
&\quad - P[(n^S)'_S + (n^F)'_F - \operatorname{grad} n^F\cdot\mathbf{w}_F] \geqslant 0.
\end{aligned} \tag{42}$$

The stress power is governed by the generalised velocity gradient

$$\bar{\mathbf{L}}_\alpha = \mathbf{L}_\alpha - \bar{\mathbf{W}}_\alpha, \tag{43}$$

while the power of the production of moment of momentum is governed by the difference angular velocity

$$\bar{\boldsymbol{\nu}}_F = \bar{\boldsymbol{\omega}}_F - \bar{\boldsymbol{\omega}}_S. \tag{44}$$

The saturation condition has been added to the entropy inequality as a constraint introducing an undetermined Lagrangian multiplier $P$. This allows for the introduction of two independent evolution equations for the volume fractions. The advantage of this procedure is that the model becomes compatible with other models either with one compressible and one incompressible constituent, so-called hybrid models (Ehlers, 1993), or with completely incompressible models. In the second case, the balances of mass are reduced to evolution equations for the volume fractions, compare Bowen (1980), Ehlers (1989), and Svendsen and Hutter (1995). Note that incompressibility refers to the effective density $\rho^{\alpha R}$ defined at the microscale. Changes of the partial density $\rho^\alpha$ are still possible due to changes of the porosity $n^F$.

On the other hand, the results obtained for the compressible model are the same, either if the Lagrangian multiplier is used in combination with two evolution equations or if the constraint is taken into account directly by use of only one evolution equation for one volume fraction (Diebels, 1997).

Based on the above assumptions and on the concept of materials of second grade, the following variables are required to describe the motion of the binary mixture under study consisting of an elastic skeleton and a viscous pore-fluid

$$\begin{aligned}
\mathcal{S} = \{&\bar{\mathbf{U}}_S,\ \nabla\bar{\mathbf{U}}_S,\ \bar{\mathbf{K}}_S,\ \nabla\bar{\mathbf{K}}_S,\ \rho^{FR},\ \operatorname{grad}\rho^{FR}, \\
&\bar{\mathbf{L}}_F,\ \operatorname{grad}\bar{\boldsymbol{\omega}}_F,\ \mathbf{w}_F,\ \bar{\boldsymbol{\nu}}_F,\ n^S,\ \operatorname{grad} n^S,\ \vartheta,\ \operatorname{grad}\vartheta\},
\end{aligned} \tag{45}$$

with $\nabla = \mathrm{G{\scriptstyle RAD}}_S$. The so-called response functions

$$\mathcal{R} = \{\psi^\alpha,\ \eta,\ \mathbf{T}^\alpha,\ \mathbf{M}^\alpha,\ \hat{\mathbf{p}}^\beta,\ \hat{\mathbf{m}}^\beta,\ (n^\alpha)'_\alpha\}, \tag{46}$$

with $\alpha \in \{S, F\}$ and $\beta \in \{F\}$ will be given by constitutive equations.

### 4.1. EVOLUTION EQUATIONS FOR THE VOLUME FRACTIONS

In the case of vanishing mass productions $\hat{\rho}^\alpha = 0$ and with view of the results of the incompressible model, evolution equations for the volume fractions are derived from the balance of mass as an a priori constitutive assumption.

In the framework of porous media theories, two effects lead to changes in the partial density $\rho^\alpha$, namely changes in the effective density $\rho^{\alpha R}$ and changes in the volume fraction $n^\alpha$. If both effects are separated, the mass balance may be re-written in the following form:

$$\rho^{\alpha R}((n^\alpha)'_\alpha + z^\alpha n^\alpha \operatorname{div} \mathbf{x}'_\alpha) + n^\alpha((\rho^{\alpha R})'_\alpha + (1 - z^\alpha)\rho^{\alpha R} \operatorname{div} \mathbf{x}'_\alpha) = 0, \tag{47}$$

where the factor $z^\alpha$ is limited to $0 \leqslant z^\alpha \leqslant 1$. The case $z^\alpha = 1$ corresponds to the incompressible limit, that is, all changes in the partial density result from changes in the volume fraction, while the effective density $\rho^{\alpha R}$ is constant. Due to the independence of the volume fraction $n^\alpha$ and the effective density $\rho^{\alpha R}$, it is a sufficient condition that both paranthesis are zero. From this assumption, the evolution equation for the volume fraction

$$(n^\alpha)'_\alpha + z^\alpha n^\alpha \operatorname{div} \mathbf{x}'_\alpha = 0 \tag{48}$$

and the balance equation for the effective density

$$(\rho^{\alpha R})'_\alpha + (1 - z^\alpha)\rho^{\alpha R} \operatorname{div} \mathbf{x}'_\alpha = 0 \tag{49}$$

are motivated. Regard that, according to this choice, the functions $z^\alpha$ replace the evolution equations $(n^\alpha)'_\alpha$ in the set of the response functions $\mathcal{R}$.

Taking into account two independent evolution equations for both constituents $\varphi^S$ and $\varphi^F$ of the saturated binary mixture, the incremental form of the saturation constraint is given by

$$\operatorname{grad} n^F \cdot \mathbf{w}_F + z^S n^S \operatorname{div} \mathbf{x}'_S + z^F n^F \operatorname{div} \mathbf{x}'_F = 0. \tag{50}$$

### 4.2. RESULTS

Following the well-known ideas proposed by Coleman and Noll (1963) for a visco-elastic continuum and adapted to the theory of porous media by Bowen (1980, 1982) and Ehlers (1989, 1993), the material time derivatives of the free energy functions $\psi^\alpha$ in the entropy inequality are replaced by the derivatives of the free energy functions with respect to the process variables according to the chain rule of differentiation. The entropy inequality (42) yields with respect to the process variables $\mathcal{S}$:

$$\bar{\mathbf{L}}_S \cdot \left[\mathbf{T}^S + z^S n^S \left(\rho^S \frac{\partial \psi^S}{\partial n^S} + P\right)\mathbf{I} - \bar{\mathbf{R}}_S \left(\rho^S \frac{\partial \psi^S}{\partial \bar{\mathbf{U}}_S} + \rho^F \frac{\partial \psi^F}{\partial \bar{\mathbf{U}}_S}\right)\mathbf{F}_S^T\right] +$$

$$+ (\nabla \bar{\mathbf{U}}_S)'_S \cdot \left[-\rho^S \frac{\partial \psi^S}{\partial \nabla \bar{\mathbf{U}}_S} - \rho^F \frac{\partial \psi^F}{\partial \nabla \bar{\mathbf{U}}_S}\right] + \nabla\nabla \bar{\mathbf{U}}_S \cdot \left[-\rho^F \frac{\partial \psi^F}{\partial \nabla \bar{\mathbf{U}}_S} \otimes \mathbf{F}_S^{-1}\mathbf{w}_F\right] +$$

$$
\begin{aligned}
&+ \operatorname{grad}\bar{\boldsymbol{\omega}}_{\mathrm{S}} \cdot \left[\mathbf{M}^{\mathrm{S}} - \bar{\mathbf{R}}_{\mathrm{S}}\left(\rho^{\mathrm{S}}\frac{\partial\psi^{\mathrm{S}}}{\partial\bar{\mathbf{K}}_{\mathrm{S}}} + \rho^{\mathrm{F}}\frac{\partial\psi^{\mathrm{F}}}{\partial\bar{\mathbf{K}}_{\mathrm{S}}}\right)\mathbf{F}_{\mathrm{S}}^{T}\right] + \\
&+ (\nabla\bar{\mathbf{K}}_{\mathrm{S}})'_{\mathrm{S}} \cdot \left[-\rho^{\mathrm{S}}\frac{\partial\psi^{\mathrm{S}}}{\partial\nabla\bar{\mathbf{K}}_{\mathrm{S}}} - \rho^{\mathrm{F}}\frac{\partial\psi^{\mathrm{F}}}{\partial\nabla\bar{\mathbf{K}}_{\mathrm{S}}}\right] + \nabla\nabla\bar{\mathbf{K}}_{\mathrm{S}} \cdot \left[-\rho^{\mathrm{F}}\frac{\partial\psi^{\mathrm{F}}}{\partial\nabla\bar{\mathbf{K}}_{\mathrm{S}}} \otimes \mathbf{F}_{\mathrm{S}}^{-1}\mathbf{w}_{\mathrm{F}}\right] + \\
&+ (\operatorname{grad}\rho^{\mathrm{FR}})'_{\mathrm{F}} \cdot \left[-\rho^{\mathrm{S}}\frac{\partial\psi^{\mathrm{S}}}{\partial\operatorname{grad}\rho^{\mathrm{FR}}} - \rho^{\mathrm{F}}\frac{\partial\psi^{\mathrm{F}}}{\partial\operatorname{grad}\rho^{\mathrm{FR}}}\right] + \\
&+ \operatorname{grad}\operatorname{grad}\rho^{\mathrm{FR}} \cdot \left[\rho^{\mathrm{S}}\frac{\partial\psi^{\mathrm{S}}}{\partial\operatorname{grad}\rho^{\mathrm{FR}}} \otimes \mathbf{w}_{\mathrm{F}}\right] + \\
&+ (\operatorname{grad} n^{\mathrm{S}})'_{\mathrm{S}} \cdot \left[-\rho^{\mathrm{S}}\frac{\partial\psi^{\mathrm{S}}}{\partial\operatorname{grad} n^{\mathrm{S}}} - \rho^{\mathrm{F}}\frac{\partial\psi^{\mathrm{F}}}{\partial\operatorname{grad} n^{\mathrm{S}}}\right] + \\
&+ \operatorname{grad}\operatorname{grad} n^{\mathrm{S}} \cdot \left[-\rho^{\mathrm{F}}\frac{\partial\psi^{\mathrm{F}}}{\partial\operatorname{grad} n^{\mathrm{S}}} \otimes \mathbf{w}_{\mathrm{F}}\right] + \dot{\vartheta}\left[-\rho\,\eta - \rho^{\mathrm{S}}\frac{\partial\psi^{\mathrm{S}}}{\partial\vartheta} - \rho^{\mathrm{F}}\frac{\partial\psi^{\mathrm{F}}}{\partial\vartheta}\right] + \\
&+ (\operatorname{grad}\vartheta)^{\bullet} \cdot \left[-\rho^{\mathrm{S}}\frac{\partial\psi^{\mathrm{S}}}{\partial\operatorname{grad}\vartheta} - \rho^{\mathrm{F}}\frac{\partial\psi^{\mathrm{F}}}{\partial\operatorname{grad}\vartheta}\right] + \\
&+ \operatorname{grad}\operatorname{grad}\vartheta \cdot \left[-\rho^{\mathrm{S}}\frac{\partial\psi^{\mathrm{S}}}{\partial\operatorname{grad}\vartheta} \otimes \mathbf{d}_{\mathrm{S}} - \rho^{\mathrm{F}}\frac{\partial\psi^{\mathrm{F}}}{\partial\operatorname{grad}\vartheta} \otimes \mathbf{d}_{\mathrm{S}}\right] + \\
&+ (\bar{\mathbf{L}}_{\mathrm{F}})'_{\mathrm{F}} \cdot \left[-\rho^{\mathrm{S}}\frac{\partial\psi^{\mathrm{S}}}{\partial\bar{\mathbf{L}}_{\mathrm{F}}} - \rho^{\mathrm{F}}\frac{\partial\psi^{\mathrm{F}}}{\partial\bar{\mathbf{L}}_{\mathrm{F}}}\right] + \operatorname{grad}\bar{\mathbf{L}}_{\mathrm{F}} \cdot \left[\rho^{\mathrm{S}}\frac{\partial\psi^{\mathrm{S}}}{\partial\bar{\mathbf{L}}_{\mathrm{F}}} \otimes \mathbf{w}_{\mathrm{F}}\right] + \\
&+ (\operatorname{grad}\bar{\boldsymbol{\omega}}_{\mathrm{F}})'_{\mathrm{F}} \cdot \left[-\rho^{\mathrm{S}}\frac{\partial\psi^{\mathrm{S}}}{\partial\operatorname{grad}\bar{\boldsymbol{\omega}}_{\mathrm{F}}} - \rho^{\mathrm{F}}\frac{\partial\psi^{\mathrm{F}}}{\partial\operatorname{grad}\bar{\boldsymbol{\omega}}_{\mathrm{F}}}\right] + \\
&+ \operatorname{grad}\operatorname{grad}\bar{\boldsymbol{\omega}}_{\mathrm{F}} \cdot \left[\rho^{\mathrm{S}}\frac{\partial\psi^{\mathrm{S}}}{\partial\operatorname{grad}\bar{\boldsymbol{\omega}}_{\mathrm{F}}} \otimes \mathbf{w}_{\mathrm{F}}\right] + (\mathbf{x}''_{\mathrm{F}} - \mathbf{x}''_{\mathrm{S}}) \cdot \left[-\rho^{\mathrm{S}}\frac{\partial\psi^{\mathrm{S}}}{\partial\mathbf{w}_{\mathrm{F}}} - \rho^{\mathrm{F}}\frac{\partial\psi^{\mathrm{F}}}{\partial\mathbf{w}_{\mathrm{F}}}\right] + \\
&+ \operatorname{grad}\mathbf{x}'_{\mathrm{S}} \cdot \left[\rho^{\mathrm{F}}\frac{\partial\psi^{\mathrm{F}}}{\partial\mathbf{w}_{\mathrm{F}}} \otimes \mathbf{w}_{\mathrm{F}}\right] + \operatorname{grad}\mathbf{x}'_{\mathrm{F}} \cdot \left[\rho^{\mathrm{S}}\frac{\partial\psi^{\mathrm{S}}}{\partial\mathbf{w}_{\mathrm{F}}} \otimes \mathbf{w}_{\mathrm{F}}\right] + \\
&+ \left((\bar{\boldsymbol{\omega}}_{\mathrm{F}})'_{\mathrm{F}} - (\bar{\boldsymbol{\omega}}_{\mathrm{S}})'_{\mathrm{S}}\right) \cdot \left[-\rho^{\mathrm{S}}\frac{\partial\psi^{\mathrm{S}}}{\partial\bar{\boldsymbol{\nu}}_{\mathrm{F}}} - \rho^{\mathrm{F}}\frac{\partial\psi^{\mathrm{F}}}{\partial\bar{\boldsymbol{\nu}}_{\mathrm{F}}}\right] + \\
&+ \operatorname{grad}\bar{\boldsymbol{\omega}}_{\mathrm{S}} \cdot \left[\rho^{\mathrm{F}}\frac{\partial\psi^{\mathrm{F}}}{\partial\bar{\boldsymbol{\nu}}_{\mathrm{F}}} \otimes \mathbf{w}_{\mathrm{F}}\right] + \operatorname{grad}\bar{\boldsymbol{\omega}}_{\mathrm{F}} \cdot \left[\rho^{\mathrm{S}}\frac{\partial\psi^{\mathrm{S}}}{\partial\bar{\boldsymbol{\nu}}_{\mathrm{F}}} \otimes \mathbf{w}_{\mathrm{F}}\right] + \mathcal{D} \geqslant 0, \qquad (51)
\end{aligned}
$$

where the dissipation $\mathcal{D}$ is given by

$$
\begin{aligned}
\mathcal{D} \;=\; \bar{\mathbf{L}}_{\mathrm{F}} \cdot \Bigg[&(1 - z^{\mathrm{F}})\rho^{\mathrm{FR}}\left(\rho^{\mathrm{S}}\frac{\partial\psi^{\mathrm{S}}}{\partial\rho^{\mathrm{FR}}} + \rho^{\mathrm{F}}\frac{\partial\psi^{\mathrm{F}}}{\partial\rho^{\mathrm{FR}}}\right)\mathbf{I} + \\
&+ n^{\mathrm{F}} z^{\mathrm{F}}\left(\rho^{\mathrm{F}}\frac{\partial\psi^{\mathrm{F}}}{\partial n^{\mathrm{F}}} + P\right)\mathbf{I} + \mathbf{T}^{\mathrm{F}}\Bigg] -
\end{aligned}
$$

$$- \mathbf{w}_F \cdot \left[ \hat{\mathbf{p}}^F + \rho^S \frac{\partial \psi^S}{\partial \rho^{FR}} \operatorname{grad} \rho^{FR} - P \operatorname{grad} n^F + \right.$$

$$\left. + \rho^F \frac{\partial \psi^F}{\partial \bar{\mathbf{U}}_S} (\nabla \bar{\mathbf{U}}_S \mathbf{F}_S^{-1})^{\overset{3}{-}} + \rho^F \frac{\partial \psi^F}{\partial \bar{\mathbf{K}}_S} (\nabla \bar{\mathbf{K}}_S \mathbf{F}_S^{-1})^{\overset{3}{-}} \right] +$$

$$+ \mathbf{M}^F \cdot \operatorname{grad} \bar{\boldsymbol{\omega}}_F - \hat{\mathbf{m}}^F \cdot \bar{\boldsymbol{\nu}}_F -$$

$$- \operatorname{grad} \vartheta \cdot \left[ \boldsymbol{\Phi}_\eta + \rho^S \frac{\partial \psi^S}{\partial \vartheta} \mathbf{d}_S + \rho^F \frac{\partial \psi^F}{\partial \vartheta} \mathbf{d}_F \right]. \tag{52}$$

Following the standard arguments, the terms connected with time and space derivatives of process variables have to vanish each for its own. The remaining dissipation inequality $\mathcal{D} \geqslant 0$ is investigated by the principle of minimum dissipation at thermodynamical equilibrium and a Taylor series expansion close to the equilibrium state. Without going into details, the following results are obtained:

1. The free energy functions $\psi^\alpha$ of both constituents do not depend on the complete set of process variables, but on the so-called equilibrium variables

$$\psi^\alpha = \psi^\alpha(\bar{\mathbf{U}}_S, \bar{\mathbf{K}}_S, \rho^{FR}, n^S, \vartheta). \tag{53}$$

2. The entropy of the mixture is given by differentiation of the free energy functions with respect to the temperature $\vartheta$

$$\rho\eta = -\rho^S \frac{\partial \psi^S}{\partial \vartheta} - \rho^F \frac{\partial \psi^F}{\partial \vartheta}. \tag{54}$$

3. The stress and couple stress of the skeleton material depend on the derivative of the free energy functions with respect to the deformation tensors:

$$\mathbf{T}^S = \bar{\mathbf{R}}_S \left[ \rho^S \frac{\partial \psi^S}{\partial \bar{\mathbf{U}}_S} + \rho^F \frac{\partial \psi^F}{\partial \bar{\mathbf{U}}_S} \right] \mathbf{F}_S^T - n^S z^S \left[ \rho^S \frac{\partial \psi^S}{\partial n^S} + P \right] \mathbf{I} \tag{55}$$

and

$$\mathbf{M}^S = \bar{\mathbf{R}}_S \left[ \rho^S \frac{\partial \psi^S}{\partial \bar{\mathbf{K}}_S} + \rho^F \frac{\partial \psi^F}{\partial \bar{\mathbf{K}}_S} \right] \mathbf{F}_S^T. \tag{56}$$

Thus, the stress tensor is influenced by the configuration pressure $\partial \psi^S / \partial n^S$ and by the undetermined Lagrangian multiplier $P$.

4. The fluid stress is governed by an equilibrium part and an extra stress, while couple stresses are only possible for nonequilibrium states

$$\mathbf{T}^F = -(1 - z^F)\rho^{FR} \left[ \rho^S \frac{\partial \psi^S}{\partial \rho^{FR}} + \rho^F \frac{\partial \psi^F}{\partial \rho^{FR}} \right] \mathbf{I} -$$

$$- n^F z^F \left[ \rho^F \frac{\partial \psi^F}{\partial n^F} + P \right] \mathbf{I} + \overset{4}{\mathbf{D}}{}^F \bar{\mathbf{L}}_F, \tag{57}$$

$$\mathbf{M}^{\mathrm{F}} = \overset{4}{\mathbf{M}}{}^{\mathrm{F}} \operatorname{grad} \bar{\boldsymbol{\omega}}_{\mathrm{F}}. \tag{58}$$

As the solid stress tensor, the fluid equilibrium stress depends on the Lagrangian multiplier $P$, which is required to guarantee saturation. From the dissipation inequality, the only restriction to the fourth-order tensors $\overset{4}{\mathbf{D}}{}^{\mathrm{F}}$ and $\overset{4}{\mathbf{M}}{}^{\mathrm{F}}$ is their positive definiteness.

5. The productions of momentum and moment of momentum are related to the free energy functions, to the Lagrangian multiplier $P$ and to the seepage velocity and the difference angular velocity, respectively:

$$\begin{aligned}\hat{\mathbf{p}}^{\mathrm{F}} &= \hat{\mathbf{p}}_0^{\mathrm{F}} + \mathbf{S}_p^{\mathrm{F}} \mathbf{w}_{\mathrm{F}} \\ &= P \operatorname{grad} n^{\mathrm{S}} - \rho^{\mathrm{S}} \frac{\partial \psi^{\mathrm{S}}}{\partial \rho^{\mathrm{FR}}} \operatorname{grad} \rho^{\mathrm{FR}} + \mathbf{S}_p^{\mathrm{F}} \mathbf{w}_{\mathrm{F}} - \\ &\quad - \rho^{\mathrm{F}} \left[ \frac{\partial \psi^{\mathrm{F}}}{\partial \bar{\mathbf{U}}_{\mathrm{S}}} (\nabla \bar{\mathbf{U}}_{\mathrm{S}} \mathbf{F}_{\mathrm{S}}^{-1})^{\underline{3}} \right]^{\underline{1}} - \rho^{\mathrm{F}} \left[ \frac{\partial \psi^{\mathrm{F}}}{\partial \bar{\mathbf{K}}_{\mathrm{S}}} (\nabla \bar{\mathbf{K}}_{\mathrm{S}} \mathbf{F}_{\mathrm{S}}^{-1})^{\underline{3}} \right]^{\underline{1}},\end{aligned} \tag{59}$$

$$\hat{\mathbf{m}}^{\mathrm{F}} = \mathbf{S}_m^{\mathrm{F}} \bar{\boldsymbol{\nu}}_{\mathrm{F}}. \tag{60}$$

The second-order tensors $\mathbf{S}_p^{\mathrm{F}}$ and $\mathbf{S}_m^{\mathrm{F}}$ must be positive definite.

6. In accordance to Fourier's law, the internal part of the heat flux is given by

$$\mathbf{q}^{i} = \mathbf{q}^{\mathrm{S}} + \mathbf{q}^{\mathrm{F}} = \mathbf{Q}^{i} \operatorname{grad} \vartheta, \tag{61}$$

with a positive definite second-order tensor $\mathbf{Q}^{i}$.

More general results of the nonequilibrium parts of the constitutive equations are possible, if a cross coupling of the variables would be taken into account. For simplicity, these effects are neglected here. Otherwise, for example the heat flux would explicitly depend on mechanical variables such as $\bar{\mathbf{L}}_{\mathrm{F}}$, $\mathbf{w}_{\mathrm{F}}$, etc. while the mechanical quantities such as the fluid stress $\mathbf{T}^{\mathrm{F}}$ and the exchange of momentum and moment of momentum would depend on grad $\vartheta$.

### 4.3. THE LAGRANGIAN MULTIPLIER

The Lagrangian multiplier $P$ governing the constitutive equations can be determined from a hydrostatic boundary-value problem. The total pressure $-\sigma_m$ acts onto the solid skeleton and onto the fluid according to the volume fractions

$$\mathbf{T}^{\mathrm{S}} = -n^{\mathrm{S}} \sigma_m \mathbf{I}, \qquad \mathbf{T}^{\mathrm{F}} = -n^{\mathrm{F}} \sigma_m \mathbf{I}. \tag{62}$$

Compared with the constitutive equation for the fluid stress (57), the thermodynamic pressure $p^{\mathrm{FR}}$ of the fluid phase and the Lagrangian multiplier $P$ are determined by

$$\sigma_m = \frac{\rho^{\mathrm{S}} \rho^{\mathrm{F}}}{n^{\mathrm{F}}} \frac{\partial \psi^{\mathrm{S}}}{\partial \rho^{\mathrm{FR}}} + \left(\rho^{\mathrm{FR}}\right)^2 \frac{\partial \psi^{\mathrm{F}}}{\partial \rho^{\mathrm{FR}}} =: p^{\mathrm{FR}}, \tag{63}$$

$$P = \rho^{\mathrm{F}} \frac{\partial \psi^{\mathrm{F}}}{\partial n^{\mathrm{S}}} + \frac{\rho^{\mathrm{S}} \rho^{\mathrm{FR}}}{n^{\mathrm{F}}} \frac{\partial \psi^{\mathrm{S}}}{\partial \rho^{\mathrm{FR}}} + \left(\rho^{\mathrm{FR}}\right)^2 \frac{\partial \psi^{\mathrm{F}}}{\partial \rho^{\mathrm{FR}}}. \tag{64}$$

In this case, due to an admissible choice of the Lagrangian multiplier $P$, the results become independent of the choice of the factor $z^{\mathrm{F}}$ which was introduced together with the evolution equation of the volume fraction $n^{\mathrm{F}}$. Especially, if $P$ is chosen in the given form, the fluid stress according to (57) does no longer depend on $z^{\mathrm{F}}$. Furthermore, the given choice leads to results which are identical with those obtained by Diebels (1997), where only one evolution equation for $n^{\mathrm{S}}$ is taken into account and $(n^{\mathrm{F}})'_{\mathrm{F}}$ is determined directly from the saturation constraint. Therefore, the factor $z^{\mathrm{S}}$ governs the evolution equation for the volume fraction $n^{\mathrm{S}}$ and the solid stress $\mathbf{T}^{\mathrm{S}}$. Following the experiments and the extended principle of effective stress proposed by Lade and de Boer (1997), $z^{\mathrm{S}}$ may be expressed in terms of the compressibility of the grains themselves and of the structural compressibility of the skeleton.

In contrast to Diebels (1997), starting from two independent evolution equations as presented in this paper, and handling the saturation condition as a true constraint by a Lagrangian multiplier makes it easier to obtain both models with one incompressible constituent, so-called hybrid models (Ehlers, 1993), and models with two incompressible constituents (Bowen, 1980; Ehlers, 1989). In the former case, the Lagrangian multiplier $P$ is determined by a constitutive equation, while in the latter case, $P$ is determined by the boundary conditions of the problem, and it can be identified to be the effective pore pressure.

## 5. Conclusions

In the first part of the paper, the kinematics of a mixture with micropolar constituents have been developed. Based on the motion and micromotion of micropolar materials, the deformation tensors and the corresponding deformation velocities have been derived. In the second part the balance equations of mass, momentum, moment of momentum, energy and entropy have been discussed by use of a master balance. It has been found that the balance of microinertia derived from a kinematic relation fits into the structure of the master balance. Applying Truesdell's metaphysical principles, a relation of the physical quantities of all constituents $\varphi^\alpha$ to the corresponding quantities of the whole mixture $\varphi$ have been given.

Based on these general results, a binary mixture has been discussed, consisting of a granular elastic skeleton saturated by a viscous pore-fluid. The volume fractions have been introduced as additional variables. Therefore, evolution equations have been required, which were motivated by the split of the balance of mass. These evolution equations and the saturation constraint have been taken into account when the entropy inequality was discussed for the binary mixture under study. As a result, it has been found that the free energy functions of the solid and of the fluid part are governed by the first Cosserat deformation tensor and by the curvature tensor of the solid, by the volume fraction of the solid, by the effective density of the fluid and by the temperature. The dependence of the response functions on the other process variables has been restricted to the nonequilibrium state. Following the well-known arguments, the general structure of the response function, especially of the stresses, couple stresses and the interactions of momentum and moment of momentum, could be derived. These results depend on a Lagrangian multiplier, which has been introduced to guarantee the saturation condition. By the investigation of a hydrostatic boundary value problem, an admissible choice of the Lagrangian multiplier has been determined. Furthermore, the resulting constitutive equations have become the same as if the saturation constraint has been incorporated directly. The advantage of the proposed procedure has been found in a compatibility of the derived model equations to so-called hybrid models, which follows directly if the corresponding quantities were dropped from the set of the process variables.

## References

Besdo, D.: 1974, Ein Beitrag zur nichtlinearen Theorie des Cosserat-Kontinuums, *Acta Mech.* **20**, 105–131.

de Boer, R.: 1982, *Vektor- und Tensorrechnung für Ingenieure*, Springer-Verlag, Berlin.

de Boer, R. and Ehlers, W.: 1986, Theorie der Mehrkomponentenkontinua mit Anwendung auf bodenmechanische Problemi, Forschungsberichte aus dem Fachbereich Bauwesen 40, Universität-GH-Essen, Essen.

de Borst, R.: 1991, Numerical modelling of bifurcation and localisation in cohesive-frictional materials, *Pageoph.* **137**, 368–390.

Bowen, R. M.: 1976, Theory of mixtures, In: A. C. Eringen (ed.), *Continuum Physics III*, Academic Press, New York, pp. 1–127.

Bowen, R. M.: 1980, Incompressible porous media models by use of the theory of mixtures, *Int. J. Engng. Sci.* **18**, 1129–1148.

Bowen, R. M.: 1982, Compressible porous media models by use of the theory of mixtures, *Int. J. Engng. Sci.* **20**, 697–735.

Coleman, B. C. and Noll, W.: 1963, The thermodynamics of elastic materials with heat conduction and viscosity, *Arch. Rat. Mech. Anal.* **13**, 167–178.

Cosserat, E. and Cosserat, F.: 1909, *Théorie des Corps Déformable*, A. Herman, Paris.

Diebels, S.: 1997, Materialtheorie für granulare Materialien, *ZAMM* (submitted).

Diebels, S. and Ehlers, W.: 1996, On basic equations of multiphase micropolar materials, *Technische Mechanik* **16**, 77–88.

Dietsche, A., Steinmann, P. and Willam, K.: 1993, Micropolar elasticity and its role in localization, *Int. J. Plast.* **9**, 813–831.

Ehlers, W.: 1989, Poröse Medien – ein kontinuumsmechanisches Modell auf der Basis der Mischungstheorie, Forschungsberichte aus dem Fachbereich Bauwesen 47, Universität-GH-Essen, Essen.

Ehlers, W.: 1993, Constitutive equations for granular materials in geomechanical context, In: K. Hutter (ed.), *Continuum Mechanics in Environmental Sciences and Geophysics*, CISM Courses and Lecture Notes No. 337, Springer-Verlag, Wien, pp. 313–402.

Ehlers, W.: 1996, Grundlegende Konzepte in der Theorie Poröser Medien, *Technische Mechanik* **16**, 63–76.

Eringen, A. C.: 1964, Simple microfluids, *Int. J. Engng. Sci.* **3**, 205–217.

Eringen, A. C. and Kafadar, C. B.: 1976, Polar field theories, In: A. C. Eringen (ed.), *Continuum Physics IV*, Academic Press, New York, pp. 1–73.

Haupt, P.: 1993, Foundations of continuum mechanics, In: K. Hutter (ed.), *Continuum Mechanics in Environmental Sciences and Geophysics*, CISM Courses and Lecture Notes No. 337, Springer-Verlag, Wien, pp. 1–77.

Hemmingsson, J., Herrmann, H. J. and Roux, S.: 1996, Vectorial cellular automaton for the stress in granular media, *J. de Physique* (submitted).

Lade, P. V. and de Boer, R.: 1997, The concept of effective stress for soil, concrete, and rock, *Geotechnique* **47**, 61–78.

Svendsen, B. and Hutter, K.: 1995, On the thermodynamics of a mixture of isotropic materials with constraints, *Int. J. Engng. Sci.* **33**, 2021–2054.

Tejchmann, J. and Wu, W.: 1993, Numerical study on patterning of shear bands in a Cosserat continuum, *Acta Mech.* **99**, 61–74.

Truesdell, C.: 1984, *Rational Thermodynamics*, Springer-Verlag, New York, p. 221.

# Propagation and Evolution of Wave Fronts in Two-Phase Porous Media

ZHANFANG LIU[1] and REINT DE BOER[2]
[1]*Chongqing University*, Department of Engineering Mechanics, Chongqing, China*
[2]*Universität Essen, FB 10-Bauwesen, Institut für Mechanik, 45117Essen, Germany*

(Received: 8 August 1997; in final form: 6 November 1997)

**Abstract.** An investigation is made into the propagation and evolution of wave fronts in a porous medium which is intended to contain two phases: the porous solid, referred to as the skeleton, and the fluid within the interconnected pores formed by the skeleton. In particular, the microscopic density of each real material is assumed to be unchangeable, while the macroscopic density of each phase may change, associated with the volume fractions. A two-phase porous medium model is concisely introduced based on the work by de Boer. Propagation conditions and amplitude evolution of the discontinuity waves are presented by use of the idea of surfaces of discontinuity, where the wave front is treated as a surface of discontinuity. It is demonstrated that the saturation condition entails certain restrictions between the amplitudes of the longitudinal waves in the solid and fluid phases. Two propagation velocities are attained upon examining the existence of the discontinuity waves. It is found that a completely coupled longitudinal wave and a pure transverse wave are realizable in the two-phase porous medium. The discontinuity strength of the pore-pressure may be determined by the amplitude of the coupled longitudinal wave. In the case of homogeneous weak discontinuities, explicit evolution equations of the amplitudes for two types of discontinuity waves are derived.

**Key words:** discontinuity waves, propagation, evolution.

## 1. Introduction

The dynamics of fluid-saturated porous media has been attracting considerable attention because of its importance in various engineering fields, for example, earthquake engineering, soil dynamics, geophysics and hydrology. The systematic investigations of wave motion in the early stage were totally based on Biot's model (Biot, 1941, 1956). In the context of this model, enormous efforts had been made in the literature (Mengi and McNiven, 1978; Plona, 1982; Zienkiewicz and Shiomi, 1984; Garg, Nafeh and Good, 1974; Dziecielak, 1986; Deresiewicz and Rice, 1986; Levy, 1979) on transient response and discontinuity waves in porous media. Biot's model has had such a remarkable influence on investigations of porous media that even today it remains among one of the most widely cited models in the field of porous media. One interesting prediction of Biot's model is the possibility of a strongly attenuated slow wave mode. Its existence gets support from the experimental measurement presented

* The author is now on research leave at Universität Essen.

by Plona (1980). A thorough insight of acoustics in porous media incorporating the model with laboratory results is embodied in a book of Bourbié, Coussy and Zinszner (1987). However, the theoretical examination of porous materials with fluid-filled pore spaces by continuum thermodynamics has elucidated the drawback of Biot's model, as pointed out by de Boer in his major review paper (de Boer, 1996). As a matter of fact, Biot's model, due to its origin along with purely intuitive derivations, cannot anchor and deepen the fundamental aspects of porous media theory such as the volume fraction concept and the closure problem.

It has been widely recognized that the mixture theory presents a solid basis for the theoretical description of porous media. However, porous media are not mixtures. Eventually, treatment upon porous media through the continuum procedure brings about additional fields in terms of the volume fractions in comparison with the mixture theory. The discussion as to how to solve this problem remains in progress.

Another issue in porous media theory is the formulation of the constitutive relations for the constituents. Their representation with the treatment of volume fractions constitutes different versions of porous media theory.

An incompressible porous media model proposed by Bowen (1980) and extended by de Boer (1996) has acquired widespread recognition among various porous media models. Incompressibility here implies that the true densities of the constituents of porous media remain unchanged during the motion, but the relevant bulk densities vary with the change of the volume fractions. This has nothing to do with the concept of incompressibility in one-component materials. An assumption of incompressibility in porous media is physically admissible and practical when we take into account the fact that the change of the true densities is rather small compared with that of the bulk densities. The applicability of this model has experienced intensive examination by applying it to the problem of uplift, friction and capillarity (de Boer and Ehlers, 1990), to the transient response (de Boer, Ehlers and Liu, 1993) and to the wave propagation (de Boer and Liu, 1994; de Boer and Liu 1995) in porous media. It was predicted that a coupled longitudinal wave and a transverse wave are realizable in incompressible porous media.

In this report we examine discontinuity waves in porous media by use of the above-mentioned incompressible porous media model. The theory of discontinuity surfaces is utilized to investigate propagation conditions and amplitude evolution of discontinuity waves. For the sake of our analysis and completeness, at first we outline concisely basic elements of propagating discontinuity surfaces and the second order conditions of compatibility. The incompressible porous media model is briefly introduced in the third section. The focus of our interest is on a linear poroelastic solid filled with an ideal liquid, such that two constituents are assumed incompressible and the whole medium is statistically homogeneous and isotropic. Possible thermal and chemical effects are excluded. In the fourth section we present propagation conditions of wave fronts in such media. We will demonstrate that the saturation condition along with the incompressibility produces a certain constraint between amplitudes of the longitudinal waves in the solid and in the fluid, where the waves in the solid and in the fluid cannot propagate arbitrarily. They are coupled and propagate with the same propagation speed. A coupled longitudinal wave and a pure transverse wave are realizable in such media. The last section is devoted to the amplitude evolution of the wave fronts, where an assumption of homogeneous weak discontinuities is suggested.

## 2. Second Order Conditions of Compatibility

This section contains a brief introduction to the singular surfaces and the compatibility conditions of the second order. We consider a moving surface of discontinuity $\sigma(t)$ in a three-dimensional Euclidian space. The parametric equation of this surface is described by (see Eringen and Suhubi, 1974)

$$\mathbf{x} = \mathbf{x}(\theta^\alpha, t), \quad \alpha = 1, 2, \tag{2.1}$$

where $\theta^\alpha$ is the curvilinear coordinate of the surface $\sigma(t)$; $t$ denotes the time, and $\mathbf{x}$ is the position vector for a reference point to the surface. For the oriented surfaces $\sigma(t)$ the tangent vector reads

$$\mathbf{x}_{,\alpha} = \frac{\partial \mathbf{x}}{\partial \theta^\alpha}, \tag{2.2}$$

where $\mathbf{x}_{,\alpha}$ is just the covariant base vector on the surface, the corresponding contravariant base vector is indicated by $\mathbf{x}^{\alpha}_{,}$. The surface metric tensor of the surface $\sigma(t)$ is given by

$$a_{\alpha\beta} = \mathbf{x}_{,\alpha} \cdot \mathbf{x}_{,\beta}, \tag{2.3}$$

where the dot denotes the scalar product. The components of the second fundamental form of the surface satisfy

$$b_{\alpha\beta} = \mathbf{n} \cdot \mathbf{x}_{,\alpha\beta} = -\mathbf{n}_{,\alpha} \cdot \mathbf{x}_{,\beta}, \tag{2.4}$$

where $\mathbf{n}$ is the unit normal vector of the surfaces $\sigma(t)$. The mean curvature $\Omega$ and the total or Gauss curvature $K$ of the surfaces $\sigma(t)$ are the first and the second invariants of the second fundamental form $b^{\alpha}_{\cdot\beta}$, written as

$$\Omega = \tfrac{1}{2} a^{\alpha\beta} b_{\alpha\beta} = \tfrac{1}{2} b^{\alpha}_{\cdot\alpha}, \tag{2.5}$$

$$K = \tfrac{1}{2}(b^{\alpha}_{\cdot\alpha})^2 - \tfrac{1}{2} b^{\alpha}_{\cdot\beta} b^{\beta}_{\cdot\alpha}. \tag{2.6}$$

The normal velocity of the surface $\sigma(t)$ is specified by

$$u_n = \mathbf{n} \cdot \dot{\mathbf{x}}, \quad \dot{\mathbf{x}} = \frac{\partial \mathbf{x}}{\partial t}(\theta^\alpha, t), \tag{2.7}$$

and the local speeds of propagation are defined by

$$U_\wp = u_n - \mathbf{x}'_\wp \cdot \mathbf{n}, \quad \wp = \mathrm{S},\ \mathrm{L}, \tag{2.8}$$

which indicates a measure of the normal velocity of the surface $\sigma(t)$ with respect to the material particles. In our case, $\mathbf{x}'_{\wp}$ denotes the velocity of the particles of the two components ($\wp = \mathrm{S}$, solid skeleton and $\wp = \mathrm{L}$, liquid) in a two-phase porous medium.

For an arbitrary time-dependent field $\phi(\mathbf{x}, t)$ confined to the surfaces, the displacement derivative is introduced in the form

$$\frac{\delta\phi}{\delta t} = \frac{\partial\phi}{\partial t} + u_n \mathbf{n} \cdot \operatorname{grad} \phi, \tag{2.9}$$

which is defined as the time derivative of $\phi$ along the normal trajectory. A useful relation here is

$$\frac{\delta\mathbf{n}}{\delta t} = -u_{n,\alpha}\mathbf{x}^{\alpha}_{,} = -u_{n,\alpha}a^{\alpha\beta}\mathbf{x}_{,\beta}, \tag{2.10}$$

which relates the displacement of the unit normal vector $\mathbf{n}$ with the propagation speed $u_n$ of the surface. In (2.10) $a^{\alpha\beta}$ is the contravariant component of the surface metric tensor $a_{\alpha\beta}$.

Being an arbitrary function, $\phi(\mathbf{x}, t)$ has a jump across the surface $\sigma(t)$ given by

$$[|\phi|] = \phi^{+} - \phi^{-} \neq 0, \tag{2.11}$$

where $\phi^{+}$ and $\phi^{-}$ are the definite limits of the function $\phi$ in front of and behind the surface $\sigma(t)$, respectively. In the following we introduce the mentioned denotations:

$$\begin{aligned} &A = [|\phi|], \qquad B = [|\operatorname{grad} \phi|] \cdot \mathbf{n}, \\ &C = [|\operatorname{grad} \operatorname{grad} \phi|] \cdot \mathbf{n} \otimes \mathbf{n}, \qquad A' = \left[\left|\frac{\partial\phi}{\partial t}\right|\right]. \end{aligned} \tag{2.12}$$

With the aid of the above-mentioned denotations (2.12) the geometrical conditions of compatibility have the following form:

$$[|\operatorname{grad} \phi|] = B\mathbf{n} + a^{\alpha\beta} A_{;\alpha}\mathbf{x}_{,\beta}, \tag{2.13}$$

$$\begin{aligned} [|\operatorname{grad} \operatorname{grad} \phi|] = {} & C\mathbf{n} \otimes \mathbf{n} + a^{\alpha\beta}(B_{;\alpha} + a^{\gamma\delta}b_{\alpha\gamma}A_{;\delta})(\mathbf{n} \otimes \mathbf{x}_{,\beta} + \mathbf{x}_{,\beta} \otimes \mathbf{n}) + \\ & + a^{\alpha\beta}a^{\gamma\delta}(A_{;\alpha\gamma} - Bb_{\alpha\gamma})\mathbf{x}_{,\beta} \otimes \mathbf{x}_{,\delta}, \end{aligned} \tag{2.14}$$

and the kinematical conditions of compatibility are as follows:

$$\left[\left|\frac{\partial\phi}{\partial t}\right|\right] = -u_n B + \frac{\delta A}{\delta t}, \tag{2.15}$$

$$\left[\left|\frac{\partial^2\phi}{\partial t^2}\right|\right] = \left(u_n C - \frac{\delta B}{\delta t} - a^{\alpha\beta} A_{;\alpha} u_{n,\beta}\right) u_n + \frac{\delta A'}{\delta t}, \tag{2.16}$$

$$\left[\left|\operatorname{grad} \frac{\partial\phi}{\partial t}\right|\right] = \left(-u_n C + \frac{\delta B}{\delta t} + a^{\alpha\beta} A_{;\alpha} u_{n,\beta}\right)\mathbf{n} + a^{\alpha\beta} A'_{;\alpha}\mathbf{x}_{,\beta}, \tag{2.17}$$

where indices following a semicolon represent covariant partial differentiation with respect to curvilinear coordinates of the surfaces $\theta^\alpha$.

## 3. Incompressible Porous Media Model

This section is concerned with an introduction of the incompressible porous media model. Interested readers can consult the articles by de Boer (1996) and Bowen (1980). With the aid of an introduction of volume fractions, liquid-saturated porous media are modelled as a binary mixture of superimposed but immiscible constituents $\varphi^\wp(\wp = \mathrm{S, L})$ with particles of solid and liquid phases, each of which is interpreted as a continuum following its own kinematics. This macroscopic treatment results in a smeared model in which at any time $t$, each spatial position $\mathbf{x}$ of the current configuration is simultaneously occupied by particles of both components and these particles proceed from different reference positions $\mathbf{X}_\wp$. Each $\varphi^\wp$ is assigned its own motion function $\boldsymbol{\chi}_\wp$:

$$\mathbf{x} = \boldsymbol{\chi}_\wp(\mathbf{X}_\wp, t), \quad \wp = \mathrm{S, L}. \tag{3.1}$$

The function $\boldsymbol{\chi}_\wp$ is postulated to be unique and uniquely invertible at any time, which requires

$$J_\wp = \det \mathbf{F}_\wp \tag{3.2}$$

to be unequal to zero. In (3.2) $\mathbf{F}_\alpha$ is the deformation gradient.

The volume fractions

$$n^\wp = n^\wp(\mathbf{x}, t) \tag{3.3}$$

are defined as the local ratios of the constituent volume elements $\mathrm{d}v^\wp$ with respect to the bulk volume element $\mathrm{d}v$. For immiscible saturated porous materials, we always have the saturation condition

$$n^\mathrm{L} + n^\mathrm{S} = 1. \tag{3.4}$$

The volume fraction of the fluid $n^\mathrm{L}$ is sometimes called the porosity of two-phase porous media, which is assumed not to vanish ($n^\mathrm{L} > 0$) by no means throughout the paper unless it is particularly mentioned. Associated with each $\varphi^\wp$ are the realistic densities $\rho^{\wp\mathrm{R}}$ and the bulk densities $\rho^\wp$. Their relation is given by

$$\rho^\wp = n^\wp \rho^{\wp\mathrm{R}}. \tag{3.5}$$

The macroscopic realistic density $\rho^{\wp\mathrm{R}}$ equals the microscopic true density only if the true density on the microscale is constant, see de Boer (1996). The implication of incompressibility in porous media is somewhat complicated. If the constituents $\varphi^\wp$ are incompressible, their real densities are kept constant during the deformation, that is, with the above assumption,

$$\rho^{\wp\mathrm{R}} = \mathrm{const.} \tag{3.6}$$

It should be mentioned that despite the constituent incompressibility, the partial bodies suffer volume change due to the structural compressibility.

We exclude mass and heat exchanges between the solid skeleton and the liquid for the problem under study, then the balance equations for the constituents necessary for the investigation are summarized as follows:

*Balance of mass*:

$$(\rho^{\wp})'_{\wp} + \rho^{\wp}\mathrm{div}\, \mathbf{x}'_{\wp} = 0, \tag{3.7}$$

where $\mathbf{x}'_{\wp}$ is the velocity of the particle $\mathbf{X}_{\wp}$. The notation $(..)'_{\wp}$ stands for the material time derivative following the constituents $\varphi^{\wp}$.

*Balance of momentum*:

$$\mathrm{div}\, \mathbf{T}^{\wp} + \rho^{\wp}(\mathbf{b}^{\wp} - \mathbf{x}''_{\wp}) + \hat{\mathbf{p}}^{\wp} = \mathbf{0}, \tag{3.8}$$

$$\hat{\mathbf{p}}^{\mathrm{S}} + \hat{\mathbf{p}}^{\mathrm{L}} = \mathbf{0}, \tag{3.9}$$

where $\mathbf{T}^{\wp}$ denotes the Cauchy stress tensor and $\hat{\mathbf{p}}^{\wp}$ the interaction force caused by the mutual influence between the individual constituents, $\mathbf{b}^{\wp}$ the external acceleration and $\mathbf{x}''_{\wp}$ the acceleration of the particle $\mathbf{X}_{\wp}$.

Considering the relation (3.5) and the incompressibility condition (3.6) we can easily write the balance of mass (3.7) as

$$(n^{\wp})'_{\wp} + n^{\wp}\mathrm{div}\, \mathbf{x}'_{\wp} = 0. \tag{3.10}$$

Obviously, in the case of incompressible porous media, the balance of mass can be reduced into the balance of volume fraction. With the aid of the saturation condition (3.4) together with the identity

$$(n^{\mathrm{L}})'_{\mathrm{S}} = (n^{\mathrm{L}})'_{\mathrm{L}} - \mathrm{grad}\, n^{\mathrm{L}} \cdot (\mathbf{x}'_{\mathrm{L}} - \mathbf{x}'_{\mathrm{S}}), \tag{3.11}$$

the balance equations of volume fraction (3.10) may be rewritten in the form

$$\mathrm{div}\, (n^{\mathrm{S}}\mathbf{x}'_{\mathrm{S}} + n^{\mathrm{L}}\mathbf{x}'_{\mathrm{L}}) = 0. \tag{3.12}$$

Thus, as a straightforward consequence of the saturation condition (3.4) and the incompressibility (3.6), it follows that the kinematics of the solid skeleton and the fluid in porous media must satisfy the relation (3.12) despite that each phase is assigned its own motion. And this relation will produce a certain constraint for the wave motion in such media. The constraint shall be further discussed in the next section.

The balance equations of momentum (3.8)–(3.9) for the solid skeleton and the pore fluid have the following forms:

$$\mathrm{div}\, \mathbf{T}^{\mathrm{S}} + \rho^{\mathrm{S}}(\mathbf{b}^{\mathrm{S}} - \mathbf{x}''_{\mathrm{S}}) - \hat{\mathbf{p}}^{\mathrm{L}} = \mathbf{0}, \tag{3.13}$$

$$\operatorname{div} \mathbf{T}^{\mathrm{L}} + \rho^{\mathrm{L}}(\mathbf{b}^{\mathrm{L}} - \mathbf{x}_{\mathrm{L}}'') + \hat{\mathbf{p}}^{\mathrm{L}} = \mathbf{0}. \tag{3.14}$$

As a consequence of the incompressibility condition (3.6) and the saturation condition (3.4), it follows from the thermodynamic examination that the stress tensors $\mathbf{T}^{\mathrm{S}}$, $\mathbf{T}^{\mathrm{L}}$ and the interaction force $\hat{\mathbf{p}}^{\mathrm{L}}$ are additively decomposed into two parts, the unknown pore-water pressure $\lambda$ and the extra or effective quantities, denoted by the index E, for which constitutive equations must be formulated, see de Boer (1996), viz.

$$\mathbf{T}^{\mathrm{S}} = -n^{\mathrm{S}}\lambda\mathbf{I} + \mathbf{T}_{\mathrm{E}}^{\mathrm{S}}, \tag{3.15}$$

$$\mathbf{T}^{\mathrm{L}} = -n^{\mathrm{L}}\lambda\mathbf{I} + \mathbf{T}_{\mathrm{E}}^{\mathrm{L}}, \tag{3.16}$$

$$\hat{\mathbf{p}}^{\mathrm{L}} = \lambda \operatorname{grad} n^{\mathrm{L}} + \hat{\mathbf{p}}_{\mathrm{E}}^{\mathrm{L}}, \tag{3.17}$$

and

$$\mathbf{T}_{\mathrm{E}}^{\mathrm{S}} = \rho_{0\mathrm{S}}^{\mathrm{S}} \frac{\partial \psi^{\mathrm{S}}}{\partial e_{\mathrm{S}}} \mathbf{I} + \mathbf{T}_{\mathrm{E}}^{\mathrm{SD}}, \tag{3.18}$$

$$\mathbf{T}_{\mathrm{E}}^{\mathrm{L}} = \rho_{0\mathrm{L}}^{\mathrm{L}} \frac{\partial \psi^{\mathrm{L}}}{\partial e_{\mathrm{L}}} \mathbf{I}, \tag{3.19}$$

$$\hat{\mathbf{p}}_{\mathrm{E}}^{\mathrm{L}} = -\mathbf{S}_{\nu}(\mathbf{x}_{\mathrm{L}}' - \mathbf{x}_{\mathrm{S}}'), \tag{3.20}$$

where $\mathbf{T}_{\mathrm{E}}^{\mathrm{SD}}$ is the devitoric part of the effective solid stress, $\psi^{\wp}$ the free Helmholtz energy, $\rho_{0\mathrm{S}}^{\mathrm{S}}$ and $\rho_{0\mathrm{L}}^{\mathrm{L}}$ denote the density functions of the constituents $\varphi^{\mathrm{S}}$ and $\varphi^{\mathrm{L}}$ in the reference state, $e_{\mathrm{S}}$ and $e_{\mathrm{L}}$ indicate the volume strains of the constituents $\varphi^{\mathrm{S}}$ and $\varphi^{\mathrm{L}}$.

Moreover, $\mathbf{S}_{\nu}$ is called the tensor of permeability. In the case of isotropic permeability, the quantity $\mathbf{S}_{\nu}$ describing the effect of permeability between the solid skeleton and the fluid, is given by

$$\mathbf{S}_{\nu} = \frac{(n^{\mathrm{L}})^2 \gamma^{\mathrm{LR}}}{k^{\mathrm{L}}} \mathbf{I} = \alpha_{\nu}\mathbf{I}, \tag{3.21}$$

where $\gamma^{\mathrm{LR}}$ is the real specific weight of the liquid and $k^{\mathrm{L}}$ the coefficient of permeability of the porous solids.

Insertion of (3.15)–(3.17) in (3.13) and (3.14) produces

$$\operatorname{div} \mathbf{T}_{\mathrm{E}}^{\mathrm{S}} - n^{\mathrm{S}} \operatorname{grad} \lambda + \rho^{\mathrm{S}}(\mathbf{b}^{\mathrm{S}} - \mathbf{x}_{\mathrm{S}}'') - \hat{\mathbf{p}}_{\mathrm{E}}^{\mathrm{L}} = \mathbf{0}, \tag{3.22}$$

$$\operatorname{div} \mathbf{T}_{\mathrm{E}}^{\mathrm{L}} - n^{\mathrm{L}} \operatorname{grad} \lambda + \rho^{\mathrm{L}}(\mathbf{b}^{\mathrm{L}} - \mathbf{x}_{\mathrm{L}}'') + \hat{\mathbf{p}}_{\mathrm{E}}^{\mathrm{L}} = \mathbf{0}. \tag{3.23}$$

The set of Equations (3.12), (3.22) and (3.23) along with the constitutive relations (3.18)–(3.20) constitutes the general statement of the boundary/initial value problem for incompressible saturated porous media.

## 4. Propagation Conditions of Wave Fronts

Consider a singular curved surface $\sigma(t)$ with the curvilinear coordinates $\theta^\alpha$ ($\alpha = 1, 2$) moving through the porous media. By definition, the propagating singular surfaces are said to be discontinuity waves if the motion itself is continuous across $\sigma(t)$, but the deformation gradient, the velocity and their higher order derivatives suffer finite jumps across $\sigma(t)$. The geometrical and the kinematical conditions of compatibility for the first and the second derivatives of a function across $\sigma(t)$ are presented in the second section. However, the field variables across $\sigma(t)$ are still restricted by the dynamical conditions of compatibility that are a part of the field equations when the discontinuity surfaces occur.

For the present problem under study, the dynamical conditions of compatibility are derived from the local mass and momentum balances. It follows that the jump field equations take the forms (de Boer and Liu, 1995):

*Balance of mass*:

$$[|\rho^{\wp}(\mathbf{x}'_{\wp} \cdot \mathbf{n} - u_n)|] = 0, \quad \text{or} \quad [|\rho^{\wp}\mathrm{U}_{\wp}|] = 0. \tag{4.1}$$

*Balance of momentum*:

$$[|\rho^{\wp}\mathbf{x}'_{\wp}(\mathbf{x}'_{\wp} \cdot \mathbf{n} - u_n)|] - [|\mathbf{T}^{\wp}\mathbf{n}|] = \mathbf{0}, \quad \text{or} \quad [|\rho^{\wp}\mathbf{x}'_{\wp}\mathrm{U}_{\wp}|] + [|\mathbf{T}^{\wp}\mathbf{n}|] = \mathbf{0}. \tag{4.2}$$

For typical porous media such as saturated soils, from physical insight we may make the assumption

$$\left|\frac{\mathbf{x}'_{\wp} \cdot \mathbf{n}}{u_n}\right| \ll 1, \tag{4.3}$$

as the propagation speed $u_n$ is several orders of magnitude higher than the particle velocity. Hence, we see from (4.1) and (4.3) that both densities suffer no jumps across the singular surfaces

$$[|\rho^{\wp}|] = 0. \tag{4.4}$$

Due to the constraint of incompressibility (3.5) and (3.6) together with the continuous mass density across $\sigma(t)$ in (4.4) we have

$$[|n^{\wp}|] = 0. \tag{4.5}$$

In turn, with the aid of (4.3) and (4.4) the jump condition (4.2) is simplified into

$$[|\mathbf{T}^{\wp}|]\mathbf{n} + \rho^{\wp}[|\mathbf{x}'_{\wp}|]u_n = \mathbf{0}. \tag{4.6}$$

For further considerations, it is convenient to substitute the velocities $\mathbf{x}'_{\wp}$ and the accelerations $\mathbf{x}''_{\wp}$ by the corresponding derivatives of the displacement vector $\mathbf{w}_{\wp}$ such that

$$\mathbf{x}'_{\wp} = \mathbf{w}'_{\wp}, \qquad \mathbf{x}''_{\wp} = \mathbf{w}''_{\wp}. \tag{4.7}$$

The investigations to follow are confined to a geometrically linear elastic porous solid filled with an ideal fluid. Thus, the constitutive relations for the stresses in the general forms (3.18) and (3.19) are reduced to

$$\mathbf{T}_{\rm E}^{\rm S} = 2\mu^{\rm S}\mathbf{E}_{\rm S} + \lambda^{\rm S} e_{\rm S}\mathbf{I}, \tag{4.8}$$

$$\mathbf{T}_{\rm E}^{\rm L} = \alpha_{\rm L} e_{\rm L}\mathbf{I}, \tag{4.9}$$

where $\mu^{\rm S}(n_{0\rm S}^{\rm S})$ and $\lambda^{\rm S}(n_{0\rm S}^{\rm S})$ are the macroscopic Lamé constants of the empty porous solids, $n_{0\rm S}^{\rm S}$ is the volume fraction of the solid in the reference state. In (4.9) $\alpha_{\rm L}$ is a material dependent quantity of the partial liquid body. In (4.8) and (4.9) the linearized Lagrangian strain tensor is introduced such that

$$\mathbf{E}_{\wp} = \tfrac{1}{2}(\text{grad}\ \mathbf{w}_{\wp} + \text{grad}\ {}^{\rm T}\mathbf{w}_{\wp}). \tag{4.10}$$

If $\mathbf{w}_{\wp}$ takes the place of $\phi$ in (2.12) and (2.15) such that $[|\mathbf{w}_{\wp}|] = \mathbf{0}$ on the shock fronts, it is easy to get

$$\left[\left|\frac{\partial \mathbf{w}_{\wp}}{\partial t}\right|\right] = -u_n \mathbf{a}^{\wp} \tag{4.11}$$

with

$$\mathbf{a}^{\wp} = [|\text{grad}\ \mathbf{w}_{\wp}|]\mathbf{n}, \tag{4.12}$$

which is defined as the amplitude vectors in the solid skeleton and in the fluid. Moreover, we further set

$$b = [|\lambda|], \tag{4.13}$$

which may be considered as a measure of the pore-water pressure discontinuity across the wave fronts.

Substitution of (3.15), (3.16), (4.7)–(4.13) into (4.6) produces

$$(\lambda^{\rm S} + \mu^{\rm S})(\mathbf{a}^{\rm S}\cdot\mathbf{n})\mathbf{n} + \mu^{\rm S}\mathbf{a}^{\rm S} - n^{\rm S} b\mathbf{n} = \rho^{\rm S} u_n^2 \mathbf{a}^{\rm S}, \tag{4.14}$$

$$\alpha_{\rm L}(\mathbf{a}^{\rm L}\cdot\mathbf{n})\mathbf{n} - n^{\rm L} b\mathbf{n} = \rho^{\rm L} u_n^2 \mathbf{a}^{\rm L}. \tag{4.15}$$

We now return to the discussion as to how the saturation condition along with the incompressibility of the constituents restricts the amplitude vectors $\mathbf{a}^{\wp}$ in (4.12). As pointed out by de Boer (1996), the saturation condition always implies a certain constraint amongst the variables and causes a pore-water pressure to exist amongst the constitutive variables. Nevertheless, for saturated porous media the saturation condition together with the incompressibility has a property that the volume occupied before deformation equals the volume after deformation, which means

$$n^{\rm S}\text{div}\ \mathbf{w}_{\rm S} + n^{\rm L}\text{div}\ \mathbf{w}_{\rm L} = 0. \tag{4.16}$$

This crucial point is examined as follows. For saturated porous media, keeping in mind that the whole medium is statistically homogeneous and isotropic, we write the saturation condition in the initial state in the form

$$n^{\mathrm{S}}_{0\mathrm{S}} + n^{\mathrm{L}}_{0\mathrm{L}} = 1. \tag{4.17}$$

In our case, the real density $\rho^{\wp\mathrm{R}}$ remains unchanged, it is easy to have

$$\begin{aligned} n^{\mathrm{S}}_{0\mathrm{S}} &= n^{\mathrm{S}} \det \mathbf{F}_{\mathrm{S}} = n^{\mathrm{S}}(1 + \mathrm{Div}\, \mathbf{w}_{\mathrm{S}}), \\ n^{\mathrm{L}}_{0\mathrm{L}} &= n^{\mathrm{L}} \det \mathbf{F}_{\mathrm{L}} = n^{\mathrm{L}}(1 + \mathrm{Div}\, \mathbf{w}_{\mathrm{L}}), \end{aligned} \tag{4.18}$$

where Div indicates a divergence with respect to the material coordinates. Thus, inserting the expressions in (4.18) into (4.17) and recalling (3.4) we obtain

$$n^{\mathrm{S}} \mathrm{Div}\, \mathbf{w}_{\mathrm{S}} + n^{\mathrm{L}} \mathrm{Div}\, \mathbf{w}_{\mathrm{L}} = 0. \tag{4.19}$$

Obviously, Equation (4.16) has the same form as Equation (4.19) in the case of infinitesimal deformations.

We then evaluate (4.16) on the wave fronts and immediately attain

$$n^{\mathrm{S}} \mathbf{a}^{\mathrm{S}} \cdot \mathbf{n} + n^{\mathrm{L}} \mathbf{a}^{\mathrm{L}} \cdot \mathbf{n} = 0, \tag{4.20}$$

where (4.5) has been utilized. Equation (4.20) indicates that the amplitudes in the solid skeleton and in the liquid satisfy a certain relation, as a matter of fact, Equation (4.20) is a constraint for the amplitudes of the coupled longitudinal waves in incompressible porous media.

With the projection of (4.14) and (4.15) in the direction of $\mathbf{n}$ and normal to $\mathbf{n}$ and upon elimination of $b$ it follows that

$$\{(n^{\mathrm{L}})^2(\lambda^{\mathrm{S}} + 2\mu^{\mathrm{S}}) + (n^{\mathrm{S}})^2 \alpha^{\mathrm{L}} - [(n^{\mathrm{S}})^2 \rho^{\mathrm{L}} + (n^{\mathrm{L}})^2 \rho^{\mathrm{S}}] u_n^2\} \mathbf{a}^{\mathrm{S}} \cdot \mathbf{n} = 0, \tag{4.21}$$

$$(\mu^{\mathrm{S}} - \rho^{\mathrm{S}} u_n^2) \mathbf{a}^{\mathrm{S}} \times \mathbf{n} = \mathbf{0}, \tag{4.22}$$

$$\mathbf{a}^{\mathrm{L}} \times \mathbf{n} = \mathbf{0}. \tag{4.23}$$

It is indicated from (4.21)–(4.23) together with (4.20) that two types of waves with two propagation speeds are realizable in incompressible porous media. The longitudinal waves (coupled P-type) can propagate either in the solid or in the liquid. Their amplitudes are associated to the relation (4.20). In other words, the shock in the fluid phase propagates to the extent that they move exactly in the opposite direction to the waves in the solid phase. The longitudinal waves stimulate two completely coupled disturbances in the incompressible porous media. These P-type waves should decay relatively fast, which will be revealed in the next section. The transverse waves (S-type) propagate only in the solid skeleton. The two propagation speeds are derived from the conditions of the existence of nontrivial amplitudes, the results read:

*For coupled P-type waves*:

$$u_{n1}^2 = \frac{(n^{\mathrm{L}})^2(\lambda^{\mathrm{S}} + 2\mu^{\mathrm{S}}) + (n^{\mathrm{S}})^2\alpha^{\mathrm{L}}}{(n^{\mathrm{L}})^2\rho^{\mathrm{S}} + (n^{\mathrm{S}})^2\rho^{\mathrm{L}}}, \tag{4.24}$$

*For S-type waves*:

$$u_{n2}^2 = \frac{\mu^{\mathrm{S}}}{\rho^{\mathrm{S}}}. \tag{4.25}$$

Because the propagation speeds are constant, it follows from (2.10) that

$$u_{n,\alpha} = 0, \qquad \frac{\partial u_n}{\partial t} = 0, \qquad \frac{\delta \mathbf{n}}{\delta t} = \mathbf{0}. \tag{4.26}$$

It is obvious from (4.26) that the wave fronts $\sigma(t)$ constitute a family of parallel surfaces.

The measure of the pore-water pressure discontinuity $b$, across the surfaces is obtained from (4.15), which takes the form

$$\begin{aligned} b &= \frac{\alpha^{\mathrm{L}} - \rho^{\mathrm{L}} u_{n1}^2}{n^{\mathrm{L}}} \mathbf{a}^{\mathrm{L}} \cdot \mathbf{n} \\ &= \frac{n^{\mathrm{S}}\rho^{\mathrm{L}} u_{n1}^2 - n^{\mathrm{S}}\alpha^{\mathrm{L}}}{(n^{\mathrm{L}})^2} \mathbf{a}^{\mathrm{S}} \cdot \mathbf{n}, \end{aligned} \tag{4.27}$$

then, the pressure discontinuity $b$ is determined by the amplitudes of the longitudinal waves. It is expected that the discontinuity of the pore-water pressure should be correlated with the jumps of the volume of two phases, and the latter has been expressed in terms of $[|\mathrm{div}\, \mathbf{w}_{\mathrm{S}}|]$ and $[|\mathrm{div}\, \mathbf{w}_{\mathrm{L}}|]$ through (4.16) and (4.20).Without exception, the pore-water pressure $\lambda$ suffers no jumps on the fronts of the transverse waves.

It is interesting to note that the velocities suffer not only the tangential jumps but also the normal jumps across the shock surfaces in incompressible porous media. The normal jumps in the solid and in the fluid unify into coupled P-type waves with the same propagation speed as indicated in (4.20). The tangential jumps occur only in the solid phase as shown through (4.21) and (4.23), which is in accordance with incompressible one-component materials where no normal jump happens across the wave fronts.

The solutions of the set of Equations (4.21)–(4.23) determine the amplitudes only to the accuracy of the direction of the amplitudes and not to the absolute value. The evolution equation of the amplitudes is presented in the following.

## 5. Evolution of the Amplitudes of Wave Fronts

In this section we investigate the evolution equations governing the amplitudes of discontinuity waves in porous media. Because our study is confined to infinitesimal deformations and to a linear poroelastic solid filled with an ideal liquid, it follows from (3.20) and (4.7)–(4.10) that the balance equations of momentum (3.22) and (3.23) may be rewritten in the form

$$(\lambda^S + \mu^S)\text{grad div}\ \mathbf{w}_S + \mu^S \text{div grad}\ \mathbf{w}_S - n^S \text{grad}\ \lambda + \\ + \rho^S \left(\mathbf{b}^S - \frac{\partial^2 \mathbf{w}_S}{\partial t^2}\right) + \alpha_\nu \left(\frac{\partial \mathbf{w}_L}{\partial t} - \frac{\partial \mathbf{w}_S}{\partial t}\right) = \mathbf{0}, \tag{5.1}$$

$$\alpha_L\ \text{grad div}\ \mathbf{w}_L - n^L \text{grad}\ \lambda + \rho^L \left(\mathbf{b}^L - \frac{\partial^2 \mathbf{w}_L}{\partial t^2}\right) - \\ - \alpha_\nu \left(\frac{\partial \mathbf{w}_L}{\partial t} - \frac{\partial \mathbf{w}_S}{\partial t}\right) = \mathbf{0}. \tag{5.2}$$

In order to derive the differential equations of the amplitudes, we need to evaluate the field equations (5.1) and (5.2) on the wave fronts. To this end, from the iterated substitution of $\phi$ in (2.14) and (2.16) by $\mathbf{w}_\wp$ into the geometrical conditions and the kinematical conditions of compatibility of the second order we have the following jumps

$$[|\text{grad div}\ \mathbf{w}_\wp|] = (\mathbf{C}^\wp \cdot \mathbf{n})\mathbf{n} + a^{\alpha\beta}[(\mathbf{a}^\wp_{;\alpha} \cdot \mathbf{n})\mathbf{x}_{,\beta} + (\mathbf{a}^\wp_{;\alpha} \cdot \mathbf{x}_{,\beta})\mathbf{n}] - \\ - b^{\alpha\beta}(\mathbf{a}^\wp \cdot \mathbf{x}_{,\alpha})\mathbf{x}_{,\beta}, \tag{5.3}$$

$$[|\text{div grad}\ \mathbf{w}_S|] = \mathbf{C}^S - b^\alpha_{\cdot\alpha}\mathbf{a}^S, \tag{5.4}$$

$$\left[\left|\frac{\partial^2 \mathbf{w}_\wp}{\partial t^2}\right|\right] = u_n^2 \mathbf{C}^\wp - 2u_n \frac{\delta \mathbf{a}^\wp}{\delta t}, \tag{5.5}$$

where the following relation

$$\mathbf{n} \cdot \mathbf{x}_{,\alpha} = 0 \tag{5.6}$$

has been used, and $b^\alpha_{\cdot\alpha}$ is the trace of the surface tensor $b^\alpha_{\cdot\beta}$, and

$$\mathbf{C}^\wp = [|\text{grad grad}\ \mathbf{w}_\wp|]\mathbf{n} \otimes \mathbf{n} \tag{5.7}$$

are introduced as unknown vectors.

Next, we first examine the amplitude evolution of the longitudinal waves in a well-known way. However, the procedure is lengthy and the details are therefore purposely neglected. After the field equations, (5.1) and (5.2) are written on the wave fronts and the relations (5.3)–(5.6) are substituted into them, upon elimination of the pore-water pressure gradient $[|\text{grad}\ \lambda|]$, we have

$$n^L(\lambda^S + \mu^S)[(\mathbf{C}^S \cdot \mathbf{n}) + a^{\alpha\beta}(\mathbf{a}^S_{;\alpha} \cdot \mathbf{x}_{,\beta})] + n^L\mu^S(\mathbf{C}^S \cdot \mathbf{n} - b^\alpha_{\cdot\alpha}\mathbf{a}^S \cdot \mathbf{n}) - \\ - n^S\alpha_L[(\mathbf{C}^L \cdot \mathbf{n}) + a^{\alpha\beta}(\mathbf{a}^L_{;\alpha} \cdot \mathbf{x}_{,\beta})] - n^L\rho^S\left(u_n^2\mathbf{C}^S \cdot \mathbf{n} - 2u_n\frac{\delta \mathbf{a}^S}{\delta t} \cdot \mathbf{n}\right) + \\ + n^S\rho^L\left(u_n^2\mathbf{C}^L \cdot \mathbf{n} - 2u_n\frac{\delta \mathbf{a}^L}{\delta t} \cdot \mathbf{n}\right) + \alpha_\nu u_n(\mathbf{a}^S \cdot \mathbf{n} - \mathbf{a}^L \cdot \mathbf{n}) = 0. \tag{5.8}$$

For simplicity we examine only the case of the so-called homogeneous weak discontinuity in our following analysis. Namely, all amplitudes tangent to the wave fronts are assumed as unchangeable quantitites, that is,

$$a^{\mathrm{S}}_{n,\alpha} = 0, \qquad a^{\mathrm{S}}_{v,\alpha} = 0, \tag{5.9}$$

where we denote

$$a^{\mathrm{S}}_{n} = \mathbf{a}^{\mathrm{S}} \cdot \mathbf{n}, \qquad a^{\mathrm{S}}_{v} = \mathbf{a}^{\mathrm{S}} \cdot \mathbf{v} \tag{5.10}$$

as the absolute amplitudes of the longitudinal and the transverse waves, $\mathbf{v}$ is chosen as a unit vector perpendicular to the unit wave normal $\mathbf{n}$ such that

$$\mathbf{v} \cdot \mathbf{n} = 0. \tag{5.11}$$

Moreover, in the scope of infinitesimal deformation, the equation for the volume fractions (3.12) are simplified into

$$\operatorname{div}\left(n^{\mathrm{S}}\frac{\partial \mathbf{w}_{\mathrm{S}}}{\partial t} + n^{\mathrm{L}}\frac{\partial \mathbf{w}_{\mathrm{L}}}{\partial t}\right) = 0. \tag{5.12}$$

In view of (2.17) it is easy to have

$$\left[\!\left[\operatorname{div}\frac{\partial \mathbf{w}_{\wp}}{\partial t}\right]\!\right] = -u_n \mathbf{C}^{\wp} \cdot \mathbf{n} + \frac{\partial \mathbf{a}^{\wp}}{\partial t} \cdot \mathbf{n} - a^{\alpha\beta} u_n \mathbf{a}^{\wp}_{;\alpha} \cdot \mathbf{x}_{,\beta}. \tag{5.13}$$

Thus, recalling (4.23) and (4.17) and upon substitution of the above expression (5.13) into (5.12) we obtain the restriction for the unknown vector $\mathbf{C}^{\wp}$ such that,

$$n^{\mathrm{S}}\mathbf{C}^{\mathrm{S}} \cdot \mathbf{n} + n^{\mathrm{L}}\mathbf{C}^{\mathrm{L}} \cdot \mathbf{n} = 0, \tag{5.14}$$

where (5.9) and (5.10) have been utilized. It should be mentioned here that the constraint (5.14) is valid only for the P-type coupled waves.

With the aid of (5.14) a rearrangement of (5.8) immediately gives rise to

$$\begin{aligned} &2[(n^{\mathrm{L}})^2\rho^{\mathrm{S}} + (n^{\mathrm{S}})^2\rho^{\mathrm{L}}]u_{n1}\frac{\delta a^{\mathrm{S}}_n}{\delta t} + \alpha_v u_{n1} a^{\mathrm{S}}_n - \\ &-[(n^{\mathrm{L}})^2(\lambda^{\mathrm{S}} + 2\mu^{\mathrm{S}}) + (n^{\mathrm{S}})^2\alpha^{\mathrm{L}}]b^{\alpha}_{\cdot\alpha} a^{\mathrm{S}}_n + \\ &+ \{(n^{\mathrm{L}})^2(\lambda^{\mathrm{S}} + 2\mu^{\mathrm{S}}) + (n^{\mathrm{S}})^2\alpha^{\mathrm{L}} - [(n^{\mathrm{L}})^2\rho^{\mathrm{S}} + (n^{\mathrm{S}})^2\rho^{\mathrm{L}}]u^2_{n1}\}\mathbf{C}^{\mathrm{S}} \cdot \mathbf{n} = 0. \end{aligned} \tag{5.15}$$

Utilizing the expression (4.24), we can eliminate the unknown dot product $\mathbf{C}^{\mathrm{S}} \cdot \mathbf{n}$ and obtain the differential equation governing the amplitude $a^{\mathrm{S}}_n$

$$\frac{\delta a^{\mathrm{S}}_n}{\delta t} - \frac{1}{2}\left[b^{\alpha}_{\cdot\alpha} u_{n1} - \frac{\alpha_v}{(n^{\mathrm{L}})^2\rho^{\mathrm{S}} + (n^{\mathrm{S}})^2\rho^{\mathrm{L}}}\right] a^{\mathrm{S}}_n = 0. \tag{5.16}$$

Analogously, we take the projection of the field equations (5.1) and (5.2) across the wave fronts in the direction of $\mathbf{v}$ and bear in mind that only the transverse waves are realizable in the solid skeleton in this case, the result is arranged in the form

$$\begin{aligned}&(\lambda^{\mathrm{S}}+\mu^{\mathrm{S}})[a^{\alpha\beta}(\mathbf{a}^{\mathrm{S}}_{;\alpha}\cdot\mathbf{n})(\mathbf{x}_{,\beta}\cdot\mathbf{v})-b^{\alpha\beta}(\mathbf{a}^{\mathrm{S}}\cdot\mathbf{x}_{,\alpha})(\mathbf{x}_{,\beta}\cdot\mathbf{v})]+\\&+\mu^{\mathrm{S}}(\mathbf{C}^{\mathrm{S}}\cdot\mathbf{v}-b^{\alpha}_{\cdot\alpha}\mathbf{a}^{\mathrm{S}}\cdot\mathbf{v})-\rho^{\mathrm{S}}\left(u_n^2\mathbf{C}^{\mathrm{S}}\cdot\mathbf{v}-2u_n\frac{\delta\mathbf{a}^{\mathrm{S}}}{\delta t}\cdot\mathbf{v}\right)+\\&+\alpha_\nu u_n\mathbf{a}^{\mathrm{S}}\cdot\mathbf{v}=0.\end{aligned}\tag{5.17}$$

In the case of the homogeneous weak discontinuity, Equation (5.8) becomes

$$2\rho^{\mathrm{S}}u_{n2}\frac{\delta\mathbf{a}^{\mathrm{S}}}{\delta t}\cdot\mathbf{v}-(\mu^{\mathrm{S}}b^{\alpha}_{\cdot\alpha}-\alpha_\nu u_{n2})\mathbf{a}^{\mathrm{S}}\cdot\mathbf{v}+(\mu^{\mathrm{S}}-\rho^{\mathrm{S}}u_{n2}^2)\mathbf{C}^{\mathrm{S}}\cdot\mathbf{v}=0,\tag{5.18}$$

where the identity

$$\mathbf{v}_{,\alpha}\cdot\mathbf{n}=-\mathbf{v}\cdot\mathbf{n}_{,\alpha}=b^{\cdot\gamma}_{\alpha}\,\mathbf{x}_{,\gamma}\cdot\mathbf{v}\tag{5.19}$$

has been utilized. With the aid of the expression (4.25), the amplitude for the transverse waves is determined by

$$\frac{\delta a_v^{\mathrm{S}}}{\delta t}-\left(\frac{1}{2}b^{\alpha}_{\cdot\alpha}u_{n2}-\frac{\alpha_\nu}{2\rho^{\mathrm{S}}}\right)a_v^{\mathrm{S}}=0.\tag{5.20}$$

With the help of (2.9), the following notation may be introduced as

$$\frac{\delta}{\delta t}=u_n\left(\mathbf{n}\cdot\operatorname{grad}+\frac{1}{u_n}\frac{\partial}{\partial t}\right)=u_n\frac{\mathrm{d}}{\mathrm{d}r},\tag{5.21}$$

where the total derivative in the direction of the normal $\mathbf{n}$ to the singular surfaces $\sigma(t)$ is represented by $\mathrm{d}/\mathrm{d}r$. Then, the substitution by means of $\mathrm{d}/\mathrm{d}r$ shows that the ordinary differential equations (5.16) and (5.20) take the forms

$$\frac{\mathrm{d}a_n^{\mathrm{S}}}{\mathrm{d}r}=(\Omega-M)a_n^{\mathrm{S}},\tag{5.22}$$

$$\frac{\mathrm{d}a_v^{\mathrm{S}}}{\mathrm{d}r}=(\Omega-N)a_v^{\mathrm{S}},\tag{5.23}$$

where definition (2.5) has been used, and the coefficients $M$ and $N$ are defined by

$$M=\frac{\alpha_\nu}{2\left[(n^{\mathrm{L}})^2\rho^{\mathrm{S}}+(n^{\mathrm{S}})^2\rho^{\mathrm{L}}\right]u_{n1}},\tag{5.24}$$

$$N=\frac{\alpha_\nu}{2\rho^{\mathrm{S}}u_{n2}}.\tag{5.25}$$

Furthermore, $\Omega$ can be rewritten in terms of a certain position, namely

$$\Omega(r) = \frac{\Omega_0 - K_0 r}{1 - 2\Omega_0 r + K_0 r^2}, \tag{5.26}$$

where $r$ is the distance of the wave fronts from their initial position. Hence, the integration of (5.22) and (5.23) together with (5.26) results in

$$a_n^S(r) = a_n^S(0)(1 - 2\Omega_0 r + K_0 r^2)^{-1/2} \exp(-Mr), \tag{5.27}$$

$$a_v^S(r) = a_v^S(0)(1 - 2\Omega_0 r + K_0 r^2)^{-1/2} \exp(-Nr), \tag{5.28}$$

where $a_n^S(0)$ and $a_v^S(0)$ are the initial values of the absolute amplitudes of waves.

It is evident from (2.55), (2.56), (2.58) and (2.59) that the growth and decay of the waves depend on the initial geometrical shapes of the wave fronts and the diffusion effect between the phases. The diffusion effect only produces the attenuation of the amplitudes of the waves but does not influence the propagation speeds.

In particular, the evolution of the amplitudes of wave fronts may reduce to the classical results in the case of nondissipation where the amplitudes rely completely on the initial geometrical property of the wave fronts.

If the initial mean curvature $\Omega_0$ is positive and the initial total curvature $K_0$ is negative, for instance in the case of the initial pseudosphere for the wave fronts, it follows from (5.27) and (5.28) that the amplitudes will reach an infinite value within a finite propagation distance. In other words, the discontinuity waves in the porous media may constitute a higher weak discontinuity. In the case of the negative initial mean curvature and the positive initial total curvature, the amplitudes of wave fronts will decay to vanish. It is of interest to consider the plane ($\Omega_0 = 0$, $K_0 = 0$), the cylindrical ($\Omega_0 = -0.5/r_0$, $K_0 = 0$) and the spherical ($\Omega_0 = -1/r_0$, $K_0 = 1/r_0^2$) waves for which the amplitudes are reduced into the following forms, respectively:

*For plane waves*:

$$a_n^S(r) = a_n^S(0) \exp(-Mr), \tag{5.29}$$

$$a_v^S(r) = a_v^S(0) \exp(-Nr); \tag{5.30}$$

*For cylindrical waves*:

$$a_n^S(r) = a_n^S(0) \left(1 + \frac{1}{r_0} r\right)^{-1/2} \exp(-Mr), \tag{5.31}$$

$$a_v^S(r) = a_v^S(0) \left(1 + \frac{1}{r_0} r\right)^{-1/2} \exp(-Nr); \tag{5.32}$$

*For spherical waves*:

$$a_n^S(r) = a_n^S(0) \left(1 + \frac{2}{r_0} r + \frac{1}{r_0^2} r^2\right)^{-1/2} \exp(-Mr), \tag{5.33}$$

$$a_v^{\mathrm{S}}(r) = a_v^{\mathrm{S}}(0)\left(1 + \frac{2}{r_0}r + \frac{1}{r_0^2}r^2\right)^{-1/2} \exp(-Nr), \tag{5.34}$$

where $r_0$ is the initial semi-diameter of the wave fronts.

## 6. Concluding Remarks

In this paper the investigation of discontinuity waves in two-phase incompressible saturated porous media is presented. The incompressible porous media model is used to describe incompressible porous materials. Our focus of interest is put on a linear elastic porous solid filled with an ideal fluid. The propagation conditions and the amplitude evolution of wave fronts are investigated by the use of the theory of singular surfaces. It is revealed that a coupled longitudinal wave and a pure transverse wave are realizable in two-phase incompressible porous media. The longitudinal waves propagate either in the solid skeleton or in the liquid with the same propagation speed. In this coupled wave mode a certain restriction between the amplitudes is required by the saturation condition together with the incompressibility of porous media. The jump of the pore pressure may be determined by the amplitudes of the coupled waves. The transverse waves propagate only in the solid phase. The two propagation speeds are determined by the material constants and the volume fractions but do not depend on the diffusion coefficient $\alpha_v$. In the case of homogeneous weak discontinuities, the differential equations governing the amplitudes of the waves of two types are then derived and the explicit solutions for the two cases are obtained. These results indicate that the amplitudes of wave fronts may either decay to vanish or grow to infinity in finite time, which depends on the geometrical property of the initial shapes of the wave fronts and the diffusion effect between the two phases.

## Acknowledgements

This work was supported by Alexander von Humboldt-Stiftung, the Deutsche Forschungsgemeinschaft (DFG), and the National Natural Sciences Foundation (China).

## References

Biot, M. A.: 1941, General theory of three dimensional consolidation, *J. Appl. Phys.* **12**(2), 155–164.

Biot, M. A.: 1956, Theory of propagation of elastic waves in a fluidsaturated porous solid: I. Low frequency range, *J. Acoust. Soc. Am.* **28**(2), 168–178.

de Boer, R.: 1996, Highlights in the historical development of the porous media theory: toward a consistent macroscopic theory, *Appl. Mech. Rev.* **49**(4), 201–262.

de Boer, R. and Ehlers, W.: 1990, Uplift, friction and capillarity – three fundamental effects for liquid-saturated porous solids, *Int. J. Solids Structure* **26**, 43–57.

de Boer, R., Ehlers, W. and Liu, Z.: 1993, One-dimensional transient wave propagation in fluid-saturated porous media, *Archive Appl. Mech.* **63**, 59–72.

de Boer, R. and Liu, Z.: 1994, Plane waves in a semi-infinite fluid saturated porous medium, *Transport in Porous Media* **16**, 147–173.

de Boer, R. and Liu, Z.: 1995, Propagation of acceleration waves in incompressible liquid-saturated porous solids, *Transport in Porous Media* **21**, 163–173.

Bourbié, T., Coussy, O. and Zinszner, B.: 1987, Acoustic of Porous media, Editions Technip, Paris.

Bowen, R. M.: 1980, Incompressible porous media models by use of the theory of mixtures, *Int. J. Engng. Sci.* **18**, 1129–1148.

Deresiewicz, H. and Rice, J. T.: 1986, The effect of boundaries on wave propagation in a liquid filled porous solid: III. Reflection of plane waves at a free plane boundary (general case), *Bull. Seism. Soc. Am.* **52**, 505–625.

Dziecielak, R.: 1986, Propagation and decay of acceleration waves in thermoconsolidation media, *Acta Mech.* **59**, 213–231.

Eringen, A. C. and Suhubi, E. S.: 1974, *Elastodynamics*, Vol. I, Academic Press, New York and London.

Garg, S. K., Nafeh, A. H. and Good, A. J.: 1974, Compressional waves in fluid-saturated elastic porous media, *J. Appl. Phys.* **45**, 1968–1974.

Levy, T.: 1979, Propagation of waves in a fluid-saturated porous elastic solid, *Int. J. Engng. Sci.* **17**, 1005–1014.

Mengi, Y. and McNiven, H. D.: 1978, Propagation and decay of waves in porous media, *J. Acoust. Soc. Am.* **64**(4), 1125–1131.

Plona, T. J.: 1980, Observation of a second bulk compressional wave in a porous medium at ultrasonic frequencies, *Appl. Phys. Lett.* **36**(4), 259–261.

Plona, T. J.: 1980, Experimental study of the two bulk compressional modes in water-saturated porous structures, *Ultrasonic Symposium, IEEE*, New York, pp. 868–872.

Plona, T. J.: 1982, Acoustics of fluid-saturated porous media, *Ultrasonic Symposium, IEEE*, New York, pp. 1044–1048.

Zienkiewicz, O. C. and Shiomi, T.: 1984, Dynamic behaviour of saturated porous media – The generalized Biot formulation and its numerical solution, *Int. J. Num. Ana. Mech. Geomech.* **8**, 71–96.

# Computer Simulation of Drying Optimal Control

S. J. KOWALSKI and A. RYBICKI
*Poznan University of Technology, Institute of Technology and Chemical Engineering, pl. Marii Skodowskiej-Curie 2, 60-965 Poznan, Poland*

(Received: 11 August 1997; in final form: 3 December 1997)

**Abstract.** The subject of this paper is the control of computer-simulated drying processes for the purpose of their optimization. The simulated processes are carried out in such a way that the drying-induced stresses never exceed the stress limit. This paper presents results of a mechanically admissible drying process on the example of a prismatic bar dried convectively, using the theory for drying of capillary-porous materials. The procedure of designing the safe drying processes, in which the destruction of dried materials no longer proceeds, is presented.

**Key words:** mechanically-admissible process, altered drying conditions, minimal drying time and energy consumption.

## 1. Introduction

Dried materials, like fluid saturated capillary-porous bodies, shrink during drying to a lesser or greater extent, and internal stresses caused by the nonuniform shrinkage arise. The deformations or crack formation occurring during drying directly influence the quality of dried products. Therefore, the development of methods of control preventing the destruction of dried materials and precisely designing the optimum drying processes in order to shorten the drying time and save energy is very desirable.

This paper is a continuation of one by Kowalski and Rybicki (1995). Its aim is to present the optimization of a computer-simulated drying process for the purpose of designing mechanically admissible drying processes with a possible maximal drying rate. One finds in this study that the reduction of the drying stresses is possible, if they rise to critical values, through altering the temperature and the moisture content of the drying medium. A question arises as to how to establish the program of alteration of these parameters in order to reduce the stresses and to not slow the drying rate.

The numerical experiment program presented in this paper allows us to calculate not only the parameters of the state of the dried material but also the evolution of these quantities in time registered on the computer monitor. The drying parameters are changed automatically by the computer when the stresses reach the critical value. One observes that the dried material responds immediately on the increase of moisture content in the drying medium but its response on the decrease of the drying medium temperature is lengthened in time. This is probably due to warming of the drying medium by the dried material for the time of stabilization of the wet bulb temperature.

Both the problem of mechanical response of dried materials to variation of drying parameters and the mechanically-admissible drying processes are demonstrated on the prismatic bar dried convectively. The theory proposed earlier by Kowalski (1987, 1990, 1994, 1997), Kowalski and Strumillo (1997) and the numerical methods elaborated by Rybicki (1993, 1994), Kowalski and Rybicki (1996) make the basis for these studies. The two-dimensional boundary-value problem described by the system of four coupled second-order differential equations is solved with the use of the finite element method for spatial derivatives and the three-point finite differences for the time derivatives. The paper presents the results in the form of diagrams drawn automatically by computer.

## 2. Model Presentation

We shall design a mechanically admissible drying process for convective drying of an infinite bar having a rectangular cross-section, Figure 1.

We shall present briefly the theory of drying based on which the mechanically admissible and optimal drying processes are designed.

The drying induced stresses $\sigma_{ij}$ are calculated from the relation (Kowalski, 1987, 1990, 1997):

$$\sigma_{ij} = M(u_{i,j}+u_{j,i})+\left[\left(A-\frac{\gamma_\Theta^2}{c_\Theta}\right)u_{l,l}-\left(\gamma_\vartheta-\gamma_\Theta\frac{c_\vartheta}{c_\Theta}\right)\vartheta-\frac{\gamma_\Theta}{c_\Theta}\rho_0\mu\right]\delta_{ij}, \quad (1)$$

where $u_i$ denotes the displacement, $\vartheta$ is the temperature, $\mu$ is the moisture potential in dried material, $M$ and $A$ are the shear and bulk modulus of the dried material, $\rho_0$ is the mass density of dry material, $\gamma_\vartheta = (2M+3A)\alpha_\vartheta$, $\gamma_\Theta = (2M+3A)\alpha_\Theta$, $\alpha_\vartheta$ and $\alpha_\Theta$ are the coefficients of linear thermal and humid expansion of the dried material

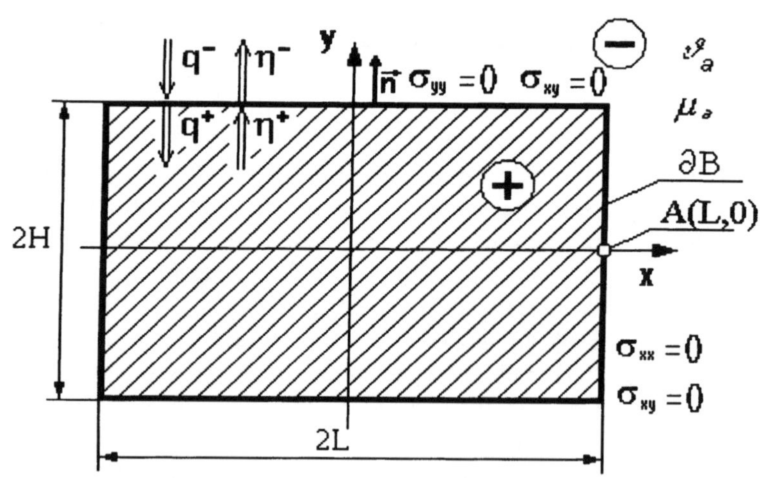


*Figure 1.* Rectangular cross-section of the drying bar.

and $c_\vartheta$ and $c_\Theta$ are temperature and moisture content coefficients in the moisture potential function.

Displacements $u_x(x, y, t)$, $u_y(x, y, t)$, temperature $\vartheta(x, y, t)$, moisture potential $\mu(x, y, t)$ are calculated from the following system of differential equations (see Kowalski and Rybicki, 1995):

$$\begin{aligned} &Lu_{x,xx} + Mu_{x,yy} + Nu_{y,xy} = P\vartheta_{,x} + R\mu_{,x}, \\ &Lu_{y,yy} + Mu_{y,xx} + Nu_{x,xy} = P\vartheta_{,y} + R\mu_{,y}, \\ &K_{\mathrm{m}}(\mu_{,xx} + \mu_{,yy}) = \mu_{,t} + \left[\gamma_\Theta(u_{x,xt} + u_{y,yt}) - c_\vartheta \vartheta_{,t}\right]\rho_0, \\ &K_{\mathrm{T}}(\vartheta_{,xx} + \vartheta_{,yy}) = \vartheta_{,t} + K_E(u_{x,xt} + u_{y,yt}) - K_\Theta \mu_{,t}, \end{aligned} \tag{2}$$

where $K_{\mathrm{m}} = \Lambda_{\mathrm{m}} c_\Theta/\rho_0^2$, $K_{\mathrm{T}} = \Lambda_{\mathrm{T}}/\overset{*}{c}_{\mathrm{v}}$, $K_E = T_{\mathrm{r}}(\gamma_\vartheta - c_\vartheta\gamma_\Theta/c_\Theta)\overset{*}{c}_{\mathrm{v}}$, $K_\Theta = T_{\mathrm{r}} c_\vartheta \rho/c_\Theta \overset{*}{c}_{\mathrm{v}}$, $L = M + N$, $N = M + A - \gamma_\Theta^2/c_\Theta$, $P = \gamma_\vartheta - c_\vartheta\gamma_\vartheta/c_\Theta$, $R = \rho_0\gamma_\Theta/c_\Theta$, $\overset{*}{c}_{\mathrm{v}} = c_{\mathrm{v}} + T_{\mathrm{r}} c_\vartheta^2/c_\Theta$, $\Lambda_{\mathrm{T}}$ and $\Lambda_{\mathrm{m}}$ are the thermal and moisture conductivity coefficients, $T_{\mathrm{r}}$ is the reference temperature, $c_{\mathrm{v}}$ is the specific heat of dried material.

The following boundary condition is assumed to be satisfied (Kowalski and Rybicki, 1994, 1995):

for the stresses

$$\sigma_{ij} n_j|_{\partial B^-} = 0; \tag{3}$$

for the moisture exchange

$$\Lambda_{\mathrm{m}}\ \mathrm{grad}\ \mu|_{\partial B^+} \cdot \mathbf{n} = \pm\alpha_{\mathrm{m}}(\mu|_{\partial B^-} - \mu_{\mathrm{a}}); \tag{4}$$

and for the heat exchange

$$\Lambda_{\mathrm{T}}\ \mathrm{grad}\ \vartheta|_{\partial B^+} \cdot \mathbf{n} = \alpha_{\mathrm{T}}(\vartheta_{\mathrm{a}} - \vartheta|_{\partial B^+}) \mp l\alpha_{\mathrm{m}}(\mu|_{\partial B^-} - \mu_{\mathrm{a}}). \tag{5}$$

The initial values are

$$\sigma_{ij}(x, y, 0) = 0, \qquad \mu(x, y, 0) = \mu_0, \qquad \vartheta(x, y, 0) = \vartheta_0. \tag{6}$$

In the above condition $\mu_{\mathrm{a}}$ and $\vartheta_{\mathrm{a}}$ denote the chemical potential and the temperature of the drying medium (air), and $l$ is the latent heat of evaporation.

The quantities on the left hand side of boundary conditions (3)–(5) are related to the inner side of boundary surface while those on the right hand side are related to the outer side of this surface. Thus, the moisture potential $\mu|_{\partial B^-}$ on the outer side of the boundary surface is the potential related to water vapour in air at the boundary, and $\mu_{\mathrm{a}}$ is the vapour potential in air at a distance from the boundary. The difference between these two potentials creates the 'force' for convective exchange of vapour between the boundary and the surrounding atmosphere.

The potentials for vapour can be written in the well known form (see, e.g., Elwell and Pointon, 1976; Strumillo, 1983, Szarawara, 1985):

$$\mu|_{\partial B^-} = \mu^\Theta(p, T_n) + RT_n \ln X_n, \tag{7}$$

$$\mu_a = \mu^{\ominus}(p, T_a) + RT_a \ln X_a, \tag{8}$$

where $p$ is the total pressure of the surrounding atmosphere, $T_n$ and $T_a$ are the absolute temperatures of the surrounding atmosphere at the boundary and at a distance from the boundary, and $X_n$ and $X_a$ are the molar vapour contents at the boundary and at a distance from the boundary, respectively. The absolute humidity $Y_a$ is related to the molar vapour content as $X_a = Y_a/(0.622 + Y_a)$. Functions $\mu^{\ominus}(p, T_n)$ and $\mu^{\ominus}(p, T_a)$ denote the potentials of pure vapour at the temperatures $T_n$ and $T_a$, respectively. Developing these functions in Taylor's series and retaining terms of first order only one can express the difference between vapour potentials at the boundary and far from the boundary as follows:

$$\begin{aligned} \mu|_{\partial B^-} - \mu_a &= 7.36(T_a - T_n) + \\ &+ 8.3\left[\left(\frac{T_a - T_n}{2}\right)\ln\frac{X_n}{X_a} - (T_a - T_n)\ln\sqrt{X_n X_a}\right]. \end{aligned} \tag{9}$$

The above system of equations fully describes the boundary-value problem. Moreover, through the control of temperature $T_a$ and moisture content $X_a$ of the drying medium, one can design the mechanically-admissible process with optimal drying parameters.

## 3. Drying Processes as an Object of Control

A relevant mathematical model of drying may serve to work out a computer program for simulation of the drying process. The complexity of the mathematical model of drying and, in particular, the coupling of the system of differential equations appearing in such a model makes it difficult to solve precisely. The only possible solution to such a complex problem is to adopt one of the numerical methods. Such a method is used in this paper.

During the numerical simulation one can analyze multiple parameters describing the state of the dried material, such as the moisture content, temperature, stresses or strains. Due to change of the temperature of the drying medium $T_a$ and its moisture content $X_a$ one can observe the mechanical response of the dried material to these changes. The numerical results obtained allow one to draw some conclusions concerning the real nature of the drying process and to put forward a proposal for its optimal control.

The notion 'control' denotes the configurational program of the drying process for the purpose of its house-keeping run. The 'object of control' is the installation through which the control proceeds. 'Identification of the control object' implies the physical properties of the object. One differentiates two groups of physical quantities which are transformed by the object of control. The first group is called the 'input quantities,' and the second one the 'output quantities'. An alteration of the input quantities means the external interaction (impulse) on the object of control. An alteration of the output quantities means the reaction (response) of the object of control on the external impulse.

The object of control in our considerations is the simulated drying process, and, precisely, the mathematical model of this process given by the Equations (2) and the boundary conditions (3)–(5). We formulate our problem as follows: to find such a program of alteration of the drying parameters (temperature $T_a$ and moisture content $X_a$), at which the drying rate is possibly maximal and the drying stresses never surpass the admissible stresses for given dried material. Thus, one must observe the tension stresses at the boundary of the dried material, which are usually the largest. These stresses must never surpass the admissible value, but their necessary reduction during the drying process, when they tend to surpass the admissible stresses, ought to be carried out in such a way that they remain close to the admissible value. Otherwise the rate of drying would be slowed too much. This is the principle of maximum drying rate (see also Leitman, 1966).

From the point of view of the control theory, Equations (2) with boundary conditions (3)–(5) for the above formulated problem can be performed by the scheme illustrated in Figure 2.

We interpret the scheme as follows. The drying medium acts on the dried material through the moisture flux $\eta$ and the heat flux $\mathbf{q}$ on the boundary. The response of the dried material to this action is manifested through the increase of the strains $\epsilon_{ij} = (u_{i,j} + u_{j,i})/2$, temperature $\vartheta$, and the decrease of moisture content $\Theta$. The fluxes $\eta$ and $\mathbf{q}$ are governed by the boundary conditions (3)–(5) and are directly influenced by the parameters of drying medium $T_a$ and $X_a$, as well as by the temperature $\vartheta$ and the moisture content $\Theta$ of the dried material on the boundary. The control unit selects the parameters of the drying medium through the analysis of tensed stresses which are calculated continually.

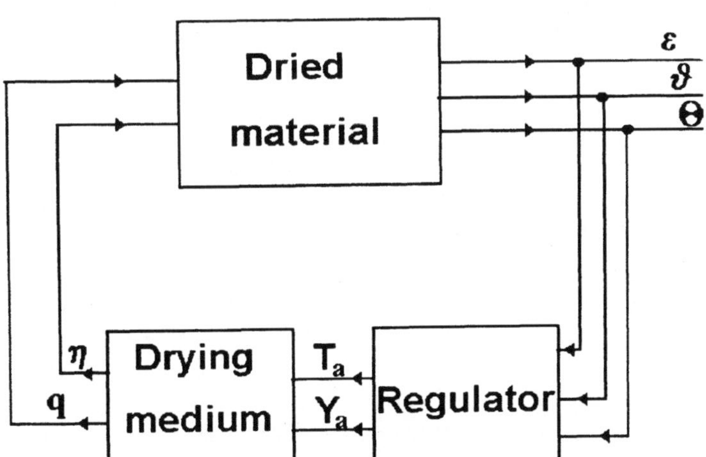


*Figure 2.* Object of control: Input and output quantities.

## 4. Simulated Drying Processes and Their Analysis

First, our task is to find functions $u_x$, $u_y$, $\vartheta$ and $\mu$ within the rectangle $(-L, L) \times (-H, H)$, (in particular in point $x = L, y = 0$), and for $t \in \mathbf{R}^+$ to satisfy the system of Equations (2), the boundary conditions (3)–(5) with the right hand side of Equation (9), and the initial condition (6).

The system of differential equation (2) is solved using the finite element method for the spatial derivatives and the three-point finite differences for the time derivatives (see Rybicki, 1994; Kowalski and Rybicki, 1994, 1995). Because of the symmetry of the problem, the calculations were confined to a quarter of the whole rectangle, namely to the domain $(0, L) \times (0, H)$ .

The drying process is carried out in the following way: The initial parameters of the drying medium are $\vartheta_{0a} = 20°C$, and the humidity $Y_{0a} = 0.015$, which is equivalent to relative humidity $\phi = 100\%$. The temperature of the dried material also equals $\vartheta_{0a} = 20°C$ and this means that no drying process proceeds at this state. Next, the temperature $\vartheta_a$ and the absolute humidity $Y_a$ are gradually increased. For each combination of these parameters, the relative humidity is $\phi < 100\%$.

The surfaces in Figure 3 illustrate the distributions of the stresses $\sigma_{yy}$, the temperature $\vartheta$, and the moisture potential $\mu$ in the quarter of bar cross-section. At the beginning of the drying process ($t = 10$ min) the temperature distribution is highly non-uniform and the thermal stresses (compressive at the boundary and tensional inside the cross-section) dominate. After some time ($t = 2.5$ h) the distribution of the temperature becomes uniform in the whole cross-section and equal to the wet bulb temperature. The thermal stresses disappear and the shrinkage stresses (tensional at the boundary and compressive inside the cross-section) caused by the non-uniform moisture distribution, or equivalently by the moisture potential distribution, start to dominate.

Figure 4 presents the evolutions of the moisture potential $\mu$ and the stresses $\sigma_{yy}$ in a quarter of the cross-section of the drying bar in the first period of drying. At the beginning, the distributions of these quantities are almost uniform in the whole cross-section. In the course of drying, the moisture potential decreases, mostly at the boundary, and the stress increases in proportion to the nonuniformity of the moisture potential.

Figure 5 presents the evolution of the moisture content $W$ [kg moisture/kg dry material] in the unit length of the bar and the maximal stresses max $\sigma_{yy}$ arising in point $x = L$, $y = 0$ of the cross-section for drying processes with drying parameters according to the programs placed in the Table I. Figure 5(a) allows one to estimate the rate of drying. It is evident that the rate of drying increases when the absolute humidity $Y_a$ decreases or the temperature $\vartheta_a$ increases. The maximal stresses, which increase in several programs in point $x = L$, $y = 0$, depend evidently on the drying rate. The greater the drying rate, the greater the resulting maximal stress. Thus, the increase of drying rate is unfortunately connected with an increase of maximal stress value. Note that programs 1 and 8 or 2 and 7 have different drying parameters, but the drying rate and the maximal stresses are almost the same. Reduction

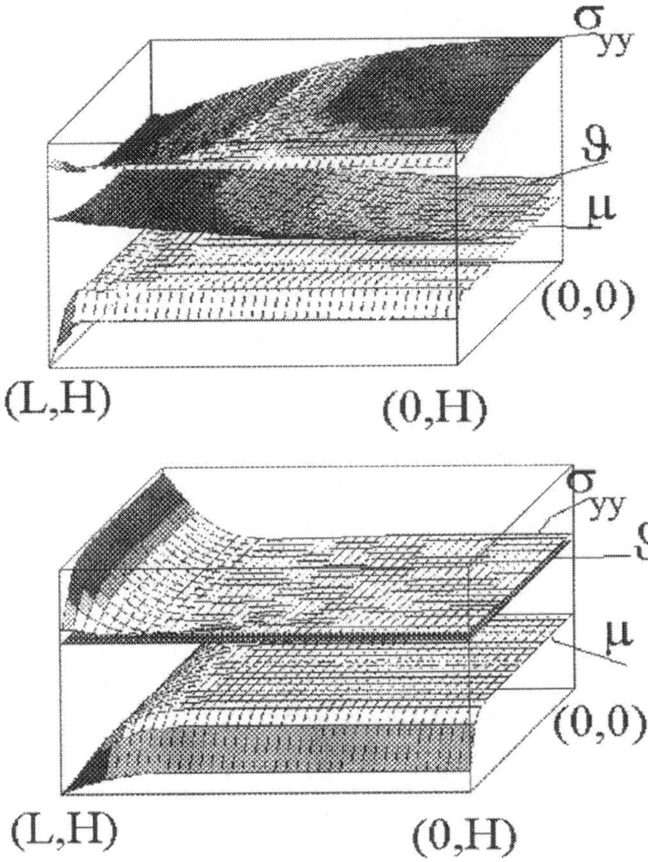


*Figure 3.* Distributions of the stresses $\sigma_{yy}$, the temperature $\vartheta$, and the moisture potential $\mu$ in the quarter of bar cross-section at time $t = 10$ min. and $t = 2.5$ h.

of the drying stresses upon reaching admissible value (say 200 KPa) is presented in Figure 6.

Figures 6(a) and 6(b) present a reaction of the drying induced stresses to alteration of drying parameters. The curves in Figure 6 are a continuation of that in Figure 5(b) for $Y_a = 0.035$ and $\vartheta_a = 50°C$. At 8th hour of the drying process the temperature of the drying medium $\vartheta_a$ is diminished about 3°C , 5°C and 10°C (Figure 6(a)) or the humidity of the drying medium $Y_a$ is increased about 10%, 15% and 20% (Figure 6(b)), respectively. However, both these operations involve a decrease of the stresses only for some time. After some time the stresses start to increase again. Therefore, the drying parameters have to be controlled all the time.

Figure 7 illustrates the sensitivity of the dried material and the drying induced stresses to alterations of the drying parameters. Observing the curves denoted by 1, one can state that the response of dried material on alteration of the drying medium

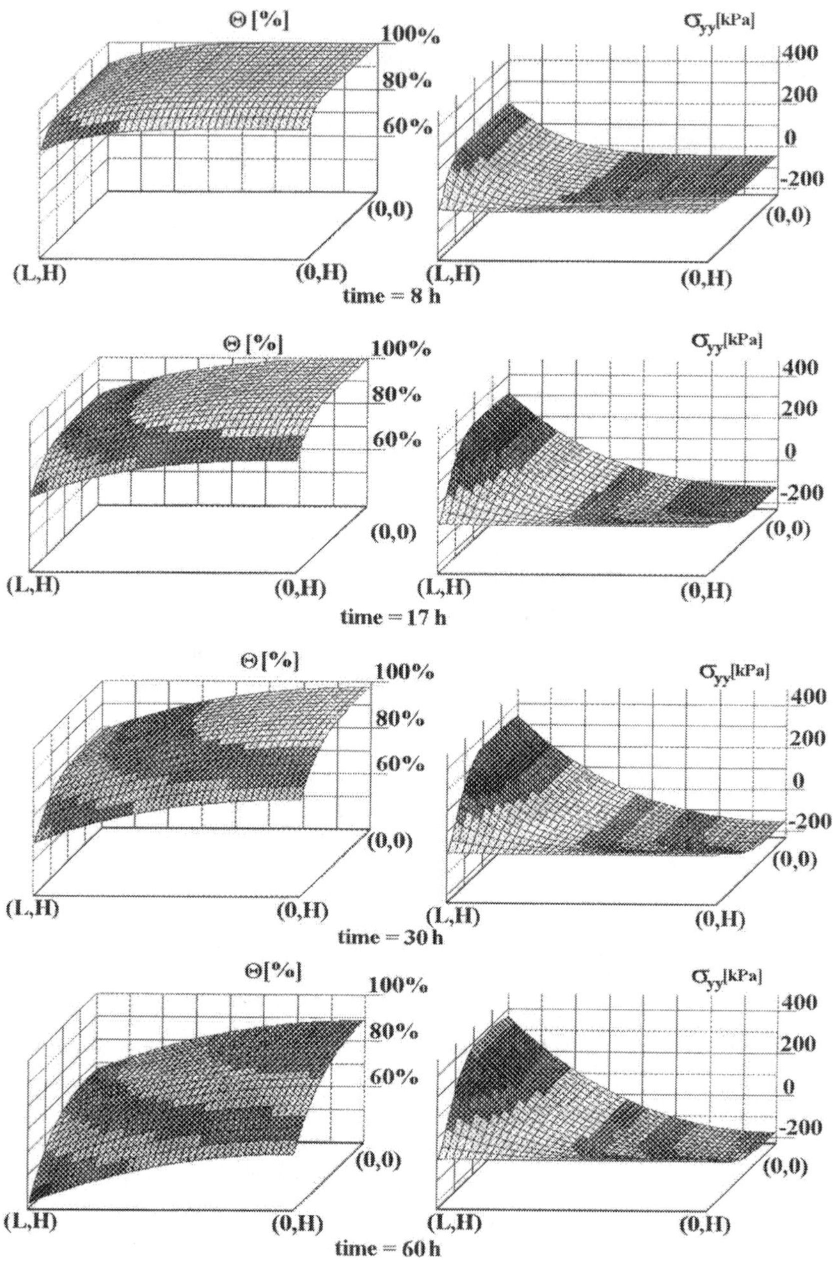


*Figure 4.* Distributions of the moisture content and the stresses in the quarter of bar cross-section during the process.

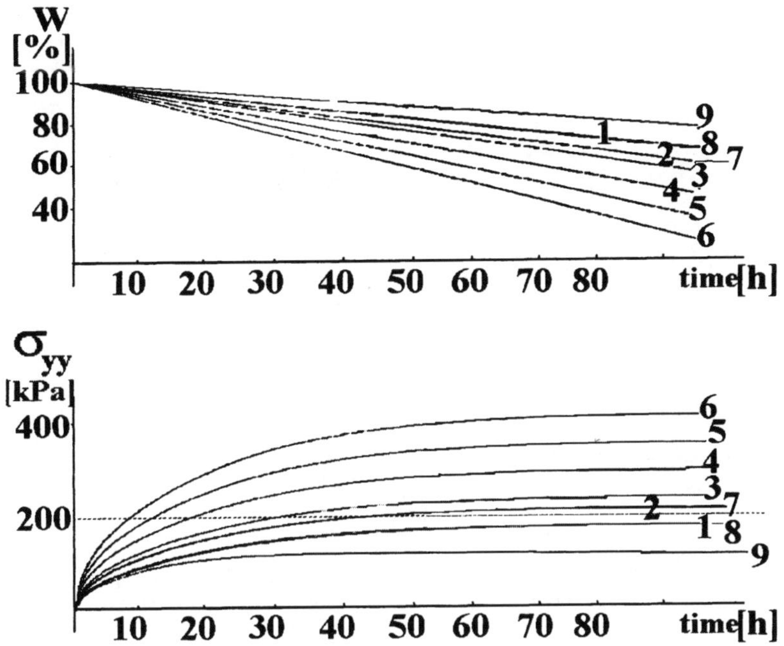


*Figure 5.* Evolution of the moisture content $W$ per unit length of the bar and the stresses in point $x = L$, $y = 0$, versus time.

*Table I.*

| Program No. | $Y_a$[kg/kg] | $\vartheta_a$ [°C] | ${}_{\max}\sigma_{yy}$ [KPa] |
|---|---|---|---|
| 1 | 0.035 | 50 | 186 |
| 2 | 0.035 | 53 | 217 |
| 3 | 0.035 | 55 | 239 |
| 4 | 0.035 | 60 | 297 |
| 5 | 0.035 | 65 | 355 |
| 6 | 0.035 | 70 | 413 |
| 7 | 0.088 | 70 | 217 |
| 8 | 0.105 | 70 | 180 |
| 9 | 0.140 | 70 | 119 |

humidity is instantaneous. The stresses decrease at once when the humidity $Y_a$ increases, and vice-versa. However, if the humidity $Y_a$ is kept constant for longer time, the stresses tend to increase again.

The response of the dried material to the temperature alterations is retarded in time, probably because of the material heat capacity. The drying rate can be increased by an increase of the drying medium temperature $\vartheta_a$ . The temperature alteration influences the stresses $\sigma_{yy}$ in a similar way to the alteration of the drying medium humidity,

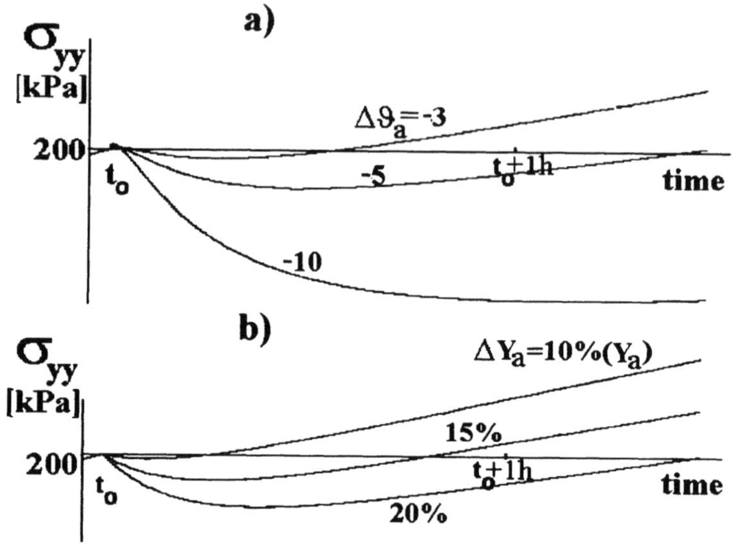


*Figure 6.* Reduction of drying stresses through changes of the drying parameters: (a) temperature of the drying medium $\vartheta_a$, (b) humidity of the drying medium $Y_a$.

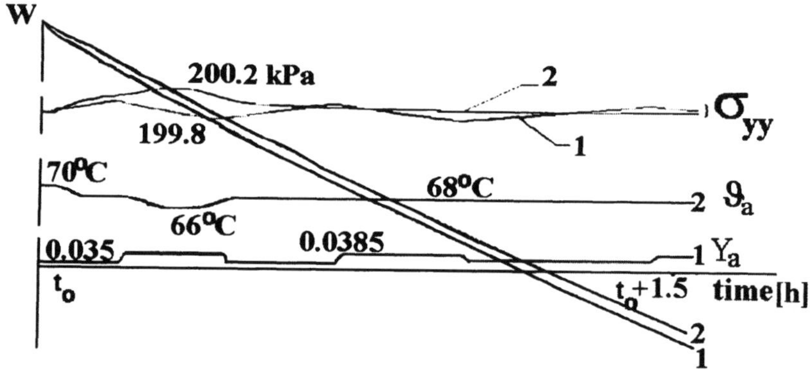


*Figure 7.* Response of dried material on alterations of the drying parameters.

but the reaction of the material to alteration of the temperature $\vartheta_a$ is a bit slower than upon alteration of the humidity $Y_a$ . Studying the above diagrams, we infer the necessity of altering the drying parameters when the stresses reach a dangerous value (in our case 200 KPa).

Figure 8 presents a controlled drying process from beginning to 11th hour through the suitable alteration of the temperature $T_a$ .

The upper part of the figure signifies the moisture content alteration W and the lower part illustrates the evolution of the stresses in point $x = L$, $y = 0$ for the following program: from the beginning of the drying process to 8.6 h the drying parameters are $T_a = 70°C$ and $Y_a = 0.035$, since during that time the humidity is

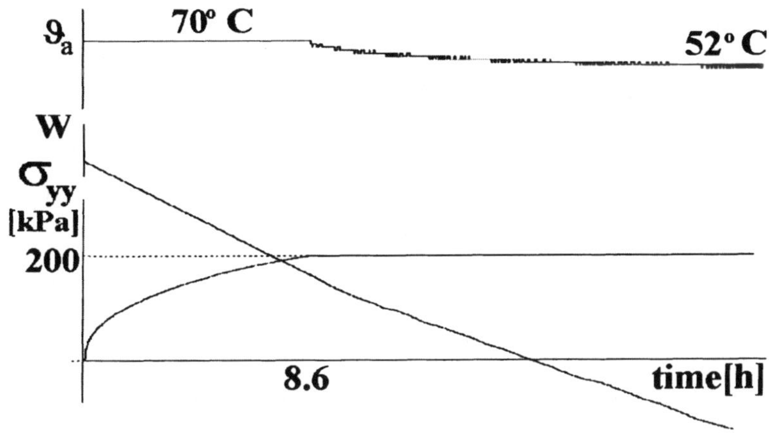


*Figure 8.* Optimal drying process.

kept constant and the temperature of the drying medium is altered by the computer automatically in such a way that the stresses never exceed the permissible stress $\sigma_{yy}$ =200 kPa. Moreover, it is seen that the stress $\sigma_{yy}$ is nearly constant and equal to 200 kPa through the whole course of drying. One expects the drying rate in this case to be higher than that in all other programs.

## 5. Final Remarks

The main achievement of this contribution is in finding that the dried materials are very sensitive to the alterations of drying conditions. This gives a possibility for the control of drying processes and for designing the optimum processes with optimal drying time and energy consumption and without violation of the dried material strength. The following conclusions result from the present studies:

1. An increase of the temperature or a decrease of the humidity of the drying medium increases considerably the drying rate;
2. The increase of drying rate causes the nonuniform distribution of the moisture content in dried material and this involves the drying induced stresses;
3. The magnitude of drying induced stresses evidently depends on the drying rate;
4. To avoid a destruction of dried material one ought to change the drying parameters in that moment when the maximal stress in the material reaches the value of permissible stress for a given material;
5. The optimal drying process seems to be when the drying rate corresponds to the maximal stress equal to the permissible stress as it is shown in Figure 8. For lower drying rates the stresses would not violate the strength of the material, but the time of drying would be very long;
6. An optimal drying process can be controlled in a fully automatic way following the given program for drying.

The above conclusions are the result of computer-simulated studies based on the physical model elaborated by the authors. The next step should be to carry on the optimal drying processes in practical applications.

## Acknowledgements

This work was carried out as a part of research project No. 3 T09C 005 12 sponsored by the Polish State Committee for Scientific Research in the year 1997.

## References

de Boer, R. and Kowalski, S. J.: 1995, Thermomechanics of fluid-saturated porous media with phase change, *Acta Mech.* **109**, 167–189.

Elwell, D. and Pointon, A. J.: 1976, *Classical Thermodynamics*, WNT-Warszawa.

Hasatani, M. and Itaya, Y.: 1992, Effect of drying process on quality control in ceramic production, In: A. S. Mujumdar (ed.), *Drying '92: Proc. 8th Int. Drying Symp. (IDS'92)*, Montreal, Quebec, Canada, 2–5 August 1992.

Kowalski, S. J.: 1987, Thermomechanics of constant drying rate period, *Arch. Mech.* **39**(3), 157–176.

Kowalski, S. J.: 1990, Thermomechanics of dried materials, *Arch. Mech.* **42**(2), 123–149.

Kowalski, S. J.: 1994, Thermomechanics of the drying process of fluid-saturated porous media, *Drying Tech.* **12**(3), 453–482.

Kowalski, S. J., Musielak, G. and Rybicki, A.: 1997, Response of dried materials on drying conditions, *Int. J. Heat Mass Transfer* **40**(5), 1217–1226.

Kowalski, S. J. and Rybicki, A.: 1994, Interaction of the thermal and moisture stresses in materials dried convectively, *Arch. Mech.* **46**(3), 251–261.

Kowalski, S. J. and Rybicki, A.: 1995, Drying induced stresses as a function of drying rate, Forschungsbericht aus dem Fachbereich Bauwesen, Beitrage zur Mechanik 66, Universität-GH Essen, pp. 237–245.

Kowalski, S. J. and Rybicki, A.: 1996, Drying stress formation induced by inhomogeneous moisture and temperature distribution, *Transport in Porous Media* **24**, 139–156.

Kowalski, S. J. and Strumillo, Cz.: 1997, Moisture transport, thermodynamics, and boundary conditions in porous materials in presence of mechanical stresses, *Chem. Engng. Sci.* **52**(7), 1141–1150.

Leitman, G.: 1966, *An Introduction to Optimal Control*, McGraw-Hill Inc., New York.

Luikov, A. V.: 1975, System of differential equations of heat and mass transfer in capillary-porous bodies (Review), *Int. J. Heat Mass Transfer* **18**, 1–14.

Rybicki, A.: 1993, Determination of drying induced stresses in a prismatic bar, *Engng. Trans.* **41**(2), 139–156.

Rybicki, A.: 1994, Problem of evolution of shrinkage stresses in dried materials, PhD Thesis, Technical University Lodz (in Polish).

Strumillo, Cz.: 1983, *Fundamentals of Drying and Its Technology*, WNT-Warszawa (in Polish).

Szarawara, J.: 1985, *Chemical Thermodynamics*, WNT-Warszawa (in Polish).

# Deformations and Stresses in Dried Wood

S. J. KOWALSKI and G. MUSIELAK
*Poznań University of Technology, Institute of Technology and Chemical Engineering, pl. Marii Skłodowskiej-Curie 2, 60-965 Poznań, Poland*

(Received: 11 August 1997; in final form: 3 December 1997)

**Abstract.** The deformations and the evolution of drying-induced stresses in wood are studied based on a model which takes into account the alteration of mechanical properties of wood in the course of drying. A two-dimensional initial-boundary value problem is solved with the help of the finite element method. An influence of wood anisotropy on the deformation and the stress distributions and evolution of maximal stresses is analysed.

**Key words:** wood drying, stress analysis, anisotropy, nonlinear model.

## 1. Introduction

Dried wood is subjected to the action of heat, moisture content and mechanical loads. The last of these is caused by drying-induced stresses, which arise when the thermal and the moisture-content fields are nonuniform. The deformation arising during drying consists generally of mechanical, shrinkage, creep, and mechano-shrinkage strains. The *mechanical strain* (elastic or elastic–plastic) is produced by the drying-induced stresses. The *shrinkage strain* arises due to the removal of moisture from wood and it is the most important of those mentioned above. The shrinkage may interact with the expansion or contraction caused by the temperature field. The *creep strain* is the time-dependent portion of the total deformation due to imposed stress history. The *mechano-shrinkage strain* is due to an interaction between changes in stress and moisture content. In tension or compression this strain is seen as a decrease or increase in the amount of shrinkage associated with change in moisture-content (Milota, 1994). The mechano-shrinkage strains seem to be proportional to the product of stress and moisture content. Such a term exists in the model considered in this paper because one assumes the material coefficients to be moisture dependent. The plastic and creep strains will be neglected in our considerations.

The main aim of this paper is to study the mechanical behaviour of dried wood based on a model which takes into account the fact that dried material changes its mechanical properties during drying. We shall use the physical relations proposed by Kowalski and Kowal (1998), established for the wetting process of wood. We assume that the hysteresis between wetting and drying is negligible and the same physical relations are valid for both of these processes. The material coefficients in these physical relations are functions of the moisture content. They were estimated for pine sapwood for the biaxial state of stress on samples cut crosswise to the wood fibres (see Kowal *et al.*, 1992; Kowal and Kowalski, 1995).

The moisture content distribution and the evolution of drying-induced stresses in wood treated as an orthotropic medium are studied using the model mentioned above and finite element methodology. The paper illustrates the convective drying of a wood bar (two-dimensional problem) in which the wood fibres run parallel to the bar axis. The influence of wood anisotropy on the stress distribution and evolution of maximal stresses is analysed.

## 2. Model Presentation

A characteristic feature of wood is the anisotropy of its physical and mechanical properties resulting from the anatomic structure of the wood. We assume wood to be orthotropic. This kind of anisotropy is characterized by three mutually perpendicular planes of symmetry, which are assigned according to the anatomic directions of wood, namely: lengthwise L; radial R; and tangential T, as it is shown in Figure 1.

The complete model of drying of deformable porous materials consists of the equations of transport phenomena (heat and mass transport equations) and the equilibrium equation together with suitable physical relations describing the material behaviour (see for instance Kowalski, 1987, 1990). For the purpose of this work we shall use simplified transport equations and focus our attention on the analysis of the mechanical behaviour of dried wood.

We use the equation of the moisture mass balance (see Kowalski, 1987) in the form

$$\varrho_0 \dot{\theta} = -w_{i,i}, \tag{1}$$

where $\theta$ is the moisture content defined as the ratio of the partial mass density of the moisture $\varrho_{\mathrm{m}}$ to the partial mass density of the dry body $\varrho_0$ and

$$w_i = -\Lambda_{\mathrm{m}ij}\mu_{,j}, \quad \Lambda_{\mathrm{m}ij} = 0 \text{ for } i \neq j, \quad i, j = \mathrm{R, T, L} \tag{2}$$

is the moisture flux, proportional to the gradient of the moisture potential $\mu_{,j}$ with $\Lambda_{\mathrm{m}ij}$ being the coefficients of moisture transport. In these equations $i$ and $j$ mean R, T and L – the aforementioned orthotropic directions of wood.

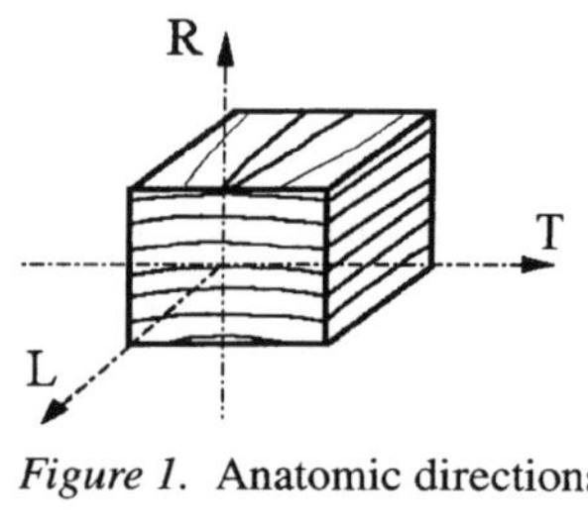


*Figure 1.* Anatomic directions of wood (L-lengthwise, R-radial, T-tangential).

Generally the moisture potential

$$\mu = \mu(\varepsilon, \vartheta, \theta) \tag{3}$$

is the function of the volumetric deformation $\varepsilon = \varepsilon_{kk}$, the relative temperature $\vartheta$ and the moisture content $\theta$. Using the chain-rule expansion for the moisture potential one gets

$$\mu_{,i} = \frac{\partial \mu}{\partial \varepsilon}\varepsilon_{,i} + \frac{\partial \mu}{\partial \vartheta}\vartheta_{,i} + \frac{\partial \mu}{\partial \theta}\theta_{,i}. \tag{4}$$

We assume small deformations for dried wood and neglect the influence of wood volume alteration on moisture transport. We assume also that the non-uniformity in the temperature distribution in wood is of minimal importance and therefore consider the influence of the temperature gradient on the moisture transport as negligible. Thus, we neglect the first two terms on the right hand side of Equation (4). Introducing the notation

$$\frac{\partial \mu}{\partial \theta} = \frac{c_\theta}{\varrho_0}, \tag{5}$$

where $c_\theta$ is the moisture content coefficient in the moisture potential function and taking into account the above assumptions, we obtain the equation of the moisture transport in the form

$$\dot{\theta} = \frac{c_\theta}{\varrho_0^2}\Lambda_{mij}\theta_{,ij}, \qquad i, j = \mathrm{R, T, L}. \tag{6}$$

The relation between the mechanical strains and the stresses has the form of a Neo-Hookean relation with material coefficients dependent on the moisture content. Since we assume wood to be orthotropic, the number of mechanical material coefficients in such a case equals 9. We introduce the engineering material constants $E_i$ (Young's moduli), $G_{ij}$ (shear moduli), $\nu_{ij}$ (Poisson's ratios) and write the physical relations for the orthotropic model as follows (see Kowalski and Kowal (1998)):

$$\begin{aligned}
\varepsilon_\mathrm{R} &= \mathring{\varepsilon}_\mathrm{R} + \frac{1}{E_\mathrm{R}}\sigma_\mathrm{R} - \frac{\nu_\mathrm{TR}}{E_\mathrm{T}}\sigma_\mathrm{T} - \frac{\nu_\mathrm{LR}}{E_\mathrm{L}}\sigma_\mathrm{L},\\
\varepsilon_\mathrm{T} &= \mathring{\varepsilon}_\mathrm{T} - \frac{\nu_\mathrm{RT}}{E_\mathrm{R}}\sigma_\mathrm{R} + \frac{1}{E_\mathrm{T}}\sigma_\mathrm{T} - \frac{\nu_\mathrm{LT}}{E_\mathrm{L}}\sigma_\mathrm{L},\\
\varepsilon_\mathrm{L} &= \mathring{\varepsilon}_\mathrm{L} - \frac{\nu_\mathrm{RL}}{E_\mathrm{R}}\sigma_\mathrm{R} - \frac{\nu_\mathrm{TL}}{E_\mathrm{T}}\sigma_\mathrm{T} + \frac{1}{E_\mathrm{L}}\sigma_\mathrm{L},\\
\varepsilon_\mathrm{RT} &= \frac{1}{2G_\mathrm{RT}}\sigma_\mathrm{RT}, \qquad \varepsilon_\mathrm{RT} = \frac{1}{2G_\mathrm{RL}}\sigma_\mathrm{RL}, \qquad \varepsilon_\mathrm{TL} = \frac{1}{2G_\mathrm{TL}}\sigma_\mathrm{TL},
\end{aligned} \tag{7}$$

where $\mathring{\varepsilon}_i$ denotes the free shrinkage strain. The material coefficients have to satisfy the following symmetry conditions:

$$\frac{\nu_\mathrm{TR}}{E_\mathrm{T}} = \frac{\nu_\mathrm{RT}}{E_\mathrm{R}}, \qquad \frac{\nu_\mathrm{RL}}{E_\mathrm{R}} = \frac{\nu_\mathrm{LR}}{E_\mathrm{L}}, \qquad \frac{\nu_\mathrm{LT}}{E_\mathrm{L}} = \frac{\nu_\mathrm{TL}}{E_\mathrm{T}}. \tag{8}$$

Both the free shrinkage strain and the material coefficients depend generally on the moisture content $\theta$ and the temperature $\vartheta$.

It is well known (see, e.g., Glijer *et al.*, 1984) that the largest shrinkage takes place in the tangential direction and the smallest in the lengthwise direction. The average shrinkage in the tangential direction is about 1.65 times greater than in the radial direction, but generally they are quantitatively comparable. The shrinkage in the lengthwise direction is about 25 times smaller than that in the tangential direction. Therefore, we shall not consider the lengthwise shrinkage in this paper. We shall also assume that the influence of radial and tangential stresses on lengthwise strains is very small ($\nu_{\mathrm{RL}} = \nu_{\mathrm{TL}} \cong 0$).

We shall consider a two-dimensional problem of the convective drying of a wood bar. So, we assume the plane state of stress. Thus, we substitute $\sigma_{\mathrm{L}} = \sigma_{\mathrm{RL}} = \sigma_{\mathrm{TL}} = 0$ in the above physical relations. Furthermore, we shall assume that the thermal strains are much smaller than the shrinkage strains and therefore assume the material coefficient to be temperature independent.

The stresses in the plane state are expressed by the relations

$$
\begin{aligned}
\sigma_{\mathrm{R}} &= \frac{E_{\mathrm{R}}}{1-\nu_{\mathrm{TR}}\nu_{\mathrm{RT}}}[(\varepsilon_{\mathrm{R}} - \mathring{\varepsilon}_{\mathrm{R}}) + \nu_{\mathrm{TR}}(\varepsilon_{\mathrm{T}} - \mathring{\varepsilon}_{\mathrm{T}})],\\
\sigma_{\mathrm{T}} &= \frac{E_{\mathrm{T}}}{1-\nu_{\mathrm{TR}}\nu_{\mathrm{RT}}}[(\varepsilon_{\mathrm{T}} - \mathring{\varepsilon}_{\mathrm{T}}) + \nu_{\mathrm{RT}}(\varepsilon_{\mathrm{R}} - \mathring{\varepsilon}_{\mathrm{R}})],\\
\sigma_{\mathrm{RT}} &= 2G_{\mathrm{RT}}\varepsilon_{\mathrm{RT}}.
\end{aligned}
\tag{9}
$$

The stresses have to fulfil the equations of balance of internal forces:

$$
\sigma_{ij,j} = 0. \tag{10}
$$

Wood changes its mechanical properties in the range from the dry state to the state of fibre saturation point. The swelling or shrinkage takes place only in this range. Therefore, the curves for free swelling or shrinkage are chosen in such a way as to express that the wood is maximally swelled in the fibre saturated point $\theta_n$ ($\cong$30%). The experimental data suggests the choice of a parabolic curve for radial free swelling strains and a trigonometrical curve for tangential swelling strains. Thus, we have

$$
\begin{aligned}
\mathring{\varepsilon}_{\mathrm{R}}(\theta) &= \mathring{\varepsilon}_{\mathrm{R}}(\theta_0) + [\mathring{\varepsilon}_{\mathrm{R}}(\theta_n) - \mathring{\varepsilon}_{\mathrm{R}}(\theta_0)](2X - X^2),\\
\mathring{\varepsilon}_{\mathrm{T}}(\theta) &= \mathring{\varepsilon}_{\mathrm{T}}(\theta_0) + [\mathring{\varepsilon}_{\mathrm{T}}(\theta_n) - \mathring{\varepsilon}_{\mathrm{T}}(\theta_0)]\tfrac{1}{2}[1 + \cos(1+X)\pi],
\end{aligned}
\tag{11}
$$

where $X = (\theta - \theta_0)/(\theta_n - \theta_0)$ is the dimensionless moisture content, $\mathring{\varepsilon}_{\mathrm{R}}(\theta_n)$ is the free swelling strain at the fibre saturation point $\theta_n$ and $\mathring{\varepsilon}_{\mathrm{R}}(\theta_0)$ is equal to the strain at the dry state $\theta_0$.

Experimental data (see Kowal *et al.*, 1992; Kowal and Kowalski, 1995) suggest a consideration of the material coefficients for wood to be the functions of moisture content. We propose the following form of these functions:

$$
\frac{1}{E_i(\theta)} = \frac{1}{E_i(\theta_0)} + B_i(\theta_0)X, \quad i = \mathrm{R}, \mathrm{T}, \tag{12}
$$

that is

$$E_i(\theta) = \frac{E_i(\theta_0)}{1 + B_i(\theta_0)E_i(\theta_0)X}, \qquad i = \mathrm{R, T}. \tag{13}$$

A similar function is proposed for the shear modulus $G_{\mathrm{RT}}$. The Poisson's ratios $\nu_{ij}$ are assumed to be constant. The coefficients $B_i(\theta_0)$ are termed as mechano-sorptive in the case of wetting processes. Similarly, we shall use them also in drying processes. They express the increase or decrease in the amount of swelling or shrinkage associated with the moisture content change with simultaneous existence of compressional or tensional stresses.

## 3. Numerical Example

A convective drying of long prismatic wooden bars having rectangular cross-sections (see Figure 2) is presented as an example.

These two bars have the same size cross-sections but they are cut out from different places of the tree trunk, so they differ from each other in the anatomic directions T-tangential and R-radial. For bar A the R-radial direction coincides with the longer side of the rectangular cross-section, and the T-tangential direction with the shorter side, and vice-versa for bar B.

Thus, a two-dimensional initial-boundary problem is solved. The initial conditions assume the moisture content of wood at the fibre saturation state:

$$\theta(x, y) = \theta_n \quad \text{for } t = 0. \tag{14}$$

The first boundary condition assumes the convective exchange of moisture vapour

$$w_i n_i = \alpha_{\mathrm{m}}(\mu - \mu_{\mathrm{a}}), \tag{15}$$

where $n_i$ is the outside normal to the boundary, $\alpha_{\mathrm{m}}$ is the coefficient of convective mass exchange and $\mu_{\mathrm{a}}$ is the moisture potential of vapour in the drying environment. The second boundary condition assumes the dried body to be free of external forces

$$\sigma_{ij} n_i = 0. \tag{16}$$

One can approximate wood as a layered body of the earlywood and latewood. The layers are parallel to the T-tangential direction and perpendicular to the R-radial one.

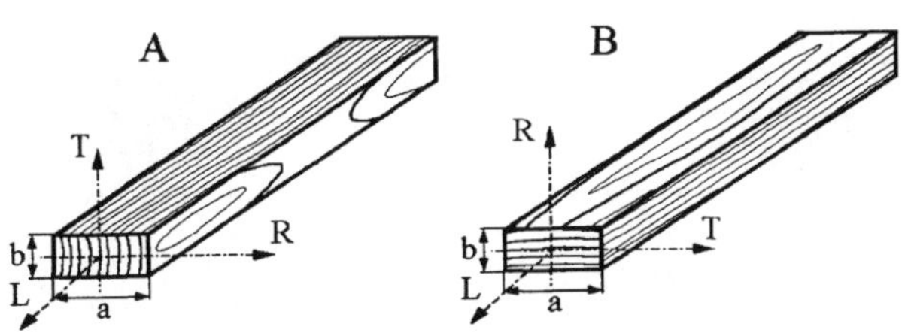


*Figure 2.* The dried bars A and B.

In such a case the coefficient of moisture transport in the T-tangential direction $\Lambda_{mT}$ is the arithmetical average of the moisture transport coefficients for the earlywood and latewood, and the coefficient in R-radial direction $\Lambda_{mR}$ is the geometrical average of these two kinds of wood. Because the arithmetical average is always greater (or equal) than the geometrical one, we have

$$\Lambda_{mT} \geqslant \Lambda_{mR}. \tag{17}$$

In our calculations we introduce $\Lambda_{mR} = 2.5 \times 10^{-9}\,\mathrm{kg\,s/m^3}$, $\Lambda_{mT} = 1.1\Lambda_{mR}$ and $\varrho_0 = 600\,\mathrm{kg/m^3}$, $c_\theta = 1.8 \times 10^4\,\mathrm{J/m^3}$, $\alpha_m = 5.0 \times 10^{-6}\,\mathrm{kg\,s/m^4}$, $\mu_a = 3\,\mathrm{J/kg}$ (see Lewis *et al.*, 1977).

Experimental tension tests in the radial direction, (see Kowalski and Kowal, (1998)), allow the estimation of Young's modulus $E_R(\theta_0) = 715\,\mathrm{MPa}$ and the mechano-sorptive coefficient $B_R(\theta_0) = 0.022\,\mathrm{MPa}$. The Poisson's ratio is assumed to be constant, $\nu_{RT} = 0.31$. Experimental compression tests in the tangential direction allow the estimation of Young's modulus $E_T(\theta_0) = 562\,\mathrm{MPa}$ and the mechano-sorptive coefficient $B_T(\theta_0) = 0.028\,\mathrm{MPa}$. The Poisson's ratio is here also assumed to be constant, $\nu_{TR} = 0.25$. Experimental biaxial tests (tension in radial direction and compression in tangential direction acting together) allow one to estimate the shear modulus $G_{RT}(\theta_0) = 245\,\mathrm{MPa}$ and the mechano-sorptive coefficient $B_{RT}(\theta_0) = 0.048\,\mathrm{MPa}$.

The method for solving the problem at hand is based on the Galerkin discretization (finite elements) method in space, and on the difference approximation of the time derivatives in the final system of ordinary differential equations (see, for instance Wait and Mitchell, 1986). The isoparametric elements consisting of eight nodes are employed. The Crank–Nicholson scheme is used for time integration.

## 4. Results

The numerical calculations were carried out for the bar of rectangular cross-section (Figure 2): $a = 0.04\,\mathrm{m}$ and $b = 0.02\,\mathrm{m}$. Due to the geometrical symmetry of the problem all graphs are presented for the quarter of the bar cross-section.

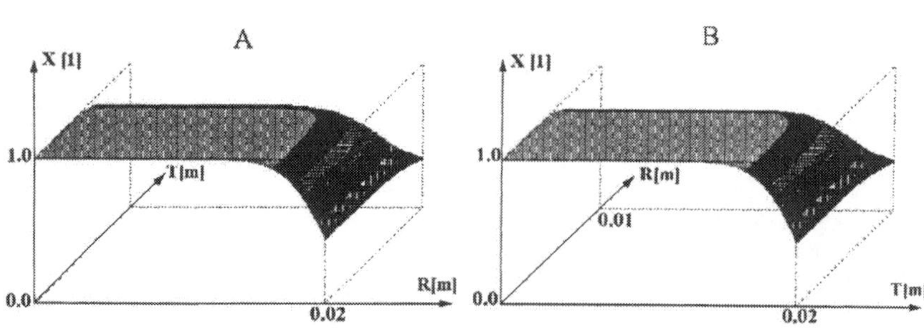


*Figure 3.* Distributions of moisture content in the cross-section of the bars A and B.

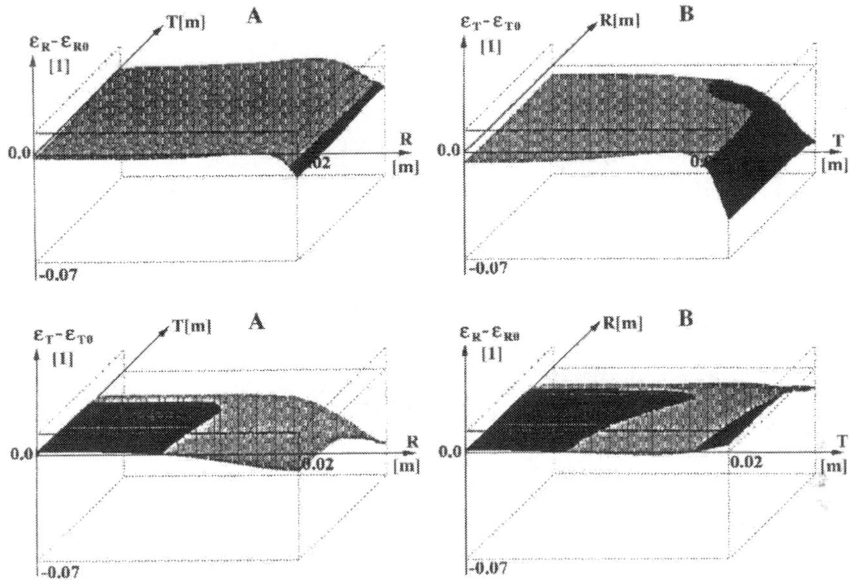


*Figure 4.* Strain distributions in the cross-section of the bars A and B.

The distribution of moisture content in the bar cross-sections after 6.5 h of the drying process is shown in Figure 3. In the middle part of the cross-section the moisture content changes slightly and is very close to the initial value. Near the boundary it decreases appreciably and tends to approach the final value. The nonuniform moisture content distribution is the reason for the nonuniform shrinkage strains inside the body and, consequently, for the generation of the internal stresses. The moisture distributions for the bars A and B differ slightly, as we have assumed a small difference in the moisture flow in the R-radial and T-tangential directions ($\Lambda_{\mathrm{mT}} = 1.1\Lambda_{\mathrm{mR}}$).

Figure 4 illustrates the difference between current and initial strains distributions $\varepsilon_i - \varepsilon_{i0}$, $i = \mathrm{R}, \mathrm{T}$ inside the bar cross-section at the same time $t = 6.5$ h. The pictures differ from each other evidently, not only quantitatively but also qualitatively. The main reason for these diversities is the anisotrophy of wood. It is seen that the shrinkage strains in the T-tangential direction are greater than those in the R-radial one. This is a natural feature of wood behaviour.

Figure 5 presents the stress distributions in the cross-sections of the two bars after 6.5 h of the drying process. The differences between them are apparent. The normal stresses $\sigma_{\mathrm{R}}$ and $\sigma_{\mathrm{T}}$ differ from each other both quantitatively and qualitatively. There would be no difference if the bars were isotropic. The shear stresses $\sigma_{\mathrm{RT}}$ differ from each other also in several bars, but only quantitatively.

In general, the normal stresses are tensional near the boundary surface and compressional in the middle part of the cross-section. The tensional values are much greater than the compressional ones.

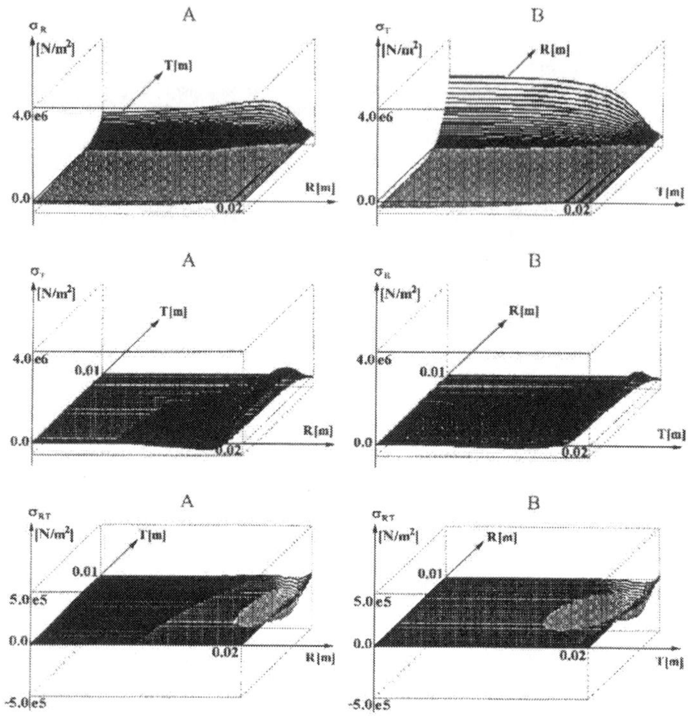


*Figure 5.* Stress distribution in the cross-section of the bars A and B.

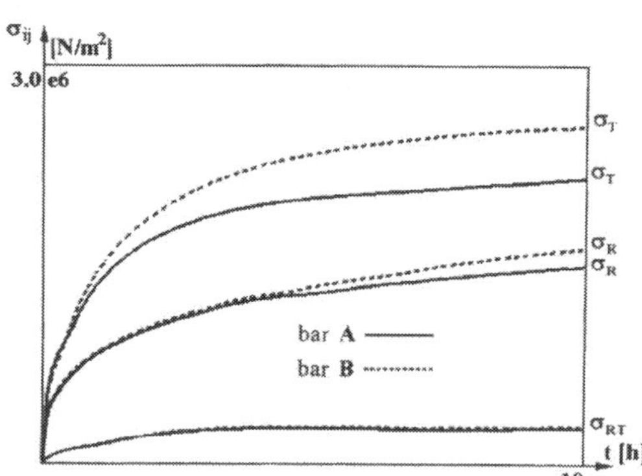


*Figure 6.* Evolution of maximal values of stresses.

If the material were isotropic, linear or nonlinear, the normal stresses would attain their maximal values at the intersection of the axis of the rectangle symmetry and the boundary surface (see Musielak, 1996; Kowalski and Musielak, 1996). This is not the case for an anisotropic material as, for example, for wood. The anisotropy, even in the case of linear material, involves the maximum of normal stress $\sigma_R$ at the boundary, but in a point found at a distance from the axis of rectangular symmetry (see the distributions of normal stresses $\sigma_R$ at the boundary surface in Figure 5).

The evolution of the maximal tensional stresses $\sigma_R$, $\sigma_T$ and shear stresses $\sigma_{RT}$ versus time are shown in Figure 6. The solid lines refer to bar A and the dashed lines to bar B. It is seen that these stresses undergo a rapid increase during the initial period of drying, and that then the increase becomes much slower. It is known (see Kowalski and Musielak, 1996) that after some time the stresses achieve their maxima and then slowly decrease. In both considered cases, the maximal stresses $\sigma_T$ reach the largest values. One may note that the structure of bar B allows production of greater stresses than the structure of bar A.

The authors also carried out numerical calculations for linearized physical relations (9). The material coefficients were considered there as constant and equal to the values suitable for dry wood. In other words the mechano-sorptive coefficients $B_i$ were taken to be zero. The results obtained are qualitatively similar to the above presented. But the values of stresses in this case are about 20 times greater than the values obtained on the basis of nonlinear physical relations.

## 5. Conclusions

Two main questions have been analysed in this paper: firstly, the influence of anisotropy on the strain and stress distributions, and secondly, the difference in evolution of maximal drying stresses for the wood structure in bars A and B. Both questions are of practical importance because one can look for the answer as to which bar, A or B, has a tendency to produce more dangerous stresses during drying. Our task here has been to present not a real example, but a general method for the analysis of stress distribution in wood during drying, taking into account the structure of wood. We have shown that the wood structure evidently influences the stress distributions and ought to be considered in the stress analysis.

We can also add that numerical analysis of stresses carried out on models which do not take into account the alteration of material properties during drying can lead to results with overstated values. This is because the Young's moduli of dry wood are much greater than that of saturated wood.

The general conclusion is that, for dried wood, the material coefficients must be moisture dependent and the model has to regard the anisotropy of wood.

## Acknowledgements

This work was carried out as a part of research project No. 3 T09C 005 12 sponsored by the Polish State Committee for Scientific Research in the year 1997.

## References

Glijer, L., Matejak, M. and Osipiuk, J.: 1984, *Theory and Technology of Drying of Wood*, PWRiL, Warsaw (in Polish).

Kowal, M. and Kowalski, S. J.: 1995, Experimental investigation of water soaked wood at uniaxial and biaxial states of stress, *Appl. Mech. Rev.* **48**, 684–688.

Kowal, M., Rybarczyk, W. and Kowalski, S. J.: 1992, Mechanical behaviour of water soaked wood at various states of stress, *Wood Sci. Technol.* **26**, 295–306.

Kowalski, S. J.: 1987, Thermomechanics of constant drying rate period, *Arch. Mech.* **39**, 157–176.

Kowalski, S. J.: 1990, Thermomechanics of dried materials, *Arch. Mech.* **42**, 123–149.

Kowalski, S. J. and Kowal, M.: 1998, Physical relations for wood at variable humidity, *Transport in Porous Media* **31**, 331–346.

Kowalski, S. J. and Musielak, G.: 1996, Modelling of mechanical properties of dried materials, In: C. Strumiłło and Z. Pakowski (eds), *Proc. 10th Int. Drying Symp. (IDS'96)*, Krakow, Poland, 30 July – 2 August 1996, vol. A, pp. 267–274.

Lewis, R. W., Strada, M. and Comini, G.: 1977, Drying-induced stresses in porous bodies, *Comp. Meth. Appl. Mech. Engng.* **11**, 1175–1184.

Milota, M. R. and Quinglin, W.: 1994, Resolution of the stress and strain components during drying of a softwood, In: V. Rudolph and R. B. Keey (eds), *Proc. 9th Int. Drying Symp. (IDS'94)*, Gold Coast, Australia, 1–4 August 1994.

Musielak, G.: 1996, Internal stresses caused by outflow of moisture and phase change inside dried material, *Drying Tech.* **14**, 289–306.

Wait, R. and Mitchell, A. R.: 1986, *Finite Element Analysis and Applications*, John Wiley & Sons, Chichester.

# Phase Transitions in Gas- and Liquid-Saturated Porous Solids

REINT DE BOER and JOACHIM BLUHM
*Institute of Mechanics, Fb-10 Bauwesen, University of Essen, D-45117 Essen, Germany*

(Received: 20 August 1997; in final form: 21 October 1997)

**Abstract.** Phase transitions in porous media consisting of a porous solid filled with liquid and gas constituents can occur, for example, due to freezing and drying processes. Although these phenomena are of certain relevance in soil mechanics and material sciences, a general thermodynamical theory is still awaited. Based on recent findings in the porous media theory, this paper is concerned with the development of thermodynamic restrictions for the constitutive relations of an elastic, incompressible porous solid, filled with an incompressible liquid and a compressible gas. The investigations show that mass conversions are related to the differences of the chemical potentials and energy transitions to the differences of temperatures. Thus, they confirm well-known results in classical thermodynamics of gases.

**Key words:** fluid-saturated elastic porous solids, first-order phase transitions, drying and freezing processes.

## 1. Introduction

Phase changes can occur in gas- and liquid-saturated porous solids under certain circumstances. For example, due to drying processes, the liquid or parts of the liquid can turn into vapour, which is a component of the gas phase, or due to a freezing process, the liquid or parts of the liquid are converted to ice, then connected with the porous solid and part of it with the same motion as the solid. These phase transitions are referred to as first-order transitions. Others, such as the conversion from superfluid liquid helium to ordinary liquid helium are called second-order transitions (see Elwell and Pointon, 1972). In this paper, exclusively first-order transitions will be treated.

Although the mentioned phase transitions, namely liquid into vapour or liquid into ice, are of considerable practical relevance, for example, in material sciences for wet porous materials in connection with drying processes, or in geomechanics for frictional materials in connection with freezing and melting processes, only few investigations have been performed within the porous media theory.

For the description of the mechanical and thermodynamical behavior of phase transitions, essentially strongly simplified constitutive models have been used in the past.

For drying processes, Luikov (1975) gives a survey of contributions mainly published by Soviet scientists in the field of the mathematical description of heat and mass transportation phenomena in rigid capillary porous solids. The article of Whitaker (1977) describes the development of the theory of drying for the same model up to 1977 and contains a broad survey of corresponding literature on this topic. Ben Nasrallah and Perre (1988) proceed also from the rigid porous solid in order to numerically calculate the temperature, moisture and pressure distribution. They obtain very clearly arranged results for these distributions in bricks for periods with constant as well as decreasing drying rates.

Whereas the articles of Luikov (1975), Whitaker (1977), as well as those of Ben Nasrallah and Perre (1988) are concerned with drying processes in rigid porous media, Lewis *et al.* (1977) transfer corresponding considerations to a deformable porous solid. Within the framework of their finite-element calculation of the temperature and moisture distribution in a porous solid they determine the deformations and stresses by means of an elastic model. Later, Lewis *et al.* (1979) extend their investigations to a viscoplastic model. However, they develop on the basic fundamental equations and the solutions for the mixture body without performing an exact analysis of the deformation and stress states incorporating all interaction effects between the individual partial bodies.

Kowalski (1987, 1990) is concerned with thermodynamics of drying materials in two extended papers. In the first article, the author presents a system of differential equations describing the distribution of strains, stresses, temperatures and moisture concentrations in a fluid-saturated capillary porous solid. The model proposed by Kowalski (1987) differs from those discussed above insofar that it considers deformations which are simultaneously taking place with the changes of the heat and moisture concentrations within the porous solid. However, the investigations are restricted to periods with constant drying rates. Kowalski (1990) extended in his second paper his investigations to periods with decreasing drying rates. Whereas in the case of constant drying rates, the evaporation surface lies in the boundary of the porous solid, this is not so in the case of decreasing drying rates. Rather, the evaporation surface can draw back from the surface at the boundary into the porous body. That means that the boundary conditions must be formulated as free boundaries (Stefan's problem). Besides, in Kowalski's (1987, 1990) papers many references are cited.

Little is known in the literature concerning the description of the mechanical and thermodynamical behavior of liquid-saturated porous solids during freezing and melting processes. In this field, fundamental investigations are obviously performed only for porous solids (concrete) in view of the pore structure and the main physical properties of the pore-water and the frost-resistance (see, e.g., Schäfer, 1964; Manns and Hartmann, 1977; Zech, 1981; Setzer *et al.*, 1994). The main results are that water in the capillary-pores behaves approximately like free water concerning the freezing-point. If, however, the water is structured in very small (micro) pores the freezing-point can be decreased considerably. Naturally, most investigations concerning freezing and melting processes are devoted to the frost-resistance of concrete as this material is very important for many constructions (see, e.g., Schäfer, 1964; Manns and Hartmann, 1977). It is stated that the frost-resistance of concrete is decisively influenced by the amount and the nature of the pores. Moreover, it is recognized that frost damages are mostly caused by a hydraulic pressure due to the extension of water in the case of freezing.

The aforementioned investigations are nearly exclusively performed without the consideration of profound mechanical or thermodynamical theories. In particular, freezing and melting processes are obviously not treated within the framework of the modern porous media theory, whereas phase transitions of water into vapour have been treated several times within this theory, although only with very simplified models.

Also, the consequences of such phase transitions within a consistent theory have not been treated in the past. For drying processes this concerns the investigation of failure criteria in order to predict cracks in drying solids. For freezing and melting processes it concerns the formulation of corresponding criteria in order to describe frost damages in such porous solids as concrete. Thus, a broad field of unresolved problems in the porous media are still awaiting search.

There have been only a few attempts to develop a thermodynamical foundation for phase transitions in the mixture and porous media theories (see, e.g., Nunziato and Walsh, 1980; Kowalski, 1987, 1990).

It seems that the first major investigations to gain criteria for phase transitions based on thermodynamical principles have been performed by de Boer (1995a, 1996) and de Boer and Kowalski (1995). In these works a ternary model with compressible and incompressible constituents have been treated. However, only the compressibility of the constituents due to the interface pressure has been taken into account. Thus, the results in the aforementioned papers are limited.

Today, a consistent macroscopic porous media theory with no mass exchange is available, see de Boer (1996, 1997) and Bluhm (1997). It is the aim of this paper to extend the thermodynamical investigations on saturated porous solids by including all supply terms in the balance equations to provide some thermodynamical background for phase changes, and in particular, to gain criteria to predict the onset of mass conversion. The results of this theory are physically and mathematically well balanced and can serve as a basis for further investigations of phase transitions, and, especially of the consequences of such conversions.

Before starting the thermodynamical investigations, the contents of this paper will be outlined. In the Preliminaries, in Section 2, the volume fraction concept, kinematics and balance equations as well as the entropy inequality, are introduced. In the sections on kinematics the new finding of a multiplicative decomposition of the deformation gradient into parts which describe the deformations of the real material and the changes of the pores will be discussed. In Section 3 restrictions will be gained for the constitutive equations. Moreover, some statements on the onset of phase conversions will be given. Finally, some conclusions will be drawn.

## 2. Preliminaries

We assume that the medium on hand consists of the following constituents:

(a) an elastic incompressible porous skeleton,
(b) an incompressible liquid and
(c) a compressible gas.

Thus, a porous medium with intercommunicating void spaces filled up with a two-component immiscible mixture of liquid and gas, called moisture, will be investigated. The special binary model, namely an elastic incompressible porous solid solely filled with an incompressible liquid or with a compressible gas are contained in the ternary model. In order to obtain this special model it is only necessary to neglect the volume fractions of the gas or of the liquid, respectively. Moreover, it is assumed that the pores, the liquid and the gas are statistically distributed over the control space, which is shaped by the solid matrix. The individual constituents will be denoted by $\varphi^\alpha$ ($\alpha = \mathrm{S}, \mathrm{L}, \mathrm{G}$, referring to the porous solid, liquid and gas).

### 2.1. VOLUME FRACTION CONCEPT

With the volume fractions (the volume fraction concept can be founded by micromechanics, see, for example, de Boer and Didwania, 1997)

$$n^\alpha = n^\alpha(\mathbf{x}, t) \tag{1}$$

at each spatial point of the control space $\mathbf{x}$ and at each time $t$. The actual volume elements $\mathrm{d}v^\alpha$ of the individual constituents are related to the volume element $\mathrm{d}v$ of the total body by

$$\mathrm{d}v^\alpha = n^\alpha \mathrm{d}v. \tag{2}$$

In the case of a saturated porous solid, the volume fractions must fulfill the saturation condition

$$n^\mathrm{S} + n^\mathrm{L} + n^\mathrm{G} = 1 \tag{3}$$

independent of the special material behavior.

According to an averaging procedure, each constituent is characterized by the volume average mass density $\rho^{\alpha\mathrm{R}}(\mathbf{x}, t)$ (macroscopic real mass density which is equal to the real mass density if the latter is constant close to the spatial point $\mathbf{x}$) and the bulk mass density $\rho^\alpha(\mathbf{x}, t)$. They are related to each other by (see, e.g., de Boer and Didwania, 1997)

$$\rho^\alpha = n^\alpha \rho^{\alpha\mathrm{R}}. \tag{4}$$

It should be mentioned that the volume fraction concept has the effect of 'smearing' the individual constituents over the control space. Thus, partial substitute constituents with smooth quantities and reduced densities are created that can be treated by continuum mechanics methods.

## 2.2. KINEMATICS

Next, the motions of the constituents are treated. These are understood as time dependent sequences of placements. The macroscopic formulation implies a model of superimposed continua where at any time $t$ each spatial point $\mathbf{x}$ is simultaneously occupied by particles $X_\alpha$ with the position vectors $\mathbf{X}_\alpha$ of the constituents $\varphi^\alpha$. Thus, each constituent is assigned its own independent motion function $\boldsymbol{\chi}_\alpha$ (mixture theory):

$$\mathbf{x} = \boldsymbol{\chi}_\alpha(\mathbf{X}_\alpha, t). \tag{5}$$

However, in the porous media theory the motion is constrained by the saturation condition (3) which can be easily proved (see, e.g., de Boer, 1996). In the following the velocity $\mathbf{x}'_\alpha$, the acceleration $\mathbf{x}''_\alpha$, and the deformation gradient $\mathbf{F}_\alpha$ will be introduced:

$$\mathbf{x}'_\alpha = \frac{\partial \boldsymbol{\chi}_\alpha}{\partial t}, \qquad \mathbf{x}''_\alpha = \frac{\partial^2 \boldsymbol{\chi}_\alpha}{\partial t^2}, \qquad \mathbf{F}_\alpha = \operatorname{Grad}_\alpha \boldsymbol{\chi}_\alpha, \tag{6}$$

where $\operatorname{Grad}_\alpha$ is the gradient with respect to $\mathbf{X}_\alpha$. It should be emphasized that the motion function $\boldsymbol{\chi}_\alpha$, the velocity $\mathbf{x}'_\alpha$ and the acceleration $\mathbf{x}''_\alpha$ as well as the deformation gradient $\mathbf{F}_\alpha$ are defined for a material particle which does not change its mass during the motion. However, due to phase transitions the mass of the particles may vary. The consequences of these phase changings on kinematic quantities are not considered in the definition for the velocity and acceleration in $(6)_{1,2}$ and on kinematic quantities derived from (6) as, for example, the rate of the volume strain. In the sequel we will consider the influence of phase transitions by separate investigations. With $(6)_1$ the material time derivative of an arbitrary scalar field $\Gamma(\mathbf{x}, t)$ for the proposed model is defined as follows:

$$\Gamma'_\alpha = \frac{\partial \Gamma}{\partial t} + \operatorname{grad} \Gamma \cdot \mathbf{x}'_\alpha, \tag{7}$$

where grad is the gradient with respect to $\mathbf{x}$. By using the inverse function of (5) the dependence of $\mathbf{x}'_\alpha$ and $\mathbf{x}''_\alpha$ on the position vectors $\mathbf{X}_\alpha$ can be replaced by the variable $\mathbf{x}$. Then, the velocity $\mathbf{x}'_\alpha$ can be expressed by

$$\mathbf{x}'_\alpha = \mathbf{x}'_\alpha(\mathbf{x}, t). \tag{8}$$

From the velocity field $\mathbf{x}'_\alpha$ the velocity gradient with respect to the actual placement can be calculated:

$$\mathbf{L}_\alpha = \operatorname{grad} \mathbf{x}'_\alpha = (\mathbf{F}_\alpha)'_\alpha\, \mathbf{F}_\alpha^{-1}. \tag{9}$$

The gradient $\mathbf{L}_\alpha$ can additively be decomposed into symmetric and skew-symmetric parts $\mathbf{D}_\alpha$ and $\mathbf{W}_\alpha$, respectively,

$$\begin{aligned} &\mathbf{L}_\alpha = \mathbf{D}_\alpha + \mathbf{W}_\alpha, \\ &\mathbf{D}_\alpha = \tfrac{1}{2}\,(\mathbf{L}_\alpha + \mathbf{L}_\alpha^{\mathrm{T}}), \qquad \mathbf{W}_\alpha = \tfrac{1}{2}(\mathbf{L}_\alpha - \mathbf{L}_\alpha^{\mathrm{T}}). \end{aligned} \tag{10}$$

With the deformation gradient $\mathbf{F}_\alpha$ the right Cauchy–Green tensor can be introduced:

$$\mathbf{C}_\alpha = \mathbf{F}_\alpha^{\mathrm{T}} \mathbf{F}_\alpha. \tag{11}$$

For some purposes it is advisable to multiplicatively decompose the deformation gradient $\mathbf{F}_\alpha$ and the right Cauchy–Green tensor $\mathbf{C}_\alpha$ into spherical and volume-preserving parts. The decomposition of the deformation gradient leads to

$$\mathbf{F}_\alpha = \overset{\circ}{\mathbf{F}}_\alpha\, \bar{\mathbf{F}}_\alpha, \tag{12}$$

where

$$\overset{\circ}{\mathbf{F}}_\alpha = (J_\alpha)^{1/3}\, \mathbf{I}, \quad J_\alpha = \det \mathbf{F}_\alpha \tag{13}$$

is the spherical part of $\mathbf{F}_\alpha$. Moreover, it is recognized that

$$\det \bar{\mathbf{F}}_\alpha = 1 \tag{14}$$

and that $\bar{\mathbf{F}}_\alpha$ is the volume-preserving part of $\mathbf{F}_\alpha$. With the multiplicative decomposition (12) in connection with (13) the right Cauchy–Green tensor can be reformulated as

$$\mathbf{C}_\alpha = (J_\alpha)^{2/3}\, \bar{\mathbf{C}}_\alpha, \quad \bar{\mathbf{C}}_\alpha = \bar{\mathbf{F}}_\alpha^{\mathrm{T}}\, \bar{\mathbf{F}}_\alpha. \tag{15}$$

In order to describe properties of the microscale by macromechanic quantities it is advisable to mulitiplicatively decompose the deformation gradient $\mathbf{F}_\alpha$ into a tensor $\mathbf{F}_{\alpha\mathrm{R}}$ which describes the deformation of the real material and into a tensor $\mathbf{F}_{\alpha\mathrm{N}}$ to describe the changes of the pores in size and shape:

$$\mathbf{F}_\alpha = \mathbf{F}_{\alpha\mathrm{N}}\, \mathbf{F}_{\alpha\mathrm{R}}. \tag{16}$$

The tensors $\mathbf{F}_{\alpha\mathrm{N}}$ and $\mathbf{F}_{\alpha\mathrm{R}}$ are in general not integrable and they are to be understood as local mappings of tangent (vector) spaces at each material point of the body. It is not possible to gain the tensors $\mathbf{F}_{\alpha\mathrm{N}}$ and $\mathbf{F}_{\alpha\mathrm{R}}$ from a motion function at the microlevel, because in the porous media theory no balance equations at the microlevel are introduced. Therefore, in the sequel constitutive equations for $\mathbf{F}_{\alpha\mathrm{R}}$ or other quantities derived from $\mathbf{F}_{\alpha\mathrm{R}}$, are formulated. Concerning the foundation of such a decomposition the reader is referred to Bluhm and de Boer (1997), de Boer (1996) and Bluhm and de Boer (1996). For further investigations some additonal kinematic quantities must be introduced. Similar to (12)–(14) the deformation tensors $\mathbf{F}_{\alpha\mathrm{N}}$ and $\mathbf{F}_{\alpha\mathrm{R}}$ are decomposed into volume-preserving and spherical parts:

$$\mathbf{F}_{\alpha\mathrm{N}} = \overset{\circ}{\mathbf{F}}_{\alpha\mathrm{N}}\, \bar{\mathbf{F}}_{\alpha\mathrm{N}}, \qquad \mathbf{F}_{\alpha\mathrm{R}} = \overset{\circ}{\mathbf{F}}_{\alpha\mathrm{R}}\, \bar{\mathbf{F}}_{\alpha\mathrm{R}}, \tag{17}$$

where

$$\begin{aligned}
&\overset{\circ}{\mathbf{F}}_{\alpha \mathrm{N}} = (J_{\alpha \mathrm{N}})^{1/3}\mathbf{I}, && \overset{\circ}{\mathbf{F}}_{\alpha \mathrm{R}} = (J_{\alpha \mathrm{R}})^{1/3}\,\mathbf{I},\\
&J_{\alpha \mathrm{N}} = \det \mathbf{F}_{\alpha \mathrm{N}}, && \bar{J}_{\alpha \mathrm{N}} = \det \bar{\mathbf{F}}_{\alpha \mathrm{N}} = 1,\\
&J_{\alpha \mathrm{R}} = \det \mathbf{F}_{\alpha \mathrm{R}}, && \bar{J}_{\alpha \mathrm{R}} = \det \bar{\mathbf{F}}_{\alpha \mathrm{R}} = 1.
\end{aligned} \tag{18}$$

In addition, the right Cauchy–Green tensor $\mathbf{C}_{\alpha \mathrm{R}}$ is introduced:

$$\mathbf{C}_{\alpha \mathrm{R}} = \mathbf{F}^{\mathrm{T}}_{\alpha \mathrm{R}}\,\mathbf{F}_{\alpha \mathrm{R}} = (J_{\alpha \mathrm{R}})^{2/3}\,\bar{\mathbf{C}}_{\alpha \mathrm{R}}, \qquad \bar{\mathbf{C}}_{\alpha \mathrm{R}} = \bar{\mathbf{F}}^{\mathrm{T}}_{\alpha \mathrm{R}}\,\bar{\mathbf{F}}_{\alpha \mathrm{R}}. \tag{19}$$

Moreover, some material time derivatives of such quantities describing effects at the microscale are necessary. The rates of $J_{\alpha \mathrm{R}}$ and $\bar{\mathbf{C}}_{\alpha \mathrm{R}}$ are given by

$$(J_{\alpha \mathrm{R}})'_{\alpha} = J_{\alpha \mathrm{R}}\,(\mathbf{D}_{\alpha \mathrm{R}} \cdot \mathbf{I}), \qquad (\bar{\mathbf{C}}_{\alpha \mathrm{R}})'_{\alpha} = 2\,\bar{\mathbf{F}}^{\mathrm{T}}_{\alpha \mathrm{R}}\,\mathbf{D}^{\mathrm{D}}_{\alpha \mathrm{R}}\,\bar{\mathbf{F}}_{\alpha \mathrm{R}} \tag{20}$$

where $\mathbf{D}_{\alpha \mathrm{R}}$ is the symmetric part of

$$\mathbf{L}_{\alpha \mathrm{R}} = (\mathbf{F}_{\alpha \mathrm{R}})'_{\alpha}\,\mathbf{F}^{-1}_{\alpha \mathrm{R}} \tag{21}$$

and where the index $(\ldots)^{\mathrm{D}}$ marks the deviatoric part of $\mathbf{D}_{\alpha \mathrm{R}}$.

### 2.3. BALANCE EQUATIONS AND THE ENTROPY INEQUALITY

In the porous media theory the balance equations – balance of mass, balance of momentum and moment of momentum as well as balance of energy – must be established for each constituent $\varphi^{\alpha}$ in consideration of all interactions and external agencies. Moreover, the sum of the partial balances of the $\kappa$ constituents $\varphi^{\alpha}$ must result in the corresponding balance laws of the mixture body. In particular, the latter point has been extensively discussed by de Boer (1995b) and Bluhm (1997). It was suggested to make some modifications in summing up the balance equations for the partial constituents in order to avoid the large formalism of kinematics and other obscure statements based on the introduction of the so-called barycentric velocity (the article of de Boer (1996), contains some misprints which have been corrected in this paper).

*Balance of mass:*
With respect to the volume element $\mathrm{d}v$, the balance equation of mass for each $\varphi^{\alpha}$ is the local statement

$$(\rho^{\alpha})'_{\alpha} + \rho^{\alpha}\operatorname{div}\mathbf{x}'_{\alpha} = \hat{\rho}^{\alpha}, \tag{22}$$

where

$$\hat{\rho}^{\alpha} = \hat{\rho}^{\alpha}(\mathbf{x}, t) \tag{23}$$

represents the mass supplied to $\varphi^\alpha$ from all other constituents that occupy $\mathbf{x}$ at time $t$. The summation of the statements (22) over all $\kappa$ constituents yields the constraint for the mass supplies

$$\sum_{\alpha=1}^{\kappa} \hat{\rho}^\alpha = 0, \tag{24}$$

that is, the sum of the local mass supply terms of all $\kappa$ constituents vanishes.

*Balance of momentum:*
In the local statement, the axiom of the balance of linear momentum for a constituent $\varphi^\alpha$ is given by

$$\operatorname{div} \mathbf{T}^\alpha + \rho^\alpha(\mathbf{b}^\alpha - \mathbf{x}''_\alpha) = \hat{\rho}^\alpha \mathbf{x}'_\alpha - \hat{\mathbf{p}}^\alpha, \tag{25}$$

where $\mathbf{T}^\alpha$ represents the partial Cauchy stress tensor of the constituent $\varphi^\alpha$ and $\mathbf{b}^\alpha$ is the external acceleration. The quantity

$$\hat{\mathbf{p}}^\alpha = \hat{\mathbf{p}}^\alpha(\mathbf{x}, t) \tag{26}$$

denotes the linear momentum supply per unit volume which can be interpreted as an interaction force and which is constrained by

$$\sum_{\alpha=1}^{\kappa} \hat{\mathbf{p}}^\alpha = \mathbf{0}, \tag{27}$$

that is, the sum of all interaction forces of all constituents $\varphi^\alpha$ vanishes.

*Balance of moment of momentum:*
The postulate (see de Boer *et al.*, 1991)

$$\mathbf{I} \times \mathbf{T}^\alpha = \hat{\mathbf{m}}^\alpha \tag{28}$$

is the local statement of the balance of moment of momentum for a constituent $\varphi^\alpha$, where the cross product, connecting the identity tensor $\mathbf{I}$ and Cauchy's stress tensor $\mathbf{T}^\alpha$ is explained in de Boer (1982) and where

$$\hat{\mathbf{m}}^\alpha = \mathbf{m}^\alpha(\mathbf{x}, t) \tag{29}$$

represents the local moment of momentum supply per unit volume, which is constrained by

$$\sum_{\alpha=1}^{\kappa} \hat{\mathbf{m}}^\alpha = \mathbf{0}. \tag{30}$$

If

$$\hat{\mathbf{m}}^\alpha = \mathbf{0} \tag{31}$$

it follows from (28) that Cauchy's stress tensor is symmetric:

$$\mathbf{T}^\alpha = (\mathbf{T}^\alpha)^{\mathrm{T}}. \tag{32}$$

In all other cases $\mathbf{T}^\alpha$ is not symmetric. However, the requirement that in (28) the right-hand side is different from zero implies in continuum mechanics the introduction of additional rotational degrees of freedom for the material point.

*Balance of energy:*
The axiom of the energy balance of the constituent $\varphi^\alpha$ is introduced in the following local form:

$$\rho^\alpha(\varepsilon^\alpha)'_\alpha - \mathbf{T}^\alpha \cdot \mathbf{L}_\alpha - \rho^\alpha r^\alpha + \operatorname{div} \mathbf{q}^\alpha = \hat{e}^\alpha - \hat{\mathbf{p}}^\alpha \cdot \mathbf{x}'_\alpha - \hat{\rho}^\alpha(\varepsilon^\alpha - \tfrac{1}{2}\mathbf{x}'_\alpha \cdot \mathbf{x}'_\alpha). \tag{33}$$

Herein, $\varepsilon^\alpha = \varepsilon^\alpha(\mathbf{x}, t)$ stands for the partial specific internal energy of $\varphi^\alpha$ per mass element $\rho^\alpha \mathrm{d}v$, $r^\alpha = r^\alpha(\mathbf{x}, t)$ is the partial external heat supply and $\mathbf{q}^\alpha = \mathbf{q}^\alpha(\mathbf{x}, t)$ is the influx of energy. The quantity

$$\hat{e}^\alpha = \hat{e}^\alpha(\mathbf{x}, t) \tag{34}$$

represents the local energy supply $\varphi^\alpha$ from the other $\kappa - 1$ constituents and is constrained by

$$\sum_{\alpha=1}^{\kappa} \hat{e}^\alpha = 0. \tag{35}$$

As aforementioned, in order to derive the constraints on the supply terms (24), (27), (30), and (34) some convenient summations must be introduced.

*Entropy inequality:*
Whereas the balance equations for a mixture were already formulated at the end of the 1950s and the beginning of the 1960s, the entropy inequality was developed first in the second half of the 1960s after extensive discussions. For a long time, it was unknown in the literature as to whether an entropy inequality had to be allowed for each constituent $\varphi^\alpha$ or whether one had to use only one entropy inequality for the mixture as a whole. Today, the opinion is that the postulate of separate entropy inequalities must be regarded as a sufficient but too restrictive requirement. The postulate of one common entropy inequality is simultaneously a necessary and a sufficient condition for the existence of dissipation mechanisms within the mixture. Proceeding from so-called metaphysical principles, Truesdell (1968) derived an inequality for the mixture as the sum of the entropy statements of the $\kappa$ constituents. In de Boer and Ehlers (1986), the following local statement of the second law of thermodynamics was derived:

$$\sum_{\alpha=1}^{\kappa} \left[ \rho^\alpha(\eta^\alpha)'_\alpha + \hat{\rho}^\alpha \eta^\alpha - \frac{1}{\Theta^\alpha} \rho^\alpha r^\alpha + \operatorname{div}\left(\frac{\mathbf{q}^\alpha}{\Theta^\alpha}\right) \right] \geqslant 0, \tag{36}$$

where $\eta^\alpha$ is the specific entropy and $\Theta^\alpha$ the local absolute temperature of the constituent $\varphi^\alpha$. Furthermore, the entropy inequality (36) can be reformulated using the partial energy balance (33). This reformulation results in

$$\sum_{\alpha=1}^{\kappa} \frac{1}{\Theta^\alpha} \Big\{ -\rho^\alpha[(\varepsilon^\alpha)'_\alpha - \Theta^\alpha(\eta^\alpha)'_\alpha] - \hat{\mathbf{p}}^\alpha \cdot \mathbf{x}'_\alpha + \mathbf{T}^\alpha \cdot \mathbf{L}_\alpha - \\ - \hat{\rho}^\alpha \left(\varepsilon^\alpha - \Theta^\alpha\eta^\alpha - \frac{1}{2}\mathbf{x}'_\alpha \cdot \mathbf{x}'_\alpha\right) - \frac{1}{\Theta^\alpha}\mathbf{q}^\alpha \cdot \operatorname{grad}\Theta^\alpha + \hat{e}^\alpha \Big\} \geqslant 0. \tag{37}$$

Using the free Helmholtz energy function

$$\psi^\alpha = \varepsilon^\alpha - \Theta^\alpha\eta^\alpha \tag{38}$$

the inequality (37) reads

$$\sum_{\alpha=1}^{\kappa} \frac{1}{\Theta^\alpha} \Big\{ -\rho^\alpha[(\psi^\alpha)'_\alpha + (\Theta^\alpha)'_\alpha\eta^\alpha] - \hat{\mathbf{p}}^\alpha \cdot \mathbf{x}'_\alpha + \mathbf{T}^\alpha \cdot \mathbf{L}_\alpha - \\ - \hat{\rho}^\alpha \left(\psi^\alpha - \frac{1}{2}\mathbf{x}'_\alpha \cdot \mathbf{x}'_\alpha\right) - \frac{1}{\Theta^\alpha}\mathbf{q}^\alpha \cdot \operatorname{grad}\Theta^\alpha + \hat{e}^\alpha \Big\} \geqslant 0\,. \tag{39}$$

The entropy inequality (39) will serve as a basis in the constitutive theory (in the next section) in order to gain valuable restrictions for the constitutive equations, in particular, for the description of the onset of phase transitions.

## 3. Constitutive Theory

It is easy to recognize from the preceding developed field equations that not all unknown mechanical and thermodynamical fields can be determined by the volume fraction concept, kinematics, and balance equations, if the external acceleration $\mathbf{b}^\alpha$ and the external heat supply $r^\alpha$ are known. Therefore, constitutive equations are needed for the unknown fields that connect these quantities with motion and temperature functions and derivations of these functions. This means that constitutive equations must be formulated for partial quantities and for several coupling mechanisms between the different constituents. Moreover, as pointed out by de Boer (1996) and Bluhm (1997), for a compressible porous medium consisting of $\kappa$ compressible constituents, $\kappa - 1$ additional constitutive relations for such quantities are necessary which describe properties of micromechanics and, in particular, to achieve closure.

The development of constitutive equations for saturated porous media is based on the same principles that govern the derivation of constitutive equations in classical continuum mechanics of one-component materials. These are: *determinism, equipresence, local action, material objectivity* and *dissipation*.

Guided by these principles, constitutive equations for a saturated porous medium consisting of

- an elastic incompressible porous solid,
- an incompressible liquid and
- a compressible gas

will be derived.

Before going into the details, the entropy inequality (39) must be reformulated for the ternary model under study, in order to be able to evaluate the dissipation principle. As has been discussed, for example, by de Boer (1996), the saturation condition (3) is recognized as a constraint. Therefore, the saturation condition must (in the rate formulation provided with a Lagrange multiplier) be considered in the entropy inequality (39). Moreover, there are two additional constraints provided with Lagrange multipliers which must be added to the entropy inequality, namely the incompressibility conditions of the solid and liquid constituents. These are:

$$\rho^{\mathrm{SR}} = \text{const.}, \qquad \rho^{\mathrm{LR}} = \text{const.} \tag{40}$$

At first, the saturation condition (3) will be formulated in the rates:

$$(n^{\mathrm{S}})'_{\mathrm{S}} + (n^{\mathrm{L}})'_{\mathrm{S}} + (n^{\mathrm{G}})'_{\mathrm{S}} = 0 \tag{41}$$

or

$$(n^{\mathrm{S}})'_{\mathrm{S}} + (n^{\mathrm{L}})'_{\mathrm{L}} + (n^{\mathrm{G}})'_{\mathrm{G}} - \operatorname{grad} n^{\mathrm{L}} \cdot (\mathbf{x}'_{\mathrm{L}} - \mathbf{x}'_{\mathrm{S}}) - \operatorname{grad} n^{\mathrm{G}} \cdot (\mathbf{x}'_{\mathrm{G}} - \mathbf{x}'_{\mathrm{S}}) = 0. \tag{42}$$

From (4) the derivations of the volume fractions are obtained in dependence of the rates of the partial and real densities:

$$(n^{\alpha})'_{\alpha} = \frac{1}{\rho^{\alpha\mathrm{R}}}\,(\rho^{\alpha})'_{\alpha} - n^{\alpha}\,\frac{1}{\rho^{\alpha\mathrm{R}}}\,(\rho^{\alpha\mathrm{R}})'_{\alpha}. \tag{43}$$

With the relations (43) and (18) the constraint (saturation condition) (42) turns out to be:

$$\begin{aligned} &-\frac{1}{\rho^{\mathrm{SR}}}\,(\rho^{\mathrm{S}})'_{\mathrm{S}} - \frac{1}{\rho^{\mathrm{LR}}}\,(\rho^{\mathrm{L}})'_{\mathrm{L}} - \frac{1}{\rho^{\mathrm{GR}}}\,(\rho^{\mathrm{G}})'_{\mathrm{G}} + \\ &+\frac{n^{\mathrm{S}}}{\rho^{\mathrm{SR}}}\,(\rho^{\mathrm{SR}})'_{\mathrm{S}} + \frac{n^{\mathrm{L}}}{\rho^{\mathrm{LR}}}\,(\rho^{\mathrm{LR}})'_{\mathrm{L}} + \frac{n^{\mathrm{G}}}{\rho^{\mathrm{GR}}}\,(\rho^{\mathrm{GR}})'_{\mathrm{G}} + \\ &+\operatorname{grad} n^{\mathrm{L}} \cdot (\mathbf{x}'_{\mathrm{L}} - \mathbf{x}'_{\mathrm{S}}) + \operatorname{grad} n^{\mathrm{G}} \cdot (\mathbf{x}'_{\mathrm{G}} - \mathbf{x}'_{\mathrm{S}}) = 0. \end{aligned} \tag{44}$$

Combined with a Lagrange multiplier $\lambda$ the constraint (44) must be added to the entropy inequality (39). Moreover, the incompressibility conditions (40) in rate formulations must also be considered as constraints in the entropy inequality (39). The corresponding Lagrange multipliers will be denoted by $\kappa^{\mathrm{SR}}$ and $\kappa^{\mathrm{LR}}$. However, at first, the entropy inequality (39) for the mixture must be reformulated as the kinematic expression $\mathbf{D}_\alpha \cdot \mathbf{I}$ contained in the stress power $\mathbf{T}^\alpha \cdot \mathbf{L}_\alpha$ in (39) is influenced by the mutual mass exchanges (see, e.g., de Boer and Kowalski, 1995) as can be seen in the mass balance equation (22) which can be written as

$$(\rho^\alpha)'_\alpha + \rho^\alpha(\mathbf{D}_\alpha \cdot \mathbf{I}) = \hat{\rho}^\alpha \tag{45}$$

or

$$\mathbf{D}_\alpha \cdot \mathbf{I} = \mathbf{L}_\alpha \cdot \mathbf{I} = -\frac{1}{\rho^\alpha}(\rho^\alpha)'_\alpha + \frac{1}{\rho^\alpha}\hat{\rho}^\alpha . \tag{46}$$

The incorporation of (46) into the entropy inequality (39) leads to

$$\sum_{\alpha=1}^{\kappa} \frac{1}{\Theta^\alpha}\Big\{ -\rho^\alpha[(\psi^\alpha)'_\alpha + (\Theta^\alpha)'_\alpha \eta^\alpha] - \hat{\mathbf{p}}^\alpha \cdot \mathbf{x}'_\alpha - \\ - \hat{\rho}^\alpha \mu^\alpha + \mathbf{T}^{\alpha \mathrm{D}} \cdot \mathbf{L}^{\mathrm{D}}_\alpha - \frac{1}{3}(\mathbf{T}^\alpha \cdot \mathbf{I})\frac{1}{\rho^\alpha}(\rho^\alpha)'_\alpha - \\ - \frac{1}{\Theta^\alpha}\mathbf{q}^\alpha \cdot \operatorname{grad}\Theta^\alpha + \hat{e}^\alpha \Big\} \geqslant 0, \tag{47}$$

where the chemical potential $\mu^\alpha$ (see Elwell and Pointon, 1972)

$$\mu^\alpha = \psi^\alpha - \frac{1}{3}(\mathbf{T}^\alpha \cdot \mathbf{I})\frac{1}{\rho^\alpha} - \frac{1}{2}\mathbf{x}'_\alpha \cdot \mathbf{x}'_\alpha \tag{48}$$

have been used and where in the stress power the index $(\ldots)^{\mathrm{D}}$ denotes the deviatoric part of corresponding tensors. The definition (48) for the chemical potential differs from that in the literature (see Elwell and Pointon, 1972) in view of the quadratic terms of the velocities.

In the sequel the entropy inequality for the mixture (47) will be rearranged for the porous medium under study, namely the ternary model with incompressible solid and liquid as well as compressible gas constituents. That means that the saturation condition (44) provided with a Lagrangian multiplier must be added to the entropy inequality (47) along with the incompressibility conditions for the solid and liquid phases. This procedure leads to

$$\frac{1}{\Theta^{\mathrm{S}}}\Big\{ -\rho^{\mathrm{S}}[(\psi^{\mathrm{S}})'_{\mathrm{S}} + (\Theta^{\mathrm{S}})'_{\mathrm{S}}\eta^{\mathrm{S}}] - \hat{\mathbf{p}}^{\mathrm{S}} \cdot \mathbf{x}'_{\mathrm{S}} - \hat{\rho}^{\mathrm{S}}\mu^{\mathrm{S}} + \\ + \mathbf{T}^{\mathrm{SD}} \cdot \mathbf{L}^{\mathrm{D}}_{\mathrm{S}} - \left[\frac{1}{3}(\mathbf{T}^{\mathrm{S}} \cdot \mathbf{I})\frac{1}{\rho^{\mathrm{S}}} + \lambda\frac{\Theta^{\mathrm{S}}}{\rho^{\mathrm{SR}}}\right](\rho^{\mathrm{S}})'_{\mathrm{S}} + \\ + n^{\mathrm{S}}\Theta^{\mathrm{S}}\left(\lambda\frac{1}{\rho^{\mathrm{SR}}} + \kappa^{\mathrm{SR}}\right)(\rho^{\mathrm{SR}})'_{\mathrm{S}} - \frac{1}{\Theta^{\mathrm{S}}}\mathbf{q}^{\mathrm{S}} \cdot \operatorname{grad}\Theta^{\mathrm{S}} + \hat{e}^{\mathrm{S}}\Big\} + \\ + \frac{1}{\Theta^{\mathrm{L}}}\Big\{ -\rho^{\mathrm{L}}[(\psi^{\mathrm{L}})'_{\mathrm{L}} + (\Theta^{\mathrm{L}})'_{\mathrm{L}}\eta^{\mathrm{L}}] - \hat{\mathbf{p}}^{\mathrm{L}} \cdot \mathbf{x}'_{\mathrm{L}} - \hat{\rho}^{\mathrm{L}}\mu^{\mathrm{L}} +$$

$$
\begin{aligned}
&+ \mathbf{T}^{\mathrm{LD}} \cdot \mathbf{L}_{\mathrm{L}}^{\mathrm{D}} - \left[ \frac{1}{3} (\mathbf{T}^{\mathrm{L}} \cdot \mathbf{I}) \frac{1}{\rho^{\mathrm{L}}} + \lambda \frac{\Theta^{\mathrm{L}}}{\rho^{\mathrm{LR}}} \right] (\rho^{\mathrm{S}})'_{\mathrm{L}} + \\
&+ n^{\mathrm{L}} \Theta^{\mathrm{L}} \left( \lambda \frac{1}{\rho^{\mathrm{LR}}} + \kappa^{\mathrm{LR}} \right) (\rho^{\mathrm{LR}})'_{\mathrm{L}} - \frac{1}{\Theta^{\mathrm{L}}} \mathbf{q}^{\mathrm{L}} \cdot \operatorname{grad} \Theta^{\mathrm{L}} + \\
&+ \lambda \Theta^{\mathrm{L}} \operatorname{grad} n^{\mathrm{L}} \cdot (\mathbf{x}'_{\mathrm{L}} - \mathbf{x}'_{\mathrm{S}}) + \hat{e}^{\mathrm{L}} \Big\} + \\
&+ \frac{1}{\Theta^{\mathrm{G}}} \Big\{ - \rho^{\mathrm{G}} [(\psi^{\mathrm{G}})'_{\mathrm{G}} + (\Theta^{\mathrm{G}})'_{\mathrm{G}} \eta^{\mathrm{G}}] - \hat{\mathbf{p}}^{\mathrm{G}} \cdot \mathbf{x}'_{\mathrm{G}} - \hat{\rho}^{\mathrm{G}} \mu^{\mathrm{G}} + \\
&+ \mathbf{T}^{\mathrm{GD}} \cdot \mathbf{L}_{\mathrm{G}}^{\mathrm{D}} - \left[ \frac{1}{3} (\mathbf{T}^{\mathrm{G}} \cdot \mathbf{I}) \frac{1}{\rho^{\mathrm{G}}} + \lambda \frac{\Theta^{\mathrm{G}}}{\rho^{\mathrm{GR}}} \right] (\rho^{\mathrm{S}})'_{\mathrm{G}} + \\
&+ n^{\mathrm{G}} \Theta^{\mathrm{G}} \left( \lambda \frac{1}{\rho^{\mathrm{GR}}} + \kappa^{\mathrm{GR}} \right) (\rho^{\mathrm{GR}})'_{\mathrm{G}} - \frac{1}{\Theta^{\mathrm{G}}} \mathbf{q}^{\mathrm{G}} \cdot \operatorname{grad} \Theta^{\mathrm{G}} + \\
&+ \lambda \Theta^{\mathrm{G}} \operatorname{grad} n^{\mathrm{G}} \cdot (\mathbf{x}'_{\mathrm{G}} - \mathbf{x}'_{\mathrm{S}}) + \hat{e}^{\mathrm{G}} \Big\} \geqslant 0.
\end{aligned} \tag{49}
$$

In the following inequality, (49) will be reformulated in order to bring the inequality in a convenient form which makes it easy to compare (49) with results in the literature (see de Boer, 1996):

$$
\begin{aligned}
&- \frac{\rho^{\mathrm{S}}}{\Theta^{\mathrm{S}}} (\psi^{\mathrm{S}})'_{\mathrm{S}} - \frac{\rho^{\mathrm{L}}}{\Theta^{\mathrm{L}}} (\psi^{\mathrm{L}})'_{\mathrm{L}} - \frac{\rho^{\mathrm{G}}}{\Theta^{\mathrm{G}}} (\psi^{\mathrm{G}})'_{\mathrm{G}} - \\
&- \frac{\rho^{\mathrm{S}}}{\Theta^{\mathrm{S}}} \eta^{\mathrm{S}} (\Theta^{\mathrm{S}})'_{\mathrm{S}} - \frac{\rho^{\mathrm{L}}}{\Theta^{\mathrm{L}}} \eta^{\mathrm{L}} (\Theta^{\mathrm{L}})'_{\mathrm{L}} - \frac{\rho^{\mathrm{G}}}{\Theta^{\mathrm{G}}} \eta^{\mathrm{G}} (\Theta^{\mathrm{G}})'_{\mathrm{G}} + \\
&+ \frac{1}{\Theta^{\mathrm{S}}} \mathbf{T}^{\mathrm{SD}} \cdot \mathbf{L}_{\mathrm{S}}^{\mathrm{D}} + \frac{1}{\Theta^{\mathrm{L}}} \mathbf{T}^{\mathrm{LD}} \cdot \mathbf{L}_{\mathrm{L}}^{\mathrm{D}} + \frac{1}{\Theta^{\mathrm{G}}} \mathbf{T}^{\mathrm{GD}} \cdot \mathbf{L}_{\mathrm{G}}^{\mathrm{D}} - \\
&- \frac{1}{\Theta^{\mathrm{S}}} \left[ \frac{1}{3} (\mathbf{T}^{\mathrm{S}} \cdot \mathbf{I}) \frac{1}{\rho^{\mathrm{S}}} + \lambda \frac{\Theta^{\mathrm{S}}}{\rho^{\mathrm{SR}}} \right] (\rho^{\mathrm{S}})'_{\mathrm{S}} - \\
&- \frac{1}{\Theta^{\mathrm{L}}} \left[ \frac{1}{3} (\mathbf{T}^{\mathrm{L}} \cdot \mathbf{I}) \frac{1}{\rho^{\mathrm{L}}} + \lambda \frac{\Theta^{\mathrm{L}}}{\rho^{\mathrm{LR}}} \right] (\rho^{\mathrm{L}})'_{\mathrm{L}} - \\
&- \frac{1}{\Theta^{\mathrm{G}}} \left[ \frac{1}{3} (\mathbf{T}^{\mathrm{G}} \cdot \mathbf{I}) \frac{1}{\rho^{\mathrm{G}}} + \lambda \frac{\Theta^{\mathrm{G}}}{\rho^{\mathrm{GR}}} \right] (\rho^{\mathrm{G}})'_{\mathrm{G}} - \\
&- \frac{1}{\Theta^{\mathrm{L}}} (\hat{\mathbf{p}}^{\mathrm{L}} - \lambda \Theta^{\mathrm{L}} \operatorname{grad} n^{\mathrm{L}}) \cdot (\mathbf{x}'_{\mathrm{L}} - \mathbf{x}'_{\mathrm{S}}) - \\
&- \frac{1}{\Theta^{\mathrm{G}}} (\hat{\mathbf{p}}^{\mathrm{G}} - \lambda \Theta^{\mathrm{G}} \operatorname{grad} n^{\mathrm{G}}) \cdot (\mathbf{x}'_{\mathrm{G}} - \mathbf{x}'_{\mathrm{S}}) - \\
&- \hat{\rho}^{\mathrm{L}} \left( \frac{\mu^{\mathrm{L}}}{\Theta^{\mathrm{L}}} - \frac{\mu^{\mathrm{S}}}{\Theta^{\mathrm{S}}} \right) - \hat{\rho}^{\mathrm{G}} \left( \frac{\mu^{\mathrm{G}}}{\Theta^{\mathrm{G}}} - \frac{\mu^{\mathrm{S}}}{\Theta^{\mathrm{S}}} \right) + n^{\mathrm{S}} \left( \lambda \frac{1}{\rho^{\mathrm{SR}}} + \kappa^{\mathrm{SR}} \right) (\rho^{\mathrm{SR}})'_{\mathrm{S}} + \\
&+ n^{\mathrm{L}} \left( \lambda \frac{1}{\rho^{\mathrm{LR}}} + \kappa^{\mathrm{LR}} \right) (\rho^{\mathrm{LR}})'_{\mathrm{L}} + n^{\mathrm{G}} \lambda \frac{1}{\rho^{\mathrm{GR}}} (\rho^{\mathrm{GR}})'_{\mathrm{G}} -
\end{aligned}
$$

$$
\begin{aligned}
&- \frac{1}{(\Theta^{S})^2}\, \mathbf{q}^{S} \cdot \operatorname{grad} \Theta^{S} - \frac{1}{(\Theta^{L})^2}\, \mathbf{q}^{L} \cdot \operatorname{grad}\Theta^{L} - \frac{1}{(\Theta^{G})^2}\, \mathbf{q}^{G} \cdot \operatorname{grad}\Theta^{G} + \\
&+ \frac{1}{\Theta^{S}\Theta^{L}}\, (\hat{\mathbf{p}}^{L} \cdot \mathbf{x}'_{S})(\Theta^{L} - \Theta^{S}) + \frac{1}{\Theta^{S}\Theta^{G}}\, (\hat{\mathbf{p}}^{G} \cdot \mathbf{x}'_{S})(\Theta^{G} - \Theta^{S}) - \\
&- \frac{1}{\Theta^{S}\Theta^{L}}\, \hat{e}^{L}(\Theta^{L} - \Theta^{S}) - \frac{1}{\Theta^{S}\Theta^{G}}\, \hat{e}^{G}(\Theta^{G} - \Theta^{S}) \geqslant 0.
\end{aligned}
\tag{50}
$$

In obtaining (50) the constraints (24), (27), and (35) have been used. The present ternary model is defined by the relation between the partial and the real mass density, the volume fraction condition, the kinematics, and the balance equations, together with the set of constitutive equations, for which restrictions will be developed from the entropy inequality.

In order to fulfill closure, constitutive equations must be formulated for the following set $\mathcal{R}$ of quantities (see the discussion in de Boer, 1996):

$$
\begin{aligned}
\mathcal{R} = \{&\psi^{\alpha}, \eta^{\alpha}, \mathbf{T}^{\alpha D}, \tfrac{1}{3}\,\mathbf{T}^{\alpha} \cdot \mathbf{I} - n^{\alpha}\Theta^{\alpha}\lambda, \hat{\mathbf{p}}^{L} - \lambda\Theta^{L}\operatorname{grad} n^{L}, \\
&\mathbf{p}^{G} - \lambda\Theta^{G}\operatorname{grad} n^{G}, \mathbf{q}^{\alpha}, \hat{\rho}^{L}, \hat{\rho}^{G}, \hat{e}^{L}, \hat{e}^{G}\}.
\end{aligned}
\tag{51}
$$

The response functions (51) are determined by the motion and temperature histories in such a way that the values of the response functions at any time $t$ and place $\mathbf{x}$ are given by the set $s\,(\mathbf{x}, t)$ of process variables. This set depends on the respective medium for which constitutive equations must be developed. For our purpose the following set of thermodynamic process variables is chosen

$$
\begin{aligned}
s = \{&\Theta^{S}, \Delta^{L} = \Theta^{L} - \Theta^{S}, \Delta^{G} = \Theta^{G} - \Theta^{S}, \operatorname{grad} \Theta^{S}, \\
&\operatorname{grad} \Delta^{L}, \operatorname{grad} \Delta^{G}, \bar{\mathbf{C}}_{S}, \rho^{S}, \rho^{L}, \rho^{G}, \rho^{GR}, \quad \mathbf{x}'_{L} - \mathbf{x}'_{S}, \mathbf{x}'_{G} - \mathbf{x}'_{S}\}.
\end{aligned}
\tag{52}
$$

The choice of the thermodynamic process variables is motivated as follows: The complete temperature state is represented by the individual temperature functions $\Theta^{S}$, $\Theta^{L}$ and $\Theta^{G}$ or $\Theta^{S}$, $\Delta^{L} = \Theta^{L} - \Theta^{S}$ and $\Delta^{G} = \Theta^{G} - \Theta^{S}$, respectively, the elastic deformations of the partial solid are described by the volume-preserving tensor $\bar{\mathbf{C}}_{S}$ and the partial density $\rho^{S}$. Thus, with the choice of $\bar{\mathbf{C}}_{S}$ and $\rho^{S}$ the complete elastic behavior of the partial solid constituent can be described. In order to describe the volumetric deformations of the liquid and gas phase the density functions $\rho^{L}$ and $\rho^{G}$ are introduced as process variables. It is to be borne in mind that the solid and liquid phase are assumed to be incompressible ($\rho^{SR}$ and $\rho^{LR}$ are constant) and the real density $\rho^{GR}$ has been introduced as an independent process variable. Thus, considering the above remarks, the volume fractions $n^{S}$, $n^{L}$ and $n^{G}$ are determined by (4) so that the volume fractions do not enter the set of process variables. Finally, the velocity differences $\mathbf{x}'_{L} - \mathbf{x}'_{S}$ and $\mathbf{x}'_{G} - \mathbf{x}'_{S}$ are related to dissipative effects. In recent findings (Lade and de Boer, 1997; de Boer, 1996; Bluhm, 1997) it has been stated that due to deviatorical deformations at the microscale a volume change of the material at the macroscale can occur. However, for many problems this volume change is small so that we will not treat this effect. As the incorporation of the reference placements

$\mathbf{X}_\alpha$ of the constituents $\varphi^\alpha$ into the set of process variables $s(\mathbf{x}, t)$ is neglected, only homogeneous porous media can be described.

The evaluation of the entropy inequality (50) for the set of response function $\mathcal{R}(s)$ leads to a laborious and lengthy formalism. The interested reader is referred to de Boer and Kowalski (1995), where a complete derivation of the thermodynamical restrictions for a similar mathematical model can be found. The use of standard arguments combined with several symmetry and skew-symmetry conditions (see de Boer and Kowalski, 1995) yields the present model to be governed by the following restrictions for the response functions in the set $\mathcal{R}(s)$ (51) (only the main results are listed for processes close to the mixture equilibrium state):

*Specific entropy functions*:

$$\eta^S = -\frac{\partial \hat{\psi}^S}{\partial \Theta^S}, \qquad \eta^L = -\frac{\partial \hat{\psi}^L}{\partial \Theta^L}, \qquad \eta^G = -\frac{\partial \hat{\psi}^G}{\partial \Theta^G}. \tag{53}$$

*Free Helmholtz energy functions*:

$$\begin{aligned} &\psi^S = \hat{\psi}^S\{\Theta^S,\, \bar{\mathbf{C}}_S,\, \rho^S\}, \qquad \psi^L = \hat{\psi}^L\{\Theta^L = \Delta^L - \Theta^S,\, \rho^L\},\\ &\psi^G = \hat{\psi}^G\{\Theta^G = \Delta^G - \Theta^S,\, \rho^G,\, \rho^{GR}\}. \end{aligned} \tag{54}$$

*Stresses*:

$$\begin{aligned} \mathbf{T}^S &= -\lambda\, \Theta^S\, n^S\, \mathbf{I} - (\rho^S)^2 \frac{\partial \hat{\psi}^S}{\partial \rho^S}\mathbf{I} + 2\,\rho^S (\mathrm{J}_S)^{-2/3}\left(\mathbf{F}_S \frac{\partial \hat{\psi}^S}{\partial \bar{\mathbf{C}}_S} \mathbf{F}_S^T\right)^D,\\ \mathbf{T}^L &= -\lambda\, \Theta^L\, n^L\, \mathbf{I} - (\rho^L)^2 \frac{\partial \hat{\psi}^L}{\partial \rho^L}\mathbf{I},\\ \mathbf{T}^G &= -\lambda\, \Theta^G\, n^G\, \mathbf{I} - (\rho^G)^2 \frac{\partial \hat{\psi}^G}{\partial \rho^G}\mathbf{I}. \end{aligned} \tag{55}$$

Moreover, it turns out that

$$\lambda = -\rho^{LR}\,\kappa^{LR} = -\rho^{SR}\,\kappa^{SR} = \frac{1}{\Theta^G}\,(\rho^{GR})^2\,\frac{\partial \hat{\psi}^G}{\partial \rho^{GR}}, \tag{56}$$

where the Lagrange multiplier $\kappa^{LR}$ and $\kappa^{SR}$ are related to the weighted incompressibility conditions $n^L(\rho^{LR})'_L = 0$ and $n^S(\rho^{SR})'_S = 0$ of the liquid and the solid.

The evaluation of the entropy inequality yields furthermore the important dissipation inequality

$$\mathcal{D}(s) \geqslant 0, \tag{57}$$

where

$$\begin{aligned}\mathcal{D}(s) = & -\operatorname{grad}\Theta^{S}\cdot(\mathbf{m}^{S}+\mathbf{m}^{L}+\mathbf{m}^{G}) - \operatorname{grad}\Delta^{L}\cdot\mathbf{m}^{L} - \operatorname{grad}\Delta^{G}\cdot\mathbf{m}^{G} - \\ & -\frac{1}{\Theta^{L}}(\hat{\mathbf{p}}^{L}-\lambda\,\Theta^{L}\operatorname{grad}n^{L})\cdot(\mathbf{x}'_{L}-\mathbf{x}'_{S}) - \\ & -\frac{1}{\Theta^{G}}(\hat{\mathbf{p}}^{G}-\lambda\,\Theta^{G}\operatorname{grad}n^{G})\cdot(\mathbf{x}'_{G}-\mathbf{x}'_{S}) - \\ & -\frac{1}{\Theta^{L}\,\Theta^{S}}(\Theta^{L}-\Theta^{S})\,(\hat{e}^{L}-\hat{\mathbf{p}}^{L}\cdot\mathbf{x}'_{S}) - \\ & -\frac{1}{\Theta^{G}\,\Theta^{S}}(\Theta^{G}-\Theta^{S})\,(\hat{e}^{G}-\hat{\mathbf{p}}^{G}\cdot\mathbf{x}'_{S}) - \\ & -\hat{\rho}^{L}\left(\frac{\mu^{L}}{\Theta^{L}}-\frac{\mu^{S}}{\Theta^{S}}\right)-\hat{\rho}^{G}\left(\frac{\mu^{G}}{\Theta^{G}}-\frac{\mu^{S}}{\Theta^{S}}\right)\end{aligned} \tag{58}$$

and

$$\mathbf{m}^{S}=\frac{1}{(\Theta^{S})^{2}}\,\mathbf{q}^{S},\qquad \mathbf{m}^{L}=\frac{1}{(\Theta^{L})^{2}}\,\mathbf{q}^{L},\qquad \mathbf{m}^{G}=\frac{1}{(\Theta^{G})^{2}}\,\mathbf{q}^{G}. \tag{59}$$

Assuming that in porous media only small temperature gradients, small relative velocities and small relative temperatures occur, only thermodynamical processes close to the equilibrium state for the mixture are considered. Thus, the nonequilibrium terms of the response functions are expanded in Taylor series close to the mixture equilibrium state with respect to the parameters $\operatorname{grad}\Theta^{S}$, $\operatorname{grad}\Delta^{L}$, $\operatorname{grad}\Delta^{G}$, $\mathbf{x}'_{L}-\mathbf{x}'_{S}$, $\mathbf{x}'_{G}-\mathbf{x}'_{S}$, $\Delta$L and $\Delta$G, where only the linear term will be taken into account.

The Taylor series expansions of the vectors $\mathbf{m}^{\alpha}$, $\hat{\mathbf{p}}^{L}-\lambda\Theta^{L}\operatorname{grad}n^{L}$, $\hat{\mathbf{p}}^{G}-\lambda\Theta^{G}\operatorname{grad}n^{G}$ and the scalars $\hat{e}^{L}-\hat{\mathbf{p}}^{L}\cdot\mathbf{x}'_{S}$ and $\hat{e}^{G}-\hat{\mathbf{p}}^{G}\cdot\mathbf{x}'_{S}$ about the mixture equilibrium state combined with the principle of frame indifference and symmetry and skew-symmetry conditions (see de Boer and Ehlers, 1986) yield

$$\begin{aligned}\mathbf{m}^{\alpha} = & -\alpha^{\alpha}_{\Theta}\operatorname{grad}\Theta^{S}-\beta^{\alpha}_{\Theta}\operatorname{grad}\Delta^{L}-\gamma^{\alpha}_{\Theta}\operatorname{grad}\Delta^{G} - \\ & -\alpha^{\alpha}_{L}(\mathbf{x}'_{L}-\mathbf{x}'_{S})-\alpha^{\alpha}_{G}(\mathbf{x}'_{G}-\mathbf{x}'_{S}),\\ \frac{1}{\Theta^{L}}(\hat{\mathbf{p}}^{L}-\lambda\Theta^{L}\operatorname{grad}n^{L}) = & -\delta^{L}_{\Theta}\operatorname{grad}\Theta^{S}-\varepsilon^{L}_{\Theta}\operatorname{grad}\Delta^{L}-\varphi^{L}_{\Theta}\operatorname{grad}\Delta^{G} - \\ & -\delta^{L}_{L}(\mathbf{x}'_{L}-\mathbf{x}'_{S})-\delta^{L}_{G}(\mathbf{x}'_{G}-\mathbf{x}'_{S}),\\ \frac{1}{\Theta^{G}}(\hat{\mathbf{p}}^{G}-\lambda\Theta^{G}\operatorname{grad}n^{G}) = & -\delta^{G}_{\Theta}\operatorname{grad}\Theta^{S}-\varepsilon^{G}_{\Theta}\operatorname{grad}\Delta^{L}-\varphi^{G}_{\Theta}\operatorname{grad}\Delta^{G} - \\ & -\delta^{G}_{L}(\mathbf{x}'_{L}-\mathbf{x}'_{S})-\delta^{G}_{G}(\mathbf{x}'_{G}-\mathbf{x}'_{S}),\\ \frac{1}{\Theta^{L}\Theta^{S}}(\hat{e}^{L}-\hat{\mathbf{p}}^{L}\cdot\mathbf{x}'_{S}) = & -\eta^{L}_{L}\Delta^{L}-\eta^{L}_{G}\Delta^{G},\\ \frac{1}{\Theta^{G}\Theta^{S}}(\hat{e}^{G}-\hat{\mathbf{p}}^{G}\cdot\mathbf{x}'_{S}) = & -\eta^{G}_{L}\Delta^{L}-\eta^{G}_{G}\Delta^{G}.\end{aligned} \tag{60}$$

The insertion of (60) into (58) yields the following expression for the dissipation mechanism:

$$\begin{aligned}
\mathcal{D}(s) = \; & (\alpha_\Theta^S + \alpha_\Theta^L + \alpha_\Theta^G)\,\mathrm{grad}\Theta^S \cdot \mathrm{grad}\Theta^S + \\
& + (\beta_\Theta^S + \beta_\Theta^L + \beta_\Theta^G + \alpha_\Theta^L)\,\mathrm{grad}\Theta^S \cdot \mathrm{grad}\Delta^L + \\
& + (\gamma_\Theta^S + \gamma_\Theta^L + \gamma_\Theta^G + \alpha_\Theta^G)\,\mathrm{grad}\Theta^S \cdot \mathrm{grad}\Delta^G + \\
& + (\alpha_L^S + \alpha_L^L + \alpha_L^G + \delta_\Theta^L)\,\mathrm{grad}\Theta^S \cdot (\mathbf{x}'_L - \mathbf{x}'_S) + \\
& + (\alpha_G^S + \alpha_G^L + \alpha_G^G + \delta_\Theta^G)\,\mathrm{grad}\Theta^S \cdot (\mathbf{x}'_G - \mathbf{x}'_S) + \\
& + (\beta_\Theta^L)\,\mathrm{grad}\Delta^L \cdot \mathrm{grad}\Delta^L + (\gamma_\Theta^L + \beta_\Theta^G)\,\mathrm{grad}\Delta^L \cdot \mathrm{grad}\Delta^G + \\
& + (\alpha_L^L + \varepsilon_\Theta^L)\,\mathrm{grad}\Delta^L \cdot (\mathbf{x}'_L - \mathbf{x}'_S) + (\alpha_G^L + \varepsilon_\Theta^G)\,\mathrm{grad}\Delta^L \cdot (\mathbf{x}'_G - \mathbf{x}'_S) + \\
& + (\gamma_\Theta^G)\,\mathrm{grad}\Delta^G \cdot \mathrm{grad}\Delta^G + (\alpha_L^G + \varphi_\Theta^L)\,\mathrm{grad}\Delta^G \cdot (\mathbf{x}'_L - \mathbf{x}'_S) + \\
& + (\alpha_G^G + \varphi_\Theta^G)\,\mathrm{grad}\Delta^G \cdot (\mathbf{x}'_G - \mathbf{x}'_S) + (\delta_L^L)\,(\mathbf{x}'_L - \mathbf{x}'_S) \cdot (\mathbf{x}'_L - \mathbf{x}'_S) + \\
& + (\delta_G^L + \delta_L^G)\,(\mathbf{x}'_L - \mathbf{x}'_S) \cdot (\mathbf{x}'_G - \mathbf{x}'_S) + (\delta_G^G)\,(\mathbf{x}'_G - \mathbf{x}'_S) \cdot (\mathbf{x}'_G - \mathbf{x}'_S) + \\
& + \eta_L^L (\Delta^L)^2 + (\eta_G^L + \eta_L^G)\,\Delta^L \Delta^G + n_G^G (\Delta^G)^2 - \\
& - \hat{\rho}^L \left( \frac{\mu^L}{\Theta^L} - \frac{\mu^S}{\Theta^S} \right) - \hat{\rho}^G \left( \frac{\mu^G}{\Theta^G} - \frac{\mu^S}{\Theta^S} \right) \geqslant 0. \qquad (61)
\end{aligned}$$

By neglecting the influence of phase transitions in the model, that is, $\hat{\rho}^\alpha = 0$, the inequality is fulfilled for arbitrary $\mathrm{grad}\Theta^S$, $\Delta^L$, $\Delta^G$, $\mathrm{grad}\Delta^L$, $\mathrm{grad}\Delta^G$, $\mathbf{x}'_L - \mathbf{x}'_S$ and $\mathbf{x}'_G - \mathbf{x}'_S$ if the following relations hold for the response parameters appearing in the constitutive equations (60):

$$\begin{aligned}
& \alpha_\Theta^S + \alpha_\Theta^L + \alpha_\Theta^G \geqslant 0, \qquad \beta_\Theta^S + \beta_\Theta^L + \beta_\Theta^G + \alpha_\Theta^L = 0, \\
& \gamma_\Theta^S + \gamma_\Theta^L + \gamma_\Theta^G + \alpha_\Theta^G = 0, \qquad \alpha_L^S + \alpha_L^L + \alpha_L^G + \delta_\Theta^L = 0, \\
& \alpha_G^S + \alpha_G^L + \alpha_G^G + \delta_\Theta^G = 0, \qquad \alpha_L^L + \varepsilon_\Theta^L = 0, \quad \alpha_G^L + \varepsilon_\Theta^G = 0, \\
& \beta_\Theta^L \geqslant 0, \quad \gamma_\Theta^G \geqslant 0, \qquad \gamma_\Theta^L + \beta_\Theta^G = 0, \\
& \delta_L^L \geqslant 0, \quad \delta_G^G \geqslant 0, \qquad \delta_G^L + \delta_L^G = 0, \\
& \eta_L^L \geqslant 0, \quad \eta_G^G \geqslant 0, \qquad \eta_G^L + \eta_L^G = 0.
\end{aligned} \qquad (62)$$

In order to ensure the validity of the entropy inequality (61) in the case of mass exchange processes the constitutive equations for $\hat{\rho}^L$ and $\hat{\rho}^G$ must be chosen in such a way that the last two terms on the left hand side of (61) are always positive. A first approach can be

$$\hat{\rho}^L = -\xi^L \left( \frac{\mu^L}{\Theta^L} - \frac{\mu^S}{\Theta^S} \right), \qquad \hat{\rho}^G = -\xi^G \left( \frac{\mu^G}{\Theta^G} - \frac{\mu^S}{\Theta^S} \right), \qquad (63)$$

where $\xi^L$ and $\xi^G$ are positive response parameters.

## 4. Conclusions

The present ternary model, consisting of an elastic incompressible porous solid and two immiscible fluids, namely an incompressible liquid and a compressible gas, reveals that phase transitions between the solid and liquid as well as gas phases, respectively, and the liquid and gas phases are governed by the differences of the chemical potentials of the corresponding phases. Furthermore, the investigations show that mass exchanges cannot occur if the chemical potentials and the temperatures of the constituent are equal. This is a well-known result in classical thermodynamics of gases.

It appears to be easy to treat freezing or drying processes in porous media with the aforementioned results. This will be done in further publications.

## Acknowledgement

The authors gratefully acknowledge the financial support of this research by the Deutsche Forschungsgemeinschaft (DFG).

## References

Ben Nasrallah, S. and Perre, P.: 1988, Detailed study of a model of heat and mass transfer during convective drying of porous media, *Int. J. Heat Mass Transfer* **31**, 957–967.

Bluhm, J.: 1997, A consistent model for saturated porous media, Forschungsberichte aus dem Fachbereich Bauwesen 74, Universität-GH Essen.

Bluhm, J. and de Boer, R.: 1996, Effective stresses – a clarification, *Arch. Appl. Mech.* **66** (Ingenieur-Archiv), 479–492.

Bluhm, J. and de Boer, R.: 1997, The volume fraction concept in the porous media theory, *Zeitschr. Angew. Math. Mech.* (ZAMM) **77**(8), 563–577.

de Boer, R.: 1982, Vektor- und Tensorrechnung für Ingenieure, Springer-Verlag, Berlin, Heidelberg, New York.

de Boer, R.: 1995a, Thermodynamics of phase transitions in porous media, *Appl. Mech. Rev.* **48**(10), 613–622.

de Boer, R.: 1995b, Some issues in the macroscopic porous media, *Report Mech.* **95**(8), Festschrift Prof. A. Siekmann, FB10/Mechanik, Universität-GH Essen.

de Boer, R.: 1996, Highlights in the development of the theory of porous media – toward a consistent macroscopic theory, *Appl. Mech. Rev.* **49**(4), 201–262.

de Boer, R.: 1997, Compressible porous media: Toward a general theory, In: N. A. Fleck and A. C. F. Cocks (eds), *Mechanics of Granular and Porous Media, Proc. IUTAM Symp. Cambridge, 1996*, Kluwer Academic Publishers, Dordrecht, Boston, London, pp. 47–56.

de Boer, R. and Didwania, A.: 1997, The effect of uplift in liquid-saturated porous solids – Karl von Terzaghi's contributions and recent findings, *Geotechnique* **47**(2), 289–298.

de Boer, R. and Ehlers, W.: 1986, Theorie der Mehrkomponentenkontinua mit Anwendung auf bodenmechanische Probleme, Teil I, Forschungsberichte aus dem Fachbereich Bauwesen 40, Universität-GH Essen.

de Boer, R., Ehlers, W., Kowalski, S. J. and Plischka, J.: 1991, Porous Media – a survey of different approaches, Forschungsberichte aus dem Fachbereich Bauwesen 54, Universität-GH Essen.

de Boer, R. and Kowalski, S. J.: 1995, Thermodynamics of fluid-saturated porous media with a phase change, *Acta Mechanica* **109**, 167–189.

Elwell, D. and Pointon, A. J.: 1972, *Classical Thermodynamics*, Penguin, Harmondsworth, Middlesex, England.
Kowalski, S. J.: 1987, Thermomechanics of constant drying rate period, *Arch. Mech.* **39**, 157–176.
Kowalski, S. J.: 1990, Thermomechanics of dried materials, *Arch. Mech.* **42**, 123–149.
Lade, P. V. and de Boer, R.: 1997, The concept of effective stress for soil, concrete and rock, *Geotechnique* **47**(1), 61–78.
Lewis, R. W., Morgan, K., Thomas, H. R. and Strada, M.: 1979, Drying-induced stresses in porous bodies – an elastoviscoplastic model, *Comput. Methods Appl. Mech. Engng.* **20**, 291–301.
Lewis, R. W., Strada, M. and Comini, G.: 1977, Drying induced stresses in porous bodies, *Int. J. Numer. Meths. Engng.* **11**, 1175–1184.
Luikov, R. W.: 1975, Systems of differential equations of heat and mass transfer in capillary-porous bodies (review), *Int. J. Heat Mass Transfer* **18**, 1–14.
Manns, W. and Hartmann, E.: 1977, Frostwiderstand von Beton, Deutscher Ausschuß für Stahlbeton, Heft 289.
Nunziato, J. W. and Walsh, E. K.: 1980, On ideal multiphase mixtures with chemical reactions and diffusion, *Arch. Rational Mech. Anal.* **73**, 285–311.
Schäfer, A.: 1964, Frostwiderstand und Porengefüge des Betons – Beziehungen und Prüfverfahren, Deutscher Ausschuß für Stahlbeton, Heft 167.
Setzer, M. J., Auberg, R. and Hartmann, V.: 1994, Bewertung des Frost-Tausalz-Widerstandes von Transportbeton, Forschungsbericht aus dem Fachbereich Bauwesen 62, Universität-GH Essen.
Truesdell, C.: 1968, Sulle basi della termodinamica delle miscele, *Rend. Lincei* **44**(8), 381–383.
Whitaker, S.: 1977, Simultaneous heat, mass, and momentum transfer in porous media: A theory of drying, *Adv. Heat Transfer* **13**, 119–203.
Zech, B.: 1981, Zum Gefrierverhalten von Wasser, Dissertation, TU München.

# A Simple Model for a Fluid-Filled Open-Cell Foam

UDO DÜNGER, HERBERT WEBER and HANS BUGGISCH
*Institut für Mechanische Verfahrenstechnik u. Mechanik, Bereich Angewandte Mechanik, Universität Karlsruhe, Germany*

(Received: 29 August 1997; in final form: 20 January 1998)

**Abstract.** A simple microstructure model is used to describe a fluid-filled open-cell foam. In the simplest case it consists of parallel elastic plates with gaps between them, which are filled with a Newtonian fluid. We assume that the load applied to this model material is uniaxial. The constitutive equation is formulated with the pressure of the fluid as an inner variable. The model yields an evolutional equation for the fluid pressure which itself is a field equation, that is a partial differential equation in time and space coordinates. This differential equation is solved for an instantaneously applied constant load and for a harmonically oscillating load. The solution of the differential equation, in combination with the constitutive equation leads to a relation between mean applied load and global strain of the test specimen. Finally, we obtain the creep compliance and the complex modulus of the foam material, respectively. The influence of different geometries of the foam and of different material behaviour of the matrix and fluid on the creep compliance and the complex modulus is discussed.

**Key words:** foam, microstructure, damping characteristic, open-cell, fluid-filled.

## 1. Introduction

In general, fluid-filled open-cell polymer foams are used as dampers of waves, impact loadings and vibrations. They can be applied in a broad range, for example damping of vibrations of machines, protection of sensitive goods against vibrations and impacts, and they are used in casual shoes and mattresses as well.

In order to construct fluid-filled open-cell polymer foams with desired damping characteristics, one needs to understand how the matrix material, the fluid properties, and the cavity structure influence the damping behaviour. The aim of this paper is to analyze the influence of the above parameters.

Fluid-filled open-cell foam is not a so-called 'simple material', that is the stress at a point $X$ does not depend only on the deformation at $X$ or the history of deformation at $X$. Since the fluid needs a finite time to flow out of high pressure regions into low pressure regions, the system will pass a nonequilibrium state. Therefore, the constitutive equation has to take into account long-range interactions. The well-known approach (Eringen, 1975) for a nonsimple material of higher-order gradients assumes that the stress tensor at a place $X$ is a function of the histories of the first $n$ gradients of deformation at $X$. This is not sufficient when we have pronounced long-range interactions. This is the case for open-cell foams with fluid flow interactions.

However, modelling the material as a 'continuum with an inner fluid state variable' also considers long-range interactions. The field equation of the inner fluid state variable is an evolutional equation. The fluid pressure $p$ can be taken as an inner state variable.

In the present paper, we will look at a model foam with as simple a microstructure as possible. Initially, we will consider the fluid flow through the hollow space of the matrix material in direction $x$. The fluid volume flow is assumed to be proportional to the loss of pressure $-\partial p/\partial x$ in this direction. For a specimen which is loaded uniaxially in $x$-direction the equation of state is given by

$$\sigma = \sigma(\varepsilon, p). \tag{1.1}$$

The fluid pressure $p$ satisfies a partial differential equation in space and time coordinates. In the following this is derived for a model foam with a simple geometry.

## 2. Mathematical Modelling of a Porous Medium

The micromechanical model foam (Figure 1) consists of linear-elastic plates which are arranged parallel to each other. The gaps between the plates are filled with an incompressible Newtonian fluid under the pressure $p$. Because of the regularity of the structure, it suffices to regard a unit cell (Figure 2). The unit cell consists of two half plates and the fluid-filled space between them. Both the distance $H$ between the middle of neighbouring plates and the width $b$ of the plates in $z$-direction are assumed to be constant. This is true if the package of plates is surrounded by rigid walls, and no shear stress occurs between the walls and the plates. The actually existing shear stress can be neglected if we consider areas with sufficient distance from the boundary.

It should be mentioned that different methods exist to gain the macroscopic constitutive equations of this special porous material. On the basis of concepts from

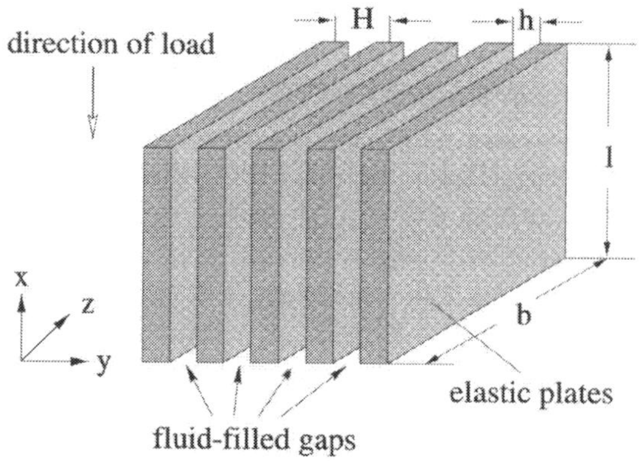


*Figure 1.* Micromechanical model foam.

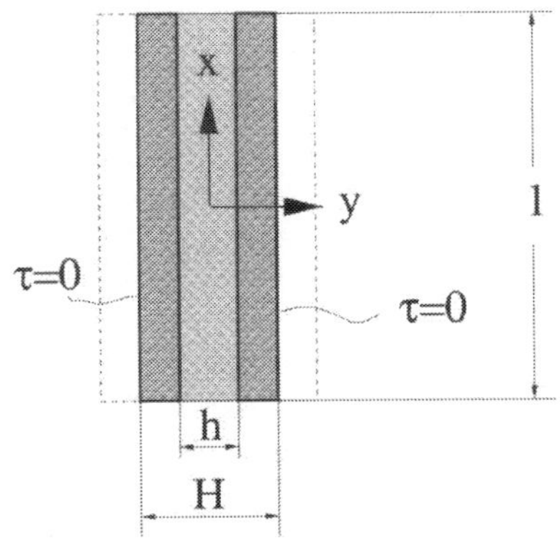


*Figure 2.* Unit cell.

continuum mechanics, Biot's theory of poroelasticity (Biot, 1956; Detournay and Cheng, 1993) could be applied, or the theory of mixtures extended by the concept of volume fractions (de Boer *et al.*, 1991). Because of the periodic microstructure, it would also be possible to apply homogenization techniques like those which are used in the mechanics of composite materials (Coussy, 1991). With the material under consideration, however, a difficulty arises when such theories are used. This is due to the fact that, in the direction perpendicular to the plates, the solid is not interconnected. Therefore, it is difficult to find the proper stress-strain relation for this direction. It seems to be easier to treat this problem in an elementary manner.

For the following considerations, we assume that inertia effects can be neglected, and that the strain is small. The shear deformation will not be taken into account. We will investigate two different cases. First, we consider the fluid flowing in the direction of the applied load. As a second case, we analyze the fluid flowing perpendicular to the direction of the applied load. In the first case the load is applied stress controlled, and, in the second one, it is applied strain controlled. This is done for the simplicity of the averaging integration of the constitutive equations.

### 2.1. FLOW PARALLEL TO THE DIRECTION OF AN APPLIED LOAD

Here, we will calculate the averaged strain $\overline{\varepsilon}$ in $x$-direction in the case when a normal stress $\sigma$ in $x$-direction is applied to the elastic plates by means of rigid end plates (Figure 3). Owing to the applied load, the fluid flows in $x$-direction.

The strain $\varepsilon_y$ in $y$-direction and the strain $\varepsilon$ in $x$-direction of the linear-elastic plates depend on both the fluid pressure $p$ within the gaps and the stress $\sigma$ in $x$-direction inside the plates.

Hooke's law for plain strain yields

$$\varepsilon = \frac{1-\nu^2}{E}\sigma + \frac{\nu(1+\nu)}{E}p, \tag{2.1}$$

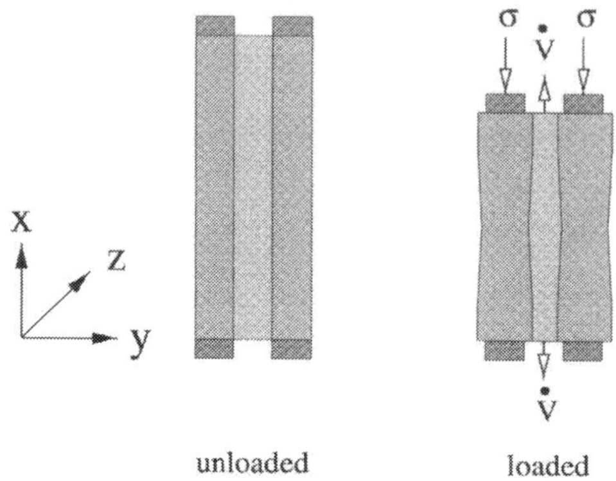


*Figure 3.* Loaded and unloaded plates, flow parallel to an applied normal stress.

$$\varepsilon_y = -\frac{1-\nu^2}{E}\,p - \frac{\nu\,(1+\nu)}{E}\,\sigma, \tag{2.2}$$

with $E$ being Young's modulus and $\nu$ Poisson's ratio of the plate material. Let $h_0$ be the distance of the unloaded plates and $\delta h$ the small change of gap width due to the applied strain. Then the gap width under load is

$$h(x,t) = h_0 + \delta h(x,t). \tag{2.3}$$

Therefore, the strain of a plate in $y$-direction is

$$\varepsilon_y = -\frac{\delta h}{H - h_0}, \tag{2.4}$$

and, with (2.2), the change $\delta h$ of the gap width can be written in the form

$$\delta h = (H - h_0)\left[\frac{1-\nu^2}{E}\,p + \frac{\nu\,(1+\nu)}{E}\,\sigma\right]. \tag{2.5}$$

The flow of fluid between two plates caused by the applied stress is assumed to be quasi-stationary. Furthermore, we assume the change of the plate thickness $\delta h$ to be small compared to the plate thickness, that is the stream lines are roughly straight-lined and parallel to each other. With these assumptions the Navier–Stokes equations lead to Darcy's law, and the averaged velocity $v$ in $x$-direction depends on the fluid viscosity $\eta$, the permeability coefficient $K$ and the loss of pressure $-\partial p/\partial x$. The velocity $v$, that is the volume flow $\dot{V}$ referred to the area $b \cdot h$, is given by

$$\frac{\dot{V}}{bh} = v = \frac{K}{\eta}\left(-\frac{\partial p}{\partial x}\right). \tag{2.6}$$

If we assume furthermore that the length $b$ of the plates is much larger than the gap width $h$, then $K$ is the permeability coefficient of two flat plates and is given by

$$K = \tfrac{1}{12}\,h^2. \tag{2.7}$$

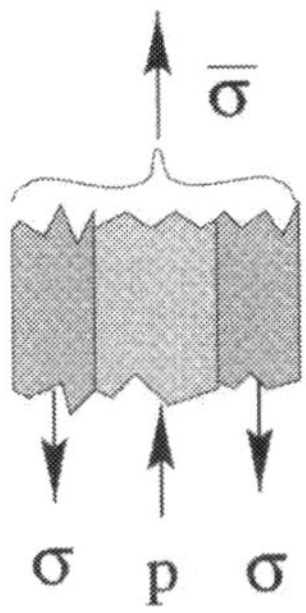


*Figure 4.* Averaged normal stress.

A change in the distance $h$ of two plates, that is a volume change of the fluid-filled space, leads to a flow of the incompressible fluid. The balance of the fluid volume yields

$$\frac{\partial(\dot{V}/b)}{\partial x} = -\frac{\partial h}{\partial t} - h\,\frac{\partial \varepsilon}{\partial t}. \tag{2.8}$$

Until now we have not taken into account that, due to the fluid flow, a shear stress $\tau_f$, interacting at the boundary between fluid and matrix material, arises. The equilibrium of forces at a unit cell (Figure 4) yields

$$(H-h)\sigma - p\,h = \overline{\sigma}\,H. \tag{2.9}$$

Here, $\overline{\sigma}$ denotes the averaged stress in $x$-direction. The pressure term in (2.9) represents the effect of the shear stress $\tau_f$. Instead of the shear stress $\tau_f$ acting at the surface area, we can imagine a volume force inside every plate. This 'pseudo volume force' has the same effect as the shear stress $\tau_f$, as can be derived from differentiating (2.9) and noting that pressure drop and shear stress are proportional:

$$\frac{\partial \sigma}{\partial x} = \underbrace{\frac{h}{H-h}\,\frac{\partial p}{\partial x}}_{\text{pseudo volume force}} \propto \frac{\tau_f}{H-h}. \tag{2.10}$$

The constitutive equation of the model foam is now obtained by substitution of the equilibrium equation (2.9) into Hooke's law (2.1), linearizing with respect to $p$, $\sigma$ and $\delta h$, and by a subsequent integration over the height of the plates:

$$\overline{\varepsilon} = A_{\parallel}\,\overline{\sigma} + B_{\parallel}\,\frac{1}{l}\int_{-l/2}^{l/2} p\,\mathrm{d}x \tag{2.11}$$

with the constants

$$A_{\parallel} = \frac{1-\nu^2}{E\,(1-h_0/H)}, \qquad B_{\parallel} = \frac{1+\nu}{E\,(1-h_0/H)}\left[\nu + (1-2\,\nu)\frac{h_0}{H}\right]. \tag{2.12}$$

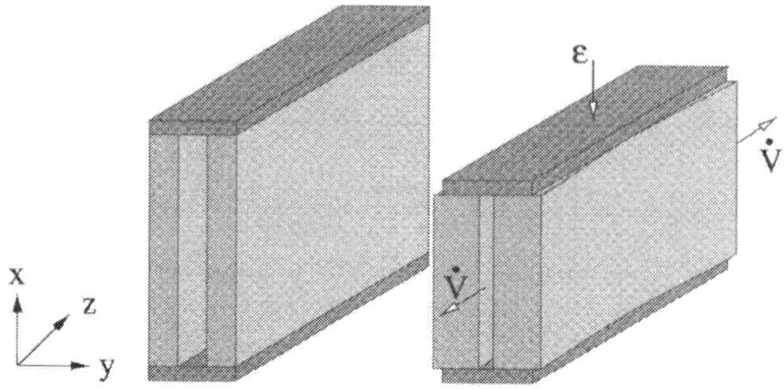


*Figure 5.* Loaded and unloaded plates, flow perpendicular to an applied strain.

Finally, we obtain the partial differential equation of the inner variable $p$ from (2.1), (2.2) and (2.5)–(2.9). Neglecting nonlinear terms results in

$$a_{\|}\,\frac{\partial^2 p}{\partial x^2} - \frac{\partial p}{\partial t} = k_{\|}\,\frac{\mathrm{d}\overline{\sigma}}{\mathrm{d}t}, \tag{2.13}$$

where

$$\begin{aligned} a_{\|} &= \frac{E\,h_0^2}{12\,\eta}\,\frac{(h_0/H)(1-h_0/H)}{(1+\nu)[1-\nu-2\,(1-2\,\nu)(h_0/H)\,(1-h_0/H)]}, \\ k_{\|} &= \frac{\nu+(1-2\,\nu)(h_0/H)}{1-\nu-2\,(1-2\,\nu)(h_0/H)(1-h_0/H)}. \end{aligned} \tag{2.14}$$

### 2.2. FLOW PERPENDICULAR TO THE DIRECTION OF AN APPLIED LOAD

We now calculate the averaged stress $\overline{\sigma}$ in $x$-direction for the case where rigid end plates apply a strain $\varepsilon$ in $x$-direction to the elastic plates and the fluid between them (Figure 5). Owing to the applied strain and the boundary conditions, the fluid flows in $z$-direction perpendicular to the applied load.

In principle, the derivation of both the constitutive equation and the partial differential equation for $p$ is analogous to the first case in 2.1, but there are some important differences.

The state of strain is now three-dimensional. Hooke's law yields

$$\varepsilon = \frac{1}{E}\,[\sigma - \nu(\sigma_z - p)], \tag{2.15}$$

$$\varepsilon_y = -\frac{1}{E}\,[p + \nu(\sigma + \sigma_z)], \tag{2.16}$$

$$\varepsilon_z = -\frac{1}{E}\,[\sigma_z - \nu(\sigma - p)]. \tag{2.17}$$

Insertion of (2.16) into (2.4) yields

$$\delta h = \frac{H - h_0}{E}[(1 - \nu^2)\ p + (\nu + \nu^2)\ \sigma + \nu E \varepsilon_z]. \tag{2.18}$$

The fluid velocity $v_z$ in $z$-direction is given by

$$\frac{\dot{V}}{lh} = v_z = \frac{K}{\eta}\left(-\frac{\partial p}{\partial z}\right). \tag{2.19}$$

Assuming that the height $l$ of the plates is much larger than the gap width $h$, the permeability coefficient $K$ again coincides with the permeability coefficient of flat plates (2.7). The balance of fluid volume yields

$$\frac{\partial(\dot{V}/l)}{\partial z} = -\frac{\partial h}{\partial t} - h\left(\frac{\partial \varepsilon}{\partial t} + \frac{\partial \varepsilon_z}{\partial t}\right). \tag{2.20}$$

The equilibrium of forces in $z$-direction at a fluid element shows that the shear stress $\tau_f$ at the boundary between fluid and matrix material is proportional to the pressure loss $-\partial p/\partial z$:

$$\tau_f = -\frac{h}{2}\frac{\partial p}{\partial z}. \tag{2.21}$$

The equilibrium of forces in $z$-direction at an element of the matrix material leads to the equation

$$\frac{\partial \sigma_z}{\partial z} = \frac{2}{H - h}\tau_f = \frac{h}{H - h}\frac{\partial p}{\partial z}. \tag{2.22}$$

If the fluid pressure $p = 0$, there will be no strain $\varepsilon_z$ in $z$-direction. This results, with (2.15) and (2.17), in the following relations:

$$\begin{aligned} \sigma &= \frac{\nu(2h - H)}{H - h}\, p + \frac{1}{1 - \nu^2}\,\varepsilon\, E, \\ \sigma_z &= \frac{h}{H - h}\, p + \frac{\nu}{1 - \nu^2}\,\varepsilon\, E. \end{aligned} \tag{2.23}$$

The linearized average stress $\overline{\sigma}$ in $x$-direction is defined as

$$\overline{\sigma} = \frac{1}{b}\int_{-b/2}^{b/2}\left[\left(1 - \frac{h_0}{H}\right)\sigma - \frac{h_0}{H}\, p\right] \mathrm{d}z. \tag{2.24}$$

Substituting (2.23) into (2.24), and neglecting nonlinear terms, we obtain the constitutive equation

$$\overline{\sigma} = A_\perp\,\overline{\varepsilon} + B_\perp\,\frac{1}{b}\int_{-b/2}^{b/2} p\,\mathrm{d}z \tag{2.25}$$

with

$$A_\perp = \left(\frac{1 - h_0/H}{1 - \nu^2}\right) E, \qquad B_\perp = -(1 - 2\nu)\frac{h_0}{H} - \nu. \tag{2.26}$$

Finally, we obtain the partial differential equation of the inner variable $p$ from (2.7), (2.15), (2.17)–(2.20) and (2.23). Neglecting nonlinear terms results in

$$a_\perp \frac{\partial^2 p}{\partial x^2} - \frac{\partial p}{\partial t} = k_\perp \frac{\mathrm{d}\overline{\epsilon}}{\mathrm{d}t}, \tag{2.27}$$

where

$$\begin{aligned} a_\perp = a_\parallel &= \frac{E\, h_0^2}{12\,\eta} \frac{(h_0/H)(1 - h_0/H)}{(1+\nu)[1 - \nu - 2(1-2\nu)(h_0/H)(1 - h_0/H)]}, \\ k_\perp &= \frac{[\nu + (1-2\nu)h_0/H](1 - h_0/H)}{(1-\nu^2)[1 - \nu - 2(1-2\nu)(h_0/H)(1 - h_0/H)]} E. \end{aligned} \tag{2.28}$$

## 3. Solution of the Partial Differential Equations

The partial differential Equations (2.13) and (2.27) are of the same type. Subsequently, the solution of (2.13) is given for both an instantaneously and a sinusoidally applied load.

### 3.1. SOLUTION FOR AN INSTANTANEOUSLY APPLIED LOAD

Let us consider the case where the load $\overline{\sigma}$ is linearly increased from 0 to $\overline{\sigma_0}$ in the time interval $[0, \delta t_0]$, and then kept constant at the value $\overline{\sigma_0}$.

The differential Equation (2.13) can then be written as

$$\begin{aligned} a_\parallel \frac{\partial^2 p}{\partial x^2} &= \frac{\partial p}{\partial t} + k_\parallel \frac{\overline{\sigma_0}}{\delta t_0} \quad \text{for} \quad 0 < t < \delta t_0, \\ a_\parallel \frac{\partial^2 p}{\partial x^2} &= \frac{\partial p}{\partial t} \quad \text{for} \quad t > \delta t_0. \end{aligned} \tag{3.1}$$

An instantaneously applied stress $\overline{\sigma}$ will lead to an abruptly increased constant fluid pressure. Therefore, the initial condition of the fluid pressure immediately after an applied jump of stress can be written as

$$p(\delta t_0) = -k_\parallel \overline{\sigma_0} \quad \text{for} \quad \delta t_0 \to 0. \tag{3.2}$$

We assume the fluid pressure at the end of the plates in $x$-direction to be zero. This gives the boundary conditions

$$p\left(x = \tfrac{1}{2} l, t\right) = p\left(x = -\tfrac{1}{2} l, t\right) = 0. \tag{3.3}$$

With these boundary conditions and the initial condition (3.2), we obtain the following Fourier series as a solution of the differential equation (2.13) for the case of a jump of stress:

$$p(x,t) = -k_{\|}\,\overline{\sigma_0}\,\frac{4}{\pi}\sum_{n=1}^{\infty}\left\{\frac{(-1)^{n+1}}{2n-1}\cdot\frac{\cos[(2n-1)\pi(x/l)]}{\exp[\pi^2(a_{\|}/l^2)(2n-1)^2 t]}\right\}. \tag{3.4}$$

## 3.2. SOLUTION FOR A SINUSOIDALLY APPLIED LOAD

Now we consider a sinusoidally applied load

$$\overline{\sigma} = \overline{\sigma_0}\sin\omega t \tag{3.5}$$

with the angular frequency $\omega$. The approach

$$p(x,t) = f(x)\sin\omega t + g(x)\cos\omega t \tag{3.6}$$

and the same boundary conditions as before (fluid pressure at the ends of the plates = ambient pressure = 0) give the following solution of the differential equation:

$$\begin{aligned} p(x,t) = {} & \frac{k_{\|}\,\overline{\sigma_0}}{\sin^2(\frac{1}{2}lK_{\|})\sinh^2(\frac{1}{2}lK_{\|}) + \cos^2(\frac{1}{2}lK_{\|})\cosh^2(\frac{1}{2}lK_{\|})} \times \\ & \times \{[-\sin^2(\tfrac{1}{2}lK_{\|})\sinh^2(\tfrac{1}{2}lK_{\|}) - \\ & - \cos^2(\tfrac{1}{2}lK_{\|})\cosh^2(\tfrac{1}{2}lK_{\|}) + \\ & + \sin(xK_{\|})\sinh(xK_{\|})\sin(\tfrac{1}{2}lK_{\|})\sinh(\tfrac{1}{2}lK_{\|}) + \\ & + \cos(xK_{\|})\cosh(xK_{\|})\cos(\tfrac{1}{2}lK_{\|})\cosh(\tfrac{1}{2}lK_{\|})]\sin\omega t + \\ & + [\sin(xK_{\|})\sinh(xK_{\|})\cos(\tfrac{1}{2}lK_{\|})\cosh(\tfrac{1}{2}lK_{\|}) - \\ & - \cos(xK_{\|})\cosh(xK_{\|})\sin(\tfrac{1}{2}lK_{\|})\sinh(\tfrac{1}{2}lK_{\|})]\cos\omega t\} \end{aligned} \tag{3.7}$$

with

$$K_{\|} = \sqrt{\frac{\omega}{2a_{\|}}}. \tag{3.8}$$

## 4. Results

### 4.1. FLOW PARALLEL TO THE DIRECTION OF AN APPLIED LOAD

#### 4.1.1. *Material Behaviour for an Instantaneously Applied Load*

The creep law, that is the relation between the averaged strain $\overline{\varepsilon}$ and an instantaneously applied load $\overline{\sigma}_0$, is obtained by inserting (3.4) into (2.11):

$$\begin{aligned}\overline{\varepsilon}(t) &= \left[A_\| - \frac{8}{\pi^2} B_\| \, k_\| \sum_{n=1}^{\infty} \frac{\exp[-\pi^2(a_\|/l^2)(2n-1)^2 t]}{(2n-1)^2}\right] \overline{\sigma_0} \\ &= D(t)\, \overline{\sigma_0}. \end{aligned} \quad (4.1)$$

The term $D(t)$ is called creep compliance. For the initial value ($t \to 0$) and for the equilibrium value ($t \to \infty$) of $D(t)$ we get

$$\begin{aligned} D(t \to 0) &= A_\| - B_\| \, k_\|, \\ D(t \to \infty) &= A_\|. \end{aligned} \quad (4.2)$$

#### 4.1.2. *Material Behaviour for a Sinusoidally Applied Load*

In the case of a sinusoidally applied load (3.5), we have to insert the solution (3.7) of the partial differential Equation (2.13) into the constitutive Equation (2.11). This gives the following expression for the averaged sinusoidal strain:

$$\overline{\varepsilon} = \overline{\sigma_0} \sqrt{C_{1\|}^2 + C_{2\|}^2}\, \sin(\omega t + \varphi) \quad (4.3)$$

with

$$\begin{aligned} C_{1\|} &= A_\| - B_\| \, k_\| + \frac{B_\| \, k_\|}{l\, K_\|} \times \\ &\times \frac{\sinh(\frac{1}{2} lK_\|)\cosh(\frac{1}{2} lK_\|) + \sin(\frac{1}{2} lK_\|)\cos(\frac{1}{2} lK_\|)}{[\sin(\frac{1}{2} lK_\|)\sinh(\frac{1}{2} lK_\|)]^2 + [\cos(\frac{1}{2} lK_\|)\cosh(\frac{1}{2} lK_\|)]^2}, \end{aligned} \quad (4.4)$$

$$C_{2\|} = \frac{B_\| \, k_\|}{l\, K_\|} \cdot \frac{\sin(\frac{1}{2} lK_\|)\cos(\frac{1}{2} lK_\|) - \sinh(\frac{1}{2} lK_\|)\cosh(\frac{1}{2} lK_\|)}{[\sin(\frac{1}{2} lK_\|)\sinh(\frac{1}{2} lK_\|)]^2 + [\cos(\frac{1}{2} lK_\|)\cosh(\frac{1}{2} lK_\|)]^2}. \quad (4.5)$$

The loss angle $\varphi$ is given by

$$\tan \varphi = \frac{C_{2\|}}{C_{1\|}}, \quad (4.6)$$

and the dynamic modulus $E_{\text{dyn}}$ is given by

$$E_{\text{dyn}} = \frac{\overline{\sigma_0}}{\overline{\varepsilon_0}} = 1/\sqrt{C_{1\|}^2 + C_{2\|}^2}. \quad (4.7)$$

The theoretical treatment predicts that the dynamic modulus $E_{\text{dyn}}$ and the tangent of the loss angle, $\tan\varphi$, are functions of the ratio $h_0/H$, Poisson's ratio $\nu$ of the linear-elastic plates and a certain dimensionless quantity. This quantity $\Pi_{\|}$ is the following function of the angular frequency $\omega$, the viscosity of the fluid $\eta$, Young's modulus $E$ of the linear-elastic plates and the ratio of the fluid path length $l$ to the gap length $h_0$:

$$\Pi_{\|} = \omega \left(\frac{l}{h_0}\right)^2 \frac{\eta}{E}. \tag{4.8}$$

For small values of the dimensionless quantity $\Pi_{\|}$, the dynamic modulus $E_{\text{dyn}}$, denoted $E_{\text{stat}}$, is equal to the reciprocal of the creep compliance for large values of $t$:

$$E_{\text{dyn}}\left(\Pi_{\|} \to 0\right) = E_{\text{stat}} = \frac{1}{D(t \to \infty)} = \frac{1}{A_{\|}}. \tag{4.9}$$

For very large values of the dimensionless quantity $\Pi_{\|}$, the dynamic modulus $E_{\text{dyn}}$ is equal to the reciprocal of the short-time creep compliance:

$$E_{\text{dyn}}(\Pi_{\|} \to \infty) = \frac{1}{D(t \to 0)} = \frac{1}{A_{\|} - B_{\|}k_{\|}}. \tag{4.10}$$

Thus, it will be sufficient to look at the material behaviour of a harmonically applied load.

The tangent of the loss angle is a measure of the energy dissipation per cycle. At low values of $\Pi_{\|}$, the tangent of the loss angle increases with the quantity $\Pi_{\|}$, due to the fluid flow relative to the matrix. The fluid flow at higher values of $\Pi_{\|}$ is more and more in phase with the applied load, and the tangent of the loss angle decreases. The fluid flow is negligible at high values of $\Pi_{\|}$, and the $\tan\varphi$ tends towards zero. This qualitative behaviour is shown in Figures 6 and 8. Furthermore, in Figure 6 it is shown that increasing values of Poisson's ratio $\nu$ cause a higher energy dissipation. The energy dissipation also increases with increasing ratio $h_0/H$ (Figure 8), which can be interpreted as the portion of the fluid volume.

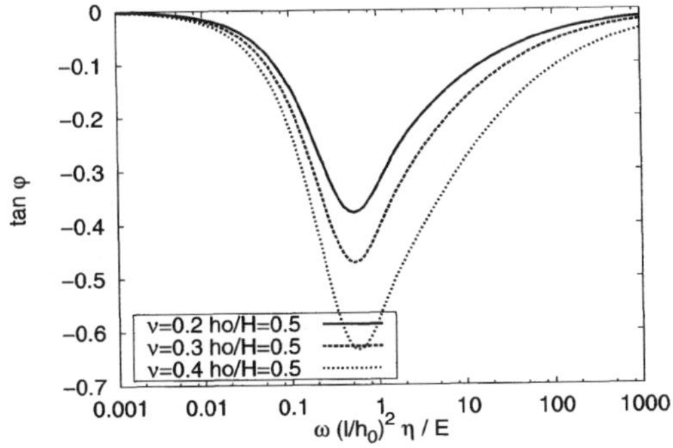


*Figure 6.* Tangent of the loss angle versus $\Pi_{\|}$, variation of $\nu$.

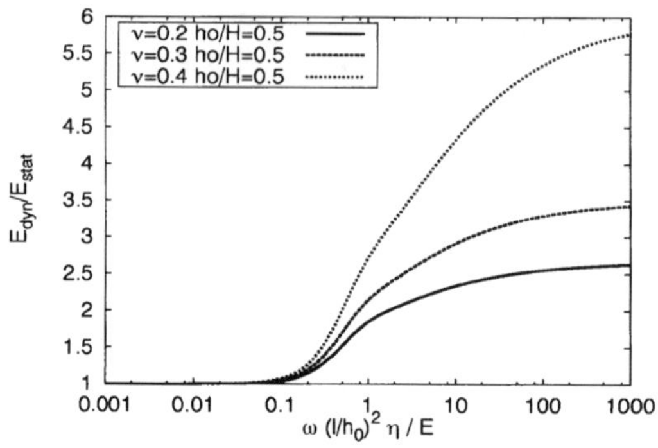


*Figure 7.* Dynamic modulus with respect to the static modulus versus $\Pi_{\|}$, variation of $\nu$.

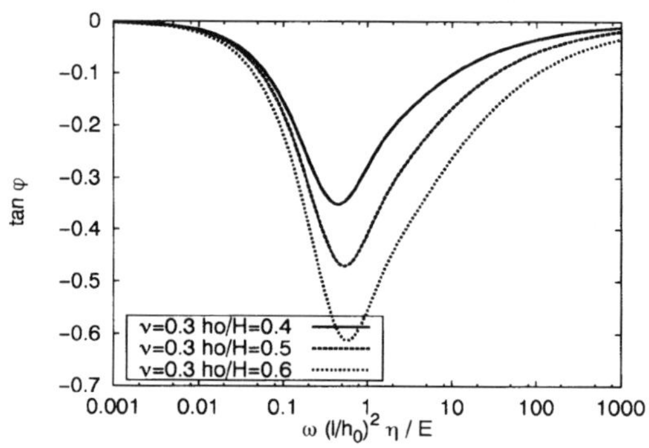


*Figure 8.* Tangent of the loss angle versus $\Pi_{\|}$, variation of $h_0/H$.

The fluid flow relative to the matrix causes the dynamic modulus to increase with the quantity $\Pi_{\|}$. At high values of $\Pi_{\|}$, the fluid flow is negligible, and the incompressible fluid acts as a hindering for the lateral strain. This means that the dynamic modulus increases with increasing values of $\Pi_{\|}$, and tends towards a maximum at high values of $\Pi_{\|}$. This behaviour is shown in Figure 7 and Figure 9. The dynamic modulus, with respect to the static modulus, increases with increasing values of Poisson's ratio $\nu$ (Figure 7), and also increases with increasing portion of the fluid volume (Figure 9).

It is worth mentioning that the Poisson's ratio has a considerable effect on the material behaviour in the case of the flow being parallel to the direction of an applied load. This effect should not be neglected. The qualitative material behaviour of our model is the same as (Rusch, 1965) predicted with his model, but he did not take into account the effect of compressible matrix material. In addition, we can omit some assumptions which (Rusch, 1965) needed to set up his model.

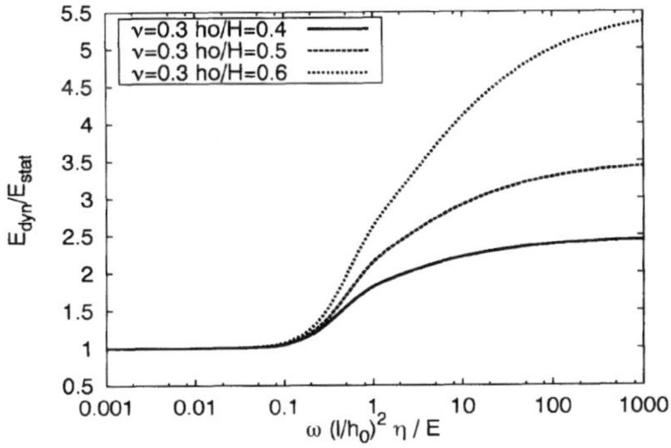


*Figure 9.* Dynamic modulus with respect to the static modulus versus $\Pi_{\|}$, variation of $h_0/H$.

## 4.2. FLOW PERPENDICULAR TO THE DIRECTION OF AN APPLIED LOAD

### 4.2.1. *Material Behaviour for a Sinusoidally Applied Strain*

In the case of a sinusoidally applied strain $\overline{\varepsilon_0} \sin \omega t$, the solution of the partial differential Equation (2.27), inserted into the constitutive Equation (2.25), gives the following expression for the averaged stress:

$$\overline{\sigma} = \overline{\varepsilon_0} \sqrt{C_{1\perp}^2 + C_{2\perp}^2} \, \sin(\omega t + \varphi) \tag{4.11}$$

with

$$\begin{aligned} C_{1\perp} &= A_\perp - B_\perp k_\perp + \frac{B_\perp k_\perp}{b K_\perp} \times \\ &\times \frac{\sinh(\frac{1}{2} bK_\perp)\cosh(\frac{1}{2} bK_\perp) + \sin(\frac{1}{2} bK_\perp)\cos(\frac{1}{2} bK_\perp)}{[\sin(\frac{1}{2} bK_\perp)\sinh(\frac{1}{2} bK_\perp)]^2 + [\cos(\frac{1}{2} bK_\perp)\cosh(\frac{1}{2} bK_\perp)]^2}, \end{aligned} \tag{4.12}$$

$$C_{2\perp} = \frac{B_\perp k_\perp}{b K_\perp} \cdot \frac{\sin(\frac{1}{2} bK_\perp)\cos(\frac{1}{2} bK_\perp) - \sinh(\frac{1}{2} bK_\perp)\cosh(\frac{1}{2} bK_\perp)}{[\sin(\frac{1}{2} bK_\perp)\sinh(\frac{1}{2} bK_\perp)]^2 + [\cos(\frac{1}{2} bK_\perp)\cosh(\frac{1}{2} bK_\perp)]^2}, \tag{4.13}$$

$$K_\perp = \sqrt{\frac{\omega}{2a_\perp}}. \tag{4.14}$$

In this case, the loss angle $\varphi$ is given by

$$\tan \varphi = \frac{C_{2\perp}}{C_{1\perp}}, \tag{4.15}$$

and the dynamic modulus $E_{\rm dyn}$ follows from

$$E_{\rm dyn} = \frac{\overline{\sigma_0}}{\overline{\varepsilon_0}} = \sqrt{C_{1\perp}^2 + C_{2\perp}^2}. \tag{4.16}$$

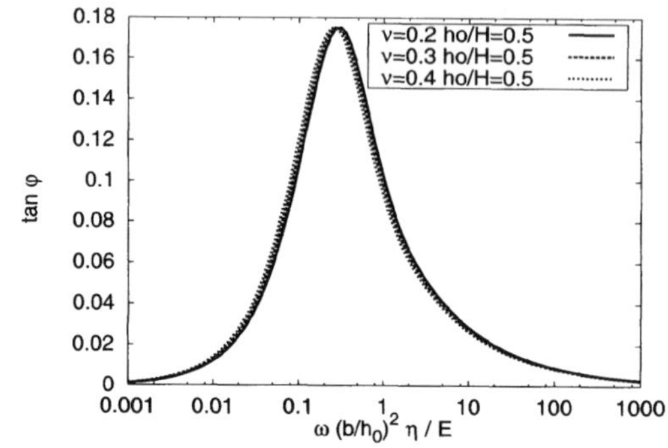


*Figure 10.* Tangent of the loss angle versus $\Pi_\perp$, variation of $\nu$.

The dynamic modulus $E_{\text{dyn}}$ and the tangent of the loss angle, $\tan\varphi$, are again functions of the ratio $h_0/H$, Poisson's ratio $\nu$ of the linear-elastic plates and a certain dimensionless quantity. This quantity $\Pi_\perp$ is analogous to the quantity $\Pi_\parallel$, but the ratio of the fluid path length to the gap length is now given by $b/h_0$:

$$\Pi_\perp = \omega \left(\frac{b}{h_0}\right)^2 \frac{\eta}{E}. \tag{4.17}$$

For small values of the dimensionless quantity $\Pi_\perp$, the dynamic modulus $E_{\text{dyn}}$, denoted $E_{\text{stat}}$, is

$$E_{\text{dyn}}\,(\Pi_\perp \to 0) = E_{\text{stat}} = A_\perp. \tag{4.18}$$

For very large values of the dimensionless quantity $\Pi_\perp$, the dynamic modulus $E_{\text{dyn}}$ is given by

$$E_{\text{dyn}}\,(\Pi_\perp \to \infty) = A_\perp - B_\perp k_\perp. \tag{4.19}$$

The qualitative material behaviour when the flow is perpendicular to the load is similar to the behaviour of the case where the flow is parallel to the load (Figures

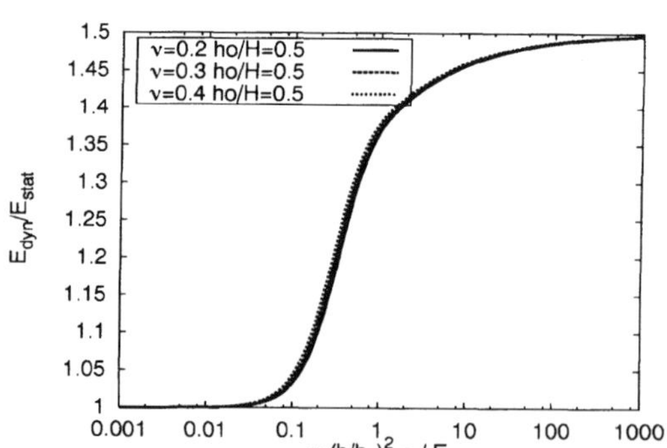


*Figure 11.* Dynamic modulus with respect to the static modulus versus $\Pi_\perp$, variation of $\nu$.

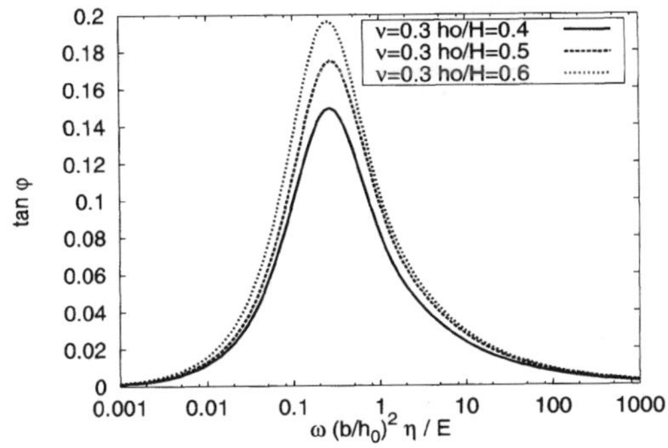


*Figure 12.* Tangent of the loss angle versus $\Pi_{\perp}$, variation of $h_0/H$.

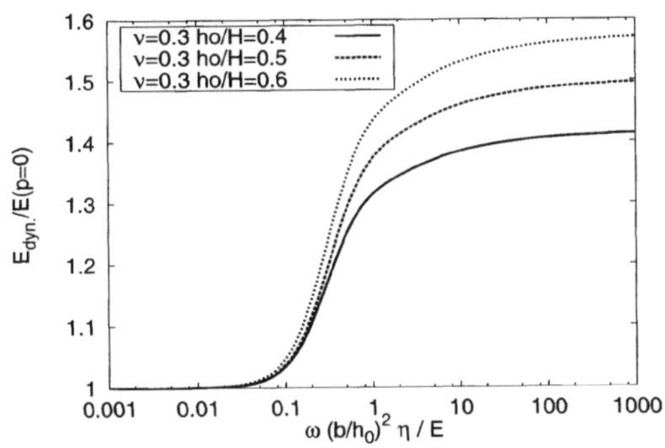


*Figure 13.* Dynamic modulus with respect to the static modulus versus $\pi_{\perp}$, variation of $h_0/H$.

10–13). The maximum energy dissipation per cycle, when the flow is perpendicular to the load, is considerably lower than the maximum energy dissipation of the first case. However, now the influence of Poisson's ratio on both, the tangent of the loss angle (Figure 10) and the dynamic modulus (Figure 11), is negligible. Increasing values of the fluid volume portion $h_0/H$ again cause a higher energy dissipation (Figure 12) and higher values of the dynamic modulus (Figure 13).

## 5. Conclusion

The presented micromechanical model of a fluid-filled open-cell foam is appropriate to describe the damping behaviour resulting from fluid flow during deformation. Furthermore, the model yields the parameters upon which the damping behaviour depends. This cannot be found by using macroscopic continuum approaches, and, with this understanding, it is possible to construct fluid-filled open-cell polymer foams with desired damping characteristics.

The parameters on which the damping behaviour depends, are the Poisson's ratio $\nu$, the fluid volume portion $h_0/H$ and a dimensionless quantity. When the load is applied parallel to the direction of the fluid flow, the quantity is given by $\omega\,(l/h_0)^2\,\eta/E$, and, when the load is applied perpendicular to the direction of the fluid flow, the quantity is given by $\omega\,(b/h_0)^2\,\eta/E$.

## References

Biot, M. A.: 1956, General solutions of the equations of elasticity and consolidation for a porous material, *J. Appl. Mech. Trans. ASME* **78**, 91–96.

Coussy, O.: 1991, *Méchanique des Milieux Poreux*, Editions Technip, Paris.

de Boer, R., Ehlers, W., Kowalski, S. and Plischka, J.: 1991, Porous media: A survey of different approaches, Forschungsberichte aus dem Fachgebiet Bauwesen 54, Univ.-GH Essen.

Detournay, E. and Cheng, A. H.-D.: 1993, *Fundamentals of Poroelasticity, Comprehensive Rock Engineering*, In: J. Hudson (ed.), Pergamon Press, Oxford.

Eringen, A. C.: 1975, *Continuum Physics*, Vol. II, Academic Press, New York.

Rusch, K. C.: 1965, Dynamic behaviour of flexible open-cell foams, PhD Thesis, The University of Akron, pp. 28–36.

# Quasi-Static and Dynamic Behavior of Saturated Porous Media with Incompressible Constituents

STEFAN BREUER
*Institute of Mechanics, Fb-10 Bauwesen, University of Essen, D-45117 Essen, Germany*

(Received: 30 August 1997; in final form: 5 January 1998)

**Abstract.** In this paper the field equations governing the dynamic response of a fluid-saturated elastic porous medium are analyzed and built up for the study of quasi-static and dynamical problems like the consolidation problem and wave propagation. The two constituents are assumed to be incompressible. A numerical solution is derived by means of the standard Galerkin procedure and the finite element method.

**Key words:** volume fraction, wave propagation, consolidation.

## 1. Introduction

Current interest in geomechanics is focused on transient phenomena occurring in earthquakes and consolidation. Since soil consists of a portion of space occupied partly by a solid phase (solid skeleton) and partly by a void space filled with fluid, it must be treated as a two-phase medium. The mechanical behavior of a fluid-saturated porous medium is governed mainly by the interaction of the solid skeleton with the fluid. This interaction occurs in quasi-static problems, like foundations, but is particularly strong in dynamic problems, for example earthquakes.

The equations governing the interaction of solid and fluid media were first established for dynamic problems by Fillunger (1936). Another approach for dynamic problems was formulated by Biot (1955), who did not proceed from the basic equations of mechanics, but introduced quite a good model in an intuitive way. At a later stage the mixture theory restricted by the volume fraction concept provided a new basis for such coupled phenomena. A survey of the historical development of the porous media theory and a discussion of inconsistencies implicit in the mixture theory has recently been given by de Boer (1996). His work ends with an introduction of the model developed by de Boer (1996) and Bluhm (1997), which is the basis for the calculation shown in this paper.

Most of the problems of the two-phase behavior of a fluid-saturated porous medium can only be predicted quantitatively by elaborate numerical computation, which, fortunately, is possible today due to the development of powerful computers. Only a few analytical solutions are available, for example de Boer *et al.* (1993). These analytical solutions have been used to verify the numerical results based on the same theory in a former investigation (see Breuer, 1997). For practical work with arbitrary boundary condition we must use the finite element method. Many investigations, based on different theories, are made in this field, for example by Prévost (1981), Zienkiewicz and Shiomi (1984), Lewis and Schrefler (1987), and Ehlers and Diebels (1994).

In this investigation we start with the basic equations of porous media theory and build them up for numerical calculation. Calculations are restricted to small deformations, thus geometric and material linear solutions are presented. In particular solutions for a two-dimensional consolidation problem and wave propagation in a two-phase porous medium with incompressible constituents will be shown. In order to simplify the problem, thermal effects and exchanges of mass between the constituents are excluded.

## 2. Governing Equations

### 2.1. KINEMATICS AND THE CONCEPT OF VOLUME FRACTIONS

Considering the kinematics of the fluid-saturated porous medium, which is an immiscible mixture of the constituents $\varphi^\alpha$ with particles $X^\alpha$ ($\alpha$ = S: solid phase, $\alpha$ = F: fluid phase), it is assumed that at any time $t$ each spatial point is simultaneously occupied by the particles $X^{\mathrm{S}}$ and $X^{\mathrm{F}}$. These particles $X^\alpha$ proceed from different reference positions $\mathbf{X}_\alpha$ at time $t = t_0$. Thus, each constituent is assigned its own independent motion function $\boldsymbol{\chi}_\alpha$. According to these remarks all unknown fields appearing in this paper are defined on the actual domain $B_0^{\mathrm{S}}$. The velocity $\mathbf{x}'_{\mathrm{S}}$ of the skeleton, its acceleration $\mathbf{x}''_{\mathrm{S}}$ and deformation gradient $\mathbf{F}_{\mathrm{S}}$ at an arbitrary position $\mathbf{X}_{\mathrm{S}} \in B_0^{\mathrm{S}}$ and an arbitrary instant of time $t$ are described by

$$\mathbf{x} = \boldsymbol{\chi}_{\mathrm{S}}(\mathbf{X}_{\mathrm{S}}, t), \quad \mathbf{x}'_{\mathrm{S}} = \mathbf{x}'_{\mathrm{S}}(\mathbf{X}_{\mathrm{S}}, t), \quad \mathbf{x}''_{\mathrm{S}} = \mathbf{x}''_{\mathrm{S}}(\mathbf{X}_{\mathrm{S}}, t), \quad \mathbf{F}_{\mathrm{S}} = \mathrm{Grad}_{\mathrm{S}}\boldsymbol{\chi}_{\mathrm{S}}, \tag{1}$$

where $\mathbf{x}$ denotes the current position of the point $\mathbf{X}_{\mathrm{S}}$ and $\mathrm{Grad}_{\mathrm{S}}$ means the derivative with respect to $\mathbf{X}_{\mathrm{S}}$.

The motion of the fluid is given in the natural Eulerian way by the velocity field $\mathbf{x}'_{\mathrm{F}}$ of the actual configuration space of point $\mathbf{x}$ and its acceleration $\mathbf{x}''_{\mathrm{F}}$ is described by

$$\mathbf{x}'_{\mathrm{F}} = \mathbf{x}'_{\mathrm{F}}(\mathbf{x}, t), \qquad \mathbf{x}''_{\mathrm{F}} = \mathbf{x}''_{\mathrm{F}}(\mathbf{x}, t). \tag{2}$$

The volume fractions $n^\alpha$ are defined as the local ratios of the constituent volumes $v^\alpha$ with respect to the bulk volume $v$ of the control space $B_{\mathrm{S}}$, which is shaped by the solid skeleton

$$n^\alpha = \frac{\mathrm{d}v^\alpha}{\mathrm{d}v}. \tag{3}$$

As a consequence of (3), we can calculate the partial volumes of each constituent by weighting the bulk volume with the volume fractions

$$v^\alpha = \int_{B_{\mathrm{S}}} n^\alpha \,\mathrm{d}v. \tag{4}$$

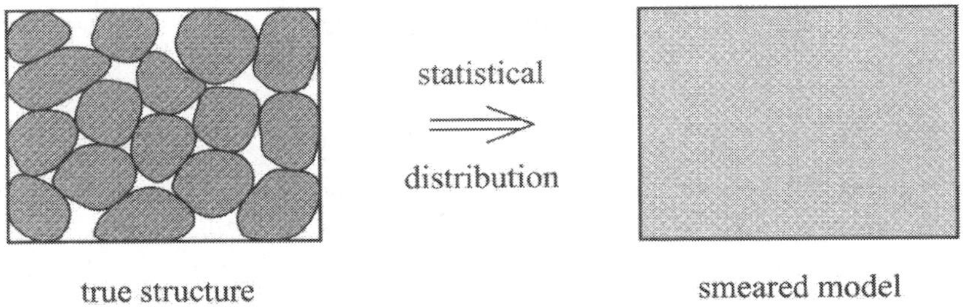


*Figure 1.* Real structure and smeared model.

With the aid of this assumption we obtain from the following identities:

$$v = \int_{B_S} \mathrm{d}v = \sum_{\alpha=1}^{\kappa} v^\alpha = \int_{B_S} \sum_{\alpha=1}^{\kappa} \mathrm{d}v^\alpha = \int_{B_S} \sum_{\alpha=1}^{\kappa} n^\alpha \, \mathrm{d}v, \tag{5}$$

the volume fraction condition

$$\sum_{\alpha=1}^{\kappa} n^\alpha = 1. \tag{6}$$

The volume fraction condition (6) plays an important role as a constraint in the constitutive theory of saturated porous media, (see Bluhm and de Boer, 1997), where this topic is discussed.

Each of the constituents $\varphi^\alpha$ have a real density $\varrho^{\alpha R}$, which is defined as the mass of $\varphi^\alpha$ per unit of $v^\alpha$. With the aid of the volume fraction concept these properties can be 'smeared' over the control space, see Figure 1, and we have the partial density

$$\varrho^\alpha = n^\alpha \, \varrho^{\alpha R}. \tag{7}$$

In Figure 1 the real structure of a porous medium is shown on the left side and on the right side the smeared model is shown.

### 2.2. FIELD EQUATIONS

Excluding mass exchanges and thermal effects the mechanical behavior of a fluid-saturated porous solid is described in porous media theory by the balance equation of mass for each individual constituent

$$(\varrho^\alpha)'_\alpha + \varrho^\alpha \ \mathrm{div} \ \mathbf{x}'_\alpha = 0, \tag{8}$$

the balance equation of momentum

$$\mathrm{div} \ \mathbf{T}^\alpha + \varrho^\alpha (\mathbf{b} - \mathbf{x}''_\alpha) + \hat{\mathbf{p}}^\alpha = \mathbf{0} \tag{9}$$

and the volume fraction condition that changes for a binary mixture into the saturation condition

$$n^{S} + n^{F} = 1. \tag{10}$$

In these equations $\mathbf{T}^{\alpha}$ is the partial Cauchy stress tensor, $\mathbf{b}$ the external acceleration and $\hat{\mathbf{p}}^{\alpha}$ the interaction force of the constituents. In addition, 'div' is the divergence operator and the symbol $(\ldots)'_{\alpha}$ defines the material time derivative with respect to the trajectory of $\varphi^{\alpha}$.

As the sum of the interaction forces must vanish, we obtain for a binary mixture

$$\hat{\mathbf{p}}^{F} + \hat{\mathbf{p}}^{S} = \mathbf{0}. \tag{11}$$

The balance equation of moment of momentum leads, excluding any moment of momentum supply, to a symmetric stress tensor

$$\mathbf{T}^{\alpha} = \mathbf{T}^{\alpha T}. \tag{12}$$

Since both constituents are incompressible, we have

$$\varrho^{\alpha R} = \text{constant}. \tag{13}$$

With this assumption, the volume fractions can be calculated from the balance equations of mass (8) and with the aid of the deformation gradient, one obtains

$$n^{S} = n^{S0}(\det \mathbf{F}_{S})^{-1}, \qquad n^{F} = 1 - n^{S}, \tag{14}$$

where $n^{S0}$ describes the initial porosity of $\varphi^{S}$.

### 2.3. CONSTITUTIVE RELATIONS

In the above-mentioned field equations, the number of unknown fields does not correspond to the number of equations, thus we have to close the problem with constitutive equations for the partial stress tensors $\mathbf{T}^{\alpha}$ and the interaction force $\hat{\mathbf{p}}^{F}$.

The constitutive relations for the solid and fluid stress tensor $\mathbf{T}^{\alpha}$ and for the interaction force $\hat{\mathbf{p}}^{F}$ consist of two terms, see Bowen (1980), Ehlers (1989), de Boer (1996) and Bluhm (1997). The former, as a result of the saturation condition, is proportional to the pore pressure $p$, while the latter represents the extra quantities, index $(\ldots)_{E}$, determined by the deformations:

$$\mathbf{T}^{\alpha} = -n^{\alpha} p\mathbf{I} + \mathbf{T}^{\alpha}_{E}, \tag{15}$$

$$\hat{\mathbf{p}}^{F} = p \text{ grad } n^{F} + \hat{\mathbf{p}}^{F}_{E}. \tag{16}$$

The viscosity of the fluid influences the interaction force $\hat{\mathbf{p}}^{\alpha}$, but with regard to the partial effective stress tensor of the fluid it can be neglected:

$$\mathbf{T}^{F}_{E} \approx \mathbf{0}. \tag{17}$$

The partial effective stress tensor of the solid can be expressed for linear elastic behavior and small deformations by Hooke's law:

$$\mathbf{T}_{\mathrm{E}}^{\mathrm{S}} = 2\mu^{\mathrm{S}}\mathbf{E}_{\mathrm{S}} + \lambda^{\mathrm{S}}(\mathbf{E}_{\mathrm{S}}\cdot\mathbf{I})\,\mathbf{I}, \tag{18}$$

where $\mathbf{E}_{\mathrm{S}} = 0.5\ (\mathrm{grad}\ \mathbf{u}_{\mathrm{S}} + \mathrm{grad}^{\mathrm{T}}\ \mathbf{u}_{\mathrm{S}})$ is Green's strain tensor, $\mathbf{u}_{\mathrm{S}}$ are the solid displacements

$$\mathbf{u}_{\mathrm{S}} = \mathbf{x} - \mathbf{X}_{\mathrm{S}}, \tag{19}$$

and $\lambda^{\mathrm{S}}$, $\mu^{\mathrm{S}}$ are the Lamé constants of the partial solid body.

The interaction between the fluid and solid constituents, caused by the motions, can be described by the extra supply term of momentum

$$\hat{\mathbf{p}}_{\mathrm{E}}^{\mathrm{F}} = -\frac{(n^{\mathrm{F}})^2\gamma^{\mathrm{FR}}}{k^{\mathrm{F}}}\mathbf{w}_{\mathrm{F}}, \tag{20}$$

with $\mathbf{w}_{\mathrm{F}} = (\mathbf{x}'_{\mathrm{F}} - \mathbf{x}'_{\mathrm{S}})$ being the seepage velocity, $\gamma^{\mathrm{FR}} = (\varrho^{\mathrm{FR}}|\mathbf{b}|)$ the real specific weight of the fluid and $k^{\mathrm{F}}$ the Darcy permeability parameter.

### 2.4. MODIFICATION OF VARIABLES

As mentioned above the control space of our porous medium is shaped by the solid skeleton. Thus, the velocity of the fluid is not of interest, but the motion relative to the solid skeleton is. Instead of the velocity $\mathbf{x}'_{\mathrm{F}}$ of the fluid, let us introduce the seepage velocity $\mathbf{w}_{\mathrm{F}}$, which is the fluid velocity relative to the motion of the solid:

$$\mathbf{w}_{\mathrm{F}} = \mathbf{x}'_{\mathrm{F}} - \mathbf{x}'_{\mathrm{S}}. \tag{21}$$

This quantity occurs in a natural way in the extra supply term of momentum, see (20). In the other equations the absolute velocity of the fluid must be replaced by

$$\mathbf{x}'_{\mathrm{F}} = \mathbf{w}_{\mathrm{F}} + \mathbf{x}'_{\mathrm{S}}. \tag{22}$$

This modification also has an advantage in describing the boundary conditions. At a moving, undrained boundary it is easier to prescribe that no fluid comes out ($\mathbf{w}_{\mathrm{F}} = 0$), than to prescribe that the velocity of the fluid is equal to the velocity of the skeleton ($\mathbf{x}'_{\mathrm{F}} = \mathbf{x}'_{\mathrm{S}}$). A disadvantage of this modification is to be seen when we want to substitute the acceleration of the fluid by the time derivations of $\mathbf{w}_{\mathrm{F}}$ and $\mathbf{x}'_{\mathrm{S}}$. In this case we have to take the convective acceleration terms into account and we get

$$(\mathbf{x}'_{\mathrm{F}})'_{\mathrm{F}} = (\mathbf{w}_{\mathrm{F}} + \mathbf{x}'_{\mathrm{S}})'_{\mathrm{S}} + \underbrace{\mathrm{grad}(\mathbf{w}_{\mathrm{F}} + \mathbf{x}'_{\mathrm{S}})\ \mathbf{w}_{\mathrm{F}}}_{\text{convective acceleration terms}}. \tag{23}$$

These convective acceleration terms in the fluid are in practice relatively insignificant in view of the shortcoming with which the permeability $k^{\mathrm{F}}$ is determined and analogous to Zienkiewicz and Shiomi (1984) these terms are to be omitted from here onwards.

### 2.5. FINAL EQUATIONS

Summing up the assumptions of the previous sections and substituting the constitutive equation in the field equations we get a final set of three equations. At first, we take the balance equation of momentum for the solid phase, inserting the assumptions (15) and (18) for the stress tensor and (16) and (20) for the interaction force. Thus, we obtain

$$\operatorname{div}(\mathbf{T}_E^S - n^S p\mathbf{I}) + \varrho^S \mathbf{b} - \varrho^S \mathbf{x}_S'' - p \text{ grad } n^F + \frac{\gamma^{FR} (n^F)^2}{k^F} \mathbf{w}_F = \mathbf{0}. \tag{24}$$

This equation describes the forces acting on the solid phase. The stress tensor is split into an effective part, determined by the motion, and the weighted pore pressure. The terms describing the weight and acceleration are similar to terms used in continuum mechanics, but this equation is extended by the interaction force. In the same manner we obtain an equation for the fluid phase, represented by

$$\operatorname{div}(-n^F p\mathbf{I}) + \varrho^F \mathbf{b} - \varrho^F (\mathbf{w}_F + \mathbf{x}_S')_S' + p \text{ grad } n^F - \frac{\gamma^{FR} (n^F)^2}{k^F} \mathbf{w}_F = \mathbf{0}. \tag{25}$$

Until here we have two vectorial equations for the unknown motions of the solid and the fluid. Another unknown is the pressure $p$. Therefore, we take the mass balance equations, considering the incompressibility and the saturation condition. Doing that, we obtain

$$\operatorname{div}(n^F \mathbf{w}_F + \mathbf{x}_S') = 0. \tag{26}$$

This equation represents the volume balance equation of the whole mixture. As a result of incompressibility, we can see, that if one part of the volume changes, for example div $\mathbf{x}_S'$, the change of the divergence of the other part is constrained.

The calculation of the mechanical behavior of a porous medium based on this theory is governed by these three equations. The motion of solid, the motion of fluid and the pore pressure appear as unknown quantities.

## 3. Numerical Solution

The system of equations introduced in the previous section have to be built up for calculation by using the standard finite element procedures and weak forms of the appropriate equations. Details of such methods are well described by Zienkiewicz (1984) or Bathe (1990). A semi-discretization procedure is used by approximating the unknown functions in the space domain. Time integration will be done by the Newmark method.

For an effective way to solve the system of equations and to match the problem to the boundary and initial conditions we have to appropriate the balance equations of momentum. Firstly, we combine the balance equations of momentum (9) of both constituents

$$\operatorname{div}(\mathbf{T}_E^S - p\mathbf{I}) + (\varrho^S + \varrho^F)\mathbf{b} - \varrho^S \mathbf{x}_S'' - \varrho^F (\mathbf{w}_F + \mathbf{x}_S')_S' = \mathbf{0}, \tag{27}$$

thus getting an equation that describes the balance equation of momentum of the whole mixture. The interaction terms vanish, the stress tensor is the addition of the partial stress tensors and the external acceleration acts on both the constituents. The sum of the inertia forces can be rearranged according to (28):

$$-\varrho^{\mathrm{S}}\mathbf{x}_{\mathrm{S}}'' - \varrho^{\mathrm{F}}\mathbf{x}_{\mathrm{F}}'' = -\varrho^{\mathrm{S}}\mathbf{x}_{\mathrm{S}}'' - \varrho^{\mathrm{F}}(\mathbf{w}_{\mathrm{F}} + \mathbf{x}_{\mathrm{S}}')_{\mathrm{S}}' = -(\varrho^{\mathrm{S}} + \varrho^{\mathrm{F}})\mathbf{x}_{\mathrm{S}}'' - \varrho^{\mathrm{F}}(\mathbf{w}_{\mathrm{F}})_{\mathrm{S}}'. \quad (28)$$

As a further equation the balance equation of momentum of the fluid is taken into account:

$$\operatorname{div} \mathbf{T}^{\mathrm{F}} + \varrho^{\mathrm{F}}(\mathbf{b} - (\mathbf{w}_{\mathrm{F}} + \mathbf{x}_{\mathrm{S}}')_{\mathrm{S}}') + p \ \operatorname{grad} n^{\mathrm{F}} - \frac{\gamma^{\mathrm{FR}}\ (n^{\mathrm{F}})^2}{k^{\mathrm{F}}}\mathbf{w}_{\mathrm{F}} = \mathbf{0}. \quad (29)$$

This equation describes the mechanical behavior of the partial fluid body. The interaction force occurs here and the stress tensor is the partial stress tensor, which is acting on the fluid.

Finally we take the volume balance equation of the whole mixture (26) into account:

$$\operatorname{div}\,(n^{\mathrm{F}}\mathbf{w}_{\mathrm{F}} + \mathbf{x}_{\mathrm{S}}') = 0. \quad (30)$$

This equation has been discussed in the previous section and till this point there is no need to change this equation.

### 3.1. WEAK FORMULATION

For numerical computations, a standard Galerkin procedure is chosen. Therefore, each of the basic Equations (27), (29) and (30) must be multiplied by a weighting function. For Equation (27) a virtual solid velocity $\bar{\mathbf{u}}_{\mathrm{S}}$ is applied, which must satisfy the geometric boundary conditions. The volume integral of a divergence can be transformed into a surface integral (see de Boer, 1982):

$$\int_B [(\mathbf{T}_{\mathrm{E}}^{\mathrm{S}} - p\mathbf{I}) \cdot \operatorname{grad} \bar{\mathbf{u}}_{\mathrm{S}} + \varrho^{\mathrm{S}}\,\mathbf{x}_{\mathrm{S}}'' \cdot \bar{\mathbf{u}}_{\mathrm{S}} + \varrho^{\mathrm{F}}\,(\mathbf{w}_{\mathrm{F}} + \mathbf{x}_{\mathrm{S}}')_{\mathrm{S}}' \cdot \bar{\mathbf{u}}_{\mathrm{S}}\,]\,\mathrm{d}v$$
$$= \int_{\partial B_t} \mathbf{t} \cdot \bar{\mathbf{u}}_{\mathrm{S}}\,\mathrm{d}a + \int_B (\varrho^{\mathrm{S}} + \varrho^{\mathrm{F}})\mathbf{b} \cdot \bar{\mathbf{u}}_{\mathrm{S}}\,\mathrm{d}v, \quad (31)$$

where $\mathbf{t}$ is the total stress vector on the surface of the mixture, consisting of the stress on the solid and on the fluid.

In Equation (29) a virtual seepage velocity $\bar{\mathbf{w}}_{\mathrm{F}}$ was taken into account; also this virtual variable must satisfy the same geometric boundary conditions like the seepage velocity, and the volume integral was transformed into a surface integral

$$\int_B \left\{(-p \operatorname{div} \bar{\mathbf{w}}_{\mathrm{F}}) + \frac{\gamma^{\mathrm{FR}} n^{\mathrm{F}}}{k^{\mathrm{F}}}\mathbf{w}_{\mathrm{F}} \cdot \bar{\mathbf{w}}_{\mathrm{F}} + \varrho^{\mathrm{FR}}\,(\mathbf{w}_{\mathrm{F}} + \mathbf{x}_{\mathrm{S}}')_{\mathrm{S}}' \cdot \bar{\mathbf{w}}_{\mathrm{F}}\right\} \mathrm{d}v$$
$$= -\int_{\partial B_t} p\bar{\mathbf{w}}_{\mathrm{F}} \cdot \mathbf{n}\,\mathrm{d}a + \int_B \varrho^{\mathrm{FR}}\mathbf{b} \cdot \bar{\mathbf{w}}_{\mathrm{F}}\,\mathrm{d}v. \quad (32)$$

Equation (30) closes the set of equations. It represents the volume balance equation of the whole mixture and was multiplied by a virtual pressure $\bar{p}$, which must satisfy the static boundary conditions:

$$\int_B \operatorname{div}\,(n^{\mathrm{F}}\mathbf{w}_{\mathrm{F}} + \mathbf{x}'_{\mathrm{S}})\ \bar{p}\,\mathrm{d}v = 0. \tag{33}$$

In this form the equation causes numerical problems, because the weighting function $\bar{p}$ is not coupled with its corresponding variable $p$. As a result of this we get a rank defect in the element martix. In order to avoid this problem we take the balance equation of momentum of the fluid (25) and solve this equation explicit for $\mathbf{w}_{\mathrm{F}}$. Next we replace $\mathbf{w}_{\mathrm{F}}$ in Equation (33) by this expression and after some transformations and changes we obtain

$$\begin{aligned}
&\int_B \operatorname{div}\left[-\frac{k^{\mathrm{F}}\,\varrho^{\mathrm{FR}}}{\gamma^{\mathrm{FR}}}\,(\mathbf{w}'_{\mathrm{F}} + \mathbf{x}''_{\mathrm{S}}) + \mathbf{x}'_{\mathrm{S}}\right]\bar{p}\,\mathrm{d}v - \\
&\quad - \int_B \frac{k^{\mathrm{F}}\varrho^{\mathrm{FR}}}{\gamma^{\mathrm{FR}}}\left[\mathbf{b} - \frac{1}{\varrho^{\mathrm{FR}}}\operatorname{grad} p\right]\cdot \operatorname{grad}\bar{p}\,\mathrm{d}v \\
&\quad = -\int_{\partial B}\left[\frac{k^{\mathrm{F}}\varrho^{\mathrm{FR}}}{\gamma^{\mathrm{FR}}}(\mathbf{w}'_{\mathrm{F}} + \mathbf{x}''_{\mathrm{S}}) + n^{\mathrm{F}}\mathbf{w}_{\mathrm{F}}\right]\cdot \mathbf{n}\ \bar{p}\,\mathrm{d}a.
\end{aligned} \tag{34}$$

This transformation was first done by Diebels and Ehlers (1996) in order to overcome these numerical problems. Thus, we get an equation where the virtual pressure $\bar{p}$ and the variable $p$ occur, thus the rank defect vanishes.

Moreover, we get a reduced system of variables and equations for quasi-static problems. In this case the acceleration terms can be neglected and we obtain the following set of equations:

$$\int_B (\mathbf{T}^{\mathrm{S}}_{\mathrm{E}} - p\,\mathbf{I})\cdot \operatorname{grad}\bar{\mathbf{u}}_{\mathrm{S}}\,\mathrm{d}v = \int_{\partial B_t} \mathbf{t}\cdot\bar{\mathbf{u}}_{\mathrm{S}}\,\mathrm{d}a + \int_B (\varrho^{\mathrm{S}} + \varrho^{\mathrm{F}})\mathbf{b}\cdot\bar{\mathbf{u}}_{\mathrm{S}}\,\mathrm{d}v, \tag{35}$$

$$\begin{aligned}
&\int_B (-p\,\operatorname{div}\bar{\mathbf{w}}_{\mathrm{F}}) + \frac{\gamma^{\mathrm{FR}}\,n^{\mathrm{F}}}{k^{\mathrm{F}}}\mathbf{w}_{\mathrm{F}}\cdot\bar{\mathbf{w}}_{\mathrm{F}}\,\mathrm{d}v \\
&\quad = -\int_{\partial B_t} p\bar{\mathbf{w}}_{\mathrm{F}}\cdot\mathbf{n}\,\mathrm{d}a + \int_B \varrho^{\mathrm{FR}}\mathbf{b}\cdot\bar{\mathbf{w}}_{\mathrm{F}}\,\mathrm{d}v,
\end{aligned} \tag{36}$$

$$\begin{aligned}
&\int_B \left[\operatorname{div}\mathbf{x}'_{\mathrm{S}}\,\bar{p} - \frac{k^{\mathrm{F}}\varrho^{\mathrm{FR}}}{\gamma^{\mathrm{FR}}}\left(\mathbf{b} - \frac{1}{\varrho^{\mathrm{FR}}}\operatorname{grad} p\right)\cdot\operatorname{grad}\bar{p}\right]\mathrm{d}v \\
&\quad = -\int_{\partial B} n^{\mathrm{F}}\mathbf{w}_{\mathrm{F}}\cdot\mathbf{n}\,\bar{p}\,\mathrm{d}a.
\end{aligned} \tag{37}$$

In the above-mentioned system of equations the balance equation of momentum of the fluid (36) becomes independent of the other two Equations (35) and (37). Thus

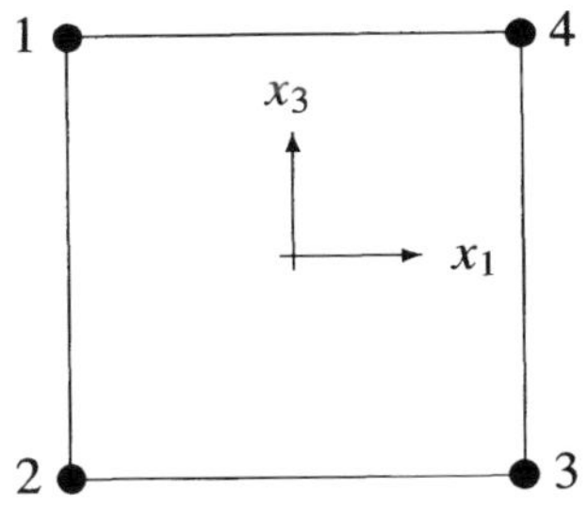


*Figure 2.* Finite element discretization.

we can solve both equations with the unknown fields ($\mathbf{u}_S$, $p$) and if we are interested in the fluid motion, we can calculate the velocity $\mathbf{w}_F$ with the help of Equation (36).

### 3.2. SOLUTION ALGORITHM

From the weak formulation one gets three scalar Equations (31), (32) and (34) with the unknown functions ($\mathbf{u}_S$, $\mathbf{w}_F$, $p$). For the discretization of the problem the unknown functions ($\mathbf{u}_S$, $\mathbf{w}_F$, $p$), as well as their time derivations and the corresponding weighting functions ($\bar{\mathbf{u}}_S$, $\bar{\mathbf{w}}_F$, $\bar{p}$) are approximated by bilinear shape functions, for example,

$$\mathbf{u}_S = \mathbf{u}_{1S} \cdot \mathbf{f}_1 + \mathbf{u}_{2S} \cdot \mathbf{f}_2 + \mathbf{u}_{3S} \cdot \mathbf{f}_3 + \mathbf{u}_{4S} \cdot \mathbf{f}_4,$$

with $\mathbf{u}_{iS}$ ($i = 1, \ldots, 4$) as the nodal values of the unknown $\mathbf{u}_S$ and $\mathbf{f}_i$ ($i = 1, \ldots, 4$) as the shape functions.

As the values of the weighting functions are not specified, the coefficient multiplied with the value of the weighting function must vanish. Now, the three scalar equations of the weak formulation are split into $n$ equations, where $n$ is the number of unknowns at each node, multiplied with the number of nodes pro element. In this case we have four nodes pro element and five unknowns ($\mathbf{u}_S$ horizontal, $\mathbf{u}_S$ vertical, $\mathbf{w}_F$ horizontal, $\mathbf{w}_F$ vertical and $p$) pro node. Thus, the discretization of the problem leads to a system of 20 equations with the unknowns ($\mathbf{u}_S$, $\mathbf{u}'_S$, $\mathbf{u}''_S$, $\mathbf{w}_F$, $\mathbf{w}'_F$, $p$). The matrix of the coefficients multiplied with the value of the discrete unknowns are determined as $\mathbf{M}$, $\mathbf{D}$ and $\mathbf{K}$. $\mathbf{M}$ means the mass matrix and is connected with the second time derivation of the unknowns. $\mathbf{D}$ represents the damping matrix and is coupled with the first time derivation of the unknowns and lastly $\mathbf{K}$ determines the stiffness matrix, which in turn is connected with the unknowns. The index M, F or K in Equation (38) means, that these coefficients are coming from the balance equation of mixture (M), the balance equation of fluid (F) or from the saturation condition (K). The second index represents the kind of unknowns, 1 stands for the motion of solid, 2 for the motion of fluid and 3 for the pressure. $\mathbf{F}$ determines the load vector of the Mixture (M), of the fluid (F) and the right-hand side vector of the volume balance equation

(K). Equation (38) shows the problem after the discretization in form of a matrix equation.

$$\begin{pmatrix} \mathbf{M}_{M1} & 0 & 0 \\ \mathbf{M}_{F1} & 0 & 0 \\ \mathbf{M}_{K1} & 0 & 0 \end{pmatrix} \begin{pmatrix} \mathbf{u}''_S \\ 0 \\ 0 \end{pmatrix} + \begin{pmatrix} 0 & \mathbf{D}_{M2} & 0 \\ \mathbf{D}_{F1} & \mathbf{D}_{F2} & 0 \\ \mathbf{D}_{K1} & \mathbf{D}_{K2} & 0 \end{pmatrix} \begin{pmatrix} \mathbf{u}'_S \\ \mathbf{w}'_F \\ 0 \end{pmatrix} + \\ + \begin{pmatrix} \mathbf{K}_{M1} & 0 & \mathbf{K}_{M3} \\ 0 & \mathbf{K}_{F2} & \mathbf{K}_{F3} \\ 0 & \mathbf{K}_{K2} & \mathbf{K}_{K3} \end{pmatrix} \begin{pmatrix} \mathbf{u}_S \\ \mathbf{w}_F \\ p \end{pmatrix} = \begin{pmatrix} \mathbf{F}^M \\ \mathbf{F}^F \\ \mathbf{F}^K \end{pmatrix}. \tag{38}$$

The Newmark method was chosen for the time integration and with the help of the result of the last time step, the problem can be converted into a $n$ x $n$ Matrix with an update of the load vector.

## 4. Examples

### 4.1. CONSOLIDATION PROBLEM

As a first example the two-dimensional consolidation problem is going to be calculated. This class of problem occurs in soil mechanics, when foundations of buildings are the point of interest, see Figure 3(a). If we take the simplest case into account, the boundary conditions are symmetric and we have only to model one half of the problem (Figure 3(b)). The same structure has been calculated by Ehlers and Diebels (1994), but by use of a different time integration and different shape functions. Therefore, an area of $10\,\text{m} \times 10\,\text{m}$ is taken into account. The material parameters are taken from de Boer *et al.* (1993) as

$$\mu^S = 5583\,\text{kN/m}^2, \qquad \lambda^S = 8375\,\text{kN/m}^2,$$
$$\varrho^{SR} = 2000\,\text{kg/m}^3, \qquad \varrho^{FR} = 1000\,\text{kg/m}^3,$$
$$n^S_{0S} = 0.67, \qquad k^F = 0.01\,\text{m/s}.$$

In order to create an FE-model for the above-mentioned problem, we discretise the plane of $10\,\text{m} \times 10\,\text{m}$ by 361 rectangle elements and prevent the vertical solid displacements and the fluid motion at the ground, as well as the horizontal solid displacements and fluid motion on both sides. At the top, the left side is unloaded

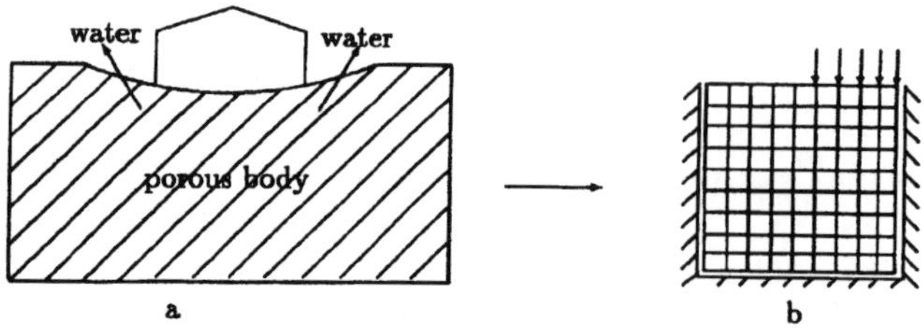


*Figure 3.* Consolidation problem.

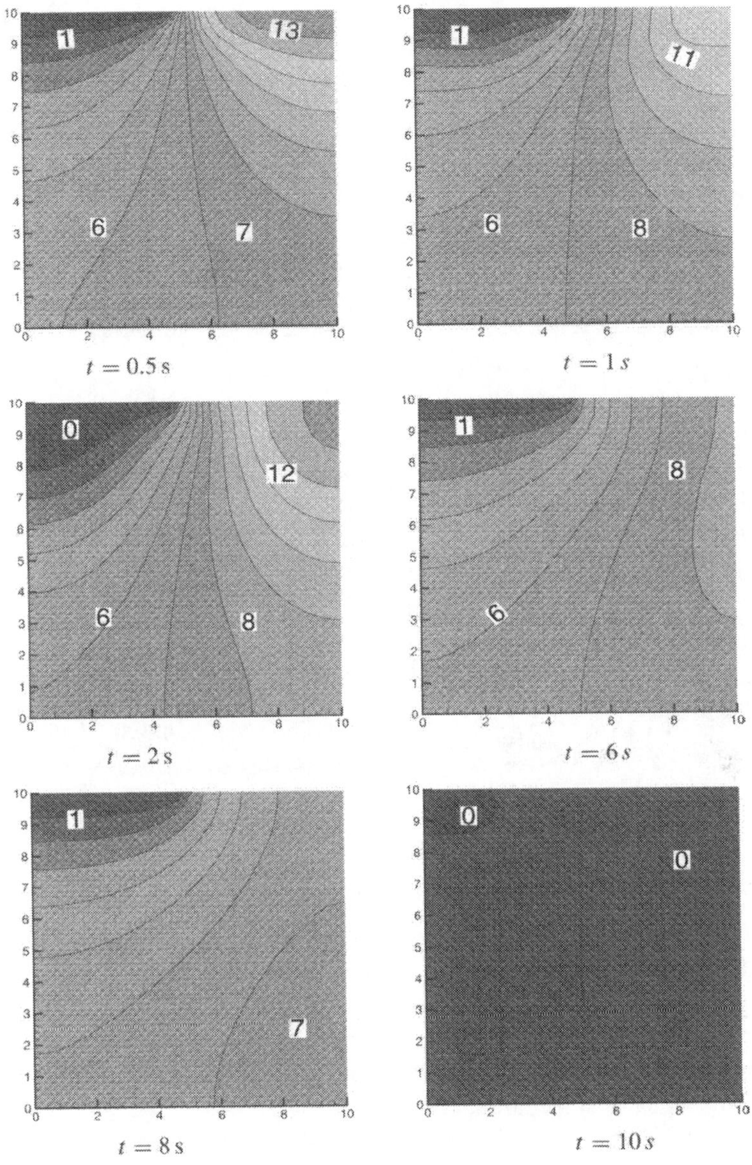


*Figure 4.* Evaluation of the pore pressure.

and drained, whereas the right side is undrained and loaded by a force which simulates the weight of a building.

Due to the constitutive parameters this consolidation process runs quite fast and after 20 s the settlement reaches its final value, while the pore pressure decreases to zero. Figure 4 shows the pore pressure at different intervals of time. The values in these figures are the pore pressures in $kN/m^2$.

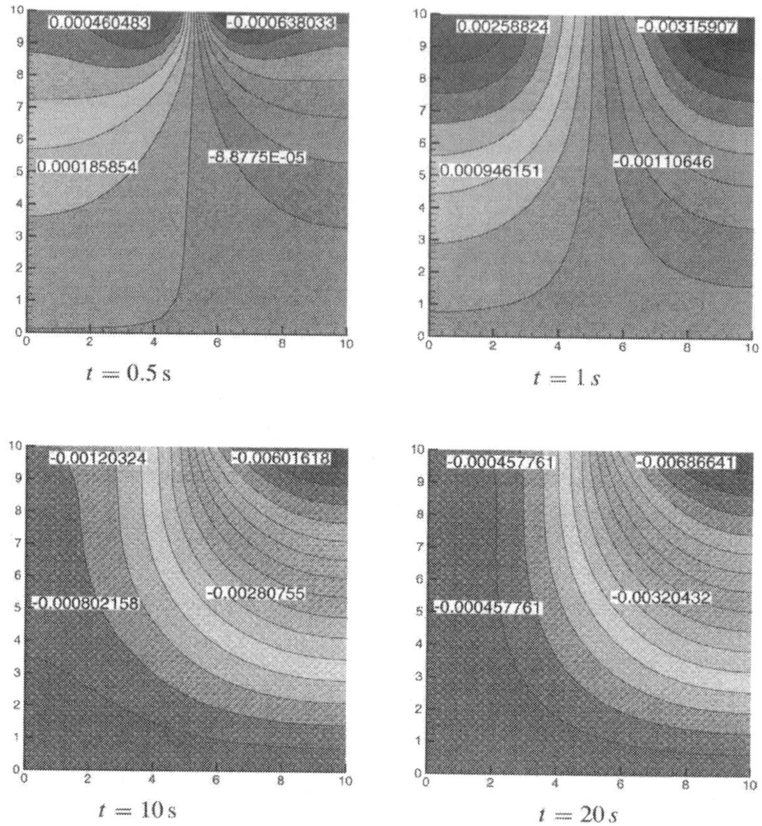


*Figure 5.* Evaluation of the solid displacement.

In Figure 4 we can see that in the beginning (time = 0.01 s) the whole external load (15 kN/m$^2$) is carried by the water body and the pore pressure under the external load moves upto 14 kN/m$^2$. With the passage of time, the pore pressure decreases and after 10 s the value of the pore pressure is nearly zero.

Figure 5 shows the evaluation of the solid displacement. The labels in these plots are the values of vertical displacements in m. At the beginning only a small amount of water is taken out of the porous body. The ground on the right side moves down and as a result of the incompressibility the ground on the left side moves up in the same way. With the passage of time the right side is still moving down whereas the left side stops moving up and starts with its settlement. During this time the water squeezes out of the porous body and after 20 s the process stops. The settlement reaches its final value and the pore pressure is zero.

In Figure 6 the direction of the fluid motion is shown. These directions do not vary, the values of the velocities start with large quantities and decreases to zero after 20 s. As an example a vector plot of the direction of the fluid motion is exhibited at time = 6 s.

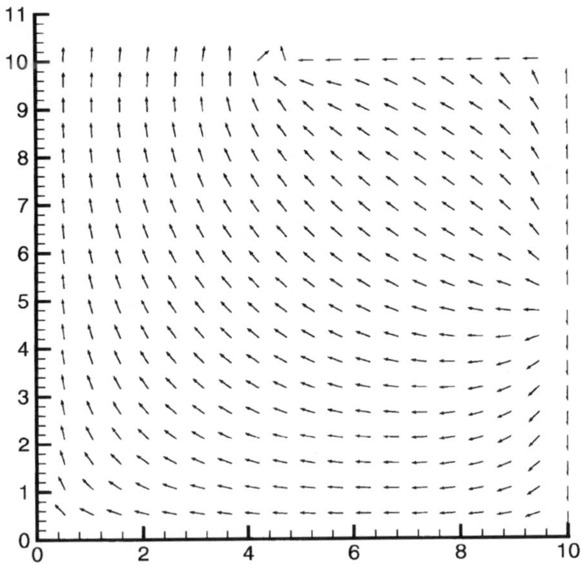

*Figure 6.* Direction of fluid motion.

## 4.2. WAVE PROPAGATION

In this section the wave propagation in porous media is taken into account. According to Biot's theory (see Biot, 1956) with two compressible constituents there are two longitudinal waves in a porous medium. One wave of dilatation is transmitted through the compressibility of the fluid and solid, the other is transmitted through the elastic structure. These two waves are coupled through the stiffness of the soil and fluid components of the system as well as through the coupling effect produced by motions of the solid and fluid, (see Richart *et al.*, 1970).

In this investigation both constituents are incompressible, thus the speed of the wave transmitted through the compressibility of the fluid and solid is infinite. The wave motion which we can observe is the motion transmitted through the elastic structure of the solid skeleton.

The calculation of longitudinal waves in a one-dimensional porous body has been the topic of a former investigation (see Breuer, 1997). In this paper the propagation of waves in a two-dimensional body is the point of interest. Therefore, we take the same structure as in the previous section (see Figure 3(b)), but here, the top is drained and unloaded. An impulse force, which acts only in the first 0.04 s in the form of a sine load:

$$F(t) = \begin{cases} 100 \ \sin \ (78.54 \cdot t)[\text{kN}] & \text{if } t \leqslant 0.04\,\text{s}, \\ 0 \quad \text{if } t \geqslant 0.04\,\text{s}, \end{cases} \tag{39}$$

is applied in the vertical direction to the center node in the middle of the structure.

Figure 7 shows the vertical solid displacements at different intervals of time. The axes determine the position of the displacements in [m]. The disturbance caused by the

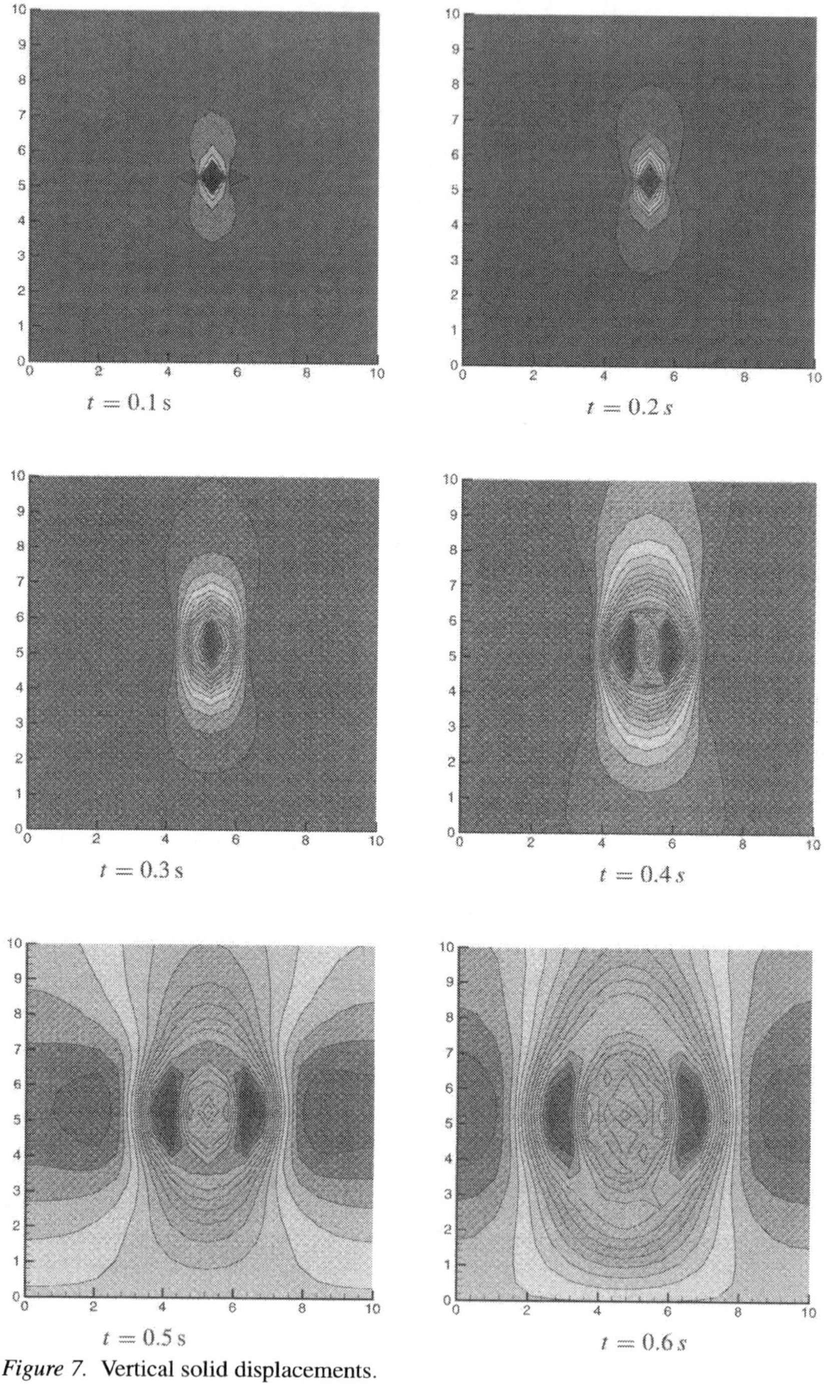


*Figure 7.* Vertical solid displacements.

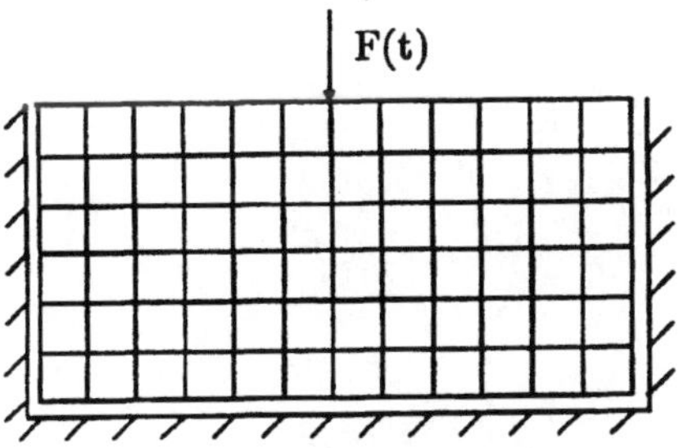


*Figure 8.* Discretization of the plane.

applied load propagates in the vertical direction faster in comparison to the horizontal direction. This causes the elliptic form of the spread of the disturbance. The reason for this phenomenon is that in the vertical direction the disturbance is transmitted through the faster longitudinal wave and in the horizontal direction it propagates with the slower speed of the transverse wave. The velocity of the longitudinal wave is exactly the same as calculated in the one-dimensional problems (see Breuer, 1997).

### 4.3. SURFACE WAVES

Next, the surface waves, which occur at a free boundary, is the point of interest. Therefore, we investigate a plane of 20 m $\times$ 10 m and load this plane in the center of the surface, see Figure 8 with the impulse force, described in Equation (39).

At the right and left boundaries of the structure the horizontal displacements and velocities are zero, as well as the vertical displacements and velocities at the bottom. The surface is drained, thus water can come out of the porous body. The pressure at the surface is zero.

Figure 8 shows the discretization of the above-mentioned plane with its applied load which vanished after 0.04 s. From this point a disturbance travels into the structure transmitted by the longitudinal and transverse waves, described in the last section. In addition to this an additional wave, called Rayleigh-wave, travels at the surface. This wave is caused by the interface between the porous body and the free boundary.

On the one side of the interface there is the porous body, as introduced in Section 2. On the other side of the interface there is the atmosphere, which is treated like a medium with negligible properties, as was done by Lord Rayleigh (1885). Thus, no pseudo Rayleigh-waves or Stoneley-waves appear.

Looking at Figure 9 we can see a qualitative plot of the horizontal displacements of the solid at different intervals of time. The first plot shows the displacements at time 0.07 s. The axes determine the position of the displacements in [m]. The propagation of a longitudinal wave cannot be observed in this plot, it causes only vertical displacements in the vertical direction, as shown in Figure 7. At the surface at point (7.5 m, 10 m) and (12.5 m, 10 m), we observe two extreme peaks travelling at

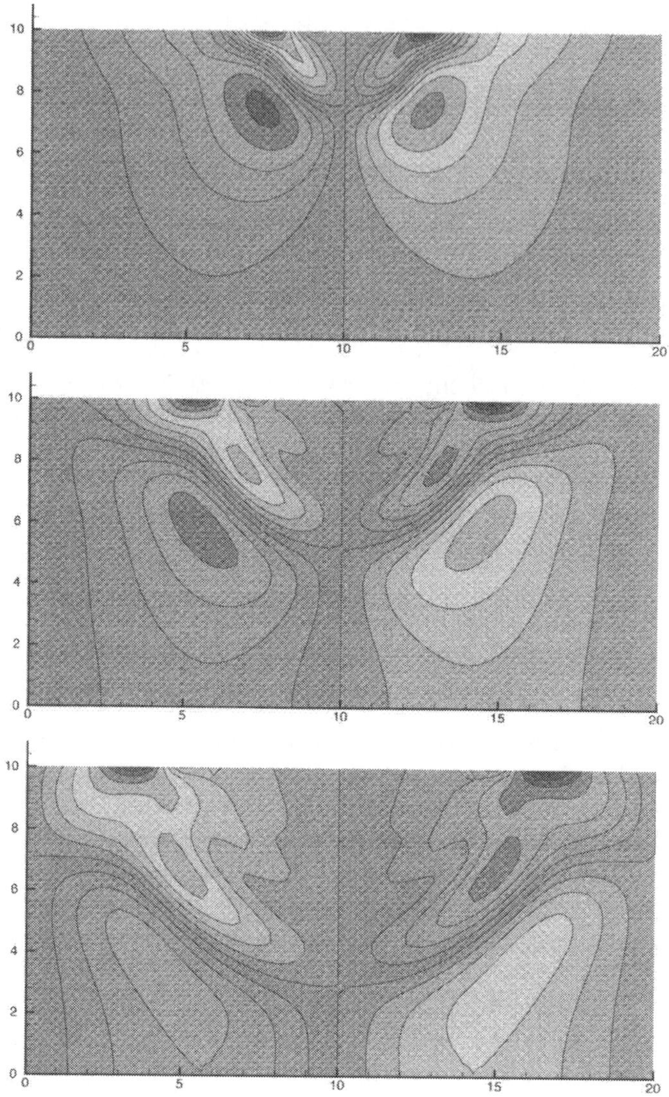

*Figure 9.* Rayleigh-waves at a free boundary.

the free boundary. These displacements are caused by the Rayleigh-wave. Looking at the next two plots at time 0.11 s and 0.15 s, we can see how these peaks propagate at the surface and how their amplitudes decrease with growing depth. The propagation-speed of Rayleigh-waves is nearly the same as that of the shear-wave (see Figure 7) and this agrees with the well known investigations about Rayleigh-waves, for example the original one of Lord Rayleigh (1885).

Figure 10 shows a plot of the deformed mesh where the solid displacements are multiplied with the factor 500, at time 0.07 s, 0.11 s and 0.15 s. We observe that the

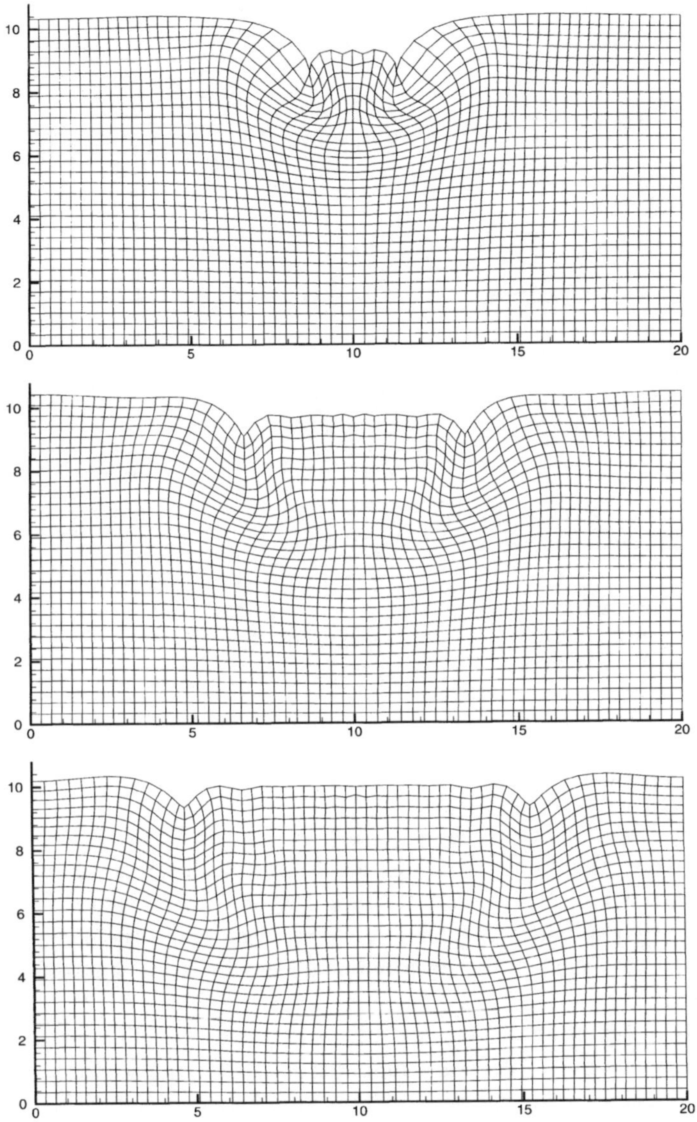


*Figure 10.* Deformed mesh (vertical and horizontal displacement scaled by factor 500).

largest deformation takes place at the surface, caused by the Rayleigh-wave. The amplitude of this wave decreases, while the disturbance travels farther away from its source. This is due to the fact that the deformation energy is constant and as more of the plane is taken under deformation, the less the amplitudes become.

We shall observe at least one particle placed at the surface at point (5 m, 10 m) while the Rayleigh-wave travels through this point. We start with small deformation, caused by the faster shear wave and observe the well known elliptic particle motion,

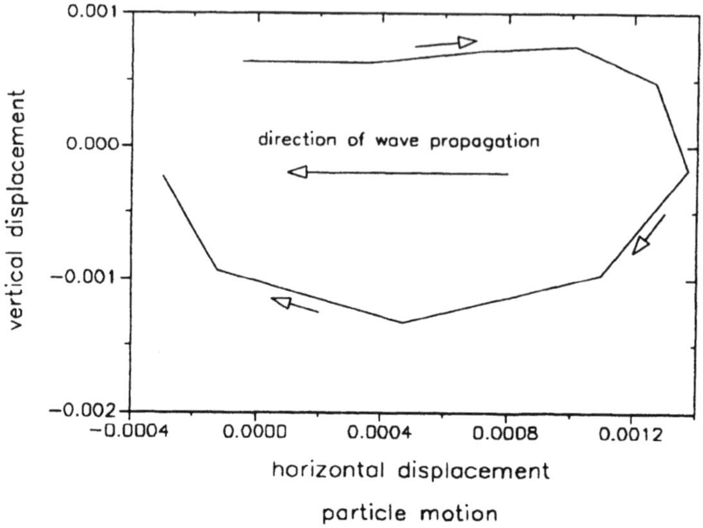


*Figure 11.* Particle motion caused by a Rayleigh-wave.

as described in the fundamental books of soil dynamics, for example Richart *et al.* (1970).

## 5. Concluding Remarks

In this investigation the elastic behavior of a fluid-saturated porous medium is taken into account. The governing equations are formulated by a model with two microscopic incompressible phases. The dynamic response is calculated for small deformation. Based on this model the classical two-dimensional consolidation process, the wave propagation in a two-dimensional plane and the propagation of surface waves are calculated.

The presented two-dimensional example of a dynamical consolidation problem shows the expected behavior of the pore pressure distribution, the seepage velocity and the solid displacement. The undrained short time behavior, where the whole external load is taken over by the fluid, as well as the drained long time behavior, where the external load is taken over by the solid skeleton, are simulated by the calculations.

The dynamic response of the second example shows the propagation of a disturbance with different velocities in different directions, caused by the transmission by different wave types.

As expected, a Rayleigh-wave occurs at the free surface and its amplitude decreases with increasing distance to the source.

The presented model is a convenient alternative to the well known Biot's theory. It presents a completely consistent procedure based on the fundamentals of mechanics.

As a result of this, further development of this model is possible and shall be the topic of further investigations.

### Acknowledgements

The author is grateful to Prof. Dr-Ing. R. de Boer and Priv.-Doz. Dr-Ing. J. Bluhm for their critical comments offered during the preparation of this paper. This work has been supported by the Deutsche Forschungsgemeinschaft (Germany).

### References

Bathe, K. J.: 1990, *Finite-Element-Methoden*, Springer-Verlag, Berlin.

Biot, M. A.: 1955, Theory of elasticity and consolidation for a porous anisotropic solid, *J. Appl. Phys.* **26**, 182.

Biot, M. A.: 1956, Theory of propagation of elastic waves in fluid-saturated porous soil: I. Low-frequency range, *J. Acoust. Soc. Am.* **28**, 168.

Bluhm, J.: 1997, A consistent model for empty and saturated porous media, Habilitation Thesis, University of Essen, Essen.

Bluhm, J. and de Boer, R.: 1997, The volume fraction concept in the porous media theory, *Zeitschr. Angew. Math. Mech. (ZAMM)* **77**(8), 563.

de Boer, R.: 1982, *Vektor- und Tensorrechnung für Ingenieure*, Springer-Verlag, Berlin.

de Boer, R.: 1996, Highlights in the historical development of the porous media theory: toward a consistent macroscopic theory, *Appl. Mech. Rev.* **49**, 201.

de Boer, R., Ehlers, W. and Liu, Z.: 1993, One-dimensional transient wave propagation in fluid-saturated incompressible porous media, *Arch. Appl. Mech.* **63**, 59.

Bowen, R. M.: 1980, Incompressible porous media models by use of the theory of mixtures, *Int. J. Engng. Sci.* **18**, 1129.

Breuer, S.: 1997, Dynamic response of a fluid-saturated elastic porous solid, *Arch. Mech.* **49**(4), 771.

Diebels, S. and Ehlers, W.: 1996, Dynamic analysis of a fully saturated porous medium accounting for geometrical and material non-linearities, *Int. J. Numerical Methods Engng.* **39**, 81–97.

Ehlers, W.: 1989, Poröse Medien – ein kontinuumsmechanisches Modell auf der Basis der Mischungstheorie, Forschungsbericht aus dem Fachbereich Bauwesen 47, Essen.

Ehlers, W. and Diebels, S.: 1994, Dynamic deformations in the theory of fluid-saturated porous solid materials, *Proc. IUTAM Symp. Anisotropy, Inhomogeneity and Nonlinearity in Solid Mechanics*, Nottingham, U.K.

Fillunger, P.: 1936, Erdbaumechanik?, Selbstverlag des Verfassers, Wien.

Lewis, R. W. and Schrefler, B. A.: 1987, The finite element method in the deformation and consolidation of a porous media, Wiley .

Prévost, J. H.: 1981, Consolidation of an elastic porous media, *J. Eng. Mech. Div. ASCE* **107** (EM 1), 169–186.

Rayleigh, L.: 1885, On waves propagating along the plane surface of an elastic solid, *Proc. Lond. Math. Soc.* **17**, 4.

Richart, F. E., Hall, J. R. and Woods, R. D.: 1970, *Vibrations of Soils and Foundations*, Prentice-Hall, Englewood Cliffs, N.J.

Zienkiewicz, O. C.: 1984, *Methode der Finiten Elemente*, Carl Hanser Verlag, München.

Zienkiewicz, O. C. and Shiomi, T.: 1984, Dynamical behaviour of saturated porous media; the generalized Biot formulation and its numerical solution, *Int. J. Num. Anal. Methods Geomech.* **8**, 71.

# A Linear Theory of Porous Elastic Solids *

PASQUALE GIOVINE
*Dipartimento di Meccanica dei Fluidi ed Ingegneria Offshore, Università degli Studi di Reggio Calabria, Via Graziella, Località Feo di vito, I-89060, Reggio Calabria, Italy*

(Received: 31 August 1997; in final form: 2 December 1997)

**Abstract.** The theory of porous elastic solids with large vacuous interstices, considered by Giovine like materials with ellipsoidal structure, includes, as a particular case, the nonlinear theory of Nunziato and Cowin of elastic materials with small spherical voids finely dispersed in the matrix.

In this paper we propose appropriate constitutive relations and then specialize the basic balance equations of Giovine to the linear theory. Also, generalizing the developments of Cowin and Nunziato, we formulate boundary-initial-value problems and examine classical applications as responses to homogeneous deformations and small-amplitude acoustic waves.

**Key words:** continua with microstructure, porous elastic solids, constitutive equations, linear theory, applications.

## 1. Introduction

The nonlinear theory of Nunziato and Cowin [1] for elastic materials with voids was, for many researchers, the basis for the study of a porous elastic solid with lacunae finely dispersed in the matrix: they considered an additional degree of kinematic freedom, representing the volume fraction of the matrix material, for which a new balance equation was postulated. But, when the pores are large, the variation of the volume fraction is not sufficient to describe the microdeformation, as observed in [2] about predictions of size effects in torsion of bars in isotropic media like compact bones or polymer foam materials.

In [3], the idea of Nunziato and Cowin was generalized to examine more complex situations. Porous solids with large vacuous interstices were considered as continua with an ellipsoidal structure for which the microdeformation is a pure strain: this means that each void has different microdeformation along principal axes, but rotates locally with the matrix of the material; alternatively, when the rotary inertia is of interest in the analysis, one is led to the Cosserat theory or, more generally, to a model of continua with affine structure (see [4]).

In this paper we propose the suitable constitutive properties for a porous elastic material which does not conduct heat, including a rate effect in the microstructural response due to inelastic surface effects associated with pore compaction and distention, and establish the restrictions imposed on these equations by thermodynamical

* Dedicated to Professor Gianfranco Capriz on the occasion of his seventy-second birthday.

considerations. Afterwards we obtain the explicit forms of the constitutive fields for isotropic materials in terms of the strain measures and apply general hypotheses on macrodisplacements and microstrains to get the linear theory, also formulating a displacement boundary-initial-value problem.

At last, we test our model in applications: (i) we consider deformation processes that are homogeneous and quasi-static and observe properties of porous solids similar to those ascribed to materials with fading memory; (ii) we study the propagation of acoustic waves and find solutions in the purely transverse and longitudinal case: in particular, we obtain not only classical elastic waves but also micro-waves propagating without affecting the elastic characteristics of the matrix material. All the results are seen to be consistent with those in [5].

## 2. Kinematics and Balance Equations

The balance equations for a porous elastic material with empty large voids, that do not diffuse through the skeleton, was obtained in [3], where we supposed that each material element of the body is able to have a microstretch different from and independent of the local affine deformation ensuing from the macromotion. This makes the proposal in [1] broader for a continuum with small spherical pores which may contract and expand homogeneously without having a gyratory movement: there we relaxed the former hypothesis by allowing distinct microstrains of the voids along principal axes of microdeformation, but maintained the absence of microrotations of the voids themselves.

The continuous material body $\mathcal{B}$ is identified with a fixed region $\mathcal{B}_*$ of the three-dimensional Euclidean space $\mathcal{E}$, the reference placement, and a generic material element of $\mathcal{B}_*$ is denoted by $x_*$; thus, the motion of $\mathcal{B}$, as a continuum with ellipsoidal structure, is a pair of smooth mappings on $\mathcal{B}_* \times \Re$ ($\Re$ is the set of real numbers),

$$x = x(x_*, \tau) \in \mathcal{E} \quad \text{and} \quad U = U(x_*, \tau) \in \mathrm{Sym}^+, \tag{1}$$

where $x$ gives the spatial position at time $\tau$ of the material point which occupied the position $x_*$ in the reference placement, $U$ is the microstructural tensor field describing the changes in the pore structure and $\mathrm{Sym}^+$ is the set of second order symmetric tensors with positive determinant. Further, $x(\cdot, \tau)$ is a bijection, for each $\tau$, with deformation gradient $F$ in $\mathrm{Lin}^+$ ($\mathrm{Lin}^+$ being the collection of all second order tensors with positive determinants):

$$F(x_*, \tau) = \frac{\partial x}{\partial x_*}(x_*, \tau) \in \mathrm{Lin}^+. \tag{2}$$

The thermal behaviour of the material is described by another smooth mapping on $\mathcal{B}_* \times \Re$, the absolute positive temperature $\theta = \theta(x_*, \tau) > 0$.

Through the inverse mapping $x_*(\cdot, \tau)$ of $x(\cdot, \tau)$, we can consider all the relevant fields in the theory as defined over the current placement $\mathcal{B}_\tau = x(\mathcal{B}_*, \tau)$ of the body $\mathcal{B}$. The local equations of balance governing an admissible thermo-kinetic process for porous solids with large voids which are nonconductors of heat are the following (see §4 of [3]):

$$\dot{\rho} + \rho \operatorname{div} \dot{x} = 0 \quad \text{(conservation law of mass),} \tag{3}$$

$$\rho \ddot{x} = \rho f + \operatorname{div} T \quad \text{(balance of linear momentum),} \tag{4}$$

$$\rho \kappa \ddot{U} = \rho B - Z + \operatorname{div} \Sigma \quad \text{(balance for micromomentum),} \tag{5}$$

$$\operatorname{skw} T = \operatorname{skw} \left( U Z^T + \operatorname{grad} U \odot \Sigma \right) \quad \text{(balance of angular momentum),} \tag{6}$$

and

$$\rho \dot{\epsilon} = T \cdot \operatorname{grad} \dot{x} + Z \cdot \dot{U} + \Sigma \cdot \operatorname{grad} \dot{U} + \rho \lambda \quad \text{(first law of thermodynamics).} \tag{7}$$

Here the reference placement $\mathcal{B}_*$ was chosen so that the reference microinertia tensor field $J_*$ has spherical value: $J_* = \kappa I$, where $I$ is the identity tensor and $\kappa$ the nonnegative microinertia coefficient depending on the reference geometric features of the pores:

$$\kappa = \tilde{\kappa}(U_*) \geqslant 0; \tag{8}$$

moreover, $\rho$ is the mass density, $f$ the vector body force, $T$ the Cauchy stress tensor, $\rho B$ and $-Z$ are the resultant symmetric tensor densities per unit volume of external and internal microactions, respectively, $\Sigma$ the microstress third-order tensor which is symmetric in the first two places, $\epsilon$ the specific internal energy per unit mass and $\lambda$ the rate of heat generation per unit mass due to radiation; at last, the superposed dot means material time derivative, while skw $A$ denotes the skew (or antisymmetric) part of a second-order tensor $A$ and $(\operatorname{grad} U \odot \Sigma)_{ij} := U_{ih,k} \Sigma_{jhk}$.

Internal microactions include interactive forces between the gross and fine structures as well as internal dissipative contributions due to the stir of the surface of pores, while the microstress is normally related to boundary microtractions, even if, in some cases, it could express weakly nonlocal internal effects; finally, external ones can be interpreted as an externally controlled pore pressure.

For a material which is a nonconductor of heat, the classical imbalance of entropy comes down to the subsequent one:

$$\rho \theta \dot{\eta} \geqslant \rho \lambda \quad \text{(Clausius–Duhem inequality),} \tag{9}$$

wherein $\eta$ is the density of entropy per unit mass; if we introduce the Helmholtz free energy per unit mass $\psi := \epsilon - \theta \eta$ and use Equation (7), we obtain a reduced version of this inequality, that is,

$$\rho \left( \dot{\psi} + \dot{\theta} \eta \right) \leqslant T \cdot \operatorname{grad} \dot{x} + Z \cdot \dot{U} + \Sigma \cdot \operatorname{grad} \dot{U}. \tag{10}$$

*Remark.* The voids theory of Nunziato and Cowin [1] can be recovered by imposing that $U$ is constrained to be spherical (see §5 of [3]).

### 3. Constitutive Prescriptions and Thermodynamic Compatibility

Our constitutive equations model a porous elastic body with homogeneous structure. We assume that the overall response of the material depends upon the deformation gradient $F$, the microstretch $U$ and its reference gradient $\nabla U$, the temperature $\theta$ and the time rate of change of the microstretch $\dot{U}$: the last one accounts for inelastic surface effects associated with changes in the deformation of the pores in the vicinity of the void boundaries.

Thus, let us call the array $\mathcal{S} := \{F, U, \nabla U, \theta\}$ of variables the elastic state of the material and, assuming equipresence, postulate as constitutive relations for constitutive quantities the following:

$$\begin{aligned} &\psi = \tilde{\psi}(\mathcal{S}, \dot{U}), \quad \eta = \tilde{\eta}(\mathcal{S}, \dot{U}), \quad T = \tilde{T}(\mathcal{S}, \dot{U}), \\ &Z = \tilde{Z}(\mathcal{S}, \dot{U}) \quad \text{and} \quad \Sigma = \tilde{\Sigma}(\mathcal{S}, \dot{U}); \end{aligned} \tag{11}$$

all the tilde functions in (11) are assumed to be twice continuously differentiable with respect to all arguments. Also, our constitutive equations have a dependence on an initial state $\{F, U, \theta\} = \{I, U_*, \theta_*\}$ of the body that we usually suppress in our notations for general considerations, but explicitly express for peculiar applications (if it is necessary).

Now, we have to check the compatibility of the constitutive prescriptions (11) with the Clausius–Duhem inequality in the reduced version (10). Making use of the chain rule of differentiation, we obtain the following equivalent relation that is valid for all values of the constitutive variables $\mathcal{S}$ and $\dot{U}$:

$$\begin{aligned} &\rho\psi_{\dot{U}} \cdot \ddot{U} + \rho\,(\eta + \psi_\theta)\,\dot{\theta} + \left(\rho\psi_F F^T - T\right) \cdot \operatorname{grad} \dot{x} \; + \\ &\quad + \left(\rho\psi_{\nabla U} \ominus F^T - \Sigma\right) \cdot \operatorname{grad} \dot{U} + (\rho\psi_U - Z) \cdot \dot{U} \leqslant 0, \end{aligned} \tag{12}$$

where the operator $\ominus$ acts filling the near index of two tensors of different dimensions, for example, $(\Phi \ominus A)_{ijl} := \Phi_{ijh} A_{hl}$.

The left-hand member of (12) is linear with respect to $\ddot{U}, \dot{\theta}, \operatorname{grad} \dot{x}$ and $\operatorname{grad} \dot{U}$, quantities that take up arbitrary values; thus the respective coefficients in the linear expression must vanish, and hence

$$\psi = \hat{\psi}(\mathcal{S}), \quad \eta = -\psi_\theta, \quad T = \rho\psi_F F^T \quad \text{and} \quad \Sigma = \rho\psi_{\nabla U} \ominus F^T. \tag{13}$$

This means that the Helmholtz free energy $\psi$, the entropy $\eta$, the Cauchy stress tensor $T$ and the microstress $\Sigma$ depend upon the elastic state of the material only; moreover, $\eta$, $T$ and $\Sigma$ are determined as soon as the constitutive equation for $\psi$ is known.

The residual inequality defines the dissipation $\mathcal{D}$ of the thermo-kinetic process

$$\mathcal{D} := H \cdot \dot{U} \leqslant 0, \tag{14}$$

where $H := \rho\hat{\psi}_U(\mathcal{S}) - \tilde{Z}(\mathcal{S}, \dot{U})$ is called the dissipation tensor and it is symmetric.

Two invariance principles need to be applied to the theory of constitutive equations: the objectivity under a rigid motion of the spatial frame of reference and the material symmetry.

With regard to the former, as $U$ transforms like the deformation gradient $F$, a convenient set of invariant geometric variables could be the following:

$$D := \tfrac{1}{2}(F^T F - I), \quad M := F^T U - U_* \quad \text{and} \quad \Gamma := F^T \ominus \nabla U \tag{15}$$

($D$ is the symmetric finite strain tensor), because they are unaffected by a rigid rotation of the body $\mathcal{B}$ and vanish if the displacement of the body from the reference placement $\mathcal{B}_*$ is rigid. Thus, if we change the elastic state $\mathcal{S}$ of the material with the new equivalent and invariant one $\mathcal{I} := \{D, M, \Gamma, \theta\}$, we can obtain a first formal consequence of the principle of material frame-indifference in terms of the constitutive relations (13), namely,

$$\begin{aligned} &\psi = \bar{\psi}(\mathcal{I}), \quad \eta = -\psi_\theta, \quad \Sigma = \rho F \ominus \psi_\Gamma \ominus F^T \quad \text{and} \\ &T = \rho F \psi_D F^T + \rho U \psi_M^T F^T + \rho \nabla U \odot (F \ominus \psi_\Gamma); \end{aligned} \tag{16}$$

moreover, the expression of the symmetric dissipation tensor $H$ changes as it follows:

$$H = \rho F \psi_M - Z, \tag{17}$$

with $F\psi_M = \psi_M^T F^T$.

For the latter invariance principle, we consider solids for which the material symmetry is of the type that possesses a center of symmetry and that are isotropic in their dependence of the constitutive fields upon the elastic state $\mathcal{I}$.

As previously observed, the free energy $\psi$ determines much of the behaviour of the porous material, thus we concentrate on dealing with it in detail firstly. We assume that the reference placement $\mathcal{B}_*$ of the body is a homogeneous placement of minimum for the free energy and for which the residual stresses $T_*$, $Z_*$ and $\Sigma_*$ vanish as well as $\psi_*$ itself.

Also, we are going to obtain the basic equations of the linear theory, hence for the free energy $\psi$ we can take the most general homogeneous, quadratic and positive semi-definite form of the joint invariants of the elastic state $\mathcal{I}$ in the case of a centrosymmetric isotropic material:

$$\begin{aligned} \rho\bar{\psi}(D, M, \Gamma, \theta) &= \tfrac{1}{2}\beta_1(\mathrm{tr} D)^2 + \tfrac{1}{2}\beta_3\left[\mathrm{tr}(\mathrm{sym}\, M)\right]^2 + \\ &+ \beta_2\,\mathrm{tr}(D^2) + \beta_4\,\mathrm{tr}[(\mathrm{sym}\, M)^2] + \beta_5(\mathrm{tr} D)\left[\mathrm{tr}(\mathrm{sym}\, M)\right] + \\ &+ \beta_6\,\mathrm{tr}[D(\mathrm{sym}\, M)] + \tfrac{1}{2}\beta_7\,\mathrm{tr}\left[(\mathrm{skw} M)^2\right] + \tfrac{1}{2}\Xi \cdot (\Gamma \otimes \Gamma) + \\ &+ \tfrac{1}{2}\gamma_1(\theta - \theta_*)^2 + \gamma_2(\theta - \theta_*)\mathrm{tr} D + \gamma_3(\theta - \theta_*)\mathrm{tr}(\mathrm{sym}\, M) \end{aligned} \tag{18}$$

(see the list (2.41) in [6]).

Here, sym $A$ denotes the symmetric part of a second-order tensor $A$, $\otimes$ is the tensor product, that is, $(\Gamma \otimes \Sigma)_{ijklmn} := \Gamma_{ijk}\Sigma_{lmn}$, and the sixth-order tensor $\Xi$ has the structure

$$\begin{aligned} \Xi_{ijklmn} &= \alpha_1(\delta_{ij}\delta_{kl}\delta_{mn} + \delta_{jk}\delta_{in}\delta_{lm}) + \alpha_2\delta_{kl}\delta_{in}\delta_{jm} + \\ &+ \alpha_3\delta_{ij}\delta_{kn}\delta_{lm} + \alpha_4\delta_{jk}\delta_{il}\delta_{mn} + \alpha_5(\delta_{jk}\delta_{im}\delta_{nl} + \delta_{ki}\delta_{jl}\delta_{mn}) + \\ &+ \alpha_6\delta_{ki}\delta_{jm}\delta_{nl} + \alpha_7\delta_{il}\delta_{jm}\delta_{kn} + \alpha_8(\delta_{jl}\delta_{km}\delta_{in} + \delta_{kl}\delta_{im}\delta_{jn}) + \\ &+ \alpha_9\delta_{il}\delta_{jn}\delta_{km} + \alpha_{10}\delta_{jl}\delta_{kn}\delta_{im} + \alpha_{11}(\delta_{ij}\delta_{km}\delta_{nl} + \delta_{ki}\delta_{jn}\delta_{lm}) \end{aligned} \tag{19}$$

(see Equation (4.5) of [7]); the twenty-one coefficients $\beta_i$ ($i = 1, \dots, 7$), $\alpha_j$ ($j = 1, \dots, 11$) and $\gamma_k$ ($k = 1, 2, 3$) in (18) and (19) are constants.

Thermal stresses are not of interest in the applications in which we are concerned in the sequel, thus we suppose that $\gamma_k = 0$ for $k = 1, 2, 3$; so, the independent coefficients in (18) reduce to eighteen.

## 4. The Linear Theory

Let $u$ be the displacement field of the continuum from the reference placement $\mathcal{B}_*$ and $V$ the change in the microstretch from the reference one (that, for the homogeneity of $\mathcal{B}_*$, is constant and can be supposed equal to the identity, that is, $U_* = I$):

$$u(x_*, \tau) := x(x_*, \tau) - x_* \quad \text{and} \quad V(x_*, \tau) := U(x_*, \tau) - I, \tag{20}$$

where $V$ is a symmetric tensor as $V \epsilon \ \mathrm{Sym}^+$.

The deformation gradient $F$ and the finite strain tensor $D$ are related to the displacement gradient $\nabla u$ by the relations

$$F = I + \nabla u \quad \text{and} \quad D = E + \tfrac{1}{2}(\nabla u)^T \nabla u, \tag{21}$$

where $E := \mathrm{sym}\,(\nabla u)$ is called the infinitesimal strain tensor. Instead the microdeformation characteristics $M$ and $\Gamma$ are related to $V$ also:

$$M = V + (\nabla u)^T + (\nabla u)^T V \quad \text{and} \quad \Gamma = \nabla V + (\nabla u)^T \ominus \nabla V. \tag{22}$$

The infinitesimal theory models physical situations in which the displacement gradient $\nabla u$, the infinitesimal microstrain field $V$, the microdeformation gradient $\nabla V$ and the microvelocity $\dot{V}$ are, in some sense, sufficiently small that their squares can be neglected. Then, we conclude from $(21)_2$ and (22) that, within an error of the second order, the following relations hold:

$$D = E, \quad M = V + (\nabla u)^T \quad \text{and} \quad \Gamma = \nabla V; \tag{23}$$

further, we have $\mathrm{sym}\, M = V + E$, $\mathrm{skw} M = W$ and $\Gamma$ symmetric in the first two places ($W := \mathrm{skw}(\nabla u)$ is the infinitesimal rotation tensor).

Here, we are interested exclusively with linear porous elastic materials. For such bodies the stresses at any time and point in a process are linear functions of the independent variables $\nabla u$, $V$, $\nabla V$ and $\dot{V}$ at the same time and place. In order to obtain their expressions, we substitute the quadratic form (18), with $\gamma_k = 0$ for each $k$, into relations $(16)_{3,4}$ and (17) and disregard second order terms in the results; thus

$$\begin{aligned} T &= [(\beta_1 + \beta_3 + 2\beta_5)\mathrm{tr} E + (\beta_3 + \beta_5)\mathrm{tr} V]\, I + \\ &\quad + 2(\beta_2 + \beta_4 + \beta_6)E + (2\beta_4 + \beta_6)V - \beta_7 W, \\ \Sigma &= \Xi\, \nabla V \quad \text{and} \\ Z &= [\beta_3 \mathrm{tr} V + (\beta_3 + \beta_5)\mathrm{tr} E]\, I + \\ &\quad + 2\beta_4 V + (2\beta_4 + \beta_6)E - \beta_7 W - H(\mathcal{I}, \dot{V}). \end{aligned} \tag{24}$$

The symmetry in the first two places of the microstress $\Sigma$ imposes restrictions on the expression (19) of the tensor $\Xi$; it must be

$$\alpha_1 = \alpha_2, \qquad \alpha_4 = \alpha_5 = \alpha_6, \qquad \alpha_7 = \alpha_{10} \quad \text{and} \qquad \alpha_8 = \alpha_9 = \alpha_{11}, \tag{25}$$

moreover, $Z$ is symmetric as well as the dissipation tensor $H$ and hence, from $(24)_3$, it results that $\beta_7 = 0$.

Finally, we need to express $H$ within the same approximation as the other terms in $(24)_3$ due to the linear dependence of internal microactions $Z$ upon infinitesimal variables; by the dissipation inequality $(14)_2$ and its smoothness, the dissipation tensor $H$ must vanish whenever $\dot{V} = 0$, then we take

$$H = -\alpha(\mathrm{tr}\dot{V})I - 2\gamma\dot{V}, \tag{26}$$

with $\alpha$ and $\gamma$ as inelastic constants. The imbalance $(14)_2$ is satisfied when

$$\gamma \geqslant 0 \quad \text{and} \quad 3\alpha + 2\gamma \geqslant 0. \tag{27}$$

The constitutive equations for the linear theory of a porous elastic material with empty large voids which do not diffuse through the matrix are the following:

$$\begin{aligned} T &= (\lambda\,\mathrm{tr}E + \omega_5\,\mathrm{tr}V)I + 2\mu E + \omega_6 V, \\ \Sigma &= \omega_1[I \otimes \mathrm{Div}\, V + \mathrm{syml}\,(\nabla(\mathrm{tr}V) \otimes I)] + \omega_8 I \otimes \nabla(\mathrm{tr}V) + \\ &\quad + 2\,\omega_2\,\mathrm{syml}\,(\mathrm{Div}\, V \otimes I) + 2\,\omega_7 \nabla V + 2\,\omega_9\,\mathrm{syml}\left[(\nabla V)^t\right], \\ Z &= (\omega_3\,\mathrm{tr}V + \omega_5\,\mathrm{tr}E + \alpha\,\mathrm{tr}\dot{V})I + 2\,\omega_4 V + \omega_6 E + 2\,\gamma\dot{V}, \end{aligned} \tag{28}$$

where we introduced the new elastic constants, related to previous ones,

$$\begin{aligned} &\lambda = \beta_1 + \beta_3 + 2\beta_5, \quad \mu = \beta_2 + \beta_4 + \beta_6, \quad \omega_1 = 2\alpha_1, \\ &\omega_2 = 2\alpha_4, \quad \omega_3 = \beta_3, \quad \omega_4 = \beta_4, \quad \omega_5 = \beta_3 + \beta_5, \\ &\omega_6 = 2\beta_4 + \beta_6, \quad \omega_7 = \alpha_7, \quad \omega_8 = \alpha_3, \quad \omega_9 = 2\alpha_8. \end{aligned} \tag{29}$$

On the third-order tensors in $(28)_2$ we used the operators left symmetrization and minor right transposition with the subsequent properties: $(\mathrm{syml}\,\Omega)_{ijl} := \frac{1}{2}(\Omega_{ijl} + \Omega_{jil})$ and $(\Omega^t)_{ijl} := \Omega_{ilj}$, respectively.

We note that in the linear theory the moment of momentum Equation (6) reduces to classical condition of symmetry of the Cauchy stress tensor $T$ and is identically satisfied by $(28)_1$.

*Remark.* When one considers thermal phenomena, linear constitutive relations $(28)_{1,3}$ have to be modified as it follows: the terms $\rho(\gamma_2 + \gamma_3)(\theta - \theta_*)I$ and $\rho\gamma_3(\theta - \theta_*)I$ are added to the right side of $(28)_{1,3}$, respectively, while, from $(16)_2$, the entropy $\eta$ is

$$\eta = -\gamma_1(\theta - \theta_*) - (\gamma_2 + \gamma_3)\mathrm{tr}E - \gamma_3\,\mathrm{tr}V. \tag{30}$$

The eleven elastic constants defined in (29), constants that completely describe the microelastic properties of the body, can be further specified if we study the restriction of the free energy density $\psi$ to be a positive semi-definite form.

Inserting (29) in (18) and neglecting terms of greater order than quadratics in $\nabla u$, $V$ and $\nabla V$, we obtain

$$\begin{aligned}\rho\psi &= \frac{3\lambda+2\mu}{6}(\mathrm{tr}\,E)^2 + \mu\,\mathrm{dev}\,E\cdot\mathrm{dev}\,E + \frac{3\omega_3+2\omega_4}{6}(\mathrm{tr}\,V)^2 + \\ &+ \omega_4\,\mathrm{dev}\,V\cdot\mathrm{dev}\,V + \frac{3\omega_5+\omega_6}{3}(\mathrm{tr}\,E)(\mathrm{tr}\,V) + \omega_6\,\mathrm{dev}\,E\cdot\mathrm{dev}\,V + \\ &+ \frac{6(\omega_1+\omega_7)+9\omega_8+2(\omega_2+\omega_9)}{18}\nabla(\mathrm{tr}\,V)\cdot\nabla(\mathrm{tr}\,V) + \\ &+ \omega_7\nabla(\mathrm{dev}\,V)\cdot\nabla(\mathrm{dev}\,V) + \frac{2(\omega_2+\omega_9)}{3}\nabla(\mathrm{tr}\,V)\cdot\nabla(\mathrm{dev}\,V) + \\ &+ \omega_2\,\mathrm{Div}\,(\mathrm{dev}\,V)\cdot\mathrm{Div}\,(\mathrm{dev}\,V) + \omega_9\nabla(\mathrm{dev}\,V)\cdot[\nabla(\mathrm{dev}\,V)]^t \geqslant 0,\end{aligned} \tag{31}$$

where dev denotes the deviatoric part of a symmetric tensor, that is, $\mathrm{dev}\,E := E - (1/3)\,(\mathrm{tr}\,E)I$. The non-negativeness of the expression in (31) assure us that $\omega_9 = 0$, while the other coefficients must resolve the following system of inequalities:

$$\begin{aligned}&\mu \geqslant 0, \quad 3\lambda+2\mu \geqslant 0, \quad (3\lambda+2\mu)(3\,\omega_3+2\,\omega_4) \geqslant (3\,\omega_5+\omega_6)^2, \\ &4\,\mu\,\omega_4 \geqslant \omega_6^2, \quad \omega_2 \geqslant 0, \quad \omega_7 \geqslant 0, \quad 3\,\omega_8+2(\omega_1+\omega_7) \geqslant 0.\end{aligned} \tag{32}$$

## 5. Equations of Motion

The general system of partial differential equations for the motion of linear elastic solids with ellipsoidal structure comes out from the insertion of constitutive relations (28), with $\omega_9 = 0$, in the balance equations (3)–(5). Then, for the homogeneous centrosymmetric isotropic materials, the displacement and microstrain equations on $u$ and $V$ are, in components,

$$\begin{aligned}&\rho = \rho_*\,(1-u_{i,i}), \\ &\rho_*\ddot{u}_j = (\lambda+\mu)u_{i,ij} + \mu u_{j,ii} + \omega_5 V_{ii,j} + \omega_6 V_{ji,i} + \rho_* f, \\ &\rho_*\kappa\ddot{V}_{jk} = (\omega_1 V_{il,il} + \omega_8 V_{ii,ll} - \omega_3 V_{ii} - \omega_5\,u_{i,i} - \alpha\dot{V}_{ii})\delta_{jk} + \\ &\qquad + \omega_1 V_{ii,jk} + 2\,\omega_7 V_{jk,ii} + \omega_2(V_{ij,ik} + V_{ik,ij}) - \\ &\qquad - 2\,\omega_4\,V_{jk} - \tfrac{1}{2}\omega_6(u_{j,k}+u_{k,j}) - 2\,\gamma\,\dot{V}_{jk} + \rho_* B_{jk}.\end{aligned} \tag{33}$$

In order to define a displacement boundary-initial-value problem we must assign a time interval $[0, \bar{\tau}]$, ten elastic and two inelastic material constants ($\lambda$, $\mu$, $\omega_i$, for $i = 1, \ldots, 8$, and $\alpha$, $\gamma$, respectively), body forces $f$ and $B$ on $\mathcal{B}\times[0,\bar{\tau}]$, initial displacement $u_0$ and microstrain $V_0$ on $\mathcal{B}$, initial velocity $v_0$ and microvelocity $\dot{V}_0$ on $\mathcal{B}$ and surface displacement $\tilde{u}$ on $\partial\mathcal{B}\otimes[0,\bar{\tau}]$ ($\partial\mathcal{B}$ is the boundary of the body $\mathcal{B}$).

Given the above data, the displacement problem consists of finding the fields $u$ and $V$ (and so $T$, $Z$ and $\Sigma$) that correspond to $f$ and $B$ and satisfy the initial conditions

$$\begin{aligned} &u(\cdot,0)=u_0, \qquad \dot{u}(\cdot,0)=v_0,\\ &V(\cdot,0)=V_0 \quad \text{and} \quad \dot{V}(\cdot,0)=\dot{V}_0 \quad \text{on} \quad \mathcal{B} \end{aligned} \tag{34}$$

and the boundary conditions

$$u=\tilde{u} \quad \text{and} \quad \Sigma\, n=0 \quad \text{on} \quad \partial\mathcal{B}\otimes[0,\bar{\tau}], \tag{35}$$

where $n$ is the outer unit normal to $\partial\mathcal{B}$.

The last condition on the microstress $\Sigma$ derives from the request of continuity of $\Sigma\, n$ across the boundary; since it is difficult to imagine a direct way to act on the pores through the boundary, except for very rare phenomena (see, for example, the case of micro-earthquakes in the basin of the caldera in the Phlegraean fields in [8]), we assume that $\Sigma$ is zero outside the continuum, thus condition $(35)_2$ follows.

## 6. Quasi-Static Homogeneous Deformations

The possibility of assigning determined linear elastic constants in a boundary-initial-value problem is connected to experimental situations involving homogeneous deformations; in fact experiments can be used to measure, with our theory, coefficients of many real materials, like rock, wood, compact bone, polymer foam, since they are isotropic and have a center of symmetry.

We consider a linear elastic porous material for which Equations (33) apply and that is subjected to a homogeneous deformation process, namely, in which the strain field is independent of position: $E\equiv\bar{E}(\tau)$; moreover, the deformation process is assumed to be quasi-static and occurs in the absence of body forces, hence the inertia terms $\rho_*\ddot{u}$ and $\rho_*\kappa\ddot{V}$ can be neglected and both $f$ and $B$ vanish.

Under these hypotheses we search solutions of the micromomentum balance $(33)_3$ in which the microstrain field $V$ is independent of position also; $(33)_3$ reduces to

$$\alpha\dot{V}_{ii}\delta_{jk}+2\gamma\dot{V}_{jk}+\omega_3V_{ii}\delta_{jk}+2\,\omega_4V_{jk}+\omega_5\bar{E}_{ii}\delta_{jk}+\omega_6\bar{E}_{jk}=0, \tag{36}$$

with the initial conditions $V_{jk}(0)=0$.

For (27), it is $\gamma\geqslant 0$ and $3\,\alpha+2\,\gamma\geqslant 0$. Now we study the most interesting case, that is, when $\gamma>0$ and $3\,\alpha+2\,\gamma>0$; in the others, that is, if one inequality, or both, are zero, we obtain solutions that are linear combinations of the strain field $\bar{E}$ plus parts of solutions of the strictly positive occurrence.

When $j\neq k$, Equation (36) is expressed as

$$\dot{V}_{jk}+\frac{\omega_4}{\gamma}V_{jk}+\frac{\omega_6}{2\gamma}\bar{E}_{jk}=0, \tag{37}$$

with the solution given by

$$V_{jk}(\tau)=-\frac{\omega_6}{2\gamma}\int_0^\tau \bar{E}_{jk}(\sigma)\mathrm{e}^{-\frac{\omega_4}{\gamma}(\tau-\sigma)}\,\mathrm{d}\sigma. \tag{38}$$

If $j = k$, we can linearly combine the three equations obtained from (36) and find the solutions for $V_{11}$ and $V_{22}$

$$V_{jj}(\tau) = V_{33}(\tau) - \frac{\omega_6}{2\gamma} \int_0^\tau (\bar{E}_{jj}(\sigma) - \bar{E}_{33}(\sigma)) e^{-\frac{\omega_4}{\gamma}(\tau-\sigma)} \, d\sigma, \tag{39}$$

for $j = 1, 2$, and an equation for $V_{33}$ only

$$(3\alpha + 2\gamma)\dot{V}_{33}(\tau) + (3\omega_3 + 2\omega_4)V_{33}(\tau) + G(\tau) = 0, \tag{40}$$

where

$$\begin{aligned} G(\tau) := {} & \frac{\gamma\omega_5 + \omega_6(\gamma + \alpha)}{\gamma} \bar{E}_{33}(\tau) + \frac{2\gamma\omega_5 - \alpha\omega_6}{2\gamma} \left[\bar{E}_{11}(\tau) + \bar{E}_{22}(\tau)\right] + \\ & + \frac{\omega_6(\alpha\omega_4 - \gamma\omega_3)}{2\gamma^2} \int_0^\tau \left[\bar{E}_{11}(\sigma) + \bar{E}_{22}(\sigma) - 2\bar{E}_{33}(\sigma)\right] e^{-\frac{\omega_4}{\gamma}(\tau-\sigma)} \, d\sigma; \end{aligned} \tag{41}$$

at last, the solution of Equation (40) is

$$V_{33}(\tau) = -\frac{1}{3\alpha + 2\gamma} \int_0^\tau G(\sigma) e^{-\frac{3\omega_3+2\omega_4}{3\alpha+2\gamma}(\tau-\sigma)} \, d\sigma. \tag{42}$$

Now, we can use previous results to give expression to the Cauchy stress tensor $T$ during an homogeneous deformation; by inserting (38), (39) and (42) in the constitutive relation $(28)_1$, we have

$$\begin{aligned} T_{jk} &= 2\mu\bar{E}_{jk}(\tau) - \frac{\omega_6^2}{2\gamma} \int_0^\tau \bar{E}_{jk}(\sigma) e^{-\frac{\omega_4}{\gamma}(\tau-\sigma)} \, d\sigma, \quad \text{if} \quad j \neq k, \quad \text{and} \\ T_{jj} &= (\lambda \operatorname{tr}\bar{E}(\tau) + 2\mu\bar{E}_{jj}(\tau)) - \frac{3\omega_5 + \omega_6}{3\alpha + 2\gamma} \int_0^\tau G(\sigma) e^{-\frac{3\omega_3+2\omega_4}{3\alpha+2\gamma}(\tau-\sigma)} \, d\sigma + \\ & + \frac{\omega_6}{2\gamma} \int_0^\tau \left[(3\omega_5 + \omega_6)\bar{E}_{33}(\sigma) - \omega_5 \operatorname{tr}\bar{E}(\sigma) - \omega_6\bar{E}_{jj}(\sigma)\right] e^{-\frac{\omega_4}{\gamma}(\tau-\sigma)} \, d\sigma. \end{aligned} \tag{43}$$

The viscoelastic behaviour in the linear range of a continuum with voids during quasi-static homogeneous motions was observed in [5]. Our relation (43) for the stress $T$ of a porous elastic solid with ellipsoidal structure generalizes the expression (4.2) of [5], thus, when $3\,\omega_3 + 2\,\omega_4 = 0$, we recover the similarity with the stress arising in linear viscoelasticity, while, in general, our material is, in the infinitesimal approximation, an isotropic simple material with fading memory of order 2 (see § 40 and 41 of [9]).

## 7. Small-Amplitude Acoustic Waves

In this application we investigate the propagation of small-amplitude acoustic waves for a material characterized by the system (33) in absence of body forces (i.e., $f = 0$ and $B = 0$).

We introduce the microdisplacement vector $y$ and then consider wave solutions of (33) of the form

$$u(x,\tau) = \zeta\, h\, \phi(x,\tau) \quad \text{and} \quad y(x,\tau) = \xi\, g\, \phi(x,\tau), \tag{44}$$

where $\phi(x,\tau) = \mathrm{Re}[\mathrm{e}^{\mathrm{i}b\tau-(a+\mathrm{i}b/c)m\cdot x}]$; $\zeta$ and $\xi$ are the wave amplitudes; $h$, $g$ and $m$ are unit vectors representing the directions of macro- and microdisplacement and of propagation, respectively; $b$ is the frequency; $a$ and $c$ are the wave attenuation and the wave speed, respectively.

The infinitesimal microstrain field $V$ is related to $y$ by the equality $V = \mathrm{sym}\,(\nabla y)$ ($=$ grad $y$ in the linear theory), thus, by substituting (44) in the system of linear Equations (33), we are led to the following:

$$\begin{aligned}
&\left\{\left[\rho_* b^2 + \mu\left(a+\mathrm{i}\frac{b}{c}\right)^2\right] h + (\lambda+\mu)\left(a+\mathrm{i}\frac{b}{c}\right)^2 (m\cdot h)m\right\}\zeta + \\
&\quad + \left(a+\mathrm{i}\frac{b}{c}\right)^2\left[\frac{\omega_6}{2}g + \left(\omega_5+\frac{\omega_6}{2}\right)(m\cdot g)m\right]\xi = 0 \quad \text{and} \\
&\left(a+\mathrm{i}\frac{b}{c}\right)\left\{\left\{\left[\omega_3 - (\omega_1+\omega_8)\left(a+\mathrm{i}\frac{b}{c}\right)^2 + \mathrm{i}\alpha b\right] I \; - \right.\right. \\
&\quad \left. - \,(\omega_1+\omega_2)\left(a+\mathrm{i}\frac{b}{c}\right)^2 m\otimes m\right\}(m\cdot g) + \\
&\quad + \left[\omega_4 - \left(\frac{\omega_2}{2}+\omega_7\right)\left(a+\mathrm{i}\frac{b}{c}\right)^2 + i\gamma b - \rho_*\frac{\kappa}{2}b^2\right] 2\,\mathrm{sym}\,(g\otimes m)\bigg\}\xi + \\
&\quad \left. + \left[\omega_5(m\cdot h)I + \omega_6\,\mathrm{sym}\,(h\otimes m)\right]\zeta\right\} = 0.
\end{aligned} \tag{45}$$

We say that the waves are purely transverse if $h$ is equal to $g$ and perpendicular to the propagation $m$. With these positions, the system (45) is equivalent to

$$\begin{aligned}
&\left[\rho_* b^2 + \mu\left(a+\mathrm{i}\frac{b}{c}\right)^2\right]\zeta + \frac{\omega_6}{2}\left(a+\mathrm{i}\frac{b}{c}\right)^2\xi = 0 \quad \text{and} \\
&\left(a+\mathrm{i}\frac{b}{c}\right)^2\left\{\left[\omega_4 - \left(\frac{\omega_2}{2}+\omega_7\right)\left(a+\mathrm{i}\frac{b}{c}\right)^2 + \mathrm{i}\gamma b - \rho_*\frac{\kappa}{2}b^2\right]\xi + \frac{\omega_6}{2}\zeta\right\} = 0.
\end{aligned} \tag{46}$$

Apart from trivial ones, we have: (i) a static solution with $b = 0$ and attenuation $a = \sqrt{(4\mu\omega_4 - \omega_6^2)/[2\mu(\omega_2+2\omega_7)]}$ (the radicand is nonnegative for $(32)_{1,4,5,6}$) in which the amplitudes are related by $\omega_6\xi = 2\mu\zeta$; (ii) if $\omega_6 = 0$, a transverse macro-wave (i.e., with $\xi = 0$) propagating without attenuation ($a = 0$) at a constant speed $c^2 = \mu/\rho_*$ and which does not affect the porosity of the material; (iii) if

$\omega_6 = 0$, a transverse micro-wave (i.e., with $\zeta = 0$) that propagates with attenuation $a = \gamma c/(\omega_2 + 2\omega_7)$ and speed given by

$$c^2 = (\omega_2 + 2\omega_7)\left[\frac{2\omega_4 - \rho_*\kappa b^2 + \sqrt{(2\omega_4 - \rho_*\kappa b^2)^2 + 4\gamma^2 b^2}}{2\gamma^2}\right]: \tag{47}$$

in the expressions of $a$ and $c$ only coefficients associated with the properties of the pores appear, thus the micro-wave propagates without modifying the elastic features of the matrix material, but is both dispersed and attenuated for the variations in the porosity which accompany the wave.

Purely elastic waves occur when the inelastic coefficient $\gamma$ vanishes; in this case we have no attenuation and obtain: (iv) a transverse micro-wave ($\zeta = 0$) that propagates with speed $c^2 = (\omega_2 + 2\omega_7)b^2/(\rho_*\kappa b^2 - 2\omega_4)$, when $\rho_*\kappa b^2 > 2\omega_4$ and $\omega_6 = 0$; (v) two distinct types of complete transverse waves, if the determinant of the coefficient matrix of the system of amplitudes $\zeta$ and $\xi$ vanish, that is, $(\rho_* - \mu/c^2)(2\omega_4 - \rho_*\kappa b^2 + (\omega_2 + 2\omega_7)b^2/c^2) + \omega_6^2/(2\rho_*) = 0$. In all elastic cases the wave speeds are functions of the frequency $b$.

When we supposed in (ii), above, that the coefficient $\omega_6$ of the nonspherical part of the microstrain in the stress tensor $T$ is null, we set ourselves in the hypotheses of Cowin and Nunziato in [5] and, in fact, we found their transverse wave solution; the others are new. Also, we observe that our micro-wave solutions in (iii), above, can be recognized in some developments of [10].

Purely longitudinal waves succeed if $h$, $g$ and $m$ are equal. Now the system (45) reduces to

$$\begin{aligned}
&\left[\rho_* b^2 + (\lambda + 2\mu)\left(a + \mathrm{i}\frac{b}{c}\right)^2\right]\zeta + (\omega_5 + \omega_6)\left(a + \mathrm{i}\frac{b}{c}\right)^2 \xi = 0,\\
&\left(a + \mathrm{i}\frac{b}{c}\right)\left\{\left[\omega_3 - (\omega_1 + \omega_8)\left(a + \mathrm{i}\frac{b}{c}\right)^2 + \mathrm{i}\alpha b\right]\xi + \omega_5\zeta\right\} = 0 \quad \text{and}\\
&\left(a + \mathrm{i}\frac{b}{c}\right)\left\{\left[2\omega_4 - \omega\left(a + \mathrm{i}\frac{b}{c}\right)^2 + 2\mathrm{i}\gamma b - \rho_*\kappa b^2\right]\xi + \omega_6\zeta\right\} = 0,
\end{aligned} \tag{48}$$

where $\omega = \omega_1 + 2\omega_2 + 2\omega_7$.

If the attenuation is null and the coefficients of the microstrain in the stress tensor vanish, we have the classical longitudinal macro-wave ($\xi = 0$) found in [5] propagating at a constant speed $c^2 = (\lambda + 2\mu)/\rho_*$ and which does not affect the porosity of the material. Instead, a purely elastic wave occurs, when the inelastic coefficients $\alpha$ and $\gamma$ are zero: it is a complete longitudinal wave with the same speed as that of the previous one that exists if the determinant of the system of the amplitudes $\zeta$ and $\xi$ vanish, that is, $(\lambda + 2\mu)(\omega_3 + 2\omega_4 - \rho_*\kappa b^2) + \rho_* b^2[2(\omega_1 + \omega_2 + \omega_7) + \omega_8] = 0$. Some other solutions of the type acoustic waves could be found with a more accurate study of our system, but this will be reported in a forthcoming work.

Finally, we want to mention a mixed case in which longitudinal and transverse features are present. We suppose that the direction of macrodisplacement $h$ is parallel to the propagation $m$, but it is perpendicular to the microdisplacement $g$. Thus, Equations (45) are reduced to

$$\begin{aligned}&\left[\rho_* b^2 + (\lambda + 2\mu)\left(a + \mathrm{i}\frac{b}{c}\right)^2\right]\zeta = 0,\\&\left(a + \mathrm{i}\frac{b}{c}\right)\left[\omega_4 - \left(\frac{\omega_2}{2} + \omega_7\right)\left(a + \mathrm{i}\frac{b}{c}\right)^2 + \mathrm{i}\gamma b - \rho_* \frac{\kappa}{2} b^2\right]\xi = 0,\end{aligned} \tag{49}$$

when the coefficients of the microstrain in the stress tensor are null. A complete elastic solution without attenuation and with amplitudes $\zeta$ and $\xi$ different from zero exists with speed $c^2 = (\omega_2 + 2\omega_7)b^2/(\rho_*\kappa b^2 - 2\omega_4)$ $(\rho_*\kappa b^2 > 2\omega_4)$, if this quantity is constant and also equal to the classical one $(\lambda + 2\mu)/\rho_*$.

So, we have a solution where the macrodisplacement is orthogonal to the microdisplacement, if the microelastic properties of the pores are, in some sense, constrained to the elastic characteristics of the matrix material.

## 8. Concluding Remarks

In [3] the theory of continua with microstructure was successfully applied to describe porous solids with very large vacuous pores viewed as materials with ellipsoidal microstructure. Hence, in this paper, we have developed in some detail the related linear theory of a nonconducting homogeneous centrosymmetric isotropic porous material by proposing the appropriate constitutive features and by specializing them with thermodynamic restrictions and invariance principles.

The voids theory in [1] is contained in the present model and the yielded results in our applications to quasi-static homogeneous deformations and to small-amplitude acoustic waves are in accordance with those previously derived in [5]. In particular, we observed that, during quasi-static homogeneous motion, the porous solid behaves like an isotropic simple material with fading memory in the linear range and it reduces to a viscoelastic medium when the microstructural variable remains spherical. Moreover, analyzing the propagation of waves, we arrived at some new solutions like a purely transverse micro-wave spreading without perturbing the elastic properties of the matrix material and a mixed complete elastic one which comes out only if the macro- and microelastic properties of the material are related.

## Acknowledgements

This research was supported by the Italian M.U.R.S.T. through 'Fondi per la ricerca scientifica 40%'.

## References


1. Nunziato, J. W. and Cowin, S. C.: A nonlinear theory of elastic materials with voids, *Arch. Rat. Mech. Analysis* **72** (1979), 175–201.
2. Lakes, R. S.: Experimental microelasticity of two porous solids, *Int. J. Solids Structures* **22** (1986), 55–63.
3. Giovine, P.: Porous solids as materials with ellipsoidal structure, In: R. C. Batra and M. F. Beatty (eds), *Contemporary Research in the Mechanics and Mathematics of Materials*, CIMNE, Barcelona, Spain, 1996, pp. 335–342.
4. Capriz, G., Podio Guidugli, P. and Williams, W.: Balance equations for materials with affine structure, *Meccanica* **17** (1982), 80–84.
5. Cowin, S. C. and Nunziato, J. W.: Linear elastic materials with voids, *J. Elasticity* **13** (1983), 125–147.
6. Smith, G. F.: On isotropic functions of symmetric tensors, skew-symmetric tensors and vectors, *Int. J. Engng. Sci.* **9** (1971), 899–916.
7. Suhubi, E. S. and Eringen, A. C.: Nonlinear theory of micro-elastic solids–II, *Int. J. Engng. Sci.* **2** (1964), 389–404.
8. Dell'Isola, F.: A possible explanation for the bradyseismic phenomena in the Phlegraean fields, Communication at *Euromech Colloquium 366 on Porous Media*, Essen, Germany, June 1997.
9. Truesdell, C. and Noll, W.: *The Non-Linear Field Theories of Mechanics*, 2nd edn, Springer-Verlag, Berlin, Germany, 1992.
10. Mindlin, R. D.: Micro-structure in linear elasticity, *Arch. Rat. Mech. Analysis* **16** (1964), 51–78.

# Derivation of Matching Conditions at the Contact Surface Between Fluid-Saturated Porous Solid and Bulk Fluid *

M. CIESZKO[1] and J. KUBIK[2]
[1] *Institute of Fundamental Technological Research PAS Mielzynskiego 27/29, 61-725 Poznan, Poland*
[2] *Department of Environmental Mechanics, Pedagogical University of Bydgoszcz*

(Received: 2 September 1997; in final form: 19 December 1997)

**Abstract.** The compatibility conditions matching macroscopic mechanical fields at the contact surface between fluid-saturated porous solid and adjacent bulk fluid are considered. Special attention is paid to the derivation of conditions for tangential components of the fluid flow velocities and to the verification of validity of the condition postulated by Beavers and Joseph. It has been shown that at the contact surface between two media, a dissipation of mechanical energy due to the fluid viscosity does exist and thus the form of a dissipation function has been proposed. It has been proven that this relation determines the form of two linear compatibility conditions derived for the tangential components of the relative fluid velocities and that these conditions describe the experimental results more precisely than the condition postulated by Beavers and Joseph.

**Key words:** velocity slip condition, compatibility conditions, fluid flow near the permeable boundary.

## List of Main Symbols

| | |
|---|---|
| $e^{\mathrm{bf}}$, $e^{\mathrm{pf}}$ | Internal energies (bulk fluid and pore fluid). |
| $f_{\mathrm{v}}$ | volume porosity. |
| $h$ | height of channel. |
| $k$ | permeability of porous skeleton. |
| $\mathbf{n}$ | unit vector normal to Γ-surface. |
| $p$ | fluid pressure. |
| $q$ | filter velocity. |
| $\mathbf{Q}$, $\mathbf{Q}^{\mathrm{T}}$ | orthogonal tensor and its transposition. |
| $\mathbf{T}$ | partial Cauchy stress tensor of bulk fluid. |
| $\overset{1}{\mathbf{T}}$, $\overset{2}{\mathbf{T}}$ | partial Cauchy stress tensors of the virtual constituents. |
| $\overset{1}{\mathbf{t}}$, $\overset{2}{\mathbf{t}}$ | partial stress vectors of the virtual constituents. |
| $\mathbf{T}^{\mathrm{s}}$, $\mathbf{T}^{\mathrm{f}}$ | partial Cauchy stress tensor of the physical constituents (skeleton and pore fluid). |
| $u_{\mathrm{s}}$ | slip velocity of the bulk fluid at permeable boundary. |
| $\mathbf{u}$, $\overset{2}{\mathbf{u}}$ | relative velocities (bulk fluid and free fluid). |

* The paper was supported by the National Scientific Committee grants No. 7T07 A02909 and No. 7T07A 1309.

| | |
|---|---|
| $\mathbf{v}$ | velocity of bulk fluid. |
| $\overset{1}{\mathbf{v}}, \overset{2}{\mathbf{v}}$ | velocities of the virtual constituents. |

*Greek Symbols*

| | |
|---|---|
| $\alpha_1, \alpha_2, \alpha_3,$ $\alpha, \alpha_2', \alpha_2'', \varepsilon$ | constant coefficients. |
| $\alpha^{\mathrm{T}}$ | tortuosity parameter. |
| $\Gamma$ | boundary surface of porous skeleton. |
| $\kappa$ | parameter of pore structure ($1/\kappa = \alpha_{\mathrm{T}}$ – tortuosity). |
| $\rho$ | mass density of bulk fluid. |
| $\rho^{\mathrm{s}}, \rho^{\mathrm{f}}$ | effective mass densities (porous skeleton and fluid). |
| $\rho^1, \rho^2$ | partial mass densities of virtual constituents. |
| $\rho_{12}$ | Biot's coupling parameter. |
| $\vartheta_e$ | sink of mechanical energy on $\Gamma$- surface. |
| $\mu$ | fluid viscosity. |
| $\mu^*$ | apparent fluid viscosity (in porous medium). |
| $\boldsymbol{\tau}$ | unit vector tangential to $\Gamma$-surface. |

## 1. Introduction

Continuum description of mechanical behaviour of fluid-saturated porous solids being in contact with a fluid flowing through the porous solid boundary is of great importance in many applications, for example in acoustics (Gogete and Munjal, 1996), in the porous bearing design (Joseph and Tao, 1966; Rhodes and Roulcau, 1966; Shir and Joseph, 1966), in glaciology (Hutter *et al.*, 1994; Wu *et al.*, 1996), in modelling of lubrication of diarthrodial joints (Ateshian, 1997; Hou *et al.*, 1989) and others.

The main open problem in this domain is the formulation of the appropriate boundary conditions to be imposed at the permeable boundary surface between a fluid-filled porous solid and the bulk fluid. A number of papers have been devoted to the problem of finding such macroscopic conditions making use of discontinuity analysis (Bear and Bahmat, 1984; Cieszko and Kubik, 1993; Hou *et al.*, 1989; Hutter *et al.*, 1994; Raats, 1972; Wilmanski, 1995) and micro-macro or semi-empirical analysis (Beavers and Joseph, 1967; Levy and Sanchez-Palencia, 1975; Taylor, 1971). In spite of great effort the problem of matching conditions for the tangential components of relative fluid velocities has not yet found a satisfactory solution.

The classical conditions at the permeable boundary of porous material filled with viscous fluid and surrounded by the bulk fluid are usually assumed to be the continuity of the fluid effective pressure and the fluid mass flux across the surface. Moreover, the tangential component of the bulk fluid velocity is assumed to be zero (Levy and Sanchez- Palencia, 1975). The first two conditions are commonly accepted, at least for the linear problems, while the adherence condition is rather restricted to very special cases of porous materials for which the migration of fluid particles across the boundary may be disregarded.

Experiments done by Beavers and Joseph (1967) have shown that the fluid mass flow rate through the channel with a permeable lower wall is much greater than that predicted for a channel with nonpermeably walls for which the adherence flow condition is valid. Therefore instead of the fluid adherence condition, these authors introduced the fluid slip velocity condition in the following form

$$\left.\frac{\mathrm{d}u}{\mathrm{d}z}\right|_{z=0} = \frac{\alpha}{\sqrt{k}}(u_s - q), \tag{1}$$

where $u_s$ is the slip velocity of the bulk fluid at the permeably boundary, $q$ is the filter velocity of a fluid flowing through the porous skeleton and $k$ is the skeleton permeability. Parameter $\alpha$ is the dimensionless material constant characterizing the skeleton pore structure within the boundary region and being independent of the fluid viscosity.

Although the condition (1) provides a qualitative agreement between the theoretical predictions and experimental measurements for the fluid flow along permeable surface this condition is not sufficiently general. It does not allow one to determine the interaction forces exerted by the flowing bulk fluid on the pore fluid and the porous skeleton. This single condition is also insufficient in solving boundary problems where the fluid flow within porous medium is described by, for example, a Brinkman (1947) type of equation.

It should be pointed out that both experimental and theoretical works have been done (see Koplik *et al.*, 1983; Levy and Sanchez-Palencia, 1975; Richardson, 1971; Saffman, 1971; Taylor, 1971) to establish the validity of the condition (1) without obtaining a satisfactory answer.

The main purpose of this work is to establish and discuss the general form of compatibility conditions matching macroscopic mechanical fields at the contact surface between fluid-saturated porous solid and adjacent bulk fluid. Special attention is paid to the derivation of conditions for the tangential components of the fluid flow velocities at the contact surface, and to the verification of validity of the condition postulated by Beavers and Joseph (1967).

Our approach to the problem of compatibility conditions is based on the assumption that the only justified conditions for mechanical fields we can formulate at the contact surface of both media are the balance equations for the mass of fluid, the linear momentum and the mechanical energy of all three components of the system. Therefore, we identify the general nonlinear compatibility conditions with the balance equations at the contact surface. Thus, the linearization provides the right form of the linear compatibility conditions.

Here we use the general compatibility conditions formulated by the authors in the paper by Cieszko and Kubik (1993) in order to prove that the macroscopic velocities of all three constituents of the system are of different values at the contact surface. This justifies the existence of dissipation of mechanical energy at such a surface due to the fluid viscosity. The general form of dissipation function is proposed, allowing one to derive two linear slip conditions for the tangential components of the relative fluid velocities. These conditions for the fluid flow within porous material described by Brinkman type equation take a form similar to the condition (1) postulated by Bevers and Joseph (1967).

## 2. Nonlinear Compatibility Conditions for Mechanical Fields

We consider a system (see Figure 1) composed of a fluid-saturated deformable porous solid and a bulk fluid being in contact at the permeable boundary $\Gamma$. The material $\Gamma$-surface is formed by particles of the skeleton. The fluid filling pores and the bulk fluid are viscous and of the same physical properties. It is assumed that the macroscopic mechanical state of the bulk fluid is characterized by three field quantities: the velocity $\mathbf{v}$, mass density $\rho$ and the Cauchy stress tensor $\mathbf{T}$, which satisfy the dynamic equations for viscous fluid (see, e.g., Landau and Lifshitz, 1986).

The mechanical macroscopic behaviour of an isotropic porous solid filled with fluid is described by the continuum immiscible mixture theory (Kubik, 1992; Cieszko and Kubik, 1996a,b) in which the geometrical pore structure is characterized by two macroparameters: (Kubik, 1986b); the volume porosity $f_v$ which represents the fluid volume fraction and the parameter $\kappa$ which is the measure of inhomogeneity of the fluid micro-velocity in its relative flow to the porous solid and is the ratio of the fluid kinetic energy expressed by average velocity to the average fluid kinetic energy. It is related to the Biot coupling parameter $\rho_{12}$ and the tortuosity parameters $\alpha$ used by Johnson (1986), as follows:

$$\frac{1}{\kappa} = 1 - \frac{\rho_{12}}{\bar{\rho}^{\mathrm{f}}} = \alpha_{\mathrm{T}},$$

where $\bar{\rho}^{\mathrm{f}}$ is the fluid partial mass density.

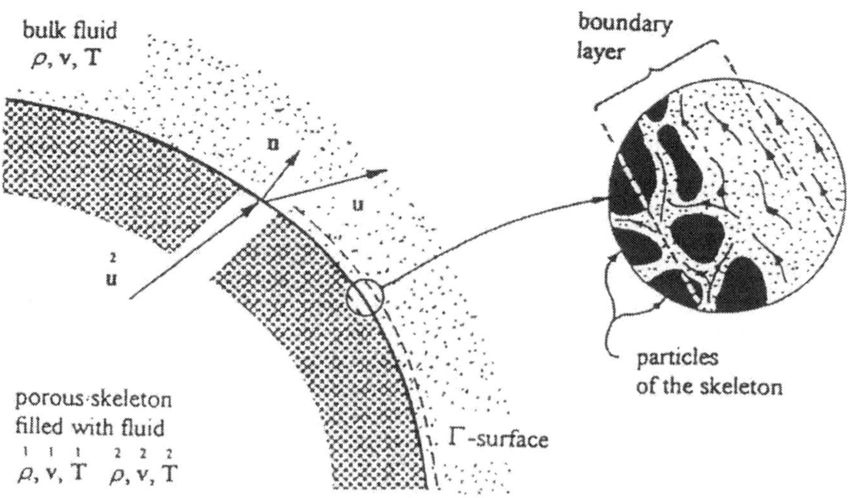


*Figure 1.* A system composed of a fluid-saturated deformable porous solid and a bulk fluid.

The characteristic feature of this theory is that the description of kinematics and dynamics of the porous solid–fluid mixture is derived by applying the micro-macro transition assuring the full linear momentum and kinetic energy representation for both components at the macrolevel (Kubik, 1986a, 1992). As a result the fluid-porous solid mixture is considered as composed of two virtual components; the first one ($|^1$) being the skeleton and the fluid associated with it moving at the skeleton velocity $\overset{1}{\mathbf{v}}$ and the other being free fluid ($|^2$) moving at the velocity $\overset{2}{\mathbf{v}}$. Between these constituents the interchange of mass during a deformation process exists, due to a change in the amount of the associated fluid. Such a description includes the dynamic coupling between pore fluid and porous skeleton (added mass effect). It should be pointed out that linear equations of this theory are equivalent to those formulated by Biot (1956).

It is worthwhile mentioning that the description of constitutive properties of the mixture is formulated for its physical constituents, that is the fluid ($|^{\mathrm{f}}$) and the porous skeloton ($|^{\mathrm{s}}$), which are chemically inert and their masses are conserved.

The partial densities $\overset{1}{\rho}.\overset{2}{\rho}$ and partial Cauchy stresses $\overset{1}{\mathbf{T}},\overset{2}{\mathbf{T}}$ of the virtual components are related to the partial densities $\bar{\rho}^{\mathrm{s}}$, $\bar{\rho}^{\mathrm{f}}$ and partial Cauchy stresses $\mathbf{T}^{\mathrm{s}}$, $\mathbf{T}^{\mathrm{f}}$ of the physical components as follows (Kubik, 1986a)

$$\overset{1}{\rho} = \bar{\rho}^{\mathrm{s}} + (1-\kappa)\bar{\rho}^{\mathrm{f}}, \qquad \overset{2}{\rho} = \kappa\bar{\rho}^{\mathrm{f}},$$

$$\overset{1}{\mathbf{T}} = \mathbf{T}^{\mathrm{s}} + (1-\kappa)\mathbf{T}^{\mathrm{f}}, \qquad \overset{2}{\mathbf{T}} = \kappa\mathbf{T}^{\mathrm{f}},$$

where

$$\bar{\rho}^{\mathrm{s}} = (1-f_{\mathrm{v}})\rho^{\mathrm{s}}, \qquad \bar{\rho}^{\mathrm{f}} = f_{\mathrm{v}}\rho^{\mathrm{f}},$$

and $\rho^{\mathrm{f}}$, $\rho^{\mathrm{s}}$ stand for the effective densities of fluid and porous skeleton, respectively.

The general compatibility conditions of mechanical fields for such a system of two adjacent media are identified with the balance equations at the $\Gamma$-surface mass of fluid, linear momentum and mechanical energy for all three components. Using the standard discontinuity analysis, they take the form (Cieszko and Kubik, 1993)

$$\rho\mathbf{u}\cdot\mathbf{n} = \overset{2}{\rho}\overset{2}{\mathbf{u}}\cdot\mathbf{n}, \tag{2}$$

$$\rho\mathbf{u}\cdot\mathbf{n}(\mathbf{u}-\overset{2}{\mathbf{u}}) = \mathbf{t} - (\overset{1}{\mathbf{t}} + \overset{2}{\mathbf{t}}), \tag{3}$$

$$\rho\mathbf{u}\cdot\mathbf{n}[\tfrac{1}{2}(\mathbf{u}^2 - (\overset{2}{\mathbf{u}})^2) + e^{\mathrm{bf}} - e^{\mathrm{pf}}] = \mathbf{t}\cdot\mathbf{u} - \overset{2}{\mathbf{t}}\cdot\overset{2}{\mathbf{u}} - \vartheta, \tag{4}$$

where

$$\mathbf{u} = \mathbf{v} - \overset{1}{\mathbf{v}} \quad \text{and} \quad \overset{2}{\mathbf{u}} = \overset{2}{\mathbf{v}} - \overset{1}{\mathbf{v}}$$

are the velocities of the bulk and free fluid flow relative to the $\Gamma$-surface, respectively, and $\mathbf{t}$, $\overset{1}{\mathbf{t}}$, $\overset{2}{\mathbf{t}}$ are the partial stress vectors for the bulk fluid and the virtual components.

They satisfy the Cauchy theorem and are related with the stress vectors $\mathbf{t}^{\mathrm{s}}$, $\mathbf{t}^{\mathrm{f}}$ for the physical components by

$$\overset{1}{\mathbf{t}} = \mathbf{t}^{\mathrm{s}} + (1-\kappa)\mathbf{t}^{\mathrm{f}}, \quad \overset{2}{\mathbf{t}} = \kappa\mathbf{t}^{\mathrm{f}}.$$

Quantities $e^{\mathrm{bf}}$ and $e^{\mathrm{pf}}$ stand for the internal energies of the bulk fluid and pore fluid, respectively.

In relation (4), the function $\vartheta$, for the decline of mechanical energy at the $\Gamma$-surface, is introduced. We will show that at this surface the dissipation of mechanical energy exists and is caused by the fluid viscosity.

## 3. Dissipation of the Mechanical Energy at the Γ-Surface

It will be shown here that at the discontinuity surface all three macroscopic velocity fields of the constituents of the considered system have different values. These discontinuities, together with the assumption that the pore and bulk fluid are viscous, promote the existence of a decline of mechanical energy on the $\Gamma$-surface.

### 3.1. DISCONTINUITIES OF THE VELOCITY FIELDS

The normal components of the macroscopic velocities of particular constituents of the system under consideration are confined by the mass continuity equation for the fluid phase. Equation (2) due to the inequality

$$\rho_{-} = \rho \neq \overset{2}{\rho} = \kappa f_{\mathrm{v}}\rho^{+} \quad (0 < \kappa f_{\mathrm{v}} \leqslant f_{\mathrm{v}} \leqslant 1),$$

shows that the normal component of the relative velocity of fluid flow through the surface of porous skeleton is discontinuous, that is,

$$0 \neq \mathbf{u}\cdot\mathbf{n} \neq \overset{2}{\mathbf{u}}\cdot\mathbf{n} \neq 0.$$

As a consequence, we have

$$\mathbf{v}\cdot\mathbf{n} \neq \overset{1}{\mathbf{v}}\cdot\mathbf{n} \neq \overset{2}{\mathbf{v}}\cdot\mathbf{n}. \tag{5}$$

The inequality (5) proves that, in general cases, the normal velocity components of particular constituents of the system differ from each other on the discontinuity surface $\Gamma$. Similar conclusion can be derived for tangential components of these velocities considering the linear momentum balance (3), and the balance of total mechanical energy (4) when the experimental results of Beavers and Joseph (1967) are taken into account.

In the case when the fluid flow along the $\Gamma$-surface takes place ($\mathbf{u}\cdot\mathbf{n} = \overset{2}{\mathbf{u}}\cdot\mathbf{n} = 0$), disregarding the decline of mechanical energy $\vartheta$, from Equations (3) and (4) we obtain

$$\mathbf{t}_{\tau} = \overset{1}{\mathbf{t}}_{\tau} + \overset{2}{\mathbf{t}}_{\tau}, \tag{6}$$

$$\mathbf{t}_\tau \cdot \mathbf{u}_\tau = \overset{2}{\mathbf{t}}_\tau \cdot \overset{2}{\mathbf{u}}_\tau, \tag{7}$$

where the subscript $\tau$ indicates vector components tangential to the $\Gamma$-surface.

When a fluid in the system is nonviscous, the conditions (6) and (7) do not impose any restriction on the tangential velocities $\mathbf{u}_\tau$ and $\overset{2}{\mathbf{u}}_\tau$. Since in such a case there are no tangential interactions, that is

$$\mathbf{t}_\tau = \overset{2}{\mathbf{t}}_\tau = \overset{1}{\mathbf{t}}_\tau = \mathbf{0},$$

the conditions (6) and (7) are satisfied identically, independently of the values of velocities $\mathbf{u}_\tau$ and $\overset{2}{\mathbf{u}}_\tau$.

For the viscous fluid, however, all the tangential stresses are not zero, and the conditions (6) and (7) confine both the values of tangential stresses as well as the tangential velocities $\mathbf{u}_\tau$ and $\overset{2}{\mathbf{u}}_\tau$.

Taking into account the experimental results of Beavers and Joseph (1967) for viscous fluid flow along a permeable boundary, which prove that the tangential component of the bulk fluid velocity is nonzero, that is,

$$\mathbf{u}_\tau \neq 0. \tag{8}$$

From Equations (6) and (7) we conclude that

$$\overset{2}{\mathbf{u}}_\tau \neq 0, \tag{9}$$

and also

$$\mathbf{u}_\tau \neq \overset{2}{\mathbf{u}}_\tau . \tag{10}$$

Otherwise the stress vectors $\mathbf{t}_\tau$ and $\overset{2}{\mathbf{t}}_\tau$ would have to be equal and, in such a case, the relation (6) would yield $\overset{1}{\mathbf{t}} = 0$. This means that there is no interaction between porous skeleton and the fluid flowing along the surface of the fluid-saturated porous medium that is physically unacceptable.

The inequalities (8)–(10), rewritten in the form

$$\mathbf{v}_\tau \neq \overset{1}{\mathbf{v}}_\tau \neq \overset{2}{\mathbf{v}}_\tau, \tag{11}$$

show that the tangential components of velocities of particular constituents of the system differ from each other on the $\Gamma$- discontinuity surface.

The above considerations allow us to state that at the $\Gamma$-surface the macroscopic velocity fields of all constituents of the system have different values. This concerns both the normal and tangential components of the velocity vectors. Accounting for the viscosity of the fluid, one can state that the relative motion of the constituents on the $\Gamma$-surface causes the viscous friction between constituents, resulting in dissipation of mechanical energy. It should be pointed out here that this dissipation process is different from the well known volume dissipation appearing in both saturated porous medium and in the viscous bulk fluid. Therefore, the function $\vartheta$ requires a separate constitutive postulate.

### 3.2. SPECIFICATION OF THE DISSIPATION FUNCTION

Taking into account the fact that the dissipation of mechanical energy on the $\Gamma$-surface results from the discontinuity of velocity fields of particular constituents of the system and assuming isotropy of the skeleton pore structure we postulate the constitutive function to be dependent on the velocities $\mathbf{v}$, $\overset{1}{\mathbf{v}}$, $\overset{2}{\mathbf{v}}$ and, additionally, on the $\Gamma$-surface orientation described by the normal vector $\mathbf{n}$. We then have

$$\vartheta = \bar{\vartheta}(\mathbf{v}, \overset{1}{\mathbf{v}}, \overset{2}{\mathbf{v}}, \mathbf{n}). \tag{12}$$

The form of constitutive function $\bar{\vartheta}$ is restricted by the requirement that it must be objective and positively definite.

The objectivity condition, Leigh (1968), when applied to Equation (12) shows that function $\bar{\vartheta}$ depends only on the relative velocities $\mathbf{u}$, $\overset{2}{\mathbf{u}}$ and the vector $\mathbf{n}$, and has to be isotropic function. We obtain the relation

$$\vartheta = \bar{\vartheta}'(\mathbf{u}, \overset{2}{\mathbf{u}}, \mathbf{n}), \tag{13}$$

and the isotropy condition for the dissipation function takes the form of identity

$$\bar{\vartheta}'(\mathbf{Qu}, \mathbf{Q}\overset{2}{\mathbf{u}}, \mathbf{Qn}) = \bar{\vartheta}'(\mathbf{u}, \overset{2}{\mathbf{u}}, \mathbf{n}) \tag{14}$$

that has to be satisfied for any orthogonal tensor $\mathbf{Q}(\mathbf{Q}^{\mathrm{T}} = \mathbf{Q}^{-1})$ and any vectors $\mathbf{u}$, $\overset{2}{\mathbf{u}}$ and $\mathbf{n}$. According to the Cauchy representation theorem for the isotropic scalar-valued functions, Leigh (1968), we find that $\bar{\vartheta}'(\mathbf{u}, \overset{2}{\mathbf{u}}, \mathbf{n})$ can be regarded as a function of the following five scalar products

$$\mathbf{I}:\ \mathbf{u}\cdot\mathbf{u},\quad \mathbf{u}\cdot\overset{2}{\mathbf{u}},\quad \overset{2}{\mathbf{u}}\cdot\overset{2}{\mathbf{u}},\quad \mathbf{u}\cdot\mathbf{n},\quad \overset{2}{\mathbf{u}}\cdot\mathbf{n}.$$

Then the relation (13) takes the form

$$\bar{\vartheta}'(\mathbf{u}, \overset{2}{\mathbf{u}}, \mathbf{n}) \equiv \bar{\vartheta}''(\mathbf{u}\cdot\mathbf{u},\ \mathbf{u}\cdot\overset{2}{\mathbf{u}},\ \overset{2}{\mathbf{u}}\cdot\overset{2}{\mathbf{u}},\ \mathbf{u}\cdot\mathbf{n},\ \overset{2}{\mathbf{u}}\cdot\mathbf{n}) = \bar{\vartheta}''(\mathbf{I}), \tag{15}$$

and for any $\bar{\vartheta}''$ the condition (14) is satisfied identically. Taking the compatibility condition (2) into account, the number of arguments in Equation (15) can be reduced by one, and we obtain

$$\vartheta = \bar{\vartheta}^{*}(\mathbf{u}\cdot\mathbf{u},\ \mathbf{u}\cdot\overset{2}{\mathbf{u}},\ \overset{2}{\mathbf{u}}\cdot\overset{2}{\mathbf{u}}, \mathbf{u}\cdot\mathbf{n}). \tag{16}$$

After the decomposition of $\mathbf{u}$ and $\overset{2}{\mathbf{u}}$ into their normal and tangential components, that is

$$\mathbf{u}_n = (\mathbf{u}\cdot\mathbf{n})\mathbf{n}, \qquad \mathbf{u}_\tau = \mathbf{u} - \mathbf{u}_n, \qquad \overset{2}{\mathbf{u}}_\tau = \overset{2}{\mathbf{u}} - (\overset{2}{\mathbf{u}}\cdot\mathbf{n})\mathbf{n}, \tag{17}$$

we get the following representation

$$\vartheta = \hat{\vartheta}(\mathbf{u}_\tau\cdot\mathbf{u}_\tau, \mathbf{u}_\tau\cdot\overset{2}{\mathbf{u}}_\tau, \overset{2}{\mathbf{u}}_\tau\cdot\overset{2}{\mathbf{u}}_\tau, \mathbf{u}\cdot\mathbf{n}). \tag{18}$$

The relation (18) is a general form of the constitutive equation for the dissipation of the mechanical energy at the interface between regions of fluid flow in the porous skeleton and the adjacent bulk fluid flow.

The only constraint for the function $\hat{\vartheta}$ results from the last requirement of its positive definiteness. This restriction takes the form of the following inequality:

$$\hat{\vartheta} = (\mathbf{u}_\tau\cdot\mathbf{u}_\tau, \mathbf{u}_\tau\cdot\overset{2}{\mathbf{u}}_\tau, \overset{2}{\mathbf{u}}_\tau\cdot\overset{2}{\mathbf{u}}_\tau, \mathbf{u}\cdot\mathbf{n}) > 0, \tag{19}$$

which should be satisfied for any nonzero values of $\mathbf{u}$, $\overset{2}{\mathbf{u}}$ and, in the case of no fluid flow ($\mathbf{u} = \overset{2}{\mathbf{u}} = 0$), it must be equal to zero

$$\hat{\vartheta}(0, 0, 0, 0) = 0. \tag{20}$$

The inequality (19) results from the entropy production principle (Wilmanski, 1985), due to the dissipation of mechanical energy on the $\Gamma$-surface.

Since, in general cases, the function $\hat{\vartheta}$ may not be an even function of the argument $\mathbf{u}\cdot\mathbf{n}$, it follows from Equation (18) that the dissipation of mechanical energy can reach different values during the inflow and outflow of fluid from the porous skeleton. However, for small values of the velocities $\mathbf{u}_\tau$, $\overset{2}{\mathbf{u}}_\tau$ and $\mathbf{u}\cdot\mathbf{n}$, conditions (19) and (20) ensure that the constitutive function $\hat{\vartheta}$ has a minimum at $\mathbf{u} = \overset{2}{\mathbf{u}} = 0$ and, therefore, in such a case, the dissipation function may be proposed in the quadratic form

$$\vartheta = \frac{\mu}{\sqrt{k}}(\alpha_1\mathbf{u}_\tau\cdot\mathbf{u}_\tau - 2\alpha_2\mathbf{u}_\tau\cdot\overset{2}{\mathbf{u}}_\tau + \alpha_3\overset{2}{\mathbf{u}}_\tau\cdot\overset{2}{\mathbf{u}}_\tau + \varepsilon(\mathbf{u}\cdot\mathbf{n})^2), \tag{21}$$

where the fluid viscosity $\mu$ and the permeability $k$ of porous skeleton have been introduced in order to make the constant coefficients $\alpha_1, \alpha_2, \alpha_3$ and $\varepsilon$ be dimensionless quantities. The coefficients $\alpha_1, \alpha_2, \alpha_3$ and $\varepsilon$ in dissipation function are macroscopic parameters characterizing the pore structure of permeable material within the boundary layer.

The choice of parameters $\mu$ and $k$ is motivated by the fact that fluid viscosity and permeability of porous material are the main properties upon which dissipation of mechanical energy appearing in the boundary region can depend. As it is shown further, the dissipation function of the form (21) leads to the Beavers–Joseph type boundary condition (see (36)), in which material parameters are independent of the fluid viscosity. Such a result is strongly supported by experimental observations (Beavers and Joseph, 1967).

Considering the fluid flow tangential and normal to the $\Gamma$- surface, it can be shown that the constitutive relation (21) will be positively definite when constant coefficients satisfy the following inequalities:

$$\alpha_1 > 0, \qquad \alpha_3 > 0, \qquad \alpha_2^2 < \alpha_1\alpha_3, \quad \varepsilon > 0. \tag{22}$$

## 4. Linear Compatibility Conditions

Equations (2)–(4) together with the dissipation function (18) form a set of the general compatibility conditions for mechanical fields on the discontinuity $\Gamma$-surface. These equations are to be used to establish the linear form of compatibility conditions which play an important role in a large number of linear problems of porous media.

The linear compatibility condition for the stress vectors of particular constituents of the system can be derived directly from the balance equation of linear momentum (3), disregarding the nonlinear terms of its left-hand side. This condition written for the normal and tangential components yields

$$\mathbf{t}_n = \overset{1}{\mathbf{t}}_n + \overset{2}{\mathbf{t}}_n, \tag{23}$$

$$\mathbf{t}_\tau = \overset{1}{\mathbf{t}}_\tau + \overset{2}{\mathbf{t}}_\tau\,. \tag{24}$$

From Equations (23) and (24), it follows that, in the linear case, the force exerted by the bulk fluid on the permeable boundary is equilibrated by forces of both constituents of saturated porous solid.

The linearization of the balance Equation (4) for mechanical energy on the $\Gamma$-surface, in contradistinction to the balance of linear momentum (3), is not straightforward. We can arrive, however, at linear relations under some assumptions. After the use of Equations (2) and (21), disregarding the left-hand side of Equation (4), we obtain

$$\left[\rho\left(\frac{t_n}{\rho} - \frac{\overset{2}{t}_n}{\overset{2}{\rho}}\right) - \frac{\mu}{\sqrt{k}}\varepsilon u_n\right] u_n$$

$$= \overset{2}{\mathbf{t}}_\tau \cdot \overset{2}{\mathbf{u}}_\tau - \mathbf{t}_\tau \cdot \mathbf{u}_\tau - \frac{\mu}{\sqrt{k}}(\alpha_1 \mathbf{u}_\tau \cdot \mathbf{u}_\tau - 2\alpha_2 \mathbf{u}_\tau \cdot \overset{2}{\mathbf{u}}_\tau + \alpha_3 \overset{2}{\mathbf{u}}{}_\tau^2). \tag{25}$$

In the above equation the components of stress vectors of bulk and free fluid are related with the corresponding relative velocities.

Since, in the linear case, the normal and tangential components of any vector field are mutually independent, Equation (25) will be satisfied if its both sides are identically equal to zero. According to the relations

$$\rho = \rho^-, \qquad \overset{2}{\rho} = \kappa f_{\mathrm{v}} \rho^+, \qquad t_n = -p^-, \qquad \overset{2}{t}_n = -\kappa f_{\mathrm{v}} p^+, \tag{26}$$

where $p^-$, $p^+$ are fluid effective pressures at both sides of $\Gamma$- surface, from Equation (25) we get

$$\rho^- \left( \frac{p^+}{\rho^+} - \frac{p^-}{\rho^-} \right) = \frac{\mu}{\sqrt{k}} \varepsilon u_n, \tag{27}$$

$$\overset{2}{\mathbf{t}}_\tau \cdot \overset{2}{\mathbf{u}}_\tau - \mathbf{t}_\tau \cdot \mathbf{u}_\tau - \frac{\mu}{\sqrt{k}} (\alpha_1 \mathbf{u}_\tau \cdot \mathbf{u}_\tau - 2 \cdot \mathbf{u}_\tau \alpha_2 \mathbf{u}_\tau \cdot \overset{2}{\mathbf{u}}_\tau + \alpha_3 \overset{2}{\mathbf{u}}{}_\tau^2) = 0, \tag{28}$$

where superscripts + and − denote values of quantities on the positive and negative side of the $\Gamma$-surface, respectively. In order to obtain the linear form of Equation (27) we assume that the fluid is barotropic, that is,

$$p = p(\rho). \tag{29}$$

Then, for the small values of $p^+ - p^-$, we can write

$$p^+ - p^- = a_0^2 (\rho^+ - \rho^-), \tag{30}$$

where

$$a_0 = \left( \frac{\mathrm{d}p}{\mathrm{d}\rho} \right)^{1/2} \Bigg|_{\rho = \rho 0}$$

is the velocity of wave front propagation in an undisturbed region of the bulk fluid. Under the above assumption, from Equation (27) we obtain the linear compatibility condition

$$p^+ - p^- = \frac{\mu}{\sqrt{k}} \frac{\varepsilon}{(1 - p_0/(\rho_0 a_0^2))} u_n. \tag{31}$$

It relates the fluid pressures on both sides of the $\Gamma$-surface and the normal component of the relative fluid velocity.

In the case where the right-hand side of Equation (31) can be omitted, the condition (31) is reduced to the classical one describing the fluid pressure continuity

$$p^+ = p^- = p. \tag{32}$$

Assumptions (29) and (30) and condition (31) allow the linearization of the fluid mass balance equation (2) at the $\Gamma$-surface. As a consequence, we obtain the compatibility condition for normal components of velocities of fluid flow across the $\Gamma$-surface

$$\mathbf{u}_n = \kappa f_{\mathrm{v}} \overset{2}{\mathbf{u}}_n \ . \tag{33}$$

To obtain the compatibility conditions for tangential components of fluid velocities at the $\Gamma$-surface we consider condition (28). Assuming that tangential forces $\mathbf{t}_\tau$ and

$\overset{2}{\mathbf{t}}_\tau$ depend linearly on the relative velocities $\mathbf{u}_\tau$ and $\overset{2}{\mathbf{u}}_\tau$, we can divide Equation (28) into two independent equations:

$$\mathbf{t}_\tau = \frac{\mu}{\sqrt{k}}(\alpha_1 \mathbf{u}_\tau - \alpha_2' \overset{2}{\mathbf{u}}_\tau), \tag{34}$$

$$\overset{2}{\mathbf{t}}_\tau = \frac{\mu}{\sqrt{k}}(\alpha_2'' \mathbf{u}_\tau - \alpha_3 \overset{2}{\mathbf{u}}_\tau), \tag{35}$$

in which

$$\alpha_2' + \alpha_2'' = 2\alpha_2.$$

Equations (34) and (35) are the two additional necessary linear compatibility conditions. Their functional form is determined by the constitutive function (21) of dissipation of mechanical energy of fluid in the boundary layer of a porous solid.

Relations (23), (24), (31) (or (32)), (33)–(35) form the complete set of the linear compatibility conditions for mechanical fields at the contact surface between fluid-saturated porous solids and bulk fluid.

Taking into account the fact that tangential stresses in the bulk and free fluid are related with the corresponding velocity gradients via the constitutive equations, the compatibility conditions (34) and (35) can be transformed into the form of conditions for the velocities only. For the one-dimensional rectilinear fluid flow along a plane surface of a porous solid, we have

$$\mathbf{t}_\tau = \mu \frac{\partial u_\tau}{\partial z} \boldsymbol{\tau} \quad \text{and} \quad \overset{2}{\mathbf{t}}_\tau = \mu^* \frac{\partial \overset{2}{u}_\tau}{\partial z} \boldsymbol{\tau},$$

where $\boldsymbol{\tau}$ is the unit vector tangential to the $\Gamma$-surface indicating the direction of fluid flow. The coordinate $z$ is directed along the vector normal to the $\Gamma$-surface. The coefficient $\mu^*$ stands for the apparent viscosity of the fluid filling the porous skeleton, corresponding to the coefficient introduced by Brinkman (1947). In such a case, from Equations (34) and (35), we obtain

$$\frac{\partial u_\tau}{\partial z} \boldsymbol{\tau} = \frac{1}{\sqrt{k}}(\alpha_1 \mathbf{u}_\tau - \alpha_2' \overset{2}{\mathbf{u}}_\tau), \tag{36}$$

$$\frac{\mu^*}{\mu} \frac{\partial \overset{2}{u}_\tau}{\partial z} \boldsymbol{\tau} = \frac{1}{\sqrt{k}}(\alpha_2'' \mathbf{u}_\tau - \alpha_3 \overset{2}{\mathbf{u}}_\tau). \tag{37}$$

Taking the above results into account, and the linear relationship between volume discharge and velocity of fluid flow through porous skeleton, we find that the Beavers–Joseph postulate (1) has the same form as that of (36). However, the Beavers–Joseph postulate is not sufficiently general, since only one condition does not allow one to solve boundary problems where the fluid flow within a porous medium is described by the second order differential equation for example the Brinkman (1947) type, of equation. Moreover, the Beavers, and Joseph way of formulating the slip-flow boundary condition does not give the possibility to determine the interaction forces exerted by the flowing bulk fluid on the pore fluid and the porous skeleton at the permeable boundary of porous solid. It results from the fact that this condition is not explicitly related with the interaction forces. In our description the relative velocities $\mathbf{u}_\tau$ and $\overset{2}{\mathbf{u}}_\tau$, used in the compatibility conditions (36) and (37), are directly related with forces appearing at the boundary surface through the formulas (34) and (35). Consequently, the use of condition (24) allows the determination of interaction force between bulk fluid and saturated porous solid.

## 5. Solution of the Beavers–Joseph Flow Problem

In this section, we use the compatibility conditions (36) and (37) to solve a problem of a viscous fluid flow along the surface of permeable porous material: the so-called the Beavers–Joseph problem. We compare this solution to that proposed by Beavers and Joseph (1967) and also with the selected results of experimental measurements performed by these authors.

Let us consider laminar flow of viscous, incompressible fluid through a channel formed by an impermeable wall and a permeable, rigid, porous halfspace filled with fluid (Figure 2). The flow induced by a constant pressure gradient $\mathrm{d}p/\mathrm{d}x$ is unidirectional and takes place in the channel as well as in the porous halfspace. We assume that the velocity field $\mathbf{u}(z)$ of fluid in the channel is given by the reduced

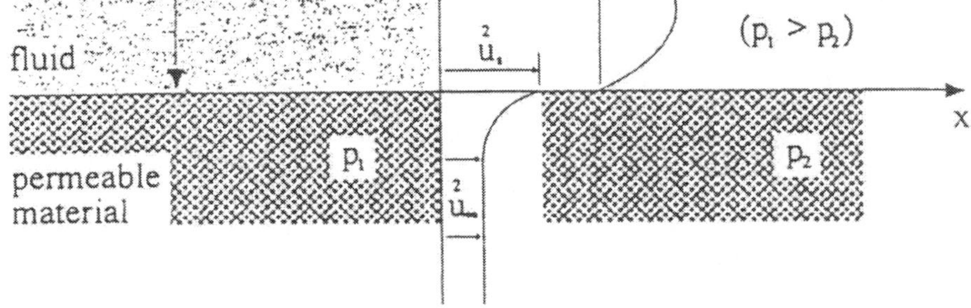


*Figure 2.* Laminar flow of viscous, incompressible fluid through a channel.

Navier–Stokes equation

$$\mu \frac{\mathrm{d}^2 u}{\mathrm{d}z^2} = \frac{\mathrm{d}p}{\mathrm{d}x} \quad \text{for} \quad 0 < z < h \tag{38}$$

and that the velocity field of fluid motion within the porous halfspace is described by an equation of the Brinkman (1947) form

$$\mu^* \frac{\mathrm{d}^2 \overset{2}{u}}{\mathrm{d}z^2} - \frac{\mu}{k} \overset{2}{u} = \kappa f_{\mathrm{v}} \frac{\mathrm{d}p}{\mathrm{d}x} \quad \text{for} \quad z < 0. \tag{39}$$

Although the fluid in the channel and the pores of the skeleton is the same, the viscosities $\mu^*$ and $\mu$ are of different values and the ratio $\mu^*/\mu$ characterizes the structure of porous material and is independent of the properties of fluid.

For the discussed problem, the boundary condition at the surface of impermeable wall is the zero fluid velocity, that is

$$u|_{z=h} = 0, \tag{40}$$

whereas at the surface of porous halfspace the compatibility conditions for velocities take the form of relations (36) and (37). Solutions of Equations (38) and (39), when conditions (40), (36) and (37) are taken into account, give the following velocity fields

$$\frac{u}{\overset{2}{u}_\infty} = \left(1 - \frac{z}{h}\right)\left(\frac{z}{h}\frac{\sigma^2}{2\kappa f_{\mathrm{v}}} + \frac{u_{\mathrm{s}}}{\overset{2}{u}_\infty}\right), \quad \text{for} \quad 0 < z < h, \tag{41}$$

$$\frac{\overset{2}{u}}{\overset{2}{u}_\infty} = \left(\frac{\overset{2}{u}_{\mathrm{s}}}{\overset{2}{u}_\infty} - 1\right) \exp\left(\frac{\sigma}{\sqrt{\mu^*/\mu}}\frac{z}{h}\right) + 1, \quad \text{for} \quad z < 0, \tag{42}$$

where $\sigma = h/\sqrt{k}$ is the dimensionless height of the channel and $u_{\mathrm{s}}$, $\overset{2}{u}_{\mathrm{s}}$ are the slip velocities of fluid at adequate sides of the surface of the porous halfspace. These velocities are linked to the parameters of the system by relations

$$\frac{u_{\mathrm{s}}}{\overset{2}{u}_\infty} = \frac{\sigma(\sigma + 2\delta_1)}{2\kappa f_{\mathrm{v}}(1 + \sigma\delta_2)}, \tag{43}$$

$$\frac{\overset{2}{u}_{\mathrm{s}}}{\overset{2}{u}_\infty} = \frac{1}{\alpha_3 + \sqrt{\mu^*/\mu}} \frac{\alpha_2''\sigma^2 + 2\kappa f_{\mathrm{v}}\sqrt{\mu^*/\mu}(1 + \alpha_1\sigma)}{2\kappa f_{\mathrm{v}}(1 + \sigma\delta_2)}, \tag{44}$$

with

$$\delta_1 = \frac{\kappa f_{\mathrm{v}} \alpha_2' \sqrt{\mu^*/\mu}}{\alpha_3 + \sqrt{\mu^*/\mu}}, \qquad \delta_2 = \alpha_1 - \frac{\alpha_2'\alpha_2''}{\alpha_3 + \sqrt{\mu^*/\mu}}. \tag{45}$$

In expressions (41)–(44) the reference velocity $\overset{2}{u}_\infty$ is introduced, that is

$$\overset{2}{u}_\infty = -\kappa f_\mathrm{v} \frac{k}{\mu} \frac{\mathrm{d}p}{\mathrm{d}x},$$

which describes the fluid flow through the porous material at a large distance from the boundary of the considered halfspace.

Results from (43) and (44) show that the fluid at both sides of the permeable surface has different velocities. It is worthwhile noting that, in the limit case where the impermeable wall approaches the surface of porous medium ($\sigma = 0$), the velocity $\overset{2}{u}_\mathrm{s}$ does not vanish and its finite value is

$$\frac{\overset{2o}{u}_\mathrm{s}}{\overset{2}{u}_\infty} = \frac{2\kappa f_\mathrm{v}\sqrt{\mu^*/\mu}}{\alpha_3 + \sqrt{\mu^*/\mu}}.$$

Such an effect is directly connected with the earlier introduced dissipation function for the boundary layer of porous material. The validity of this effect needs to be justified by experimental measurements.

Using the velocity field (41) and the velocity field for the fluid flow through the channel between two impermeable walls, we can define the fractional increase

$$\Phi = \frac{M - M_0}{M_0}$$

in the mass flow rate through the channel with a permeable lower wall ($M$), over what it would be if the wall were impermeable ($M_0$). We obtain

$$\Phi = \frac{3(\sigma + 2\delta_1)}{\sigma(1 + \delta_2\sigma)}. \tag{46}$$

The above relation is more general than that derived by Beavers and Joseph (1967), that is

$$\Phi = \frac{2(\sigma + 2\alpha)}{\sigma(1 + \alpha\sigma)}, \tag{47}$$

where the compatibility condition (1) was used. Expression (46) contains two parameters $\delta_1$ and $\delta_2$ that, according to (45), are functions of the parameters, characterizing pore structure of the interior and the surface layer of porous material, while the expression (47) contains only one parameter $\alpha$, introduced by the compatibility condition (1).

The experimental measurements performed by Beavers and Joseph have shown the qualitative agreement of the experimental data with the theoretical predictions of relation (47). In some cases, however, the obtained quantitative agreement is not satisfactory. As an example, in Figure 3 are shown the results of measurements performed by Beavers and Joseph for the oil flow over specimens of granular material

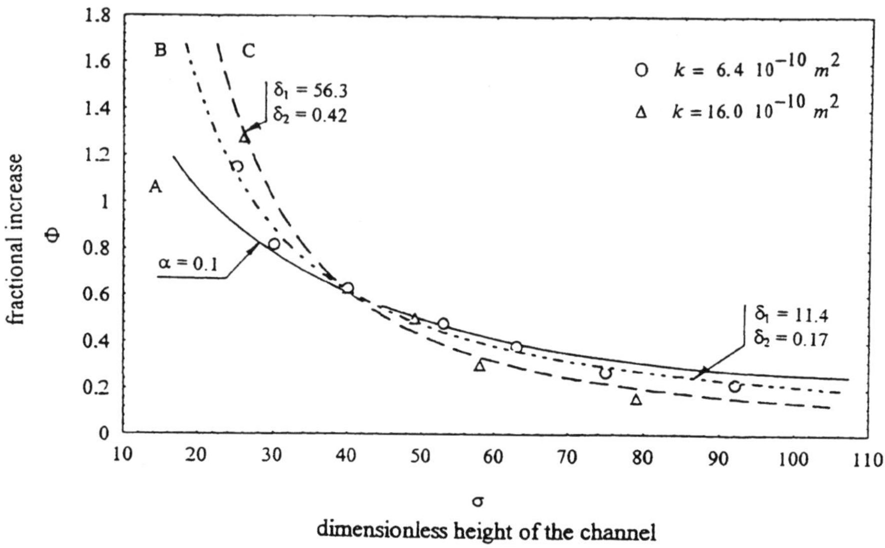


*Figure 3.* The results of measurements performed by Beavers and Joseph (1967).

(aloxite) with two various permeabilities and the adequate theoretical curves obtained from (46) and (47). The solid line A represents the theoretical predictions obtained by Beavers and Joseph from relation (47) ($\alpha = 0.1$) for both kinds of porous specimens. It is seen that a considerable difference between theoretical and experimental results occurs. The dashed lines *B* and *C* in this figure are graphs obtained from the function (46) for parameters $\delta_1$ and $\delta_2$ determined for each porous specimens by the use of optimization procedure. For specimens with permeability $k = 6.45 \times 10^{-10}\,\mathrm{m}^2$ and $k = 16.0 \times 10^{-10}\,\mathrm{m}^2$, these parameters take the following values, $\delta_1 = 11.4$, $\delta_2 = 0.17$ and $\delta_1 = 56.3$, $\delta_2 = 0.42$, respectively.

The above results show that the compatibility conditions (36) and (37) derived in this paper assure a more precise description of fluid flow along a permeable wall of porous material, in comparison to the condition (1) postulated by Beavers and Joseph.

## 6. Final Remarks

In this paper the problem of compatibility conditions for macroscopic mechanical fields at the contact surface between the fluid-saturated porous solid and the bulk fluid has been analyzed. Special attention has been paid to the derivation of conditions for the tangential components of the fluid flow velocities at the contact surface. This has allowed the verification of the accuracy of the condition postulated by Beavers and Joseph.

The analysis of the general form of compatibility conditions, that is the balance equations for mass, linear momentum and mechanical energy at the discontinuity surface has proven the existence of discontinuity in the macroscopic relative fluid velocities, justifying the postulate of the existence of mechanical energy dissipation at this surface due to the fluid viscosity. It has turned out that the introduction of the term responsible for this effect into the balance equation of mechanical energy has been the crucial point for the explanation of the validity of the Beavers–Joseph type conditions.

The linearization of the balance equations at the contact surface have provided the complete set of linear compatibility conditions and have allowed us to obtain the generalized solution of the Beavers–Joseph flow problem, leading to an excellent agreement with experimental results.

## References

Ateshian, G. A.: 1997, A theoretical formulation for boundary friction in articular cartilage, *J. Biomech. Engng.* **119**, 81–86.

Bear, J. and Bahmat, Y.: 1984, Transport phenomena in porous media basic equation, In: J. Bear and M. Y. Corapcioglu (eds), *Fundamentals of Transport Phenomena in Porous Media*, Martinus Nijhoff, Dordrecht, pp. 5–61.

Biot, M. A.: 1956, Theory of propagation of elastic waves in a fluid saturated porous solid, *J. Acoust. Soc. Am.* **28**(2), 168–191.

Beavers, G. S. and Joseph, D. D.: 1967, Boundary conditions at a naturally permeable wall, *J. Fluid. Mech.* **30**, 197–207.

Brinkman, H. C.: 1947, A calculation of the viscous force exerted by a flowing fluid on a dense swarm of Particles, *Appl. Sci. Res. A* **1**, 27–34.

Cieszko, M. and Kubik, J.: 1993, On the compatibility conditions in the fluid–fluid saturated porous solid contact problems, *Arch. Mech.* **45**(1), 77–91.

Cieszko, M. and Kubik, J.: 1996a, Constitutive relations and internal equilibrium condition for fluid-saturated porous solid. Nonlinear theory, *Arch. Mech.* **48**(5), 893–910.

Cieszko, M. and Kubik, J.: 1996b, Constitutive relations and internal equilibrium condition for fluid-saturated porous solid. Linear description, *Arch. Mech.* **48**(5), 911–923.

Gogete, G. R. and Munjal, M. L.: 1996, Sound propagation in ducts with bulk reacting lining in the presence of laminar mean flow, *J. Acoust. Soc. Am.* **99**(3), 1779–1782.

Hou, J. S., Holmes, M. H., Lai, W. H. and Mow, V. C.: 1989, Boundary conditions at the cartilage-synovial fluid interface for joint lubrication and theoretical verifications, *J. Biomech. Engng.* **111**, 78–87.

Hutter, K., Johnk, K. and Svendsen, B.: 1994, On interfacial transition conditions in two phase gravity flow, *ZAMP* **45**, 746–762.

Johnson, D. L.: 1986, Recent developments in the acoustic properties of porous media, In: *Frontiers in Physical Acoustics*, XCIII Corso, Soc. Italiana di Fisica Bologna.

Joseph, D. D. and Tao, L. N.: 1966, Lubrication of a porous bearing Stockes' solution, *J. Appl. Mech.* **33**, 753–760.

Koplik, J., Levine, H. and See, A.: 1983, Viscosity renormalization in the Brinkman equation, *Phys. Fluids* **26**, 10.

Kubik, J.: 1986a, On the internal coupling in dynamic equations of fluid-saturated porous solid, *Int. J. Engng. Sci.* **24**(6), 981–989.

Kubik, J.: 1986b, A macroscopic description of geometrical pore structure of porous solid, *Int. J. Engng. Sci.* **24**(6) 971–980.

Kubik, J.: 1992, Pore structure in dynamic behaviour of saturated materials, *Transport in Porous Media* **9**, 15–24

Landau, D. D. and Lifshitz E. M.: 1986, *Hydromechanics* (in Russian), Nauka, Moskva.
Leigh, D. C.: 1968, *Nonlinear Continuum Mechanics*, McGraw-Hill, New York.
Levy, T. and Sanchez-Palencia, E.: 1975, On the boundary conditions for fluid flow in porous media, *Int. J. Engng. Sci.* **13**, 923–940.
Raats, R. A. C.: 1972, The role of inertia in the hydrodynamics of porous media, *J. Rational Mech. Anal.* **44**, 267–280.
Rhodes, C. A. and Rouleau, W. T.: 1966, Hydrodynamic lubrication of partial porous metal bearings, *J. Basic Engng. Trans. ASME D* **88**(1), 53–60.
Richardson, S.: 1971, A model for the boundary condition of porous material, Part II, *J. Fluid Mech.* **49**, 327–336.
Saffman, P. G.: 1971, On the boundary condition at the surface of a porous medium, *Studies in Appl. Mech. L* **2**, 93–101.
Shir, C. C. and Joseph, D. D.: 1966, Lubrication of a porous bearing Reynolds' solution, *J. Appl. Mech.* **33**, 761–767.
Taylor, G. I.: 1971, A model for the boundary condition of a porous material, Part I, *J. Fluid Mech.* **49**, 319–326.
Wilmanski, K.: 1985, Phenomenological thermodynamics, In: *Engineering Mechanics*, Vol. 1, *Foundations of Mechanics* (in Polish), PWN, Warsaw.
Wilmanski, K.: 1995, Lagrangian model of two-phase porous material, *J. Non-Equil. Thermodyn.* **20**, 50–77.
Wu, T., Johnk, K., Svendsen, B. and Hutter, K.: 1996, On the gravity-driven shear flow of an ice-till mixture, *Ann. of Glaciol.* **23**, 124–128.

**SPRINGER NATURE**

# GPSR Compliance

*The European Union's (EU) General Product Safety Regulation (GPSR) is a set of rules that requires consumer products to be safe and our obligations to ensure this.*

*If you have any concerns about our products, you can contact us on ProductSafety@springernature.com*

In case Publisher is established outside the EU, the EU authorized representative is:

Springer Nature Customer Service Center GmbH
Europaplatz 3
69115 Heidelberg, Germany

FSC
www.fsc.org
MIX
Papier | Fördert gute Waldnutzung
FSC® C083411

Kolibri 360 GmbH
Ferdinand-Jühlke-Straße 7
99095 Erfurt, Deutschland
produktsicherheit@kolibri360.de